# ZEBRAFISH

This is Volume 29 in the

FISH PHYSIOLOGY series
Edited by Anthony P. Farrell and Colin J. Brauner
Honorary Editors: William S. Hoar and David J. Randall

*A complete list of books in this series appears at the end of the volume*

# ZEBRAFISH

Edited by

Dr. STEVE F. PERRY
*Department of Biology*
*University of Ottawa*
*Ottawa, ON, Canada*

Dr. MARC EKKER
*Department of Biology*
*University of Ottawa*
*Ottawa, ON, Canada*

Dr. ANTHONY P. FARRELL
*Faculty of Land and Food Systems & Department of Zoology*
*The University of British Columbia*
*Vancouver, British Columbia*
*Canada*

COLIN J. BRAUNER
*Department of Zoology*
*The University of British Columbia*
*Vancouver, British Columbia*
*Canada*

AMSTERDAM • BOSTON • HEIDELBERG • LONDON • OXFORD
NEW YORK • PARIS • SAN DIEGO • SAN FRANCISCO
SINGAPORE • SYDNEY • TOKYO
Academic Press is an imprint of Elsevier

Academic Press is an imprint of Elsevier
32 Jamestown Road, London NW1 7BY, UK
30 Corporate Drive, Suite 400, Burlington, MA 01803, USA
525 B Street, Suite 1800, San Diego, CA 92101-4495, USA

First edition 2010



**British Library Cataloguing-in-Publication Data**
A catalogue record for this book is available from the British Library

**Library of Congress Cataloging-in-Publication Data**
A catalog record for this book is available from the Library of Congress

ISBN: 978-0-12-374983-3
ISSN: 1546-5098



Typeset by MPS Limited, a Macmillan Company, Chennai, India
www.macmillansolutions.com

Printed in the United States of America
Transvered to Digital Printing, 2011

# CONTENTS

# CONTRIBUTORS

*The numbers in parentheses indicate the pages on which the authors' contributions begin.*

LEILA ABBAS *(123), MRC Centre for Developmental and Biomedical Genetics, Department of Biomedical Science, University of Sheffield, Sheffield, S10 2TN, UK*

MARIE-ANDRÉE AKIMENKO *(1), Center for Advanced Research in Environmental Genomics, Department of Biology, University of Ottawa, Ottawa, ON, Canada*

BRIAN BAGATTO *(289), Department of Biology, University of Akron, Ohio 44325, USA*

LAURE BALLY-CUIF *(25), Zebrafish Neurogenetics Group, Laboratory of Neurobiology & Development (NED), UPR 3294, CNRS, Institute of Neurobiology Alfred Fessard, Bâtiment 5E, Avenue de la Terrasse, 91198 Gif-Sur-Yvette Cédex, France*

ELLEN R. BUSBY *(173), Department of Biology, University of Victoria, Victoria, B.C. V8W 3N5, Canada*

SARAH J. CHILDS *(249), Department of Biochemistry and Molecular Biology, University of Calgary, 3330 Hospital Drive NW, Calgary, AB, Canada, T2N 4M6*

MARC EKKER *(1), Department of Biology, University of Ottawa, Ottawa, ON, Canada*

JARED.V. GOLDSTONE *(367), Biology Department, MS #32, Woods Hole Oceanographic Institution, Woods Hole, MA 02543, USA*

MARK. E. HAHN *(367), Biology Department, MS #32, Woods Hole Oceanographic Institution, Woods Hole, MA 02543, USA*

PUNG-PUNG HWANG *(311), Institute of Cellular and Organismic Biology, Academia Sinica, Nankang, Taipei, Taiwan 11529, ROC*

JOHN D. MABLY *(249), Departments of Pediatrics and Genetics, Children's Hospital Boston and Harvard Medical School, 320 Longwood Avenue, Boston, MA, 02115, USA*

STEPHAN C. F. NEUHAUSS *(81), University of Zurich, Institute of Molecular Life Sciences, Neuroscience Center Zurich and Center for Integrative Human Physiology, Winterthurerstrasse 190, CH-8057 Zurich, Switzerland*

BERND PELSTER *(289), Institut für Zoologie, Universität Innsbruck, Austria; Center for Molecular Biosciences, Innsbruck, Austria*

STEVE F. PERRY *(311), Department of Biology, University of Ottawa, Ottawa, ON, Canada*

GRAEME J. ROCH *(173), Department of Biology, University of Victoria, Victoria, B.C. V8W 3N5, Canada*

NANCY M. SHERWOOD *(173), Department of Biology, University of Victoria, Victoria, B.C. V8W 3N5, Canada*

JOHN. J. STEGEMAN *(367), Biology Department, MS #32, Woods Hole Oceanographic Institution, Woods Hole, MA 02543, USA*

PHILIPPE VERNIER *(25), Development and Neurobiology Laboratory, CNRS UPR3294, Institute of Neurobiology Alfred Fessard, Avenue de la Terrasse, F-91198 Gif-sur-Yvette, France*

TANYA T. WHITFIELD *(123), MRC Centre for Developmental and Biomedical Genetics, Department of Biomedical Science, University of Sheffield, Sheffield, S10 2TN, UK*

CONG XU *(345), Harvard Medical School, Boston, MA 02115, USA*

LEONARD I. ZON *(345), Stem Cell Program and Division of Hematology/Oncology, Children's Hospital Boston and Dana-Farber Cancer Institute, Howard Hughes Medical Institute, Harvard Stem Cell Institute, Harvard Medical School, Boston, MA 02115, USA*

# PREFACE

First, we really hope that you enjoy this book. Second, a confession – the idea to devote an entire volume of *Fish Physiology* to a single species, zebrafish, was not ours. Thus, although a precedent may have been set, surely Marc and I cannot be held responsible for future volumes bearing titles such as "Eelpout" or "the Gudgeon". In actual fact, Colin Brauner approached me at a conference several years ago with the suggestion that zebrafish might make a good topic for an upcoming volume. I enthusiastically endorsed his idea but nevertheless felt somewhat deflated when he next asked me if I knew anyone who studied zebrafish who could serve as a guest Editor; I think his exact words may have been "credible guest Editor". Naturally, I immediately thought of my colleague Marc Ekker who began his research on zebrafish in 1989, a year that saw fewer than 20 zebrafish papers published. Just 20 years later, an astounding 1800 scientific publications on zebrafish biology were produced.

So, here we are a few years after that fateful meeting with Colin – considerably grayer but immensely pleased to introduce this latest volume of *Fish Physiology*. This book presents something more than just ten chapters in key areas of physiology by the world's leading authorities. It provides, for the first time, a compilation of comprehensive reviews outlining the physiology of zebrafish, the tools used to study zebrafish physiology and the applicability of zebrafish for studies of human disease and aquatic toxicology. As such we hope you and your research programs will benefit.

Needless to say, we are extremely grateful to our authors for their dedication to this project and to Colin and to Tony Farrell for their words of wisdom. We can't say enough about the helpful professionals at Elsevier, Pat Gonzalez and Kristi Gomez, who patiently guided us along the way and kept us from wandering too far off course. And, of course, we are eternally grateful to Dave Randall and Bill Hoar for conceiving of this wonderful monograph series that has so beautifully chronicled the evolution of *Fish Physiology* over the past 40 years.

Editing a book is most definitely a learning experience; in this case we have gained a new appreciation of the remarkable variation in the use and interpretation of three of the simplest adjectives used in modern language – soon, nearly and almost.

Steve F. Perry<br>Marc Ekker<br>Ottawa, Ontario – 31 January 2010

# 1

# GENETIC TOOLS

*MARC EKKER*
*MARIE-ANDRÉE AKIMENKO*



The production of a large collection of mutations that affect almost every aspect of embryonic development has largely contributed to the rise in popularity of the zebrafish as an experimental system in vertebrate biology. The development of novel genetic methods to manipulate the zebrafish or its genome further widened the spectrum of possible biological questions that can be asked in this species. Forward genetics approaches allow us to randomly affect genes with chemical mutagens or the insertion of retroviral DNA sequences. In contrast, reverse genetics approaches take advantage of the wealth of information provided by the zebrafish genome characterization to perform the functional analysis of specific genes, mainly through loss of function studies. Morpholino oligonucleotides are by far the most widely used method to knock down gene function in zebrafish. Novel reverse genetics approaches such as TILLING or the use of zinc finger nuclease facilitate the loss of function analysis of those genes that are less amenable to morpholino knockdown. The characterization of regulatory DNA sequences that enable targeting of genes to specific cell types provided



*Zebrafish: Volume 29*
FISH PHYSIOLOGY

DOI: 10.1016/S1546-5098(10)02901-8

novel experimental strategies such as the targeted misexpression of specific gene products. They also provide genetic means to ablate specific cell types.

## 1. INTRODUCTION

The emergence of the zebrafish in the last two decades as one of the most widely used experimental animals in biology is largely attributable to the characteristics of this fish that enable embryological manipulations and large-scale genetics studies. As the use of this teleost expanded, a number of novel genetic approaches were developed or adapted for this species. This, combined with a number of genetic resources, including a characterized and sequenced genome, further contributed to the popularity of the zebrafish. Most of the early investigators using zebrafish were interested in embryonic development and, thus, many of the initial techniques were aimed at facilitating embryological studies. However, it became clear that this came with limitations and this explains why some approaches were developed in recent years that are more appropriate to biological events that take place later than embryogenesis (2–3 days post fertilization) or the early larval stages and during adulthood.

The initial rise in zebrafish use also resulted from large-scale genetic screens for mutations that affected all aspects of development, and of most of the biology of these fish (Driever et al. 1996; Haffter et al. 1996). The screens somewhat mimicked similar efforts that had previously been carried out in invertebrate species such as the fruit fly *Drosophila* and the nematode *C. elegans*. As screens for mutations, referred to as forward genetics, produce a wealth of exciting phenotypes but often do not easily provide information about the affected genes, the genome of the zebrafish had to be characterized. Genetic maps (meiotic maps) were made as these are essential to position mutations on one of the 25 chromosomes of the zebrafish genome. Further tools such as radiation hybrid maps (Geisler et al. 1999; Hukriede et al. 1999, 2001) and collections of expressed sequence tags (ESTs) were important milestones in the characterization of the zebrafish genome whose "sequencing" can be considered almost completed. The genome characterization efforts led to the identification of a number of genes, most with orthologs in other vertebrate species. Gene identification enabled "reverse genetics" approaches in which gene function is assessed, generally through loss-of-function and sometimes gain-of-function experiments.

We will now review some of the most widely used genetic approaches to which the zebrafish is amenable, starting, for historical reasons, with a brief overview of forward genetics screens, followed by reverse genetics approaches and transgenic methods. Finally, methods for the ablation of specific cell types will be described.

## 2. FORWARD GENETICS: FROM LARGE SCREENS TO TARGETED SCREENS

The rise in the popularity of the zebrafish as an experimental system in biology is largely attributable to the large screens for mutations that took place in the 1990s. The first two large screens (Driever et al. 1996; Haffter et al. 1996) used the chemical mutagen ethylnitrosourea because of its high mutation efficiency. More than 6600 mutations were identified. Families of mutagenized fish were screened for visible abnormalities by examination under a dissection microscope of selected tissues and organs at 3–5 developmental time points. Simple behavior such as motility and response to touch (Haffter et al. 1996) as well as optokinetic behavior (Brockerhoff et al. 1995) were examined. More elaborate screening procedures also allowed the identification of mutations affecting retino tectal projections (Baier et al. 1996). Some of the mutants exhibited a specific phenotype but as many as 70% were considered non-specific, meaning that the mutants had general growth defects, extensive cell death. The mutants that were maintained and characterized probably correspond to about 400–600 loci. Importantly, most of the mutations in the screens produced a lethal phenotype in homozygous fish during the embryonic or larval stages.

The molecular identification of the gene affected by a particular mutation requires positional cloning. This was initially a very demanding endeavor but characterization of the zebrafish genome, including dense genetic and physical maps and a nearly complete genome sequence, has facilitated this task to a significant extent. More than 150 genes affected by mutations obtained in the chemical screens have been so far identified.

Chemical mutagenesis is carried out by exposing male zebrafish to ENU several weeks before mating and mutations taking place in the spermatogonia are passed on to the next generations. ENU usually causes point mutations that could result in altered protein sequence, truncated proteins, impaired splicing of the precursor mRNA or mRNA degradation through non-sense mediated mRNA decay. Point mutations affecting *cis*-acting regulatory elements were also reported. Thus, chemical mutagens can create hypomorphs, gain of function mutations, and dominant negative alleles, in addition to the null (complete loss of function) alleles. In this sense, chemical mutagenesis offers a wider phenotypic variety than some of the reverse genetics approaches described in the following sections, such as morpholino oligonucleotides. It is believed that all loci in the genome are equally accessible to the point mutations introduced by ENU but specific experiments aimed at investigating this property of ENU and the relative frequency of alleles in the large screens suggest that some chromosome loci are more frequent targets of ENU than others.

In parallel to the screens based on ENU, insertional mutagenesis was also carried out on a large scale in one laboratory (Gaiano et al. 1996; Amsterdam et al. 1999) using a retrovirus as an insertional mutagen. Although the efficiency of mutagenesis by the insertional method is estimated to be about seven times lower than that achieved by ENU, the method offers the important advantage that the retroviral sequences can be used as a "tag" to rapidly recover the genomic sequences flanking the insertion site. Combined to the genomic tools described above, it becomes relatively simple to identify the gene affected by the insertion. Thus, of about 520 mutations recovered in the insertional screen, 335 of the corresponding loci have been identified. Interestingly, the sets of genes obtained with the two approaches, chemical and insertional mutagenesis, overlap extensively.

Both chemical and insertional mutagenesis screens require considerable resources when the project is carried out on a large scale. As phenotype examination, mutant recovery and maintenance represent a significant proportion of this effort, more focused screens for very specific phenotypes, although still challenging, constitute an interesting alternative. Examples of such recent focused screens include screen for visual behavior mutants (Muto et al. 2005; Wehman et al. 2005), heart atrioventricular cushion and valve development (Beis et al. 2005), and mutations causing genome instability or increased cancer susceptibility (Shepard et al. 2005). Somewhat surprisingly, the phenotype of mutants obtained in these more focused screens was also often lethal at the embryonic or larval stages.

## 3. REVERSE GENETICS: MORPHOLINO KNOCKDOWNS

Antisense technologies have been used in numerous species, both in whole animals and in cells, to create loss-of-function phenotypes. These antisense RNA technologies include RNA interference (RNAi) and morpholinos (MO). Although RNAi is used in a number of species, widespread off-target effects (Oates et al. 2000) have prevented their use in zebrafish. On the other hand, MOs are now so widely used that, when the approach is applicable, it is almost unthinkable to see characterization of gene function in zebrafish without a morpholino-based loss of function study.

### 3.1. Structure and Design of Morpholino Oligonucleotides

Morpholinos (MOs) are synthetic oligonucleotides, in which the ribose or deoxyribose sugar has been replaced by a morpholine ring (Summerton and Weller 1997). The size of the MOs is generally around 25 nucleotides. The structure of morpholinos allows them to participate in Watson–Crick pairings with cellular nucleic acids. The presence of the morpholine ring confers on MOs resistance to endogenous nucleases and makes them particularly stable. MOs are either designed so that they hybridize with the mRNA of the gene of interest near the translation initiation codon or at a splice junction. They are referred to as translation-blocking and splice-blocking MO, respectively. Contrarily to some of the other antisense methods, MOs do not function through RNAse-mediated degradation. Translation-blocking MOs, as their name suggests, inhibit translation of their target mRNA whereas splice-blocking morpholinos prevent interaction of the mRNA with the mRNA splicing machinery, resulting in a mature mRNA of altered size that will code for a truncated or abnormal protein.

Morpholinos are injected into fertilized zebrafish eggs at the one- to eight-cell stage and are more or less uniformly distributed in the various cells as the embryo develops (Nasevicius and Ekker 2000; Bill et al. 2009). A dose–response curve is generally carried out and specific phenotypes are generally obtained in at least 50% of the injected embryos with doses of 5 ng or less. Efficacy of the morpholinos is considered to last until about five days post-fertilization.

### 3.2. Determining the Efficacy of Morpholinos

The ideal test of the efficacy of a translation-blocking MO necessitates measurement of the protein amount or activity for the gene under study. As protein activity measurements are not always possible, determination of protein amounts necessitates a suitable antibody. Although generation of this antibody is expensive and time-consuming, it is often essential. A second best alternative is to use a tagged version of the MO target, which can be obtained, for example, by creating a synthetic mRNA where the MO target is linked to the coding sequences of the green fluorescent protein (GFP) gene (Collart et al. 2005). Loss of GFP activity in embryos that received both the tagged mRNA and the MO is already an indication that the MO is able to bind its target in an in vivo context but does not necessarily guarantee that the translation of the endogenous, bona fide, target is as efficiently blocked.

When using a translation-blocking MO, the need for a suitable antibody is circumvented when using splice-blocking morpholinos as their efficacy can be measured by the relative amounts of the native and altered transcripts, as determined by reverse-transcription PCR (RT-PCR).

### 3.3. Determining the Specificity of the Morpholinos

All antisense technologies are subject to non-specific or off-target effects and it is especially important to assess the specificity of the phenotype obtained with MO in addition to its efficacy. The wide use of MOs and the numerous publications that have resulted have led to refinements and some standardization in the controls. We briefly describe a few of these controls here. For a comprehensive review, the reader is referred to Eisen and Smith (2008).

#### 3.3.1. Multiple Morpholinos

The use of a second morpholino targeted against the same mRNA is the simplest control for specificity of the morphant phenotype and has become increasingly required from authors. The second morpholino should be non-overlapping in sequence with the first and produce the same phenotype. Having one of the two morpholinos blocking translation and the second blocking splicing is a particularly desirable combination. Finally, a particularly convincing result is obtained following co-injection of the two morpholinos, each in a dose that, when used individually, does not produce a morphant phenotype.

#### 3.3.2. Control and Mismatched Morpholinos

GeneTools, the only supplier of custom MOs for research use, also provides a standard morpholino directed against the human β-globin mRNA that will not be found in zebrafish embryos and larvae. This standard morpholino serves as a control for the use of MOs in general but is not necessarily a good control for every specific MO as it does not exclude the possibility that the MO of interest is also able to recognize a closely related target sequence in a gene other than its target. Therefore, a more specific control MO will correspond to a 25-mer that differs from the original MO by five nucleotides. Some control morpholinos with a smaller number of mismatches (e.g. four nucleotides) were found to mimic the effects of the original MO when used in high concentrations (Cornell and Eisen 2002). Theoretical calculations based on the *Xenopus tropicalis* genome sequence predicted a non-negligible probability (0.3) that a translation-blocking MO will recognize another target sequence near its translation start site with four mismatches or less. However, there are reasons to expect that the binding of a MO to such a mismatched sequence will have weaker phenotypic consequences than binding of the morpholino to its true target. Nevertheless, such considerations highlight the importance of using multiple controls in MO experiments.

#### 3.3.3. Rescue of the Phenotype

The rescue of the morphant phenotype by injection of synthetic mRNA is another useful evidence for the specificity of the MO effect. Of course, the synthetic mRNA must not be a target for the MO itself. In the case of translation blocking MO, this can be achieved by using an mRNA that lacks 5′-untranslated sequences, in cases where the MO is entirely directed against them. If the MO is partially complementary to the protein-coding region near the translation initiation site, it may become necessary to mutagenize the gene sequence in vitro to introduce mismatches that are conservative, that is, that do not alter the amino acid sequence. The order of injection of the MO and the synthetic mRNA used for rescue may also be important and some sources recommend injection of the synthetic mRNA first. Finally, rescue with a synthetic mRNA may not always be possible. This problem is particularly likely for genes expressed at low levels or in restricted cell types as the injected mRNA will be expressed ubiquitously and at levels that could be considered non-physiological. Thus, the ectopic or overexpression could have phenotypic consequences on its own and mask any rescuing effects. It is clearly recommended to first test the synthetic mRNA by itself and carry out a dose–response. Typical amounts of injected mRNA range between 50 pg to 1 ng.

#### 3.3.4. The p53 Apoptosis Pathway

Off-target effects of MOs often share common characteristics. This was particularly noticed in a large screen of translation-blocking MOs. It was later determined that some of these off-target effects are attributable to the activation of the p53 apoptosis pathway. Co-injection of a translation-blocking MO targeted against the p53 mRNA along with the MO under study often eliminated the non-specific apoptosis. Note that non-MO-induced cell death can be blocked with the p53MO. Nevertheless, co-injection of the p53 MO may be desirable when non-specific cell death is suspected and the gene whose function is being studied is not involved in p53-dependent biological mechanisms.

### 3.4. New Types and Uses of MOs

#### 3.4.1. Morpholinos Targeting MicroRNAs

Micro-RNA genes generate short (about 22 nucleotides) RNAs involved in the control of expression of a number of target genes by blocking the translation or by causing the degradation of their messenger RNAs. Thus, microRNAs play important roles in development and, probably, in most cellular and physiological pathways. It is possible to design MOs that will interfere with micro-RNA function. These MOs will act as target protectors by preventing the maturation of the microRNA and, thus, their ability to interact with their target messenger RNAs.

#### 3.4.2. Delivery of Morpholinos by Electroporation

MOs can also be delivered to specific cells or adult tissues using electroporation (Cerda et al. 2006). This approach has been successfully used for regenerating zebrafish fins and retina (Thummel et al. 2006, 2008; Hoptak-Solga et al. 2008; Thatcher et al. 2008) and should be applicable to groups of cells in embryos such as individual regions of the developing central nervous system.

### 3.4.3. Photoactivable MOs

Photoactivable MOs have been recently commercialized. Although injected into 1–8-cell stage embryos as conventional MOs and thus, distributed throughout the embryo, photoactivable MOs will only become active when hit by a focused laser beam. Therefore, we can activate the MO in a restricted region of the embryo at a given time in development (Shestopalov et al. 2007).

## 4. REVERSE GENETICS: TILLING

Despite their simplicity, low cost and generally high success rate, morpholino oligonucleotides have their limitations, especially for genes expressed late in development. The availability of a "complete" sequence and the identification of a large fraction of the genes that constitute the zebrafish genome enabled additional reverse genetics approaches; one of them is TILLING. Initially applied to the plant *Arabidopsis thaliana* (Colbert et al. 2001; Till et al. 2003), TILLING takes advantage of the ability of performing random chemical mutagenesis in those species, as is also the case for zebrafish (Wienholds et al. 2002, 2003). A TILLING screen starts with the generation of a large number of individuals, each carrying several chemically induced mutations. The mutagen used in zebrafish is ethylnitrosourea (see Section 2) which introduces point mutations in the genome. As for traditional mutagenesis screens, males are exposed in order to create mutations in the spermatogonia. The exposure regimen to the mutagen is designed in such a way as to provide an estimated number of hits in the genome. Families produced by breeding the exposed male fish are either kept in the fish-holding facility or the sperm from one individual is frozen for later use (Fig. 1.1). DNA is prepared from the fish and is used as a template in the TILLING procedure. The purpose of TILLING is to look for specific mutations in specific genes in a collection of DNA samples prepared from the chemically mutagenized individuals. Identification of mutations is done by either one of two methods. The first consists of the amplification by PCR of a specific region of the gene of interest, generally one of the first coding exons, followed by sequencing of the amplified fragment. This is done in each of the samples of the mutagenized fish DNA collection. The second method, based on the CEL-I enzyme, isolated from celery, also begins by PCR amplification of a region of the gene of interest.

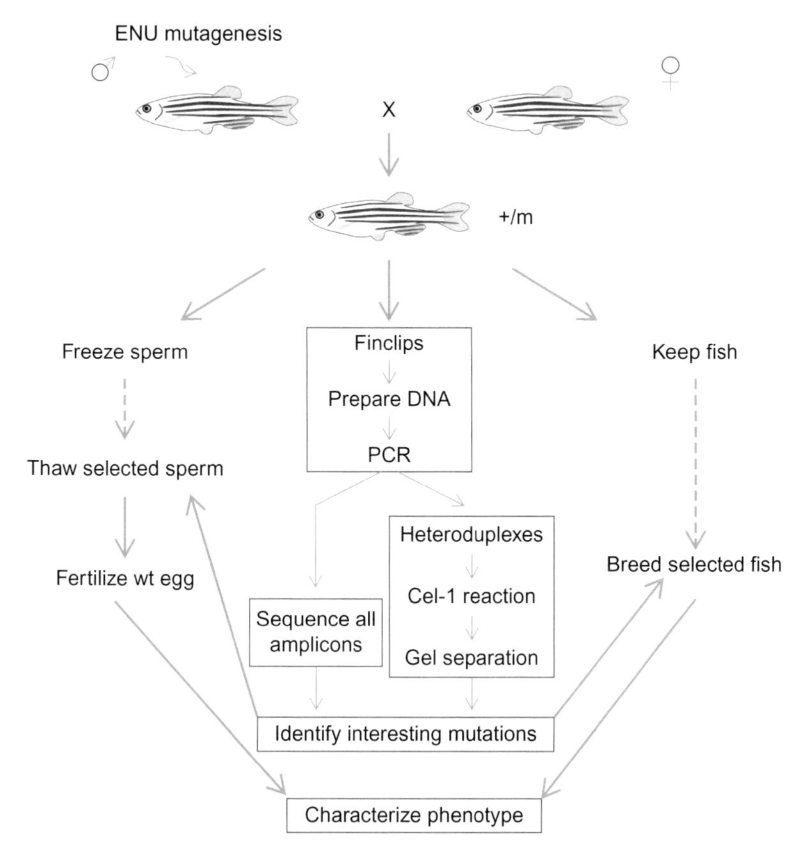


**Fig. 1.1.** Identification of mutations in specific genes by TILLING. Male zebrafish are mutagenized with ethylnitrosourea (ENU) and mated to wild-type females. Individuals from the F1 generation are either kept in the aquatic facility (right) or their sperm is frozen (left). Fin clips are used to prepare DNA that is amplified by PCR at specific gene loci. The amplicons are screened for mutations by sequencing (center left) or by detection of heteroduplexes with CEL-I endonuclease (center right). Once interesting mutations are identified, the fish are mated (right) or the sperm is thawed (left). After breeding the mutation to homozygosity, the phenotype caused by the mutation is analyzed. For details, see text.

The two DNA strands of the amplicon are melted by raising the temperature of the DNA solution. The two strands are then allowed to reanneal. As zebrafish are diploids and the mutation will be present in a heterozygous state, a cycle of melting and re-annealing will result in a mix of homoduplexes, where the two strands of the same allele will find each other, and heteroduplexes where one strand of the wild-type allele anneals with the complementary strand of the mutant allele. The CEL-I enzyme is able to recognize and cleave heteroduplexes, generating fragments that are smaller than the original amplicon. The DNA fragments are subjected to electrophoresis and the presence of lower size DNA fragments is indicative of a mutation. The size of the fragment also gives an indication of the position of the mutation which can then be confirmed by sequencing. Overall, the "sequencing" approach is simpler, easier to carry out on a large scale but is more costly than the "CEL-I" approach.

Regardless of the method used to "TILL" for the mutation, the next step is to determine if the identified mutation is likely to have an impact on the gene product. Mutations introducing stop codons are generally preferred and their finding in the first coding exons will produce severely truncated proteins. However, some mutations may also be silent if they fall in the third nucleotide of a codon and do not change the amino acid sequence of the protein. Others will result in amino acid substitutions. When interesting mutations are found, the next step is to thaw the frozen sperm from this particular sample and use it to fertilize wild-type eggs (Fig. 1.1). If the mutagenized fish were kept in the zebrafish-holding facility, they are immediately bred. Once zebrafish homozygous for the mutation have been obtained, their phenotype is examined throughout development and even in late larvae or adults. The phenotype of heterozygous individuals can also be observed, which can reveal haploinsufficiency for a particular gene. Heterozygous individuals carrying the mutation can be maintained. All these are characteristics of the TILLING approach that are not possible with the acute morpholino knock-down method. As mentioned earlier, the chemical mutagenesis that initiates a TILLING screen offers a wider phenotypic variety than morpholino knock-downs.

In the initial zebrafish TILLING publication, Wienholds and colleagues (2003) screened for mutations in 16 genes in a library of 4608 mutagenized animals at the F1 generation. They used the CEL-I approach. They were able to identify a total of 255 mutations, 14 of which corresponded to premature stop codons, seven to alterations of splicing donor/acceptor sites and 119 to amino acid changes. Their estimate is that potential null mutants had been obtained for 13 of the 16 genes tested (Wienholds et al. 2003).

## 5. REVERSE GENETICS: ZINC FINGER NUCLEASES

The limitations in the use of morpholinos and the randomness of the TILLING approach created the need for alternative methods to impair function of specific genes. Engineered zinc finger nucleases provide such ability to create specific mutations in zebrafish genes of interest (Doyon et al. 2008; Meng et al. 2008). Zinc finger nucleases (ZFNs) are molecular chimeras that consist of three or four DNA binding domains of the zinc finger type and the restriction endonuclease domain of the enzyme *Fok*1 (Fig. 1.2). ZFNs will induce a double-strand break in the DNA molecule at a site whose specificity is conferred by the combined DNA-binding preferences of the multiple zinc fingers (Fig. 1.2A). As each zinc finger recognizes a triplet of DNA base pairs (bp), a ZFN containing 3–4 zinc

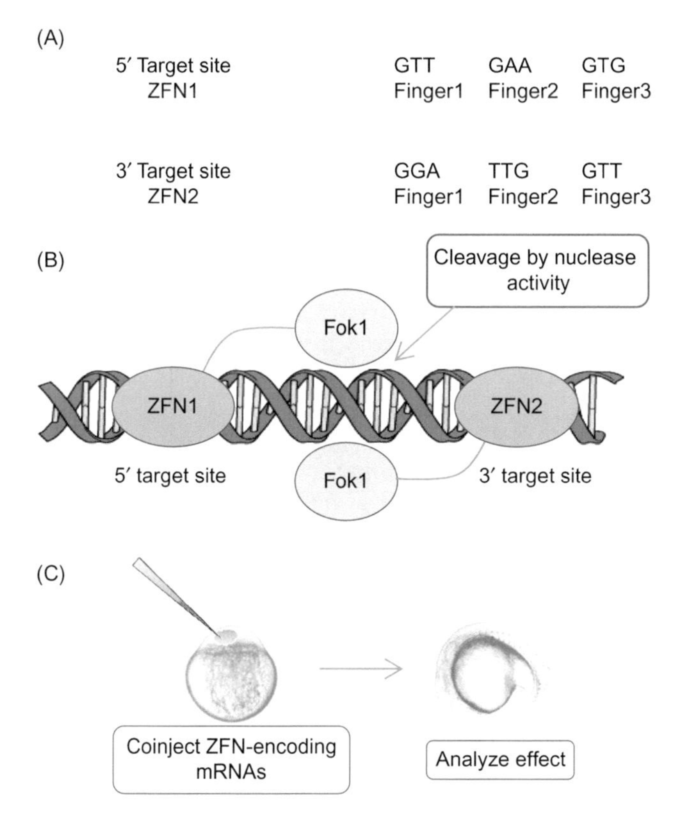


**Fig. 1.2.** Mutations at specific zebrafish gene loci using zinc finger nucleases. (A) The cleavage specificity of each zinc finger nuclease (ZFN) is conferred by the combined binding preferences of the three zinc fingers. Two ZFNs are designed for a targeted locus. (B) Mechanism of action of the ZFNs. The two ZFNs each bind their target site, bringing the two *Fok1* nuclease domains in close proximity. Cleavage takes place and the lesion is repaired by the non-homologous end joining machinery. This will create small mutations. (C) Synthetic mRNA encoding each of the two ZFNs are produced in vitro and injected into zebrafish embryos at the one-cell stage. Lesions will take place early in development using the mechanism shown in (B). The phenotype-resulting mutations caused by ZFN-mediated cleavage and subsequent repair can be passed to future generations or can be observed in primary zebrafish, if both alleles of the gene are mutated.

fingers will specifically cleave DNA at a recognition sequence of 9–12 bp. When two ZFNs bind the target sequence, the *Fok*1 nuclease domains are brought together and this creates a double-strand DNA break (Fig. 1.2B), which is subsequently repaired by one of the two endogenous DNA repair mechanisms: non-homologous end joining and homologous recombination. DNA repair through non-homologous end joining often introduces the insertion or deletion of a few base pairs and this property is used to create mutations at specific sites in the genome. The strategy is thus to design two ZFNs that will recognize a specific sequence in the gene of interest and to introduce the mRNA encoding such ZFNs into fertilized zebrafish eggs (Fig. 1.2C). Efficacy of the ZFNs can be monitored by phenotypic examination of the embryos (or of their progeny) and a molecular screening for mutations by PCR. The mutations introduced by the ZFNs can be passed to the progeny of the injected fish with a high probability (30–50%) (Doyon et al. 2008; Meng et al. 2008). Because of the length of the sequence recognized by the pair of ZFNs, it is expected that the activity of these enzymes will be highly specific and this has indeed been verified experimentally. Depending on the amounts of ZFN mRNA that are injected, one or both copies of the target gene will be cleaved. The crucial step in targeted mutagenesis with ZFNs, and the main obstacle to the wide applicability of the method, at least to date, is the design, selection and validation of the ZFNs. Although a few well-validated targets are known for specific zinc finger domains, the design of a modular ZFN that will recognize a specific sequence remains a difficult task and the enzymatic properties of ZFNs designed in silico have to be verified using an experimental test that is performed either in microorganisms and/or that involve testing multiple candidate ZFNs in zebrafish embryos to determine which one functions with the highest efficacy and specificity. Overall, this approach remains difficult for laboratories experimenting with this system for the first time. It is possible that commercial services may in the future perform this service.

Proof of concept for this method was provided by the phenocopy of phenotypes of mutants obtained by more traditional means (e.g. ENU mutagenesis) for genes such as golden (*gol*), no tail (*ntl*) and the vascular endothelial growth factor-2 receptor gene *kdr* (Doyon et al. 2008; Meng et al. 2008).

Finally, it must be noted that variations on the procedure are currently under development that would allow DNA cleavage at specific sequence with ZFNs while favoring DNA repair through the homologous recombination pathway. This would provide greater control of the mutations that are induced and would open the door to a wide variety of manipulations of the genome analogous to those carried in mouse embryonic stem cells such as gene deletions or replacements.

## 6. GAIN OF FUNCTION: TARGETED MISEXPRESSION OF GENES

Most of the phenotypes resulting from the genetic alterations described in the previous sections of this chapter result from the complete or partial loss of gene function. Gain-of-function phenotypes can also be informative in characterizing the role of specific genes and a number of methods are now available to do this in zebrafish. By far the simplest method, the microinjection into fertilized zebrafish eggs of mRNA synthesized in vitro has been widely used to ectopically express or to overexpress the product of specific genes. Some control on the amounts of protein produced can be achieved by varying the dose of the injected mRNA. The external development of fish and amphibians makes them particularly amenable to this methods and it has been widely used, including for the rescue of the phenotypes induced by morpholinos (see Section 3.3.3). However, this approach suffers from multiple drawbacks. First, it can be assumed that, for most mRNAs, the injected transcripts will be more or less uniformly distributed as the embryo develops, resulting in ubiquitous or nearly ubiquitous expression. This can lead, for some gene product, to pleiotropic effects that render phenotype analysis in specific cell populations difficult. Second, mRNAs differ vastly in their stability and ectopic/overexpression in cell populations that differentiate later in development may be difficult to achieve. Therefore, transgenic approaches that rely on DNA constructs where the gene under study is placed under the control of tissue-specific and/or inducible regulatory elements offer advantages despite the extra experimental efforts that they require.

Recent years have seen the identification of a number of *cis*-regulatory DNA sequences, including promoters and enhancers, which regulate gene expression in spatial and temporal manner. Such sequences were used in DNA constructs to target expression of genes in developing zebrafish and in adults. For a recent review, see MacDonald and Ekker 2010. Tissue/stage-specific regulatory elements can also be used as part of a dual transgene system such as the GAL-UAS that offers a wide variety of applications for the ectopic expression of genes (Scheer and Campos-Ortega 1999; Davison et al. 2007; Asakawa et al. 2008; Distel et al. 2009). In such a dual system the first construct, called the driver, encodes the gene for the yeast GAL4 DNA-binding protein placed under the control of tissue/stage-specific regulatory elements. The GAL4 gene is further modified such that it makes a chimeric protein with the VP 16 transcription activation domain, producing a strong transcriptional activator. The GAL4-VP16 transgene is thus expressed in specific zebrafish cell populations of established transgenic lines (Fig. 1.3). A large number of independent transgenic driver lines can be produced, each

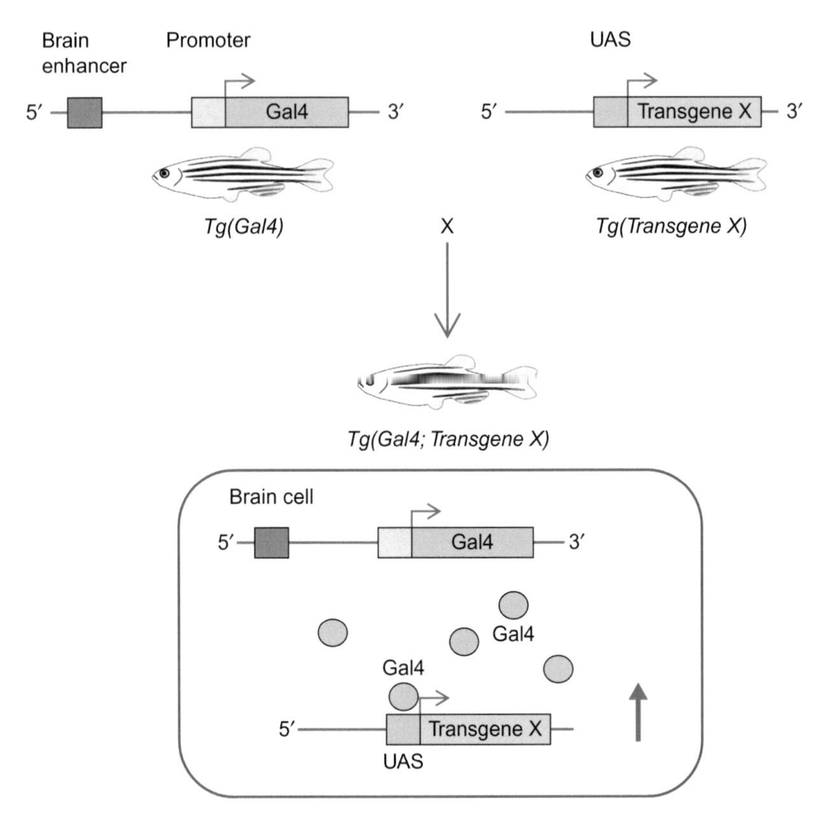


**Fig. 1.3.** The GAL4-UAS dual system for tissue-specific expression of transgenes. Two transgenic lines are necessary. The first contains the coding region for the GAL4-VP16 transcriptional activator under the control of tissue specific *cis*-regulatory elements (e.g. brain-specific enhancer). The second transgenic line contains a construct in which the gene of interest (e.g. gene X) is placed under the control of the UAS sequence, a GAL4 target site. When individuals of the two transgenic lines are mated, the GAL4-VP16 protein, expressed in brain cells of the progeny, will activate expression of gene X by binding to the UAS sequence. For details, see text.

with a distinct pattern of GAL4-VP16 expression. The second construct of the dual system encodes a gene of interest placed under the control of a promoter containing multiple copies of the upstream activator sequence (UAS), the recognition sequence for the GAL4 transcriptional activator (Fig. 1.3). Again, a number of independent transgenic lines, each with a different gene of interest can thus be produced. Once zebrafish from a GAL4 driver line are mated with individuals of a UAS-gene X line, gene X will be specifically expressed where and when the GAL4 protein is produced. The large numbers of mating combinations give this dual transgene approach its high versatility, compared to the individual production of transgenic fish with specific construct where the gene of interest is placed under the control of specific *cis*-regulatory elements.

In addition to the above transgenic lines in which the GAL4-VP16 gene is placed under the control of characterized *cis*-regulatory elements, the GAL4-VP16 system can also be used as enhancer traps to obtain a wider variety of driver expression patterns or potentially target cell populations for which no *cis*-regulatory elements have yet been characterized.

Inducible systems: Some applications of transgenesis necessitate the ability to induce expression of a transgene via an exogenous control mechanism. Inducible systems used for controlled expression generally consist of two components: an activator, usually a transcription factor whose expression is controlled in a tissue- or time-specific manner, and a second gene that contains a promoter region that is responsive to the activator. Examples of inducible systems that have been tested in zebrafish include the tetracycline repressor system and the Cre-lox system.

The Cre recombinase is a protein isolated from the bacteriophage P1. It recognizes and catalyzes DNA recombination between recognition sequences referred to as loxP sites (Hamilton and Abremski 1984; Hoess and Abremski 1984). Any DNA sequence flanked by loxP sites (or "floxed") will be excised by recombination if Cre is present in the cell. Therefore, the use of Cre recombinase to produce tissue-specific and/or inducible transgene expression requires independent lines of transgenic animals produced with the following transgenes. The first transgene contains the gene of interest flanked by loxP sites in such a way that they prevent its expression. The second transgene contains the coding sequences for the Cre recombinase protein placed under the control of tissue-specific regulatory elements. Mating of transgenic animals containing each transgene will result in activation of the gene of interest only in those cells that express the Cre recombinase. A number of reports using and optimizing the Cre-loxP system in zebrafish have been published in recent years (Langenau et al. 2005; Pan et al. 2005; Thummel et al. 2005; Le et al. 2007; Hans et al. 2009). As in the above GAL4-UAS system the availability of lines of transgenic animals with the various constructs provides versatility to this approach.

The activation of heat shock proteins by cellular stress (e.g. an increase in temperature) is a highly conserved process shared between prokaryotes and eukaryotes (Bardwell and Craig 1984; Liu et al. 1997). The heat shock response can be utilized as an inducible system, and has been tested in zebrafish using heterologous heat shock promoter regions from frog and mouse (Adam et al. 2000. It is possible to target the heat shock response to specific cells using physical methods. Thus, Halloran et al. (2000) induced expression of a reporter gene driven by a heat shock promoter in single cells using a sublethal laser microbeam. In another study, Hardy and colleagues (Hardy et al. 2007) modified a soldering iron to induce transgene misexpression in a small area of the embryo. These methods were found to be successful in inducing transgene expression in small groups of cells in a temporally controlled manner. Heat shock induction also allows temporal control of Cre recombinase activity (Le et al. 2007).

Heat shock promoters can also be used as part of transgenes to target expression of dominant negative forms of proteins or constitutively active forms of receptors. Expression of these altered proteins is achieved by causing heat shock at the appropriate time and, if desired, at the appropriate location using one of the above methods. This strategy has been particularly used to study regeneration of adult zebrafish tissues (Lee et al. 2005, 2009; Stoick-Cooper et al. 2007).

## 7. GENETIC ABLATION OF SPECIFIC CELL POPULATION

The reverse genetics approaches described in the above sections enable the functional analysis of individual genes. The combined characterization of multiple genes allows tracing of the pathways responsible for the proper development or physiological role of specific cell populations. However, the ability to ablate specific cell populations is also a powerful tool to study developmental or physiological processes. Physical techniques exist in zebrafish to ablate cells either by physical surgery or using laser irradiation. However, these techniques are labor-intensive and not really amenable to the treatment of a large number of individuals. Their reproducibility can also be variable. Genetic cell ablation tools that function in a spatially controllable manner, can be induced temporally, are reversible, and can be transmitted through the germline to a large progeny, are desirable. The development of such methods in the zebrafish encountered the same pitfalls experienced in other systems. For example, a simple genetic strategy to ablate a specific cell type is to target the expression of a gene encoding a toxin such as the diphtheria toxin A-chain to specific cells by constructing a transgene in which the toxin is placed under the control of *cis*-regulatory elements (promoter, enhancers) that are exclusively active in the cells one aims to ablate. The transgene is injected into fertilized zebrafish eggs and a stable transgenic line is established. A frequently encountered problem with such an approach or with its variations is the "leakiness" of the regulatory elements used to drive transgene. This leakiness results in off-target ablations if not in general toxicity that impairs the ability to generate the stable transgenic lines or cause undesirable side effects before the cell population under study has formed.

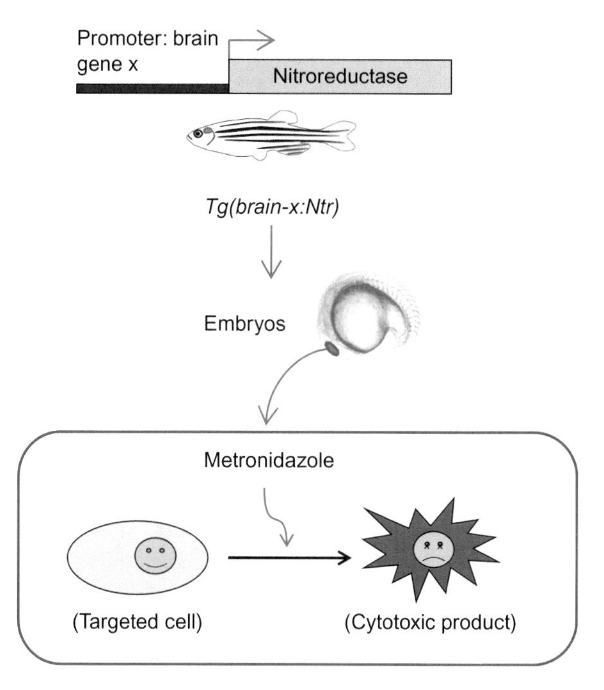


**Fig. 1.4.** Targeted ablation of specific cell types with the Ntr-Mtz strategy. The nitroreductase (*Ntr*) gene is placed under the control of cell-specific *cis*-regulatory sequences and this construct is used to produce a line of transgenic zebrafish that will express the Ntr protein in the cells of interest. When fish are administered the prodrug metronidazole (Mtz), it is metabolized into a cytotoxic compound only in those cells that express the Ntr protein. Specific ablation of these cells is thus accomplished in a time-controlled manner.

Recently, separate groups developed the use of the bacterial enzyme nitroreductase (NTR) that converts a prodrug, metronidazole (Mtz), into a cytotoxic DNA cross-linking compound (Curado et al. 2007; Pisharath et al. 2007). As in the above strategy, the NTR gene is part of a transgene in which it is controlled by cell-specific *cis*-regulatory elements (Fig. 1.4). However, the transgene has no toxic effects by itself until animals are exposed to the prodrug. The time of administration of the prodrug Mtz provides temporal control. The cytotoxic agent produced by the conversion of Mtz by the NTR enzyme is toxic to all cell types, including non-proliferating, which provides an advantage to the Ntr/Mtz system over analogous approaches based on the thymidine kinase/ganciclovir combination which inhibits the activity of DNA polymerase and is thus only active in proliferating cells.

The Ntr/Mtz system has been tested to ablate cells in the pancreas, heart, and liver. Both Pisharath and colleagues (2007) and Curado and colleagues (2007) targeted the NTR protein to the β cells of the pancreas with a transgene construct that contained the insulin promoter (Huang et al. 2001). Exposure of the transgenic fish to the prodrug resulted in specific ablation of the β cells without affecting the α or δ cells, although the loss of β cells perturbed the arrangement of the α cells. When pancreatic tissue of zebrafish expressing Ntr were exposed to Mtz for 18–24 hours, ablation of the targeted cell population was not complete, at least when assessed by marker gene expression or by TUNEL labeling for apoptotic cells (Curado et al. 2007). It is possible that a differential sensitivity of the cells in the tissue to the drug or a high turnover of the tissue (see below) may explain this observation.

Zebrafish possess a remarkable ability to regenerate cells after injury and this property can be studied with the Ntr/Mtz system. It was demonstrated that the cytotoxin produced by Ntr conversion of Mtz can be cleared from the organism following Mtz removal from the medium. The Ntr/Mtz system is thus reversible. Thus the authors of both studies that targeted zebrafish pancreatic β cells observed regeneration of these cells, as evidenced by transgene or insulin expression, when the fish were allowed to recover in the absence of the prodrug (Curado et al. 2007; Pisharath et al. 2007). Thus, zebrafish can serve as an experimental system to study the mechanisms of pancreatic β cell regeneration. Pisharath and colleagues (2007) provided evidence that the exocrine pancreas was not necessary for β cell recovery following Ntr/Mtz-mediated ablation. They used a morpholino directed again the *ptf1a* transcript to prevent exocrine pancreas development (Lin et al. 2004; Zecchin et al. 2004) and observed comparable β cell recovery in the absence or presence of *ptf1a* function.

## 8. CONCLUDING REMARKS

The identification of genes and the characterization of their function contribute largely to the understanding of the physiology of cells, organs, and systems. The large number of genes that were uncovered by the genome projects that took place in the last decade, although quite helpful, has also given us a flavor of the scope of what remains to be discovered. The characteristics of experimental animals such as the zebrafish enable some acceleration of our discovery of how the information provided by DNA sequences is converted into biological function. The evolutionary conservation of a large proportion of vertebrate genes and of the pathways in which they participate adds some significance to the findings made in zebrafish as they can be translated, with due care, not only to other fish species but also to other classes of vertebrates.

Although the gene complements of the zebrafish and of other vertebrates overlap to a large extent, the zebrafish, as well as other teleosts may have, overall, comparatively larger gene numbers due to an additional round of genome duplication that took place early after the separation of the evolutionary branches that led to actinopterygians (ray-finned fish) and to sarcopterygians (lobe-finned fish). This proposed whole genome duplications resulted in zebrafish (and other teleosts) having often two paralogs for a single gene found in other classes of vertebrates, such as mammals (Amores et al. 1998). As the two zebrafish paralogs will sometimes show partial functional redundancy, this may complicate phenotype analysis in both forward and reverse genetics projects. For example, it may explain why screens for mutations failed to provide some phenotypes or phenotypes that were subtle to a point that they escaped the rapid examination procedures inherent to large screens. Models have been proposed to explain the fate of duplicated genes including the Duplication-Degeneration-Complementation (DDC) model that proposes that complementary changes in *cis*-regulatory elements that control the expression of the two paralogs may contribute in large part to their subfunctionalization and their retention throughout evolution (Force et al. 1999). Thus, although the existence of duplicated genes in zebrafish may complicate somewhat the elucidation of their function, their existence may also provide interesting tools to understand the evolution of gene families and the evolution of gene function.

Although the global loss of function of a gene still remains the fastest route to the understanding of its role, refinements that allow tissue-specific or inducible loss of function enhance possible studies by potentially avoiding the pleiotropic effects of the global loss of function. Similarly, the search for random mutations in specific genes will provide, in addition to null phenotypes, hypomorphs or gain of function phenotypes that will shed further light on how genes work, how they interact with other genes in a pathway, and how they contribute to the proper operation of biological systems.

## ACKNOWLEDGMENTS

Work in the authors' laboratories is supported by grants from the Natural Sciences and Engineering Research Council of Canada and from the Canadian Institutes of Health Research.

## REFERENCES

Adam, A., Bartfai, R., Lele, Z., Krone, P. H., and Orban, L. (2000). Heat-inducible expression of a reporter gene detected by transient assay in zebrafish. *Exp. Cell. Res.* **256**(1), 282–290.

Amores, A., Force, A., Yan, Y.-L., Joly, L., Amemiya, C., Fritz, A., Ho, R. K., Langeland, J., Prince, V., Wang, Y.-L., et al. (1998). Zebrafish *hox* clusters and vertebrate genome evolution. *Science* **282**, 1711–1714.

Amsterdam, A., Burgess, S., Golling, G., Chen, W., Sun, Z., Townsend, K., Farrington, S., Haldi, M., and Hopkins, N. (1999). A large-scale insertional mutagenesis screen in zebrafish. *Genes Dev.* **13**(20), 2713–2724.

Asakawa, K., Suster, M. L., Mizusawa, K., Nagayoshi, S., Kotani, T., Urasaki, A., Kishimoto, Y., Hibi, M., and Kawakami, K. (2008). Genetic dissection of neural circuits by Tol2 transposon-mediated GAL4 gene and enhancer trapping in zebrafish. *Proc. Natl Acad. Sci. USA* **105**(4), 1255–1260.

Baier, H., Klostermann, S., Trowe, T., Karlstrom, R. O., Nusslein-Volhard, C., and Bonhoeffer, F. (1996). Genetic dissection of the retinotectal projection. *Development* **123**, 415–425.

Bardwell, J. C., and Craig, E. A. (1984). Major heat shock gene of Drosophila and the Escherichia coli heat-inducible dnaK gene are homologous. *Proc. Natl Acad. Sci. USA* **81** (3), 848–852.

Beis, D., Bartman, T., Jin, S. W., Scott, I. C., D'Amico, L. A., Ober, E. A., Verkade, H., Frantsve, J., Field, H. A., Wehman, A., et al. (2005). Genetic and cellular analyses of zebrafish atrioventricular cushion and valve development. *Development* **132**(18), 4193–4204.

Bill, B. R., Petzold, A. M., Clark, K. J., Schimmenti, L. A., and Ekker, S. C. (2009). A primer for morpholino use in zebrafish. *Zebrafish* **6**(1), 69–77.

Brockerhoff, S. E., Hurley, J. B., Janssen-Bienhold, U., Neuhauss, S. C., Driever, W., and Dowling, J. E. (1995). A behavioral screen for isolating zebrafish mutants with visual system defects. *Proc. Natl Acad. Sci. USA* **92**(23), 10545–10549.

Cerda, G. A., Thomas, J. E., Allende, M. L., Karlstrom, R. O., and Palma, V. (2006). Electroporation of DNA, RNA, and morpholinos into zebrafish embryos. *Methods* **39**(3), 207–211.

Colbert, T., Till, B. J., Tompa, R., Reynolds, S., Steine, M. N., Yeung, A. T., McCallum, C. M., Comai, L., and Henikoff, S. (2001). High-throughput screening for induced point mutations. *Plant Physiol.* **126**(2), 480–484.

Collart, C., Verschueren, K., Rana, A., Smith, J. C., and Huylebroeck, D. (2005). The novel smad-interacting protein smicl regulates chordin expression in the Xenopus embryo. *Development* **132**(20), 4575–4586.

Cornell, R. A., and Eisen, J. S. (2002). Delta/Notch signaling promotes formation of zebrafish neural crest by repressing neurogenin 1 function. *Development* **129**(11), 2639–2648.

Curado, S., Anderson, R. M., Jungblut, B., Mumm, J., Schroeter, E., and Stainier, D. Y. (2007). Conditional targeted cell ablation in zebrafish: A new tool for regeneration studies. *Dev. Dyn.* **236**(4), 1025–1035.

Davison, J. M., Akitake, C. M., Goll, M. G., Rhee, J. M., Gosse, N., Baier, H., Halpern, M. E., Leach, S. D., and Parsons, M. J. (2007). Transactivation from GAL4-VP16 transgenic insertions for tissue-specific cell labeling and ablation in zebrafish. *Dev. Biol.* **304**(2), 811–824.

Distel, M., Wullimann, M. F., and Koster, R. W. (2009). Optimized GAL4 genetics for permanent gene expression mapping in zebrafish. *Proc. Natl Acad. Sci. USA* **102**, 13365–13370.

Doyon, Y., McCammon, J. M., Miller, J. C., Faraji, F., Ngo, C., Katibaj, G. E., Amora, R., Hocking, T. D., Zhnag, L., Rebar, E. J., et al. (2008). Heritable targeted gene disruption in zebrafish using designed zinc-finger nucleases. *Nat. Biotechnol.* **26**, 702–708.

Driever, W., Solnica-Krezel, L., Schier, A. F., Neuhauss, S. C. F., Malicki, J., Stemple, D. L., Stainier, D. Y. R., Zwartkruis, F., Abdelilah, S., Rangini, Z., et al. (1996). A genetic screen for mutations affecting embryogenesis in zebrafish. *Development* **123**, 37–46.

Eisen, J. S., and Smith, J. C. (2008). Controlling morpholino experiments: Don't stop making antisense. *Development* **135**(10), 1735–1743.

Force, A., Lynch, M., Pickett, F. B., Amores, A., Yan, Y.-L., and Postlethwait, J. (1999). Preservation of duplicate genes by complementary, degenerative mutations. *Genetics* **151**, 1531–1545.

Gaiano, N., Amsterdam, A., Kawakami, K., Allende, M., Becker, T., and Hopkins, N. (1996). Insertional mutagenesis and rapid cloning of essential genes in zebrafish. *Nature* **383**(6603), 829–832.

Geisler, R., Rauch, G.-J., Baier, H., van Bebber, F. V., Bro, L., Dekens, M. P. S., Finger, K., Fricke, C., Gates, M. A., Geiger, H., et al. (1999). A radiation hybrid map of the zebrafish genome. *Nature Genet.* **23**, 86–89.

Haffter, P., Granato, M., Brand, M., Mullins, M. C., Hammerschmidt, M., Kane, D. A., Odenthal, J., van Eeden, F. J. M., Jianmg, Y.-J., Heisenberg, C.-P., et al. (1996). The identification of genes with unique and essential functions in the development of the zebrafish, *Danio rerio*. *Development* **123**, 1–36.

Halloran, M. C., Sato-Maeda, M., Warren, J. T., Su, F., Lele, Z., Krone, P. H., Kuwada, J. Y., and Shoji, W. (2000). Laser-induced gene expression in specific cells of transgenic zebrafish. *Development* **127**, 1953–1960.

Hamilton, D. L., and Abremski, K. (1984). Site-specific recombination by the bacteriophage P1 lox-Cre system. Cre-mediated synapsis of two lox sites. *J. Mol. Biol.* **178**(2), 481–486.

Hans, S., Kaslin, J., Freudenreich, D., and Brand, M. (2009). Temporally-controlled site-specific recombination in zebrafish. *PLoS ONE* **4**(2), e4640.

Hardy, M. E., Ross, L. V., and Chien, C. B. (2007). Focal gene misexpression in zebrafish embryos induced by local heat shock using a modified soldering iron. *Dev. Dyn.* **236**(11), 3071–3076.

Hoess, R. H., and Abremski, K. (1984). Interaction of the bacteriophage P1 recombinase Cre with the recombining site loxP. *Proc. Natl Acad. Sci. USA* **81**(4), 1026–1029.

Hoptak-Solga, A. D., Nielsen, S., Jain, I., Thummel, R., Hyde, D. R., and Iovine, M. K. (2008). Connexin43 (GJA1) is required in the population of dividing cells during fin regeneration. *Dev. Biol.* **317**(2), 541–548.

Huang, H., Vogel, S. S., Liu, N., Melton, D. A., and Lin, S. (2001). Analysis of pancreatic development in living transgenic zebrafish embryos. *Mol. Cell. Endocrinol.* **177**(1-2), 117–124.

Hukriede, N., Fisher, D., Epstein, J., Joly, L., Tellis, P., Zhou, Y., Barbazuk, B., Cox, K., Fenton-Noriega, L., Hersey, C., et al. (2001). The LN54 radiation hybrid map of zebrafish expressed sequences. *Genome Res.* **11**, 2127–2132.

Hukriede, N. A., Joly, L., Tsang, M., Miles, J., Tellis, P., Epstein, J. A., Barbazuk, W. B., Li, F. N., Paw, B., Postlethwiat, J. H., et al. (1999). Radiation hybrid mapping of the zebrafish genome. *Proc. Natl Acad. Sci. USA* **96**, 9745–9750.

Langenau, D. M., Feng, H., Berghmans, S., Kanki, J. P., Kutok, J. L., and Look, A. T. (2005). Cre/Lox-regulated transgenic zebrafish model with conditional myc-induced T cell acute lymphoblastic leukemia. *Proc. Natl Acad. Sci. USA* **102**(17), 6068–6073.

Le, X., Langenau, D. M., Keefe, M. D., Kutok, J. L., Neuberg, D. S., and Zon, L. I. (2007). Heat shock-inducible Cre/Lox approaches to induce diverse types of tumors and hyperplasia in transgenic zebrafish. *Proc. Natl Acad. Sci. USA* **104**(22), 9410–9415.

Lee, Y., Grill, S., Sanchez, A., Murphy-Ryan, M., and Poss, K. D. (2005). Fgf signaling instructs position-dependent growth rate during zebrafish fin regeneration. *Development* **132**(23), 5173–5183.

Lee, Y., Hami, D., De Val, S., Kagermeier-Schenk, B., Wills, A. A., Black, B. L., Weidinger, G., and Poss, K. D. (2009). Maintenance of blastemal proliferation by functionally diverse epidermis in regenerating zebrafish fins. *Dev. Biol.* **331**(2), 270–280.

Lin, J. W., Biankin, A. V., Horb, M. E., Ghosh, B., Prasad, N. B., Yee, N. S., Pack, M. A., and Leach, S. D. (2004). Differential requirement for ptf1a in endocrine and exocrine lineages of developing zebrafish pancreas. *Dev. Biol.* **274**(2), 491–503.

Liu, X. D., Liu, P. C., Santoro, N., and Thiele, D. J. (1997). Conservation of a stress response: Human heat shock transcription factors functionally substitute for yeast HSF. *EMBO. J.* **16**(21), 6466–6477.

MacDonald, R., and Ekker, M. (2010). Spatial and temporal regulation of transgene expression in fish. In: *Aquaculture Biotechnology* (G.L. Fletcher and M.L. Rise, eds). John Wiley & sons.

Meng, X., Noyes, M. B., Zhu, L. J., Lawson, N. D., and Wolfe, S. A. (2008). Targeted gene inactivation in zebrafish using engineered zinc-finger nucleases. *Nat. Biotechnol.* **26**, 695–701.

Muto, A., Orger, M. B., Wehman, A. M., Smear, M. C., Kay, J. N., Page-McCaw, P. S., Gahtan, E., Xiao, T., Nevin, L. M., Gosse, N. J., et al. (2005). Forward genetic analysis of visual behavior in zebrafish. *PLoS Genet.* **1**(5), e66.

Nasevicius, A., and Ekker, S. C. (2000). Effective targeted gene "knockdown" in zebrafish. *Nat. Genet.* **26**, 216–220.

Oates, A. C., Bruce, A. E., and Ho, R. K. (2000). Too much interference: Injection of double-stranded RNA has nonspecific effects in the zebrafish embryo. *Dev. Biol.* **224**(1), 20–28.

Pan, X., Wan, H., Chia, W., Tong, Y., and Gong, Z. (2005). Demonstration of site-directed recombination in transgenic zebrafish using the Cre/loxP system. *Transgenic Res.* **14**(2), 217–223.

Pisharath, H., Rhee, J. M., Swanson, M. A., Leach, S. D., and Parsons, M. J. (2007). Targeted ablation of beta cells in the embryonic zebrafish pancreas using *E. coli* nitroreductase. *Mech. Dev.* **124**(3), 218–229.

Scheer, N., and Campos-Ortega, J. A. (1999). Use of the GAL4-UAS technique for targeted gene expression in the zebrafish. *Mech. Dev.* **80**(2), 153–158.

Shepard, J. L., Amatruda, J. F., Stern, H. M., Subramanian, A., Finkelstein, D., Ziai, J., Finley, K. R., Pfaff, K. L., Hersey, C., Zhou, Y., et al. (2005). A zebrafish bmyb mutation causes genome instability and increased cancer susceptibility. *Proc. Natl Acad. Sci. USA* **102**(37), 13194–13199.

Shestopalov, I. A., Sinha, S., and Chen, J. K. (2007). Light-controlled gene silencing in zebrafish embryos. *Nat. Chem. Biol.* **3**(10), 650–651.

Stoick-Cooper, C. L., Weidinger, G., Riehle, K. J., Hubbert, C., Major, M. B., Fausto, N., and Moon, R. T. (2007). Distinct Wnt signaling pathways have opposing roles in appendage regeneration. *Development* **134**(3), 479–489.

Summerton, J., and Weller, D. (1997). Morpholino antisense oligomers: Design, preparation, and properties. *Antisense Nucleic Acid Drug Dev.* **7**, 187–195.

Thatcher, E. J., Paydar, I., Anderson, K. K., and Patton, J. G. (2008). Regulation of zebrafish fin regeneration by microRNAs. *Proc. Natl Acad. Sci. USA* **105**(47), 18384–18389.

Thummel, R., Bai, S., Sarras, M. P., Jr., Song, P., McDermott, J., Brewer, J., Perry, M., Zhang, X., Hyde, D. R., and Godwin, A. R. (2006). Inhibition of zebrafish fin regeneration using in vivo electroporation of morpholinos against fgfr1 and msxb. *Dev. Dyn.* **235**(2), 336–346.

Thummel, R., Burket, C. T., Brewer, J. L., Sarras, M. P., Jr., Li, L., Perry, M., McDermott, J. P., Sauer, B., Hyde, D. R., and Godwin, A. R. (2005). Cre-mediated site-specific recombination in zebrafish embryos. *Dev. Dyn.* **233**(4), 1366–1377.

Thummel, R., Kassen, S. C., Montgomery, J. E., Enright, J. M., and Hyde, D. R. (2008). Inhibition of Muller glial cell division blocks regeneration of the light-damaged zebrafish retina. *Dev. Neurobiol.* **68**(3), 392–408.

Till, B. J., Reynolds, S. H., Greene, E. A., Codomo, C. A., Enns, L. C., Johnson, J. E., Burtner, C., Odden, A. R., Young, K., Taylor, N. E., et al. (2003). Large-scale discovery of induced point mutations with high-throughput TILLING. *Genome Res.* **13**(3), 524–530.

Wehman, A. M., Staub, W., Meyers, J. R., Raymond, P. A., and Baier, H. (2005). Genetic dissection of the zebrafish retinal stem-cell compartment. *Dev. Biol.* **281**(1), 53–65.

Wienholds, E., Schulte-Merker, S., Walderich, B., and Plasterk, R. H. A. (2002). Target-selected inactivation of the zebrafish rag1 gene. *Science* **297**, 99–102.

Wienholds, E., van Eeden, F., Kosters, M., Mudde, J., Plasterk, R. H., and Cuppen, E. (2003). Efficient target-selected mutagenesis in zebrafish. *Genome Res.* **13**(12), 2700–2707.

Zecchin, E., Mavropoulos, A., Devos, N., Filippi, A., Tiso, N., Meyer, D., Peers, B., Bortolussi, M., and Argenton, F. (2004). Evolutionary conserved role of ptf1a in the specification of exocrine pancreatic fates. *Dev. Biol.* **268**(1), 174–184.

# 2

# ORGANIZATION AND PHYSIOLOGY OF THE ZEBRAFISH NERVOUS SYSTEM

*LAURE BALLY-CUIF*
*PHILIPPE VERNIER*





*Zebrafish: Volume 29*
FISH PHYSIOLOGY


DOI: 10.1016/S1546-5098(10)02902-X

The zebrafish is now a well-established animal model popularized mostly by developmental biologists for its easy observation and manipulation during embryogenesis. Since the zebrafish is also amenable to genetic approaches and molecular manipulation in vivo, it followed the fate of other standardized laboratory animals by becoming also a model for adult functions and behaviors in vertebrates. This prompted a wealth of studies about the neural physiology and neurobiology of behaviors in zebrafish, in seeking comparison with human or other vertebrate neurophysiology. The emerging picture shows the zebrafish as retaining most of the basic features of other vertebrates, confirming it as a pertinent model for comparison with humans, provided its phylogenetical relationships and natural way of life are readily understood to prevent overstatement, misinterpretation or even disregard of zebrafish data. Given the success of the zebrafish to study the genetics of vertebrate development, extension of its use as a model for physiology and behavior promises to be very successful in the next few years. This chapter provides a brief overview of our current knowledge about the development of neural functions, basic neurophysiology and the control of behavior in zebrafish.

## 1. INTRODUCTION

### 1.1. Zebrafish Phylogeny

Knowledge of the phylogenetic relationships as well as of the conserved and derived characteristics of the zebrafish (*Danio rerio*) with other vertebrates, particularly humans, is instrumental for understanding the extent to which it can be a "model" for studying neural development and functions. *Danio rerio* belongs to the teleosts, comprising 99% of actinopterygian species (ray finned fishes, ~ 27 000 species), the sister group of sarcopterygians (~ 27 000 species also), suggesting that the two main groups of Gnathostomes (jawed vertebrates; see Fig. 2.1) have undergone similar expansion (Nelson 2006). Teleosts emerged approximately 350 million years ago and began to diversify significantly during the Triassic period, about 200 million years ago (Metscher and Ahlberg 1999), probably favored by a whole genome duplication closely preceding this diversification (Ravi and Venkatesh 2008). More precisely, danios are cyprinids, which comprise more than 3600 different species, including well-studied fresh-water fishes (the carp *Cyprinus carpio*, or the goldfish *Carassius auratus*). Cyprinids are members of an evolutionarily rather successful group, the Ostariophysi (over 6000 members), which conserved some ancestral traits of Actinopterygians, but also diverged significantly from the morphology – and probably from the physiology – of the common actinopterygian/sarcopterygian ancestor (Fig. 2.1). Among conserved characters, it is interesting to mention the retention of functional gills, lateral line organs, or median fins, which have been lost or greatly modified in tetrapods (Metscher and Ahlberg 1999). The body plan of zebrafish is also close to the general osteichthyan morphotype, and the sensory organs or the general organization of the nervous system are those shared by most Gnathostomes. However, it is important to state that the zebrafish does not resemble a primitive vertebrate, and that many characteristics of zebrafish are indeed true novelties, which cannot be found in any other vertebrates. Similar remarks hold true for *Xenopus*, chicken or mouse, the other widespread lab models (Metscher and Ahlberg 1999). For example, as derived characters of all Ostariophysi, zebrafish emit the Schreckstoff (alarm pheromone), possess unculi and breeding tubercles (keratinous structures in the skin), and modifications in the anterior vertebrae making the Weberian apparatus, a set of four bones connecting the inner ear to the swim bladder for transmitting and amplifying sounds. They also exhibit peculiarities in skull bone pattern, dentition, pattern of the fin endoskeletons, and tail and gut morphology which are very different from the ancestral condition deduced from comparative studies with tetrapods.

### 1.2. Natural History of Zebrafish

To better understand the physiology and behavioral characteristics of zebrafish as a model organism for integrative biology and not only for developmental studies, some knowledge of zebrafish natural history is necessary. Zebrafish are found in freshwater streams and slow-moving waters from South and East Asia (Pakistan, Burma, India, and Nepal). Danio lives in shallow water at temperatures ranging from 24°C to 38°C, conductivities of 10 to 271 μS and pH 6.0–8.0 (Engeszer et al. 2007 and references therein). They prefer still water with vegetation, but can also accommodate turbid environments. They presumably feed on mosquito larvæ and other insects, and compete for food with minnows (Cyprinidae). They are themselves hunted by larger carnivorous fish such as the snakeheads of the genus *Channa* or the knifefish, *Notopterus notopterus*. They are communal and seasonal breeders (from April to August, depending on the climatic

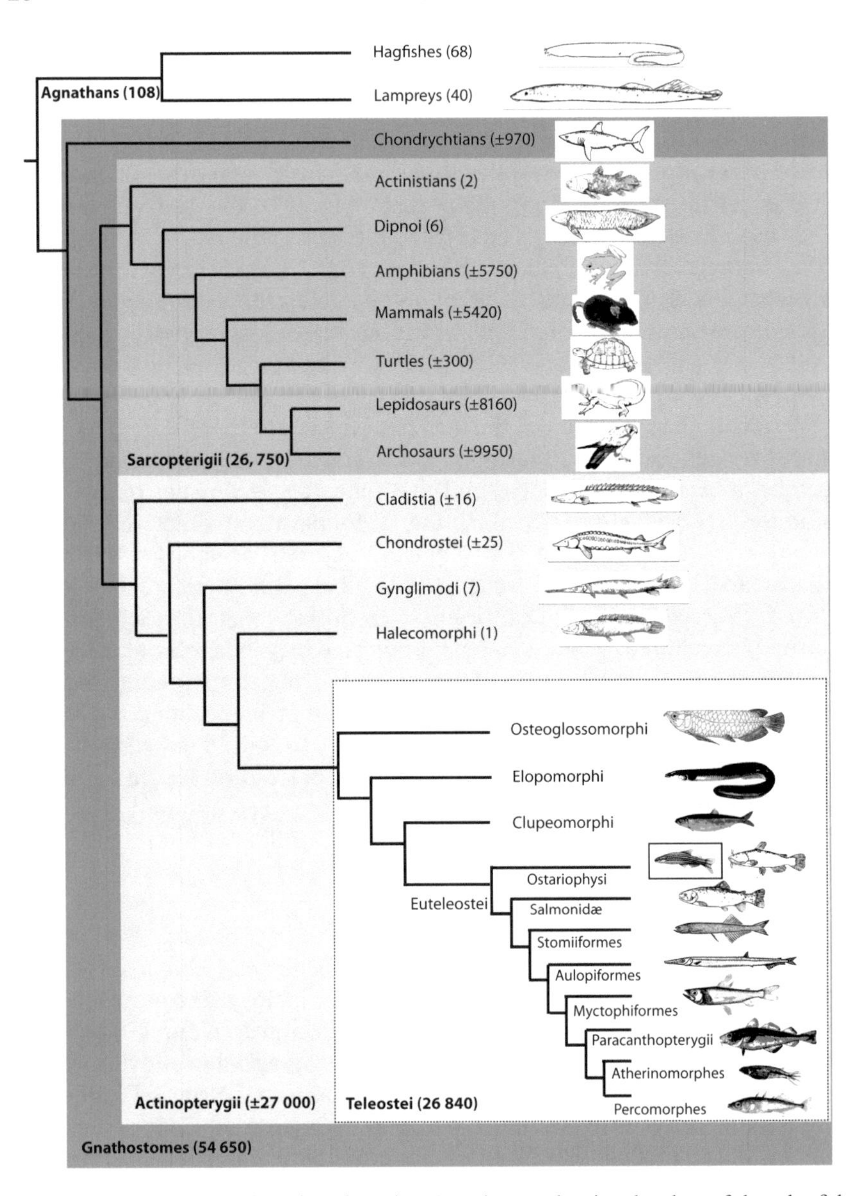


**Fig. 2.1.** Phylogenetic relationships of craniates/vertebrates, showing the place of the zebrafish within this class. The number of species known to date in each of the delineated group of species is indicated in parentheses. The dark gray rectangle encompasses the jawed vertebrates (gnathostomes) divided into Sarcopterigii (middle gray rectangle) and Actinopterygii (pale gray rectangle). Note that the two groups contain about the same number of species. The group of teleosts is proportionally enlarged (white box with dotted line) and the zebrafish, a member of Ostariophysi, is boxed. The zebrafish is distantly related to medaka (an Atherinomorph), the

conditions) and they generally lay eggs in still-water, flooded areas, such as rice paddies. Zebrafish larvæ and juveniles grow there before moving back into the river streams when waters decline in ponds and paddies. Spawning behavior itself has not been described in the wild where zebrafish often take up to 5–6 months to reach reproductive maturity. In the laboratory, embryos may hatch within 3 days post-fertilization and can mature within 2–3 months in good conditions. Finally, variations also exist in the wild for many phenotypic traits, and their adaptive nature needs to be better addressed in the future. Indeed, pigmentation patterns, which affect visual recognition and mating behavior, lateral line morphology and the shape of jaws and teeth, are likely to affect the predator–prey relationships and behaviors of zebrafish. The genetics of such phenotypic variations is not well understood and its relationships to laboratory populations should be better evaluated.

## 2. NERVOUS SYSTEM DEVELOPMENT AND GENERAL ORGANIZATION

### 2.1. Neural Induction, Neurulation and Neurogenesis

The first step in building the nervous system is the specification of the neural plate, a process called neural induction. It takes place at a very early developmental stage, preceding or concomitant to gastrulation. Like in other vertebrates, neural plate specification within the zebrafish embryo integrates antagonistic activities from different signaling pathways, such as meso- and endoderm inducers (Nodal/TGFβ) (Jia et al. 2009), neural inducers (FGFs, Wnts), and ventralizing (BMPs) versus dorsalizing (BMP antagonists) agents (reviewed in Wilson and Houart 2004). Little is known of the immediate transcriptional events that result from these interactions and regulate neural fate. The SoxB protein Sox3 is one of the very few factors known to be expressed throughout most of the neural plate (Okuda et al. 2006; Dee et al. 2007), in a complementary pattern to non-neural markers such as Bmp2b, ΔNp63 or Gata2, and is both necessary and sufficient for neural fate induction (Dee et al. 2008). SoxB proteins generally maintain the neural precursor fate, and downstream events in other systems involve activation of Geminin (Rogers et al. 2009) and counteraction of proneural proteins (Bylund et al. 2003).

---

other main teleost animal model, about as distant as the human is from the birds. Based on these relationships, it can be hypothesized that shared characteristics between zebrafish and mammalian species have been inherited from their common gnathostome ancestor and should be basal to all members of the group.

The next process by which the zebrafish neural tube is formed from the neural plate, called neurulation, involves the formation of a neural rod followed by the establishment of a lumen. In spite of this two-step process, the morphogenetic movements and molecular mechanisms involved are very similar to those occurring in other vertebrates (Geldmacher-Voss et al. 2003, reviewed in Lowery and Sive 2004). The convergence of dorsal tissues is driven in a large part by the Planar Cell Polarity (PCP) pathway, mediated by non-canonical Wnt signaling (Jessen et al. 2002; reviewed in Wada and Okamoto 2009). This process is critical for the formation of a single neurulation center, as several neural tubes can form in mutants where convergence is delayed (Tawk et al. 2007). Folding and apposition of the left and right halves of the neural plate requires the definition of hinge points, in part through the activity of *zic* genes on cytoskeletal organization (Nyholm et al. 2009), and cadherin-mediated cell adhesion (Lele et al. 2002). The superficial and deep layers of the neural plate undergo radial intercalation (Hong and Brewster 2006), and factors involved in the maintenance of epithelial structures or their polarized behavior, such as N-cadherin (Hong and Brewster 2006), are important. Other cellular behaviors at the neural rod stage are epithelialization and the establishment of apico-basal polarity, as well as the translocation of daughter cells to the opposite side of the neural rod upon cell division (reviewed in Clarke 2009). These processes integrate PCP signaling and the Par3/Par6/APKc pathway (Ciruna et al. 2006; Tawk et al. 2007; Munson et al. 2008) and are crucial for the formation of the future neurocoel. They likely permit the cytoskeletal rearrangements and junction barrier formation necessary for midline separation, and the shaping of the neural tube ventricles and their inflation by cerebrospinal fluid (CSF) (reviewed in Lowery and Sive 2009). In zebrafish, formation of the embryonic CSF (eCSF) requires the $Na^{+}K^{+}$ATPase ion pump (Lowery and Sive 2005), which creates an osmotic gradient regulating fluid flow. The eCSF contains proteoglycans, growth factors and signaling factors, and is likely secreted by neuroepithelial cells at stages preceding formation of the choroid plexus. Its function in zebrafish has not been directly studied. Analyses in other vertebrates however, demonstrated its implication in the control of neuroepithelial proliferation and survival, identifying the eCSF as a crucial component of embryonic CNS physiology (reviewed in Gato and Desmond 2009).

The commitment of neuroepithelial progenitors towards neuronal or glial differentiation is initiated concomitantly to neural plate formation and neurulation. The first specified neuronal precursors, identified by their expression of proneural genes such as *neurog1*, *ascl1* or *coe2*, are recognized as early as the tailbud stage (neural plate, 0 somites) (Allende and Weinberg 1994; Blader et al. 1997; Bally-Cuif et al. 1998). Their organization in clusters ("proneural clusters") follows a pre-patterning of the neural plate in neurogenesis-competent and -incompetent territories, which is achieved by the early complementary expression of neurogenesis-promoting and -inhibitory factors in response to positional cues. The former comprise members of the Iro family, and the latter a subgroup of Hairy/Enhancer of split-related (Her) transcription factors (reviewed in Bally-Cuif and Hammerschmidt 2003; Stigloher et al. 2008). Neurogenesis-incompetent territories, often located at neural tube boundaries, are maintained in an undifferentiated state over a long time and can persist until adulthood, where they likely play crucial roles in brain physiology through the maintenance of neural stem cells (Chapouton et al. 2006). The early proneural clusters, in contrast, are rapidly recruited to build the first larval neuronal scaffold, which will permit autonomous larval behavior. This early phase of neurogenesis, also called "primary neurogenesis", is controlled by the "lateral inhibition" process. During lateral inhibition, proneural expression in committed progenitors enhances expression of the Notch ligands Delta, triggering Notch signaling in adjacent progenitors and their transient inhibition from neurogenesis (Dornseifer et al. 1997; Haddon et al. 1998; Takke et al. 1999; Itoh et al. 2003) (reviewed in Appel and Chitnis 2002). A major effector of Notch signaling in the early neural plate is the transcription factor Her4 (Takke et al. 1999; Yeo et al. 2007).

"Secondary neurogenesis" will continue later on to increase neuron numbers and diversify neuronal subtypes, and recent studies suggest that the same proneural and neurogenic pathways are reiterated at these later stages. Hence, *neurog1* and members of the Notch pathway are expressed in the larval brain (Mueller and Wullimann 2003), and Notch-mediated lateral inhibition controls formation of later motor- or interneurons. Specifically, at these late stages, Notch signaling is involved in the maintenance of some neuronal progenitors (Yeo et al. 2007), the generation of glial cells (oligodendrocytes and radial glia) from neural progenitors (Park and Appel 2003; Park et al. 2005; Kim et al. 2008), and in binary neuronal identity choices in the final neurogenic divisions of neuronal progenitors (Shin et al. 2007).

### 2.2. Establishment of the Bauplan of Central Neuronal Identities and Connections

The precise organization of the neurogenesis pattern follows positional cues that are first established during gastrulation, concomitant to neural induction, and refined later on, in zebrafish like in other vertebrates. The embryonic organizer is the source of dorsalizing and posteriorizing information, such as Fgf and Wnt, while gastrulation movements and Wnt antagonists from the anterior neural ridge ensure "protection" of the anterior neural plate and head formation (Houart et al. 2002, reviewed in Wilson and Houart 2004; Stern 2006). These cues together pattern the neural plate along the anteroposterior axis. They are helped in this process by the progressive definition of "secondary organizers", themselves responding to general positional cues and acting as local sources of signal to refine patterning. Such organizers include the zona limitans intrathalamica (a source of Hh signaling) (Scholpp et al. 2006, 2009), the midbrain–hindbrain boundary (Wnt and Fgf) (Wilson et al. 2002), and in zebrafish rhombomere 4 (Fgf) (Maves et al. 2002). The dorsoventral axis is likewise patterned through opposing gradients generated from the axial mesendoderm, relayed by the floor plate, and the non-neural ectoderm, relayed by the roof plate. These cues involve signaling from the Hh, Nodal, BMP and Wnt pathways and have been extensively reviewed (Schier and Talbot 2005). Together, these signaling processes subdivide the early neural plate and tube into discrete areas according to a grid of anterior–posterior (A–P) and dorsal–ventral (D–V) coordinates, which have two major outcomes: (i) they define brain subdivisions (tel-, di-, mes- and rhombencephalon, along the A–P axis, and the roof, alar, ventral and basal plate along the D–V axis) which prefigure the major organization of the adult brain, and, within each, identify transient units of organization of the embryonic brain characterized by specific molecular identities and fate, the "neuromeres" (Puelles and Rubenstein 2003), and (ii) they regulate expression of the neurogenesis competence and inhibitory factors, hence positioning the proneural clusters.

Striking features follow from the co-regulation of patterning and neurogenesis. The first is the coincidence of the first neuronal clusters with neuromere centers in both the basal and alar plates, while neuromere boundaries are generally inhibitory. It results in a stereotyped organization of the primary neuronal scaffold (Fig. 2.2A). The second is the possibility to attribute specific characteristics to the neurons and neural progenitors of the different CNS subdivisions. For example, each neuromere of the rhombencephalon (rhombomere, –r–) harbors primary motorneurons, but these differ in their projection patterns and function following expression of a distinct transcription factor code in a rhombomere-specific fashion (hence, motorneurons from r1 form the trochlear nerve, from r2 the trigeminal nerve, from r4 the facial and vestibuloacoustic, etc). The third is the very similar organization of the body plan (Bauplan) between zebrafish and other vertebrates at this mid-embryogenesis stage, a property proposed to be "phylotypic", that is, common to all vertebrates (Duboule 1994). Several interpretations have been proposed for this phenomenon, of which the most likely is the existence of temporal and spatial developmental constraints in the generation and differentiation of a basic set of neuronal identities and connections common to all vertebrates (Wullimann and Vernier 2006). Some differences can be noted between species, such as, in zebrafish, the absence of serotonergic neurons from r3 (and not r4) (Lillesaar et al. 2007) and the diencephalic (and not mesencephalic) localization of dopaminergic neurons with ascending projections (Wullimann and Rink 2001; Rink and Wullimann 2001). The potential functional consequences of these differences have not been assessed.

## 2.3. The Primary Neural Scaffold and The Escape Response

The primary neuronal scaffold (Fig. 2.2A) is largely composed of cholinergic motorneurons, both cranial and spinal (reviewed in Chandrasekhar 2004), interneurons of various subtypes (glutamatergic, glycinergic, GABAergic) (Higashijima et al. 2004a,b) and large spinal cord glutamatergic sensory neurons, the Rohon-Beard cells. It is primarily involved in controlling the escape response (Grunwald et al. 1988). Within the first 5–6 days of life, it is rapidly complemented by modulatory neurons (McLean and Fetcho 2004a,b) to ensure the basic functions of the autonomous zebrafish larva, including visually guided behavior, feeding, sleep, refined locomotion, mechanosensation and escape. The specific circuits mediating the escape response have been most extensively studied (see Gahtan and Baier 2004, McLean and Fetcho 2008 for reviews) (Fig. 2.2B), although recent work is shedding important light on swimming (McLean et al. 2007; 2008, Orger et al. 2008; Liao and Fetcho 2008), olfaction (Yaksi et al. 2009; Miyasaka et al. 2009), visual perception (Ramdya and Engert 2008; Sumbre et al. 2008) and visually guided behaviors such as prey capture (Gahtan et al. 2005; Orger et al. 2008; Roeser and Baier 2003) (Fig. 2.2B) (see also Chapter 3 "Zebrafish Visual System"). The natural stimulus triggering escape response is a stimulation of the somatic sensory column, including the mechanosensory cells of the lateral line or located within the skin, and the acousticovestibular system of the inner ear (Kohashi and Oda 2008; McHenry et al. 2009). It can be evoked by touch or water flow. This stimulation is sensed by somatosensory neurons such as the Rohon-Beard cells, trigeminal neurons, or neurons of the VIIIth nerve, which connect to the reticulospinal sensory system. This system is extremely sensitive, as the generation of a single spike of activation in a single Rohon-Beard neuron at 24 h is sufficient to trigger escape (Douglass et al. 2008). Second-order projections transmitting information from the lateral line also connect to other targets, such as the contralateral hindbrain, the contralateral midbrain torus semicircularis, gaze- and posture-controlling centers in the midbrain and thalamic nuclei (Fame et al. 2006; Sassa et al. 2007) and are involved in other behaviors. A major reticulospinal target of somatosensory neurons is the Mauthner cell, a huge neuron located in r4 (reviewed in Korn and Faber 2005). Mauthner cells

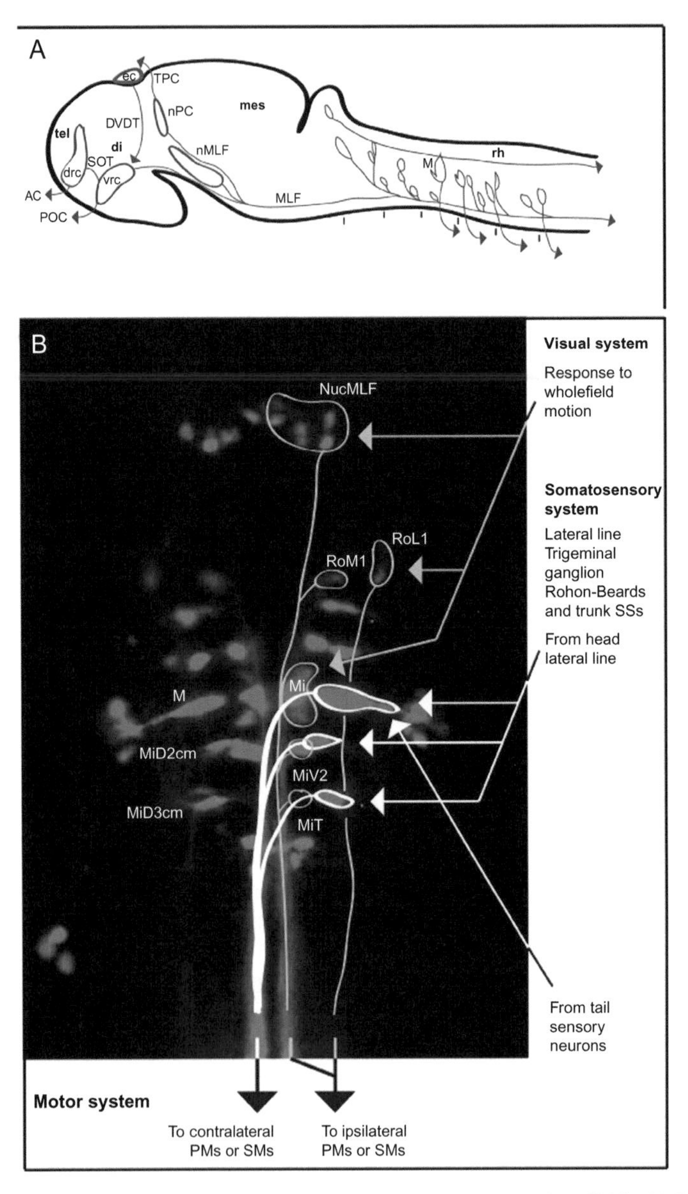


**Fig. 2.2.** Organization and selected functions of the primary neuronal scaffold. (A) Schematic representation of the early scaffold on an embryo at 36–48 hpf (anterior to the left). Neuronal clusters (circled) and tracts (arrowed) are represented. The position of rhombomere boundaries are indicated by small vertical bars. For a more precise visualization of the reticulospinal system, see (B). AC: anterior commissure; di: diencephalon; drc: dorso-rostral cluster; DVDT: dorso-ventral diencephalic tract; ec: epiphyseal cluster; mes: mesencephalon; MLF: medial

synapse directly onto contralateral interneurons and motorneurons in the spinal cord via axon collaterals (Fetcho and Faber 1988). Because Mauthner cells simultaneously excite contralateral motorneurons and inhibitory glycinergic commissural interneurons that target the opposite Mauthner cell (Satou et al. 2009), the sensory signal elicits unilateral muscle contraction, ensuring that the animal turns opposite to the stimulus. Although stimulation of the Mauthner neuron is sufficient to elicit the escape response, it is in itself dispensable for this behavior provided other reticulospinal neurons, such as MiD2 cm and MiD3 cm, are functional (Liu and Fetcho 1999) (Fig. 2.2B). Hence, reticulospinal "functional groups" rather than individual neurons are involved in sensory–motor connection. Their identity and localization within the brainstem renders them sensitive to distinctly localized stimuli (such as occurring at the level of the head or the otic vesicle in the hindbrain) (Kohashi and Oda 2008), and they are distinct from the reticulospinal groups that mediate swimming or struggling (Liao and Fetcho 2008; Orger et al. 2008; Ritter et al. 2001). The Mauthner cell also receives inhibitory input from several classes of interneurons.

## 3. GENERAL ORGANIZATION OF THE ZEBRAFISH NERVOUS SYSTEM, SUBDIVISIONS AND MAIN FUNCTIONS

Since the development of the zebrafish nervous system follows the vertebrate Bauplan, the general organization of the zebrafish central nervous system is indeed very similar to that of other vertebrates (Fig. 2.3). Along the antero-posterior axis, the central nervous system is divided into four parts. The spinal cord is the most caudal part of the CNS, protected by the bony column of vertebrae. It is continued anteriorly by the rhombencephalon, the most caudal part of the encephalon, within the skull. The forebrain is the most rostral and largest area of the brain, separated from the rhombencephalon by the intervening mesencephalon. These four main regions of the CNS are

longitudinal fascicle; nMLF: nucleus of the medial longitudinal fascicle; nPC: nucleus of the posterior commissure; POC: post-optic commissure; rh: rhombencephalon; SOT: superior optic tract; tel: telencephalon; TPC: tract of the posterior commissure; vrc: ventro-rostral cluster. (B) Dorsal view of the reticulospinal system in a 5 dpf larva (anterior to the top) with indications – on one side of the hindbrain – of individual neurons identified to be involved in escape (bold white circles and tracts) and visual-guided swimming (light gray circles and tracts). Afferent (right) and efferent (bottom) circuits are also indicated. The reticulospinal neurons are highlighted by the retrograde transport of rhodamine dextran applied at spinal levels. Abbreviations: M: Mauthner cell; Mi: middle reticulospinal neurons; NucMLF: nucleus of the medial longitudinal fascicle; PMs: primary motorneurons; Ro: rostral reticulospinal neurons; SMs: secondary motorneurons; SSs: somatic sensory neurons.

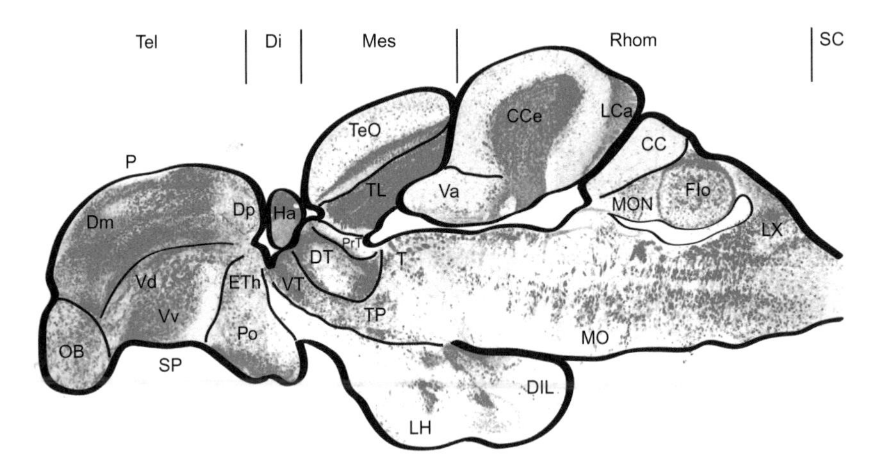


**Fig. 2.3.** The main cytoarchitechtonic divisions of the zebrafish brain. Sagittal section of an adult zebrafish brain showing the major brain areas, some of them delineated by black lines to facilitate their recognition. Neuronal cell bodies appear black. Di: diencephalon; CC: crista cerebellaris; CCe: corpus cerebelli; DIL: diffuse nucleus of the inferior lobe of the hypothalamus; Dm: medial domain of the dorsal telencephalon; Dp: posterior domain of the dorsal telencephalon; ETh: Eminentia thalami; Flo: facial lobe; Ha: habenula; LCa: lobus caudalis cerebelli; LH: lateral hypothalamic nucleus; LX: vagal lobe; Mes: mesencephalon; MO: medulla oblongata; MON: medial octavolateralis nucleus; P: pallium; Po: pre-optic nucleus; PrT: pretectum; OB: olfactory bulb; SC: spinal cord; SP: subpallium; T: tegmentum; Tel: telencephalon; TL: torus longitudinalis; TP: posterior tuberculum; Vd: ventral domain of the telencephalon, dorsal part; Vv: ventral domain of the telencephalon, ventral part; Va: valvula cerebelli; VT: ventral thalamus.

subdivided into morphological and functional areas, the main ones of which are shown on Figure 2.3. In particular, the forebrain, or prosencephalon, comprises the caudally located diencephalon and the anterior "secondary prosencephalon". The latter gives rise ventrally to the hypothalamus (corresponding to the anterior most tip of the basal neural tube) first intermingled with the eye field (which is then externalized dorsolaterally) and to the telencephalon, dorsally. The general organization of the zebrafish CNS will now be described from anterior to posterior according to this general scheme.

### 3.1. Telencephalon

The telencephalon of the zebrafish develops by eversion as in the other ray-fined fishes, instead of evagination as in other vertebrates, altering the topology of the pallial areas, as compared to mammals or other amniotes.

The telencephalon is divided into a ventral subpallium (septum/striatum) and a dorsal pallium, a region that takes multiple shapes and neuronal arrangements in vertebrates. Nevertheless, it can be considered homologous as a field to the cerebral cortex of mammals. Due to the eversion process, the homology between the four main pallial divisions depicted in amniotes (dorsal, medial, lateral and ventral pallium) and the various areas of the fish telencephalon defined cytoarchitectonically (medial –Dm–, dorsal –Dd–, lateral –Dl–, posterior –Dp–, and central –Dc– regions) are still debated (Wullimann and Mueller 2004). In addition, the organization and connections of the zebrafish pallium are more complex than in early radiating actinopterygians. Thus, it represents one of the most derived characters of the nervous system and it cannot be easily compared to that of tetrapods.

Nevertheless, Dl, Dc and Dm zones of the zebrafish pallium are somewhat comparable to the dorsal pallium/neocortex, since Dl receives visual input, Dl and Dm receive auditory projections, Dl, Dc and Dm are targets of the mechanosensory input from the lateral line, and Dm receives a gustatory input, making these pallial areas multimodal as is also the case in most of the association cortex of mammals. Based on these neuronal connections and the analysis of morphogenetic movements, it is very likely that the zebrafish homolog of the medial pallium lies laterally (which receives multisensory inputs from the nuclei of the posterior tuberculum, but not from the olfactory bulb), while the lateral pallium is medially located. The latter is targeted by afferents from the olfactory bulb, mostly in the posterior zone, and it could be homologous to the olfactory cortex of mammals. More detailed and specific descriptions of the pallial areas are not yet available for the zebrafish (Wullimann and Mueller 2004). This is also true for the neurochemical anatomy of the pallium, for which very limited description has been made. It contains many GABAergic neurons, likely to be interneurons, the origin of which is not yet understood. Similarly, other probable interneurons express somatostatin and dopamine. In addition, the pallium receives unevenly distributed afferents immunoreactive for several peptides including somatostatin, neuropeptide Y, cholecystokinin, GnRH and TRH (Castro et al. 2009 and references therein).

The subpallium of the zebrafish can be divided into two domains, Vd/Vc and Vv/Vl, which have been proposed to be homologous respectively to the striatum intermingled with the pallidum, and to the septal areas. Similarities of the zebrafish Vd with the striatum of tetrapods include the presence of substance P-immunoreactive neurons and of high densities of GABA-containing neurons (Mueller et al. 2006), also expressing *glutamate decarboxylase* (*GAD*) mRNA (Mueller and Guo 2009), as well as *dopamine receptor D2* mRNA (Boehmler et al. 2004) and *D1 receptor* mRNA (Li et al. 2007). Thus, at least one of the two classes of GABA-containing striatal output neurons described in tetrapods seems to exist in Vd. The second class of GABAergic output neurons should contain enkephalin, as described in tetrapods. This has not been studied yet in zebrafish although it has been described in other teleosts (Vecino et al. 1995) or in Polypterus (Reiner and Northcutt 1992). These GABA output neurons are likely to be modulated by dopamine afferents coming from the posterior tuberculum, as is the case in tetrapods. Whether these dopaminergic afferents play a major role in motor programming as they do in amniotes is not known yet (Vernier et al. 2004). In addition, as suggested in other teleost species, pallidal-like neurons may also be present among the striatal-like neurons of the zebrafish Vd, without being segregated as a separate pallidal formation as in tetrapods (Wullimann and Rink 2002). However, a significant difference of the zebrafish and teleost Vd with the striatum of tetrapods and agnathans is the absence of cholinergic interneurons (see below). Finally, the basal ganglia circuitry involved in motor control or motivational behaviors in zebrafish is not very well known, as in other teleosts.

The second subpallial structure, the ventral subpallial nuclei (Vv, Vl), exhibits critical features identifying them as septal formations. One of the most prominent features is that Vv is the only telencephalic region where cholinergic neurons have been detected (Mueller et al. 2004), as is the case in other teleosts (Ekstrom 1987; Perez et al. 2000). These cholinergic cells are likely to be homologous to the cholinergic septal neurons found in the amniote basal forebrain (the basal nucleus of Meynert). In addition, the Vl area has extensive ascending projections to the teleostean pallium (Dm, Dc, Dl; Murakami et al. 1983), in agreement with a role of cholinergic pallial innervation similar to that of amniote basal forebrain cholinergic neurons. This contrasts with the absence of striatal cholinergic interneurons in Vd, the zebrafish homolog of striatal areas of tetrapods and cyclostomes. Although a rigorous outgroup comparison needs to be carried out, it is tempting to speculate that teleosts have conserved cholinergic neurons in the basal (septal) forebrain as in tetrapods, but they lost ancestral striatal cholinergic components. The second major argument for identifying the ventral subpallial nuclei (Vv, Vl) as septal formation is their massive descending output to the tuberal midline hypothalamus, another characteristic feature of the septal formation in amniotes.

### 3.2. Hypothalamus

In the zebrafish, as in other vertebrates, the hypothalamus consists of the ventral part of the secondary prosencephalon. It is still often presented as the basal diencephalon, especially in fish, but this is a topological assumption only, not corresponding to its developmental morphogenesis.

It comprises an alar and a basal component. The preoptic area is generally associated with the hypothalamus, although it lies at an intervening position between the hypothalamus per se and the subpallium. At odds with other vertebrates including mammals, the third ventricle extends into the hypothalamus to form a hypothalamic ventricle. Among the various differentiations of the ventricular epithelium, part of the hypothalamic ventricle protrudes ventrally to form the saccus vasculosus. Although its function remains hypothetical, it may contain sensory receptors for depth chemical detection. Other probable functions are osmoregulation, ionic transport, skeletal growth and tooth regeneration, although other homeostatic functions are not excluded. The saccus is innervated by the tractus sacci vasculosi, which terminates in the nucleus sacci vasculosi in the posterior tuberculum of the diencephalon.

In the preoptic area, the most anterior nucleus is the nucleus preopticus parvocellularis. Anterior and caudal to it, the nucleus preopticus magnocellularis is found. Above it, just below the prethalamus, lies the dopaminergic nucleus preopticus parvocellularis anterior. The preoptic area also contains CRF-expressing neurons that are part of the stress axis and which project onto the anterior pituitary gland.

The alar hypothalamus includes mainly the suprachiasmatic, supraoptic and paraventricular hypothalamic nuclei. Located just above the optic chiasm, as suggested by its name, the suprachiasmatic nucleus plays a major role in the clock of the circadian rhythms. The two other nuclei secrete vasotocin and isotocin and project on the pars nervosa of the hypophysis (see below).

The basal hypothalamus is divided into dorsal and ventral parts, where are respectively located several nuclei such as the retrochiasmatic, ventromedial, and mammillary hypothalamic nuclei, or the nuclei tuberis (ventralis, lateralis and anterior) made of large neurons. It also comprises the subthalamic nucleus. More caudally, the hypothalamus protrudes in an inferior lobe, into which the ventricle extends. This lobe contains the nucleus diffusus and a nucleus centralis. Generally speaking, the clear correspondence (homology) between these fish nuclei and that of tetrapods including mammals is poorly known.

Beside projections on the hypophysis, the hypothalamic areas have numerous interconnections with many brain areas. They include widespread interconnections between the preoptic area and the hypothalamus, which may play a role in feeding behavior, courtship, and reproduction (Bass and Grober 2001), and ascending input from the medial hypothalamus to the subpallium and possibly the pallium (Rink and Wullimann 2002), which may regulate food intake and aggressive behaviors. In contrast, the lateral hypothalamus (the periventricular, central, and diffuse nuclei of the inferior lobe) receives only minor pallial and subpallial input and does not send ascending fibers to the telencephalon. Thus, the lateral and medial hypothalamic divisions have very different telencephalic inputs and outputs, with the medial hypothalamus strongly resembling the amniote condition (Wulliman 1998). Other demonstrated roles of the hypothalamic areas include the regulation of sleep, heart rate and blood pressure, and the control of body temperature. Nevertheless, much remains to be understood in the hypothalamic functions in the zebrafish.

### 3.3. The Pituitary Gland and the Urophysis

As in the other jawed vertebrates, the pituitary gland comprises two main parts. The posterior hypophysis or pars nervosa corresponds mostly to the nerve terminals of the hypothalamic neurons of the supraoptic and paraventricular nuclei secreting vasotocin and isotocin, respectively homologous to vasopressin and oxytocin in mammals. Vasotocin increases diuresis by acting on the kidney glomeruli and also increases blood pressure by causing contraction of the smooth-muscles of the vessels, together with isotocin, which also acts on ovary and oviducts to promote egg laying. The neurohypophysis also has terminals containing melanin concentrating hormone synthesized from duplicated genes in several neurons from the hypothalamus (Berman et al. 2009). This hormone acts on melanophores to modify the colored pattern of the skin. The anterior hypophysis is made of two lobes, the *pars distalis* and the *pars intermedia*. The latter contains mostly corticotrope and melanotrope cells, which synthesize the α-melanocyte stimulating hormone (α-MSH) also acting on the melanophores of the skin.

The *pars distalis* of the anterior pituitary gland harbors several hormone-secreting cell types, such as corticotropes producing adrenocorticotropin (ACTH), which is part of the stress axis, lactotropes producing prolactin (PRL), which plays a major role in the control of osmolarity and plasma sodium homeostasis, thyrotropes producing thyrotropin (TSH), which controls the secretion of the thyroid gland and plays a major role in metabolic regulation, somatotropes producing growth hormone (GH), which regulates cell and animal growth via glucose and protein metabolism, and gonadotropes that secrete the gonadotropins (LH, FSH) controlling gamete synthesis and reproduction. Like other teleosts, the zebrafish has somatolactin-producing cells with terminals located both in the anterior and posterior pituitary gland (reviewed in Takei and Loretz 2006).

In the caudal-most spinal cord a group of neurosecretory neurons, the Dahlgren cells, is located ventral to the central canal. They project onto the urophysis, a neurohemal organ secreting urotrophin and acetylcholine and located ventral to the most caudal part of the spinal cord (Parmentier et al.

2006). This system closely resembles the neurohypophysis, and it plays several roles in regulating smooth muscles activity and blood osmolarity.

### 3.4. The Diencephalon Including the Posterior Tuberculum

The diencephalon corresponds to the three first prosomeres of the forebrain. They are from posterior to anterior the pretectum and, anterior to it, the epithalamic structures (epiphysis and habenula), dorsal thalamus (or thalamus) with its anterior, dorsal posterior, central posterior nuclei and ventral thalamus (or prethalamus) with the ventrolateral and ventromedial nuclei.

The posterior tuberculum can be defined as the ventral part of prosomeres 2 and 3, and could eventually be extended to the ventral portion of prosomere 1, which is the region of the nucleus of the medial longitudinal fascicle (Vernier and Wullimann 2009). The periventricular area of the posterior tuberculum exhibits three components, the nucleus of the posterior tuberculum (TPp), the paraventricular organ (PVO), mainly ependymal cells, which, in anamniotes, exhibit a specific differentiation in continuity with the hypothalamic periventricular wall, and a third nucleus made of larger cells, the nucleus tuberis posterior or posterior tuberal nucleus. All these nuclei, which contain dopaminergic neurons, are housed in the basal part of prosomere 3. In the zebrafish, only TH-positive (small and large) cells of the TPp give rise to ascending projections toward the subpallial area, as demonstrated in the adult zebrafish (Rink and Wullimann 2002). This evidence as well as the known mechanisms of the differentiation of these dopamine cells suggests that they are homologous to the VTA/SN of other vertebrates (Blin et al. 2008). The dopamine neurons of the nucleus tuberis posterior are projecting to the spinal cord and are therefore probable homologs of the diencephalo–spinal pathway of tetrapods.

In zebrafish, as in other teleosts, several migrated (preglomerular) nuclei are present in the lateral posterior tuberculum, and display a large anatomical complexity and interspecific variability. These nuclei provide the major diencephalic input to the pallial zones of the area dorsalis telencephali. Specific sensory preglomerular nuclei exist for the auditory, lateral line mechanosensory, lateral line electrosensory (when present), and gustatory systems, and for components of the somatosensory and visual systems.

The predominant diencephalic targets of teleostean ascending sensory projections are in the preglomerular region located in the lateral periphery of the posterior tuberculum rather than in the dorsal thalamus as in amniotes. Although the homology of sensory pathways between teleosts and tetrapods is not always certain, the functional similarities between the teleostean preglomerular region and the amniote dorsal thalamus are striking; both make up a large proportion of the diencephalon, are subdivided into many nuclei associated with specific sensory systems, and, further, most of them have reciprocal connections with the pallium.

### 3.5. The Mesencephalon Including the Tectum

It comprises three main components, as in other vertebrates, i.e., the cerebral peduncles, the tegmentum and the tectum. The first one contains fiber bundles from the telecephalon. The tegmentum area contains the red nucleus, cranial nerve nuclei and parts of the reticular formation, but nothing equivalent to the substantia nigra or ventral tegmental area, a major difference to most other vertebrate groups.

The tectum plays major roles in controlling eye movement, fine motor programs and sensory–motor coupling. The layering of the zebrafish tectum as for other teleosts is six tangential layers. These layers exhibit different fiber and cell composition. From the pia mater to the ventricle, the layers are named stratum fibrosum marginale (SM), stratum opticum (SO), stratum fibrosum et griseum superficiale (SFGS), stratum griseum centrale (SGC), stratum album centrale (SAC), and stratum griseum periventriculare (SPV). The SPV is predominantly composed of cells (pyriform neurons), while unmyelinated axons are predominantly found in the SM. The other layers contain cells scattered among neurites from different sources.

The tectum afferent inputs come from different sensory systems and nuclei. The unmyelinated axons of the SM originate from the torus longitudinalis. The SO and the SFGS receive the bulk of the retinal terminals. Mechanoreceptive information reaches the layers SGC and SAC of the tectum from the torus semicircularis receiving lateral line and vestibular nerve input. The telencephalon projects to the SGC (ipsilaterally and possibly contralaterally). Collaterals from the nerve terminalis (the most rostral cranial nerve) may also terminate within the tectum. From the diencephalon, numerous pretectal, thalamic, and hypothalamic nuclei contribute inputs to the tectum. Other afferents come from the pretectal nuclei, the torus semicircularis, the torus longitudinalis, the nucleus isthmi, and the mesencephalon (tegmental nucleus) as well as a rostral rhombencephalic reticular nucleus.

As far as outputs are concerned, areas of the pretectum, the mesencephalon, and the rhombencephalon receive tectal inputs, forming direct and indirect visual circuitry and visuo-cerebellar circuitry. Commissural efferents exit to the contralateral tectum and the torus longitudinalis. Descending projections go to the torus semicircularis, nucleus isthmi, a tegmental dorsolateral nucleus, lateral and medial reticular formation, medial reticular formation, and in the rhombencephalon, the cerebellar nucleus lateralis valvulae and the facial and vagal lobes.

A few studies have combined anatomical identification of tectal neurons in zebrafish and other teleosts with a description of their response properties. As expected, some cells are light-responsive with large, center-surround receptive fields, possess sustained and transient responses to light stimulation, and/or are bimodally sensitive. The optic tectum of teleosts seems to be constructed and function for orientation and spatial motor tasks. However, the type and direction of movement depends upon stimulation locus in both depth and visuotopic position for eye movements and body orientation. Local stimulation of the anteromedial tectum evokes convergent eye movements, while stimulation of the anteromedial, medial and caudal zones evokes conjugate eye movements of different sorts. Similarly, stimulation of different regions of the tectum in freely moving or restrained fishes resulted in body reorientation, turning, or rolling to one side or the other, and some escape movements.

### 3.6. The Rhombencephalon or Brainstem

The zebrafish rhombencephalon is characterized as in all vertebrates by its association with the majority of cranial nerves and their primary motor and sensory centers, i.e., the trochlear (IV), trigeminal (V), abducens (VI), facial (VII), otic (VIII), glossopharyngeal (IX) and vagal (X) nerves, as well as the lateral line nerves (including a mechano- and an electroreceptive component). It also comprises the reticular formation, the raphe nuclei and many ascending or descending fiber tracts, all of which are discussed elsewhere in this chapter.

### 3.7. The Vascular System and the Cerebrospinal Fluid

Finally, little is known on the precise organization of the vascular system of the zebrafish brain, but as in the other vertebrates, it is likely to represent a major energy supplier for the neurons. The cerebrospinal fluid circulates in the ventricular system, which is significantly modified as compared to tetrapods. As in all other vertebrates, the CSF is produced by the choroid plexus saccus vasculosus.

In some localized areas, the ependymal lining of the ventricles is differentiated into specialized structures, the circumventricular organs, some of which secrete hormones and other neuroactive substances into the CSF. Part of these organs is sensitive to neuroactive substances circulating in the CSF, which acts as a chemical transport system within the CNS. These circumventricular organs include the area postrema in the dorsal rhombecephalon, the subcommissural organ, the pineal at the dorsal midline of the diencephalon, the median eminence and the posterior pituitary, and the dopaminergic paraventricular organ of the posterior tuberculum.

## 4. THE PERIPHERAL NERVOUS SYSTEM

The peripheral nervous system (PNS) of vertebrates encompasses the sensory neurons (somatic nervous system, providing information about muscle and limb position and receiving external stimuli) and the autonomic nervous system, comprising the sympathetic, parasympathetic and enteric divisions and controlling heart and smooth muscles and exocrine glands.

### 4.1. Development of the Peripheral Nervous System

The large majority of these neurons is derived from the neural crest, a population of multipotent cells identified at the junction between the neural plate and the non-neural ectoderm during gastrulation (Le Douarin and Dupin 2003). The presumptive neural crest field is defined by the activities of BMPs, Wnts and Fgfs (Ragland and Raible 2004; Lewis et al. 2004; reviewed in Barembaum and Bronner-Fraser 2005; Raible 2006), driving combinatorial expression of several growth factors (Liedtke and Winkler 2008) or transcription factors including Neurog1 and 2, Dlx3, Msx, Sox9 and 10, Foxd3, Tfap2a and Prdm1 (Hernandez-Lagunas et al. 2005; Arduini et al. 2009 and refs therein; reviewed in Huang and Saint-Jeannet 2004) and restricting Olig3 expression to the lateral aspects of the neural plate (Filippi et al. 2005). It is found in close proximity or in partial overlap with the presumptive fields of Rohon-Beard neurons and placodes–ectodermal specializations that will contribute sensory cells and sensory neurons to the PNS, although the individual precursor cells for the three lineages seem to be different (Cornell and Eisen 2000). The choice between Rohon-Beard and neural crest cell fate is mediated by Notch signaling, which favors the latter fate (Cornell and Eisen 2002; Jiang et al. 1996; Schier et al. 1996), at least in part through repression of Olig3 expression (Filippi et al. 2005).

### 4.2. Cranial Sensory Ganglia

As in other anamniotes, the Rohon-Beard cells, located within the dorsal hindbrain and spinal cord, are the first sensory neurons of the zebrafish. These cells, however, are eliminated by apoptosis after the first few days of life (Reyes et al. 2004) and are progressively replaced by neurons of the cranial sensory ganglia in the hindbrain and of the dorsal root ganglia (DRG) in the spinal cord, which will extend peripheral axons underneath the skin to detect a range of mechanical thermal and chemical stimuli. DRG arise from ventrally migrating neural crest cells that settle next to the ventral neural tube. They are specified through the action of Sox10 (Carney et al. 2006) and Foxd3 (Lister et al. 2006). Within this sensory lineage, expression of Neurog1 (Andermann et al. 2002; Cornell and Eisen 2002) specifies the neuronal versus the glial fate (McGraw et al. 2008). Using a *neurog1:EGFP* transgenic line, the formation of DRGs could be followed in vivo. By 36 hpf, the nascent DRG neurons could be seen extending axons centrally to the spinal cord and out to peripheral targets (McGraw et al. 2008). The number of neurons per DRG is initially no more than a few. It increases to about 15 by 2 weeks, and more than 100 by 4 weeks, in part through the division of mitotically active differentiated neurons (An et al. 2002).

Formation of the cranial sensory ganglia of the zebrafish, like in other vertebrates, involves the dual contribution of neural crest cells and placodes. Sensory ganglia are associated with six cranial nerves (1, 5, 7, 8, 9, 10), of which nerve 5 (trigeminal) is totally crest-derived, nerves 1 (olfactory) and 8 (octaval, or statoacoustic) are totally placode-derived, and the others are mixed. The formation of placodes and their contribution to the cranial sensory PNS has been extensively described (Haddon and Lewis 1996; Kobayashi et al. 2000; Raible and Kruse 2000; Andermann et al. 2002). The otic placode will form the otic vesicle (which will generate the inner ear and associated sensory hair cells) but also the neurons of the 8th nerve ganglion, which will transduce sensory information involved in the escape response. The lateral line placodes give rise to migrating primordia at the origin of sensory neuromasts distributed along the head and body and containing mechanosensory hair cells that detect water flow (Ghysen and Dambly-Chaudiere 2007). They are innervated by sensory ganglia and RB cells. Finally, the epibranchial placodes, positioned dorsally to the posterior pharyngeal pouches, generate visceral sensory neurons innervating the pouches through the facial, glossopharyngeal and vagal nerves. They will transmit information such as heart rate, blood pressure, and the status of other visceral organs. Expression of Neurog1 is found in all placodes and is instrumental to the differentiation of their derived sensory neurons (Andermann et al. 2002). In addition, local signals contribute to the differentiation of specific placodes, such as BMP (Holzschuh et al. 2005) or Fgf (Nechiporuk et al. 2005) for epibranchial placodes. Less is known about the establishment of the cranial sensory circuitry itself, although a recent study reports a cell-autonomous requirement for Cdh2 in cranial sensory ganglia neurons for axon guidance towards central and peripheral targets (LaMora and Voigt 2009).

The development of the trigeminal ganglion (gV), in contrast, only involves neural crests. Ganglion assembly, first visible at 11 hpf, involves cell migration, Cxcr4b-mediated chemokine signaling and E- and N-cadherin-mediated cellular interactions (Knaut et al. 2005). By 1 day, the ganglion comprises some 30 neurons and can transduce mechanical stimuli (Saint-Amant and Drapeau 1998). An ever-increasing number of trigeminal neurons is generated until at least 4 days, and the date of origin of these neurons determines in part their nociceptive or mechanosensory phenotype (Caron et al. 2008).

### 4.3. The Sympathetic Nervous System

Sympathetic neurons also originate from ventrally migrating neural crest cells, which initially settle lateral to the dorsal aorta (Raible et al. 1992). As for DRGs, proper migration is controlled by Erbb2/Erbb3 signaling (Honjo et al. 2008). Sympathetic neurons differentiate later than DRGs, between 2 to 8 dpf with an anterior to posterior sequence (An et al. 2002). Condensation of the ganglia takes place progressively thereafter, and is followed by progressive differentiation and expression of markers of the adrenergic phenotype, such as tyrosine hydroxylase (TH) and dopamine β-hydroxylase (DBH). This process is paralleled by a continuous increase in the number of sympathetic neurons per cervical and trunk ganglion to reach approximately 50 cells by 4 weeks of age.

### 4.4. The Parasympathetic Nervous System

The parasympathetic system is largely derived from mesencephalic neural crests that coalesce into the ciliary ganglion. These cells transiently express noradrenergic properties before acquiring their final cholinergic phenotype, and there is less information about their development in zebrafish than for the sympathetic system. Cholinergic innervation of the heart is functional at 5 dpf (Schwerte et al. 2006). Among other functions, the parasympathetic system is involved at late stages in the controlled inflation of the swim bladder through direct innervation of swim bladder muscles and control of gas secretion (Robertson et al. 2007).

### 4.5. The Enteric Nervous System

Enteric neurons are derived in the majority from vagal neural crests that migrate to the intestine. Differentiated enteric neurons can be first detected at the anterior end of the intestine at 3 dpf, and they cover the entire length of the intestine by 4 dpf. Major factors that control the specification, differentiation or proliferation of these cells in zebrafish include endoderm-derived cues (Pietsch et al. 2006), such as Hh, mesoderm factors (Reichenbach et al. 2008), and GDNF. Blocking GDNF function or the GDNF receptor c-Ret and co-receptors GFL-alpha through injection of antisense morpholinos impaired both enteric precursors differentiation and migration (Shepherd et al. 2001; Shepherd et al. 2004; Heanue and Pachnis 2008). In addition, several transcription factors (Phox2b, Sox10) have been implicated (Elworthy et al. 2005), and ongoing forward genetic screens promise to uncover yet-unknown enteric determinants (Kuhlman and Eisen 2007). The latter study further demonstrated a clear correlation between the number of enteric neurons and gastrointestinal motility in the 5.5 dpf larva, demonstrating the fundamental importance of this innervation. The circuitry of the zebrafish enteric nervous system, however, remains largely unknown.

## 5. SENSORY SYSTEMS

In zebrafish, most sensory pathways have been described and traced from peripheral sense organs and primary sensory centers into the telencephalon (reviewed in Wullimann and Mueller 2004). As in most other vertebrates, the zebrafish telencephalon receives essentially non-overlapping information from all sensory systems, at odds with the contention that the fish pallium is mostly multimodal.

Although the homology of sensory pathways between zebrafish and tetrapods is not certain in each single case, the degree of similarity is nevertheless of great functional and evolutionary interest. Vision reaches the telencephalon via a direct retino-thalamic and an indirect retino-tecto-thalamic pathway. In contrast to amniotes, the major telencephalic target is the subpallium. However, tectofugal visual information reaches the pallium via the preglomerular region, a complex of migrated nuclei lateral to the posterior tuberculum. The sensory systems which ascend multisynaptically in the lateral longitudinal fascicle (audition, lateral line mechanoreception and electroreception) to the mesencephalon reach the pallium mostly via nuclei in the preglomerular complex and, possibly, dorsal thalamus (only audition). Gustation in teleosts reaches the pallium via a medullary secondary gustatory nucleus and diencephalic preglomerular (and hypothalamic) nuclei. Finally, teleosts possess a direct and indirect spinal ascending somatosensory system reaching the dorsal thalamus and preglomerular region. Thus, the predominant diencephalic targets of zebrafish ascending sensory projections are in the preglomerular region located in the lateral periphery of the posterior tuberculum, not in the dorsal thalamus as in amniotes. The zebrafish sensory preglomerular nuclei receive inputs from the auditory, the lateral line mechanosensory, the electrosensory, the gustatory, the somatosensory, and the visual systems. These preglomerular nuclei have a high degree of cytoarchitectonic differentiation and they provide the main diencephalic input to the pallial zones of the area dorsalis telencephali, together with dorsal thalamic inputs. Thus, the functional similarities between the zebrafish preglomerular region and the amniote dorsal thalamus are striking: both make up a large proportion of the diencephalon, and are subdivided into many nuclei associated with specific sensory systems and, further, most of them have reciprocal connections with the pallium.

### 5.1. Chemoreception (Smell and Taste)

As in other vertebrates, odorant compounds bind to a large set of olfactory receptors located in the terminals of the olfactory nerve in the epithelium of the nostril, a derivative of the olfactory placode. Odor information is then conveyed to the olfactory bulb, the first central processing center, organized in glomeruli. Each glomerulus receives convergent input from sensory neurons expressing the same odorant receptor, resulting in the formation of combinatorial topologic maps. However, odors sharing certain molecular properties preferentially activate glomeruli within defined areas, as shown in various vertebrate species. The processing of odors in the olfactory bulb includes direct input onto mitral cells, which also transfer output information of the olfactory bulb to higher brain centers, and regulatory interaction with GABA and dopamine interneurons (periglomerular and granule cells). The interneurons mediate spatially organized inhibitory lateral interactions between mitral cells, resulting in dynamic changes of olfactory bulb output activity patterns during an odor response.

In rodents and zebrafish, primary molecular properties (e.g., characteristic functional groups) are mapped onto relatively large domains, whereas secondary molecular features (e.g., chain length) are mapped onto subregions within these domains. Chemotopic maps are therefore organized hierarchically such that "fine" maps of secondary features are nested within "coarse" maps of primary features. Within a given region, however, not all glomeruli respond to all stimuli with the associated feature, and a given stimulus usually activates glomeruli in more than one region. Compared to topological maps in other sensory systems, the chemotopy of glomerular activity patterns therefore appears rough and fractured, possibly as a consequence of reducing a high-dimensional molecular feature space onto two spatial dimensions. These network interactions create temporal patterns of activity on multiple time scales and result in dynamic changes of OB output activity patterns during an odor response. It is, however, unclear how this circuitry processes chemotopically organized patterns of glomerular input. Secondary olfactory input reaches a limited pallial territory (i.e., the homolog of the lateral pallium or olfactory cortex) and most subpallial areas.

Taste buds are chemosensory organs consisting of modified epithelial cells. These cells are used to test for edible substances and potential food. They are located on the lips, in the mouth, in the oropharyngeal cavity, on the two pairs of barbels, and on the ventral and dorsal surface of the head. Gustatory information is conveyed by the gustatory components of the teleosts reaching the diencephalon and telencephalon via a medullary secondary gustatory nucleus, comparable to the parabrachial nuclear region of mammals.

### 5.2. Vision and Visual Processing

The fish visual system exhibits a direct retino-thalamofugal and an indirect retino-tecto-thalamofugal system with synaptic relays in the dorsal thalamus, both of which may terminate in the subpallium and not in the pallium. The optic tectum of zebrafish, like the superior colliculus of mammals, provides a common, body-centered framework for multisensory integration and sensory–motor transformations, participating in the translation of the sensory inputs coded in spatial coordinates into a temporal signal in the brainstem motor generators, and is, thus, crucial for generating actions within an egocentric frame of spatial reference (see Section 3.5 and Chapter 3).

### 5.3. Audition, Vestibular Sense

Peripheral receptors from the inner ear are exquisitely sensitive to changes in movements (equilibration, encoded in the semi-circular canal) and sound. Sound perception depends mostly on macular organs, the saccule and utricle, containing an otolith (reviewed in Nicolson 2005). Zebrafish are hearing specialists, due to the presence of the Weberian ossicles that connect the swim bladder to the saccular organs, a derived character of Ostariophysi (see Introduction). They use this to detect frequencies between 10 and 4000 Hz. The organization and morphology of the inner ear neuroepithelium in fish resembles that found in higher vertebrates. The sensory cells are connected to the axons of the eighth cranial nerve, which convey the auditory systems to the medial and lateral zones of the dorsal pallium. In this pathway following the lateral longitudinal fascicle it firstly reaches the mesencephalic torus semicircularis, then the diencephalon. Here, the vestibular and part of the auditory components mostly reach the preglomerular complex of the posterior tuberculum (anterior and lateral nuclei). However, the acoustic pathways relay mostly through the dorsal thalamus from where they reach the subpallium and the pallium.

### 5.4. Mechanosensation (the Lateral Line system) and Electroreception

The lateral line is a sensory system made of sensory receptors, the neuromasts arranged on the body surface of the zebrafish in a specific pattern. Neuromasts comprise a core of mechanosensory hair cells, surrounded by support cells, and are innervated by sensory neurons that are localized in a cranial ganglion. The lateral line is sensitive to changes in the water motion and is involved in a large variety of behaviors, from detection of other individuals (predators, prey, congeners, etc.), swimming coordination and sexual courtship (reviewed in Ghysen and Dambly-Chaudière 2004). Similar to the auditory system, the lateral line mechanoreceptive inputs ascend multisynaptically in the lateral longitudinal fascicle via the mesencephalic torus semicircularis and the diencephalic preglomerular complex to the pallium; i.e., they are very similar to the lateral lemniscal system of tetrapods.

### 5.5. Somatosensation

Zebrafish possess a direct spinal ascending somatosensory system similar to the mammalian anterolateral (protopathic) system in addition to indirect spinal ascending projections which are relayed at the obex level, comparable to the mammalian medial lemniscal (epicritic) system.

## 6. MOTOR SYSTEMS AND INTEGRATIVE CENTERS

The functional organization of the major integrative centers next to the telencephalon, namely optic tectum and cerebellum, is surprisingly similar between fishes and tetrapods. It is beyond the scope of this contribution to review this information (for details, see Wulliman 1998). The cytoarchitectonic and modular organization of the craniate optic tectum, its segregated multimodal input, and the topographical representation of this input and output to the reticular formation appear to be well designed for integrative orientation tasks, such as object identification and location, and coordinated motor control, as discussed above. In zebrafish, the cerebellum is rather large and exhibits the typical three-layered cortex with comparable cell types and internal circuits to most other gnathostomes. Also the afferent and efferent connections of the zebrafish cerebellum are similar to tetrapods' and this suggests that the cerebellum may have ancestral functions in motor learning and coordination in all gnathostomes.

It remains to be analyzed how the fish brain manages to access the efferent structures, i.e., the primary motor nuclei of brain and spinal cord, for displaying a particular behavior. The fish, as with all the known non-mammalian species, have no palliospinal and palliopontine tracts. However, the motor (spinal and cranial nerve motor nuclei) and premotor systems of fishes including zebrafish resemble those of tetrapods. As in mammals, descending spinal projections (reviewed in Wulliman 1998) originate in all divisions of the reticular formation, in the caudal (inferior) raphe region (but not in the superior raphe), in vestibular and sensory trigeminal nuclei, and even in a nucleus ruber. Furthermore, the nucleus of the medial longitudinal fascicle is an ancestral craniate premotor system descending to medullary and spinal levels.

As previously noted, both optic tectum and cerebellum act on various premotor centers, in particular on the reticular formation. However, the forebrain motor centers of the zebrafish that control the spinal descending systems are less well understood. Even more than in the case of the ascending sensory systems, further studies are needed to understand the precise role of the multisynaptically descending (extrapyramidal) systems.

## 7. MONOAMINERGIC AND RELATED REGULATORY SYSTEMS

The vast majority of the neurons contributing to the sensory, motor and higher-order circuits described above use either excitatory neurotransmitters, such as glutamate and aspartate, or inhibitory neurotransmitters such as GABA and glycine. With the exception of GABA (Mueller et al. 2006), these networks remain to be described in detail.

A more complete picture may be provided for the neuromodulatory circuits using monoamines, such as the catecholamines dopamine (DA) and norepinephrine, the indolamine serotonin (5HAT), and related compounds such as histamine, and acetylcholine. These neuronal systems control and modify the sensory, motor and integrative circuits to match the inner state of the animal to its interaction with the environment. Accordingly, they are the main substrate of behavioral processes such as reward, motivation and emotions, awareness, aggression, sleep, feeding or reproductive behaviors. The neuromodulatory systems are particularized by their synthesis in a small number of neurons generally clustered in nuclei, which project in a highly divergent manner to most of the brain areas where they modulate their target in a non-synaptic manner by the so-called "volume transmission" in the ascending modulatory systems. They act on target cells by activating membrane receptors, which mostly belong to several classes of G protein-coupled receptors, and mediate different types of neuronal and glial responses (Callier et al. 2003; Kapsimali et al. 2003).

Neurons synthesizing catecholamines (mostly dopamine and norepinephrine in craniates) were essentially studied by immunohistochemistry or in situ hybridization of tyrosine hydroxylase (TH), the rate-limiting enzyme of catecholamine synthesis, sometimes supplemented by analyzing the distribution of other components of the biosynthetic pathways, such as the enzyme aromatic amino acid decarboxylase, the vesicular monoamine transporter or the plasma membrane dopamine transporter DAT. The direct analysis of dopamine or norepinephrine distribution has been also reported (for reviews, see Kaslin and Panula 2001; Smeets and Gonzalez 2000).

In the telencephalon, dopaminergic cells are mostly found in the olfactory bulb, and TH expression is highly plastic, as it disappears upon olfactory deafferentation. Other dopamine cells include amacrine cells of the retina, where dopamine acts mostly on D2-like receptors likely to increase discrimination for the two sensory pathways as in amniotes. There are also scattered subpallial dopamine cells as in most other vertebrate groups. Preoptic zebrafish dopamine cells may partially correspond to the mammalian group A14, as there are strong preoptic projections to the ventral telencephalon (Rink and Wullimann 2004). Other preoptic dopamine cells in teleosts project on the GnRH-producing cells of the ventral hypothalamus where they probably exert a mostly inhibitory effect on gametogenesis and ovulation via $D_2$ receptors (Dufour et al. 2005).

There are many TH-expressing cells of the posterior tuberculum, some of them being homologous to the amniote substantia nigra/ventral tegmental area (mammalian A9/A10; groups 1, 2, 4 in Rink and Wullimann 2001; Fig. 2.4). Posterior tubercular zebrafish dopamine neurons project to telencephalic ventral (septum) and dorsal (basal ganglia) divisions, but some projections from the posterior tuberculum may also reach dorsal (pallial) areas, as is the case in mammals and other amniotes. The action of dopamine on these structures is mediated by receptors of the D1 and D2 classes. In subpallial structures, two receptor subtypes are mainly found, the $D_{1A}$ and $D_2$ subtypes, which are located on different populations of neurons (Kapsimali et al. 2003). They likely mediate integration of sensorimotor cues in automatic programs of movements (dorsal striatum), as in other gnathostomes. In contrast, the $D_{1B}$ receptors are clearly present in an area located at the Dm–Dl junction, which has been proposed to be homologous to the mammalian hippocampus (Kapsimali et al. 2000; Salas et al. 2003). In addition, some dopaminergic cells in the zebrafish posterior tubercular area project to the spinal cord (McLean and Fetcho 2004a) and, thus, may correspond to A11. Clearly, teleostean posterior tubercular and hypothalamic dopaminergic populations are more numerous (groups 3, 5 to 7, 9 to 11; Fig. 2.4) than those of amniotes and they include three distinct cell types. Liquor-contacting cells (zebrafish groups 3, 5, 7, 10) are absent in mammals (but present in all other vertebrates). Zebrafish large pear-shaped dopaminergic cells are long-distance projections neurons (see above). Thus, only small round dopaminergic cells remain as possible candidates for an A12 (mammalian hypothalamic dopamine cells) homolog. Accordingly, numerous nuclei in the ventral and dorsal hypothalamic regions are targets of dopaminergic neurons. In some teleosts, the $D_{1A}$ and $D_{1B}$ receptor transcripts have been detected in preoptic nuclei and in the dorsal and ventral periventricular hypothalamic areas. In addition, the $D_{1C}$ receptor, a dopamine receptor subtype which has been lost in mammals, is found in a few restricted areas of the dorsal hypothalamus,

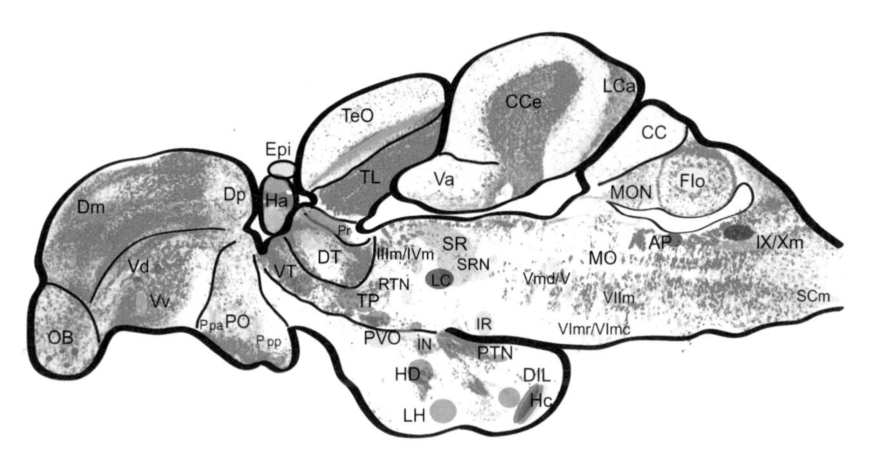


**Fig. 2.4.** Schematic representation of the main neuromodulatory pathways in the adult zebrafish brain. Dopaminergic nuclei and cells are in red; note that no dopamineric cells are found posterior to the diencephalon. Noradrenergic cells are only present in the locus cœruleus and medulla oblongata. Serotoninergic cells are in green and they are present in all brain divisions as are also the cholinergic nuclei and cells (yellow). Histaminergic cells (in blue) are only found in the caudal hypothalamic area. IIIm: oculomotor nerve nucleus; IVm: trochlear nerve nucleus; Vmd/V: dorsal/ventral trigeminal nerve motor nucleus; VImr/VImc: rostral/caudal abducens nerve motor nucleus; VIIm: facial nerve motor nucleus; IX/Xm: glossopharyngeal/vagal nerve motor nucleus; AP: area postrema; CC: crista cerebellaris; CCe: corpus cerebelli; DIL: diffuse nucleus of the inferior lobe of the hypothalamus; Dm: medial domain of the dorsal telencephalon; Dp: posterior domain of the dorsal telencephalon; DT: dorsal thalamus; Epi: epiphysis; Flo: Facial lobe; Ha: habenula; Hc/HD: caudal and dorsal periventricular hypothalamus; IN: intermediate nucleus; IR: inferior raphe; LC: locus cœruleus; LCa: lobus caudalis cerebelli; LH: lateral hypothalamic nucleus; MO: medulla oblongata; MON: medial octavolateralis nucleus; OB: olfactory bulb; PVO: paraventricular organ; PO: pre-optic nucleus; Ppa: anterior parvocellular preoptic nucleus; Ppp: posterior parvocellular preoptic nucleus; Pr: pretectum; PTN: posterior tubercular nucleus; RTN: rostral tegmental nucleus; SCm: spinal cord motoneurons; SR: superior raphe; SRN: superior reticular nucleus; TeO: optic tectum; TL: torus longitudinalis; TP: posterior tuberculum; Vd: ventral domain of the telencephalon, dorsal part; Vv: ventral domain of the telencephalon, ventral part; Va: valvula cerebelli; VT: ventral thalamus. See color plate section

including the liquor-contacting cells. Pretectal dopaminergic cells project to tectal layers where $D_{1A}$ receptors are found. Other diencephalic zebrafish dopamine cells include a ventral thalamic group corresponding to mammalian zona incerta (A13; group 0 of Rink and Wullimann 2001).

Two TH genes (*TH1* and *TH2*) have been reported to exist in the genome of some teleost fishes, *TH1* being orthologous to the mammalian *TH* gene (Candy and Collet 2005). The distribution of *TH1* and *TH2* transcripts revealed that *TH1* and *TH2* are differentially expressed in the zebrafish adult brain, as often observed for duplicated genes. Particularly we found that *TH2* transcripts were much more abundant than *TH1* in the hypothalamus, and that the *TH2* cells along the periventricular zone are devoid of TH immunoreactivity (THir). Although these neurons have not been considered to be dopaminergic in previous studies, the expression of other monoaminergic markers such as *AADC*, *DAT*, and *vMAT* suggests that these *TH2* cells are likely to be previously non-described dopaminergic cells (Yamamoto et al. 2010). Finally, all the dopaminergic cells which have been described also express the dopamine transporter.

The zebrafish adrenergic system (Ma 1994; Kaslin and Panula 2001) includes medullary cells close to the viscerosensory column/area postrema (comparable to mammalian groups A1/A2; Smeets and Gonzalez 2000) and a locus coeruleus (mammalian A6; Fig. 2.4). The other adrenergic neurons corresponding to the A3 to A5 and A7 groups of mammals are not distinguishable separately in the zebrafish brain. Neurons of mammalian A1/2 exert local control on the respiratory pacemaker and related functions (e.g., swallowing, response to pH changes). These mixed dopaminergic/adrenergic neurons are highly conserved in vertebrates and probably induce very similar cell responses in a broad range of species. The axons of the zebrafish locus coeruleus mainly project anteriorly, with a smaller contingent going to the hindbrain and spinal cord (Ma 1997). Anterior projections reach virtually all midbrain and forebrain structures as they do in other gnathostomes. In mammals, the locus coeruleus is crucial for two basic components of behaviors, namely arousal (as opposed to sleep or resting states) and awareness, the latter being necessary for focusing on specific aspects of sensory perceptions. Three receptor classes mediate the effect of norepinephrine, $a_1$, $a_2$ and b, each of which comprising typically 3–5 subtypes, highlighting the large variety of cellular actions promoted by this neurotransmitter. No precise distribution of all receptors are available yet in zebrafish, although $a_{2A}$ and $b_2$ receptors seem to be more concentrated in anterior pallial areas and $b_1/b_2$ receptors in pretectal and cerebellar target areas (Zikopoulos and Dermon 2005). The remaining tyrosine hydroxylase-positive cells rostral to the locus coeruleus in the zebrafish brain are dopaminergic (Ma 1994; Kaslin and Panula 2001).

Turning now to serotoninergic zebrafish brain populations, there are also some striking correspondences to amniotes (Kaslin and Panula 2001; Fig. 2.4). However, there are three tryptophan hydroxylases in the zebrafish, instead of only two in mammals, one of them being the result of the additional whole genome duplication specific to teleost fish (Bellipanni et al. 2002). They exhibit complementary distribution in the zebrafish brain. *tph1b* is mostly expressed in the epiphysis. In contrast, *tph1a* expression is detected in retinal amacrine cells and in restricted preoptic and posterior tubercular nuclei within the basal diencephalon. *tph1a* is also found in the caudal hypothalamus and in branchial arches-associated neurons. All sites of *tph1a* or *1b* expression within the anterior central nervous system are also immunoreactive for 5-HAT. Although the five classes of serotoninergic receptors, which have been isolated in mammals, also exist in teleosts, very few studies have addressed the relationship of serotonin projections and receptor localization. The large serotoninergic population in the superior raphe, which has a telencephalic projection (Bellipanni et al. 2002; Lillesaar et al. 2009), is almost certainly homologous to mammalian dorsal and central superior raphe nuclei (Lillesaar et al. 2009). The main target areas of these raphe neurons are probably hypothalamic nuclei where a high number of receptor-binding sites have been identified, as well as striatal and pallial areas. Serotoninergic cells in the zebrafish inferior raphe and in the more caudolaterally located reticular formation (both with spinal projections; Wulliman 1998) may correspond to mammalian nucleus raphes magnus (B3) and nuclei raphe pallidus/obscurus (B1/2), respectively. There is a distinct population of serotoninergic cells in the posterior tuberculum, and two more in the hypothalamus of the zebrafish (Fig. 2.4). The situation in the amphibian posterior tuberculum and hypothalamus is similar (Dicke et al. 1997). Sauropsids – but not mammals – also have serotoninergic cells in the posterior tuberculum (Smeets and Steinbusch 1988; Challet et al. 1996). The zebrafish hypothalamic intermediate nucleus exhibits exclusively serotoninergic cells, while the paraventricular organ, as well as the caudal hypothalamus, contains both dopaminergic and serotoninergic cells in the zebrafish. Although absent in amniotes, serotoninergic cells seen in the teleost pretectum may be ancestral for vertebrates, as they occur in amphibians (Dicke et al. 1997), chondrichthyans and lampreys (see below), but those in the pineal stalk may be unique to actinopterygians. As in tetrapods, histaminergic cell populations in the zebrafish brain are exclusively present in the most caudal hypothalamus (Kaslin and Panula 2001; Fig. 2.4).

The cholinergic system of the zebrafish is highly similar to the amniote pattern (Fig. 2.4; see Mueller et al. 2004). First, all motor cranial nerve nuclei are cholinergic, as expected. Second, there are cholinergic subpallial as well as brain stem neurons possibly corresponding to amniote cholinergic basal forebrain (Fig. 2.4) and ascending reticular (i.e., pedunculopontine-laterodorsal tegmental) systems. Also, the zebrafish secondary gustatory nucleus is at least partially cholinergic and projects to the hypothalamus. Furthermore, an isthmic nucleus (comparable to the mammalian parabigeminal nucleus) projects to the optic tectum.

## 8. EMOTIONS AND COGNITION

The recent development of assays assessing cognitive and emotional functions revealed that the zebrafish could exhibit highly complex behaviors that clearly bring it close to classical animal behavioral models such as the laboratory mouse or rat. These findings highlight the evolutionary conservation of such behaviors. Overall, however, we understand relatively little of the complex neuroanatomical bases sustaining these functions. Comparative functional neuroanatomy, and hence understanding the circuitry underlying emotions and cognition in zebrafish, becomes in this context all the more interesting.

### 8.1. Reward, Emotions and Motivation

Associative learning paradigms are widely used to screen for rewarding substances or study the mechanisms of reward. Such assays, for example the conditioned place preference test, are based on associating an initially neutral (usually visual) cue with a positive stimulus, and measuring the progressively acquired preference of the animal for the neutral cue. These tests in fact jointly measure reward and some learning capabilities of the subject, reflecting also the intricate relationship of these two behavioral components under normal conditions. Various compounds have been found to be rewarding in zebrafish adults, such as alcohol (Gerlai et al. 2000; Kily et al. 2008), classical drugs of addiction (cocaine, amphetamines, opiates) (Darland and Dowling 2001; Ninkovic and Bally-Cuif 2006; Ninkovic et al. 2006; Bretaud et al. 2007), food (Williams et al. 2002; Colwill et al. 2005), and the presence of conspecifics (Al-Imari and Gerlai 2008; Saverino and Gerlai 2008).

The neurotransmitter pathways involved have been studied largely through pharmacological approaches, based on known pathways implicated in emotions, reward and motivation in mammals. The dopaminergic system is a prime component and, in zebrafish larvae, the treatment with antagonists of dopamine receptors attenuates morphine-conditioned place preference (Bretaud et al. 2007). Their effect on the reinforcing properties of direct modulators of dopamine levels, such as cocaine and amphetamine, which primarily inhibit the dopamine transporter DAT, has not been studied. Antagonists of opioid receptors – such as naloxone or the kappa-opioid antagonist nor-binaltorphimide – abolished the reinforcing effects of morphine or salvinorin A, respectively (Lau et al. 2006; Braida et al. 2007; Bretaud et al. 2007). In addition, the effects on reward behavior of a few mutations have been studied. Hence, larvae carrying the genetic mutation *too few* – affecting the zinc finger protein Fezl – fail to experience the reinforcing properties of morphine (Bretaud et al. 2007; Lau et al. 2006). This might be linked to the altered development of diencephalic dopaminergic and serotonergic clusters in this animal (Levkowitz et al. 2003). Interestingly, preference for food in this mutant is preserved, suggesting that other transmitters and/or peptides predominate in this behavior (see Volkoff and Peter 2006). In addition, increased levels of acetylcholine are associated with decreased amphetamine-conditioned place preference in adults carrying the AChE mutation (Ninkovic et al. 2006). Several drugs of addiction, such as EtOH, cocaine and amphetamine, have been directly used in genetic screens with the hope to unravel more of the genetic and neuroanatomical pathways underlying reward and motivation (Gerlai et al. 2000; Darland and Dowling 2001; Ninkovic and Bally-Cuif 2006; Webb et al. 2009). These mutations, however, could not be cloned to date.

At this point, the territories of the brain involved in mediating reward and motivation in zebrafish have been tentatively identified from comparative neuroanatomical studies, and from mapping of the main modulatory neurotransmitters and their circuitry (see above, the presentation of the subpallium and of the posterior tuberculum). Two subpopulations of dopaminergic neurons from the posterior tuberculum, namely a group of small cells in the periventricular nucleus of the posterior tuberculum, and an adjacent group of large, pear-shaped cells, were found to project towards the ventral telencephalon (subpallium, Vv and Vd) (Rink and Wullimann 2002). The latter domain exhibits immunoreactivity for substance P and GABA in other teleosts, and expresses the D1 dopamine receptor. This makes it comparable to the mammalian striatum (Wullimann and Rink 2002), a major reward center that receives dopaminergic projections from the midbrain ventral tegmental area (mesolimbic pathway). In goldfish, further ascending dopaminergic projections have been described from the posterior tuberal nucleus (PTN, group C) towards the dorsal telencephalon (pallium) (Wullimann et al. 1991). They are possibly equivalent to the mammalian mesocortical pathway, another crucial player of motivational and emotional responses. Whether efferent projections from Vv can further compare to the rest of the mammalian basal ganglia circuitry, which controls reward and motivation, is still unknown. Adding to the circuitry possibly modulating reward, Facchin et al. (2009) demonstrate in a recent article that perturbation of epithalamic asymmetry (through manipulation of the Nodal pathway) has consequences on the initiation of movement in larvae, reflecting a potential role of the parapineal/habenular circuitry in motivation (Facchin et al. 2009). Finally, our microarray analysis of the adult amphetamine-insensitive mutant *no addiction* revealed a subset of genes abnormally regulated by the drug and co-expressed in the neurogenesis zone (ventricular zone) of the adult telencephalon (Webb et al. 2009). We further showed that amphetamine treatment of wild-type fish strongly impaired expression of some of these genes at the telencephalic ventricular zone (Webb et al. 2009) and was associated with altered neurogenesis (K. Webb and L. Bally-Cuif, unpublished). These results suggest that modulations of telencephalic adult neurogenesis (see below) might play a role in the establishment of an amphetamine-conditioned place preference behavior. Overall, experiments are crucially needed in zebrafish to determine which neuronal populations are recruited in behavioral tasks involving reward and motivation and to precisely map the relevant circuitry.

### 8.2. Learning and Memory

In mammals there are several kinds of learning and memory systems, such as relational processes (e.g., spatial learning based on a series of visual cues), implicit learning processes (e.g., the conditioning of simple motor reflexes) or emotional processes (e.g., avoidance learning). These are in large part neuroanatomically distinct, primarily involving the hippocampus, the cerebellum and the amygdala, respectively. To date, learning in zebrafish has been mostly assessed in a number of rewarded tests for place choice. The most widely used reward is food (Colwill et al. 2005; Williams et al. 2002) but fish can also learn in response to drugs (see references above) or to a computed image of conspecifics (Pather and Gerlai 2009). In a few instances (Darland and Dowling 2001; Peitsaro et al. 2003), completing a T-maze was indirectly rewarded through the target arm ending in a deeper, more comfortable container. These tests overall make it difficult to distinguish between learning, memory and reward, although, interestingly, some mutations have permitted the dissociation of these functions. For example, AChE mutants learn significantly faster than wild-type fish to find a food-rewarded target in a T-maze, although their sensitivity to the reinforcing effect of psychostimulants is lowered (Ninkovic et al. 2006). Additionally, a one-trial inhibitory avoidance test was developed where an electric shock coupled to a white compartment leads to avoidance of this compartment (Blank et al. 2009). Memory is assessed in the same tests after some latency (e.g., Levin and Chen 2004). In contrast, thorough tests have been performed in the goldfish, which we will mention here as a major contribution to the field. In a test for spatial memory, goldfish can be trained to find a target in a four-arm maze surrounded by a display of distant visual cues, and accomplish the task irrespective of their start position (reviewed in Salas et al. 2006). Likewise, they can locate a baited feeder within a 25-feeder matrix bordered by an arrangement of visual information (Rodriguez et al. 2005). Thus, they are capable of establishing allocentric spatial representations (so-called cognitive maps). Motor conditioning in goldfish can be measured by the eye-retraction movement induced by a predictive stimulus (e.g., light or sound) initially associated with an unconditioned stimulus (e.g., air puff or electric shock). Finally, emotional memory testing involves avoidance learning paradigms, such as a shuttle box where fish are conditioned to change side in an aquarium in response to a colored light paired to an electric shock (Portavella et al. 2004), or measurements of fear heart rates (Yoshida et al. 2004; Rodriguez et al. 2005). In both instances, the time separating the two stimuli can be varied (delay versus trace conditioning). The territories involved in these different behavioral assays have been mapped through focal ablations in the goldfish brain, and, to some extent, through mapping of transcriptionally active neurons. It was found that lesions of the lateral pallium (Dl) strongly impaired spatial memory as well as performance in trace conditioning assays, demonstrating the role of this domain in a spatial, relational and temporal memory system (Portavella et al. 2003). In support of this finding, protein synthesis is increased in Dl following spatial learning training (Vargas et al. 2000). In contrast, lesions of the medial pallium (Dm) affected performance in emotional memory tests. Finally, the cerebellum was found to participate in both spatial and emotional cognition in delay conditioning assays (Yoshida et al. 2004; Rodriguez et al. 2005). Taking into account the everted morphology of the teleost telencephalon, Dl is in a homologous location to the hippocampus and Dm to the amygdala. Analyses of adult neurogenesis, documenting active proliferation and neuronal production in Dl like in the hippocampus, further support these homologies (Adolf et al. 2006; Grandel et al. 2006) (see below). Together, the functional results obtained in goldfish can be placed in direct parallel to what is known of the neuroanatomical control of memory in mammals (reviewed in Salas et al. 2006).

The behavior of zebrafish AChE mutants points to ACh as a major neurotransmitter pathway involved in the cognitive functions of the adult zebrafish, as observed in mammals (Ninkovic et al. 2006). Cholinergic neurons, as revealed by their expression of the choline acetyltransferase (ChAT) enzyme, are widely distributed through the adult zebrafish brain (Mueller et al. 2004). The role of ACh is further corroborated by the observed positive effects of low doses of nicotine – which activates the nAChR – on memory (Levin and Chen 2004) and learning (Levin et al. 2006). This effect is reversed when the nAChR antagonist mecamylamine is given shortly before testing, and is correlated with an increase in the levels of DOPAC, the primary dopamine metabolite (Eddins et al. 2009), pointing to a link between nicotine effects on learning and dopamine signaling. Another relevant neurotransmitter is histamine. Zebrafish adults treated with the histidine decarboxylase inhibitor alpha-FMH display decreased brain histamine levels and a reduced number of histaminergic fibers (Peitsaro et al. 2003). In a T-maze, these animals learn and remember the task over one day as efficiently as control fish; however, they forget faster and need a longer time to find the target again after four days (Peitsaro et al. 2003). These interesting results point to a selective role of histamine in long-term memory. Histaminergic neurons are concentrated in the caudal hypothalamus (Eriksson et al. 1998) but innervate profusely the entire brain with, in particular, extensive connections to other aminergic systems in the telencephalon, hypothalamus, and raphe (Kaslin and Panula 2001). In the telencephalon, major innervation is seen in the pallium (with lower levels in Dm), and around the pallium/subpallium junction (Peitsaro et al. 2003). Finally, in both zebrafish and goldfish, administration of N-methyl-D-aspartic acid (NMDA) receptor antagonists prevents acquisition of avoidance conditioning (Davis and Klinger 1994; Xu et al. 2003; Blank et al. 2009). NMDA-binding sites are present in the goldfish telencephalon, and, as previously mentioned, NMDA causes LTP in the adult zebrafish telencephalon (Nam et al. 2004), a process that has been implicated in memory formation and consolidation. As in mammals (Yu et al. 2006), the capacity for adaptive associations and more generally cognitive functions are impaired with aging in zebrafish.

### 8.3. Anxiety

Anxiety is classically estimated in mammals as an inverse measure of boldness in open field tests. In a neutral aquarium and when the fish is allowed to swim freely, boldness is measured as the time spent in the center of the tank, as opposed to swimming along the sides, or as time spent at intermediate water levels (Bencan et al. 2009). In a two-colored tank, zebrafish naturally prefer dark, and the time spent in the bright compartment is also used as a measure of boldness (Serra et al. 1999; Ninkovic et al. 2006). Other studies measured the proportion of time spent away from a friendly environment, such as conspecifics (Gerlai et al. 2000), or the latency to enter an enriched chamber (Swain et al. 2004). The intensity of these behavioral indices is paralleled by the levels of whole-body cortisol, reflecting the role of this hormone in physiological stress responses (Egan et al. 2009). Differences can be observed in these responses between species under normal conditions, suggesting a genetic modulation (Ninkovic and Bally-Cuif 2006). In addition, anxiety and stress can be triggered by a number of stimuli, such as introduction into a new tank (Bencan et al. 2009), exposure to alarm pheromone, acute exposure to caffeine (Egan et al. 2009), or withdrawal from drugs of addiction (Lopez-Patino et al. 2008). Most of these behavioral indices can be mimicked by treatment with an anxiogenic benzodiazepine inverse agonist, and lowered upon treatment with the anxiolytic benzodiazepine diazepam (Lopez-Patino et al. 2008). The serotonergic (5HT1A receptor) agonist buspirone, as well as the 5HT reuptake inhibitor fluoxetine, also act as anxiolytics, e.g. in the novel tank test (Bencan et al. 2009; Egan et al. 2009), implicating the 5HT pathway in anxiety behavior. Finally, nicotine was also found to have an anxiolytic effect, in a manner that could be blocked by the nAChR antagonist mecamylamine when administered together with, but not later than, nicotine (Levin et al. 2007). Hence, a burst activation of nAChR seems sufficient to trigger the anxiolytic response. The use of nAChR subtype-specific antagonists further demonstrated that both the alpha7 and alpha4beta2 receptors were involved in the anxiolytic effect of nicotine (Bencan and Levin 2008). The neuroanatomical areas involved in mediating anxiety behavior in zebrafish have not been mapped.

### 8.4. Aggression

Aggressive behavior can be observed in zebrafish in the wild as well as in the laboratory. In the latter case, aggression can result from the establishment of a dominant–subordinate relationship, when several fish of the same sex are hosted in the same tank, for example (Larson et al. 2006). Another classical set-up uses a mirror placed at an angle along one side of the tank and where the image of the swimming fish mimics an intruder; aggressive behavior is measured as the frequency of attacks or aggressive postural displays towards the mirrors (Gerlai et al. 2000). The neuroanatomical or genetic bases for this behavior remain to be precisely defined. The neuropeptide arginine vasotocin, the teleostean homolog of arginine vasopressin, is differently expressed depending on the status of the individual in a group; dominant individuals express vasotocin in large neurons of the magnocellular preoptic area, while positive neurons in subordinate zebrafish are of small size and located in the parvocellular preoptic area (Larson et al. 2006). The dominance status and aggressiveness of zebrafish can be reverted upon exposure to the synthetic estrogen 17alpha-ethinylestradiol (Colman et al. 2009), highlighting the roles of estrogen levels in shaping zebrafish agonistic behavior. Finally, aggressive behavior shows a clear laterality, whereby the right eye is preferentially used to view a predator (Miklosi and Andrew 1999), although the neuroanatomical correlates of this observation have not been identified. In another teleostean fish, the mudskipper, expression of the neuronal activity marker *cfos* was used to monitor neuronal activation changes following an aggressive episode (Wai et al. 2006). No change in *cfos* expression compared to control fish was observed in the telencephalon and pituitary. In contrast, *cfos* was significantly increased in the diencephalon in scattered cells throughout the thalamus and hypothalamus, in some identified nuclei of the pons (including the lateral and medial longitudinal fascicle and the trigeminal sensory and motor nuclei), and in the medulla oblongata. The relevance of these neurons in the generation of aggression (as opposed to associated behaviors such as locomotion, for example) remains to be determined.

## 9. NERVOUS SYSTEM PLASTICITY

### 9.1. Neuronal and Synaptic Plasticity

To date, neuronal and synaptic plasticity have been best studied during zebrafish development, although it is likely that they also play a role in the adult. Behavioral plasticity in adults is evident for instance from the successful performance of teleost fish in learning paradigms (see above), although its molecular or cellular correlates remain largely unexplored. Ependymins, glycoproteins secreted by meningeal fibroblasts and found in correlates the teleostean extracellular brain fluid, have been implicated in memory consolidation in goldfish (Piront and Schmidt 1988) and zebrafish (Pradel et al. 1999). They also support neurite outgrowth and fasciculation (Schmidt et al. 1991), suggesting a role in neuronal and/or synaptic plasticity, which may be relevant to the very rapid effect of the intracerebral ventricular injection of anti-ependymin antibodies (Pradel et al. 1999). Anti-HNK1 antibodies have the same effect, and recognize HNK1-bearing ependymins in brain extracts. The presence of ependymins in the CSF and in the "ependymal zone" of the adult brain (Pradel et al. 1999) would also be in support of an effect on adult neurogenesis, perharps for longer-term behavioral modulations. Injection of antibodies directed against the adhesion molecule L1.1 (but not against Tenascin-C) also prevents memory consolidation (Pradel et al. 2000). Such an effect has been observed in various other vertebrates, and L1.1 is involved in synapse stabilization. LTP can be induced in the zebrafish telencephalon by a high concentration of KCl in a manner dependent on AMPA receptors, calcium and protein kinases (Nam et al. 2004), but the relevance of this process in zebrafish memory behavior has not been studied.

Plasticity of the locomotor circuitry has been documented in the larva. While projections of individual motorneurons and development of their specific terminal fields does not seem to involve either activity- or competition-dependent mechanisms during larval development (Liu and Westerfield 1990), second-order sensory projections, in particular from the lateral line, initially involve excess innervations followed by retraction between the larval and adult stages (Fame et al. 2006). *Deadly seven* mutants provide a further example of plasticity occurring during the construction of this system, where defective Notch1a signaling leads to the production of supernumerary Mauthner neurons (Gray et al. 2001). All Mauthner cells (between 3 and 8 cells per side instead of 1) are integrated in the escape circuit and active during the escape response; however this excess is compensated for by a decrease in the number of axonal collaterals per Mauthner cell and their regulated distribution to non-overlapping motorneuron populations, visible at 4 dpf (Liu et al. 2003). Finally, mutants for the glial glycine transporter GlyT1 (*shocker*), for example, which have excess extracellular glycine and increased inhibitory synaptic potentials, initially display severe motility and escape response defects, but these gradually recover as the locomotor system matures such that normal locomotion is restored by 90–120 hpf (Cui et al. 2004, 2005). This adaptation to high glycine levels, likely triggered by activity, was associated with decreased post-synaptic expression of glycine receptors, in particular on motorneurons (Mongeon et al. 2008).

Such adaptive capacities have to be assessed for each larval circuitry. The visual system has been rigorously analyzed, and synaptic plasticity at photoreceptor terminals is detectable by the on and off development of spinules depending on illumination (Biehlmaier et al. 2003). In contrast, intraretinal projections are largely pre-patterned (e.g., see Mumm et al. 2006; Godinho et al. 2005, for the retina). The tectal neuropil is organized into molecularly distinct sublaminae receiving the majority of the retinal ganglion cell axons. In this system, NMDA receptor-mediated activity, modifications of calcium levels through activation or inhibition of CamKII, or altering PKC activity, controls the branching level and stabilization of retinal arbors in the larva between 2 and 7 dpf (Schmidt et al. 2004). Similar effects have been observed in the regenerating retinotectal projections in the adult goldfish (Schmidt 1994). In contrast, modification of ganglion cells' activity at 5 dpf through rearing zebrafish larvae in the dark, application of an agonist of the metabotropic glutamate receptor mGluR6 – present on the dendritic terminals of some ganglion cells – or blockade of chemical synaptic transmission by Botulinum toxin B, largely preserved IPL sublamination (Nevin et al. 2008). Although these experiments suggest hardwiring in the sorting of retino-tectal projections, it remains possible that they were conducted outside the critical period(s) and that manipulations at an earlier developmental stage would reveal plasticity. Finally, in addition to lamination, retinotectal projections display somatotopy whereby the dorsal, ventral, temporal and nasal regions of the retina innervate non-overlapping regions of the tectum. Activity of retinal ganglion cells is not required for the establishment of the retinotopic map (Stuermer et al. 1990) but controls its refinement between 4 and 6 dpf, after complete tectal innervation is first achieved (Gnuegge et al. 2001).

### 9.2. Adult Neurogenesis

As in other teleosts, continuous neurogenesis is a prominent feature of the adult zebrafish brain (Fig. 2.5). Although the function of adult-born neurons remains virtually unknown, they are likely an important part of adult CNS plasticity.

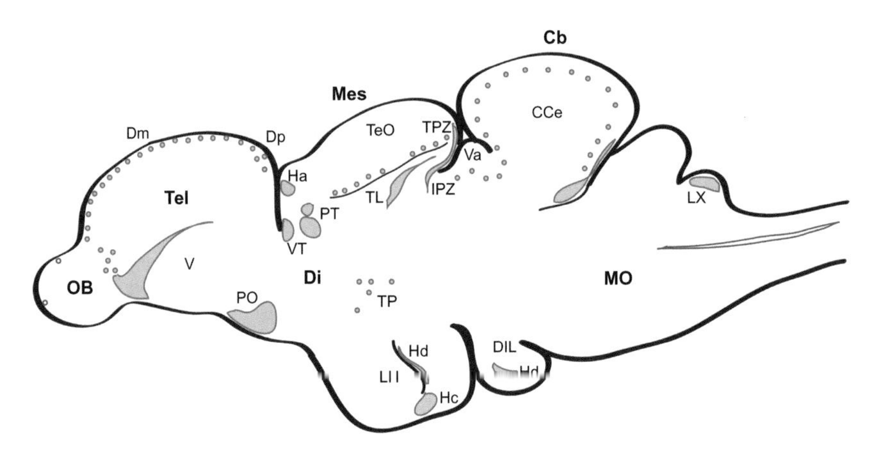


**Fig. 2.5.** Neurogenesis domains in the adult zebrafish brain. Proliferation domains are schematically indicated as gray zones or dots on a representation of the adult brain at a midsagittal level (anterior left). All these domains contain label-retaining progenitors and are neurogenic, except for proliferating cells in the OB (empty dots) and MO (empty zone). Cb: cerebellum; CCe: corpus cerebelli; Di: diencephalon; DIL: diffuse nucleus of the inferior lobe of the hypothalamus; Dm: medial domain of the dorsal telencephalon; Dp: posterior domain of the dorsal telencephalon; Ha: habenula; Hc: caudal zone of periventricular hypothalamus; Hd: dorsal zone of periventricular hypothalamus; IPZ: isthmic proliferation zone; LH: lateral hypothalamic nucleus; LX: vagal lobe; Mes: mesencephalon; MO: medulla oblongata; OB: olfactory bulb; PO: pre-optic area; PT: posterior thalamus; Tel: telencephalon; TeO: optic tectum; TL: torus longitudinalis; TP: posterior tuberculum; TPZ: tectal proliferation zone; V: ventral domain of the telencephalon; VT: ventral thalamus.

Adult-born neurons in the zebrafish were first described in the olfactory bulb (Byrd and Brunjes 1998, 2001), where BrdU-positive neurons can be found in the internal cell layer three to four weeks after labeling. Meanwhile, a total of 12 proliferation zones, covering all brain subdivisions, were identified through BrdU incorporation (Zupanc et al. 2005; Grandel et al. 2006; Pellegrini et al. 2007) (Fig. 2.5). Most of these zones are ventricular (except in the cerebellum), and BrdU tracing followed by staining for the postmitotic neuronal malker HuC/D indicates that they predominantly generate neurons (Adolf et al. 2006; Grandel et al. 2006). Whether these neurons become integrated and which circuitries they complement have not been determined, but at least a subset of them can persist for months in the adult brain (Zupanc et al. 2005). Analyses of label retention indicate the presence of slow-proliferating, self-renewing, radial glia-like progenitors in all proliferation zones (Adolf et al. 2006; Chapouton et al. 2006; Kaslin et al. 2009; Lam et al. 2009), and it is likely that the active maintenance of these potential neural stem cells accounts for the high neurogenic capacity of the adult zebrafish brain in comparison to, for example, the adult mammalian brain. The molecular mechanisms underlying this sustained activation of neurogenesis remain only partially understood. Expression of FGF receptors and Fgf signaling targets has been demonstrated in non-proliferating ventricular radial glial cells (Topp et al. 2008), most of which are also positive for the progenitor markers Nestin and Sox2 (Mahler and Driever 2007; Lam et al. 2009), suggesting that Fgf signaling plays a role in the maintenance of quiescence or neural stem cell properties. Fgf8 was recently shown to control production of adult-born granule neurons in the cerebellum (Kaslin et al. 2009). Adult neurogenic domains of the zebrafish brain are also strongly positive for expression of Aromatase B (Cyp19a1b) (Pellegrini et al. 2007; Kim et al. 2009), suggesting a potential involvement of steroids in adult neurogenesis control. Finally, expression of Enhancer-of-Split (Her) factors in adult long-lasting progenitors (Chapouton et al. 2006) suggests an implication of Notch signaling in the maintenance of adult zebrafish neural stem cells.

A number of neuronal subtypes can be generated in the adult zebrafish brain, including TH-positive and GABAergic neurons in the olfactory bulb and ventral telencephalon (subpallium), dopaminergic and serotonergic neurons in the ventral diencephalon, glycinergic neurons in the midbrain, or glutamatergic neurons in the cerebellum (Adolf et al. 2006; Grandel et al. 2006; Chapouton et al. unpublished). It is believed that adult neuronal production, for a large part, underlies brain growth and is involved in matching brain size and processing capacity with the increased peripheral sensory input that follows from the addition of new sensory cells and neurons throughout life. The generality of this hypothesis remains to be assessed. Intense neuronal production is observed during adulthood in both the optic tectum and retina for example, but the functional connection of adult-born neurons in the two systems has not been studied. It is also of note that some brain areas, such as the primary olfactory center in the dorsal telencephalon (Dl), are devoid of neuronal production during adulthood (Lillesaar et al. 2009). This contrasts strikingly with the intense neuronal production into the olfactory bulb, likely from a ventricular domain located at the junction of Vv and Vd in the subpallium and proposed to be equivalent to the rodent subependymal zone (SEZ) in the lateral ventricle.

The exact spatial contribution of adult neurogenesis to brain renewal also remains to be studied, an analysis that will necessitate permanent labeling of adult-born cells. From the few studies of adult neurogenesis that involved very long tracing times (Zupanc et al. 2005; Grandel et al. 2006), it is likely that a high number of adult-born neurons survive over long periods. Their integration gives evidence of the functional plasticity of zebrafish brain circuits. Finally, data on the contribution of adult neurogenesis to modulatory behaviors such as learning and memory or motivation are also lacking. Along these lines, a highly proliferating area has been identified in the zebrafish adult telencephalon in the most lateral aspect of Dl. This domain generates neurons that populate the adjacent parenchyma and, using ablation experiments, has been implicated in spatial memory in goldfish (Portavella et al. 2004; Salas et al. 2006). It might therefore be equivalent to the second neurogenic domain of the adult mammalian brain, the sub-granular zone (SGZ) of the dendate gyrus of the hippocampus. Hippocampal adult-born neurons in mammals are believed to play a role in spatial learning as well as modulations of mood.

### 9.3. Post lesional Plasticity and Regeneration

The capacity for regenerating injured CNS axons is prominent in a number of anamniotes, including zebrafish. Following spinal cord transsection in zebrafish, the spinal cord-projecting neurons located proximal to the section will re-grow their axons, leading to locomotion recovery in a majority of cases (Becker et al. 1997). Interestingly, this capacity seems however to decrease with increasing age of the animal. In contrast, the terminal fields of ascending, transected spinal fibers are generally not re-established, either due to a lack of regeneration or to the failure of axons to re-innervate their original target. The contribution of de novo neurogenesis to the re-innervation process seems at best minimal. In detail, there is however a rather high heterogeneity in the capacity of distinct neuronal populations to regenerate their transected axon. For example, the nucleus rubber, the nucleus of the lateral lemniscus, the tangential nucleus, and, prominently, the Mauthner cell, have a low regeneration capacity (Becker et al. 1997, 2005). Regeneration, when successful, is associated with the up-regulation of expression of the growth-associated protein GAP-43 and adhesion molecules such as L1.1 and L1.2 but not NCAM in regenerating neurons, and of L1.2 in glial cells surrounding the area of axonal regrowth (Becker et al. 1998). Inhibition of L1.1 function upon application of L1.1 morpholinos to the lesioned axons was shown to inhibit regrowth and synapse formation, demonstrating the functional relevance of this adhesion molecule in the regeneration process (Becker et al. 2004). In the retinal system, an experimental lesion of the optic nerve is also followed by regeneration and a high degree of visual functional recovery (McDowell et al. 2004). Recently, this system was used in a microarray experiment (Munderloh et al. 2009) to recover hundreds of candidate genes transcriptionally regulated during the regeneration process (Veldman et al. 2007). At the juvenile stage (5–7-day-old animals), the regeneration of a severed Mauthner cell axon can be boosted by intracellular application of cyclic AMP, suggesting that this compound may also contribute to the cell's intrinsic ability for axonal regeneration (Bhatt et al. 2004). cAMP not only influences growth but also the capacity for proper pathfinding. Studies of axonal regeneration will need to solve the mechanisms guiding the regenerating axons to navigate and establish the proper connections in a mature environment.

Recent studies also illustrate that neurons, in addition to axons, can be generated in response to lesions. Upon a spinal cord lesion, proliferation of ventricular Olig2-positive ependymo-radial glial cells located adjacent to the lesion is induced, and a subset of these progenitors generates Isl1- and/or HB9-positive motorneurons (Reimer et al. 2008). These neurons show signs of integration into the spinal circuitry, such as post-synaptic contacts and connection to their muscle targets. Given that the spinal cord normally exhibits very little proliferation and neurogenesis, these results illustrate the high plasticity of spinal ventricular zone cells in response to environmental changes. Likewise, a crush of the optic nerve, as well as mechanical or light-induced retinal lesions, is followed by the loss of expression of specific glial markers by Müller glial cells (Thummel et al. 2008), their re-initiation of proliferation (Yurco and Cameron 2005; Fausett and Goldman 2006) and their generation of neurons of most retinal subtypes (Fausett and Goldman 2006; Sherpa et al. 2008; Thummel et al. 2008).

## REFERENCES


Adolf, B., Chapouton, P., Lam, C. S., Topp, S., Tannhauser, B., Strähle, U., Götz, M., and Bally-Cuif, L. (2006). Conserved and acquired features of adult neurogenesis in the zebrafish telencephalon. *Dev. Biol.* **295**, 278–293.

Al-Imari, L., and Gerlai, R. (2008). Sight of conspecifics as reward in associative learning in zebrafish (Danio rerio). *Behav. Brain Res.* **189**, 216–219.

Allende, M. L., and Weinberg, E. S. (1994). The expression pattern of two zebrafish achaete-scute homolog (ash) genes is altered in the embryonic brain of the cyclops mutant. *Dev. Biol.* **166**, 509–530.

An, M., Luo, R., and Henion, P. D. (2002). Differentiation and maturation of zebrafish dorsal root and sympathetic ganglion neurons. *J. Comp. Neurol.* **446**, 267–275.

Andermann, P., Ungos, J., and Raible, D. W. (2002). Neurogenin1 defines zebrafish cranial sensory ganglia precursors. *Dev. Biol.* **251**, 45–58.

Appel, B., and Chitnis, A. (2002). Neurogenesis and specification of neuronal identity. *Results Probl. Cell Differ.* **40**, 237–251.

Arduini, B. L., Bosse, K. M., and Henion, P. D. (2009). Genetic ablation of neural crest cell diversification. *Development* **136**, 1987–1994.

Bally-Cuif, L., Dubois, L., and Vincent, A. (1998). Molecular cloning of Zcoe2, the zebrafish homolog of Xenopus Xcoe2 and mouse EBF-2, and its expression during primary neurogenesis. *Mech. Dev.* **77**, 85–90.

Bally-Cuif, L., and Hammerschmidt, M. (2003). Induction and patterning of neuronal development, and its connection to cell cycle control. *Curr. Opin. Neurobiol.* **13**, 16–25.

Barembaum, M., and Bronner-Fraser, M. (2005). Early steps in neural crest specification. *Semin. Cell Dev. Biol.* **16**, 642–646.

Bass, A. H., and Grober, M. S. (2001). Social and neural modulation of sexual plasticity in teleost fish. *Brain Behav. Evol.* **57**, 293–300.

Becker, C. G., Lieberoth, B. C., Morellini, F., Feldner, J., Becker, T., and Schachner, M. (2004). L1.1 is involved in spinal cord regeneration in adult zebrafish. *J. Neurosci.* **24**, 7837–7842.

Becker, T., Lieberoth, B. C., Becker, C. G., and Schachner, M. (2005). Differences in the regenerative response of neuronal cell populations and indications for plasticity in intraspinal neurons after spinal cord transection in adult zebrafish. *Mol. Cell. Neurosci.* **30**, 265–278.

Becker, T., Wullimann, M. F., Becker, C. G., Bernhardt, R. R., and Schachner, M. (1997). Axonal regrowth after spinal cord transection in adult zebrafish. *J. Comp. Neurol.* **377**, 577–595.

Becker, T., Bernhardt, R. R., Reinhard, E., Wullimann, M. F., Tongiorgi, E., and Schachner, M. (1998). Readiness of zebrafish brain neurons to regenerate a spinal axon correlates with differential expression of specific cell recognition molecules. *J. Neurosci.* **18**, 5789–5803.

Bellipanni, G., Rink, E., and Bally-Cuif, L. (2002). Cloning of two tryptophan hydroxylase genes expressed in the diencephalon of the developing zebrafish brain. *Gene. Expr. Patterns* **2**, 251–256.

Bencan, Z., and Levin, E. D. (2008). The role of alpha7 and alpha4beta2 nicotinic receptors in the nicotine-induced anxiolytic effect in zebrafish. *Physiol. Behav.* **95**, 408–412.

Bencan, Z., Sledge, D., and Levin, E. D. (2009). Buspirone, chlordiazepoxide and diazepam effects in a zebrafish model of anxiety. *Pharmacol. Biochem. Behav.* **94**, 75–80.

Berman, J. R., Skariah, G., Maro, G. S., Mignot, E., and Mourrain, P. (2009). Characterization of two melanin-concentrating hormone genes in zebrafish reveals evolutionary and physiological links with the mammalian MCH system. *J. Comp. Neurol.* **517**, 695–710.

Bhatt, D. H., Otto, S. J., Depoister, B., and Fetcho, J. R. (2004). Cyclic AMP-induced repair of zebrafish spinal circuits. *Science* **305**, 254–258.

Biehlmaier, O., Neuhauss, S. C., and Kohler, K. (2003). Synaptic plasticity and functionality at the cone terminal of the developing zebrafish retina. *J. Neurobiol.* **56**, 222–236.

Blader, P., Fischer, N., Gradwohl, G., Guillemont, F., and Strähle, U. (1997). The activity of neurogenin1 is controlled by local cues in the zebrafish embryo. *Development* **124**, 4557–4569.

Blank, M., Guerim, L. D., Cordeiro, R. F., and Vianna, M. R. (2009). A one-trial inhibitory avoidance task to zebrafish: Rapid acquisition of an NMDA-dependent long-term memory. *Neurobiol. Learn. Mem.* **92**, 529–534.

Blin, M., Norton, W., Bally-Cuif, L., and Vernier, P. (2008). NR4A2 controls the differentiation of selective dopaminergic nuclei in the zebrafish brain. *Mol. Cell. Neurosci.* **39**, 592–604.

Boehmler, W., Obrecht-Pflumio, S., Canfield, V., Thisse, C., Thisse, B., and Levenson, R. (2004). Evolution and expression of D2 and D3 dopamine receptor genes in zebrafish. *Dev. Dyn.* **230**, 481–493.

Braida, D., Limonta, V., Pegorini, S., Zani, A., Guerini-Rocco, C., Gori, E., and Sala, M. (2007). Hallucinatory and rewarding effect of salvinorin A in zebrafish: Kappa-opioid and CB1-cannabinoid receptor involvement. *Psychopharmacology (Berl.)* **190**, 441–448.

Bretaud, S., Li, Q., Lockwood, B. L., Kobayashi, K., Lin, E., and Guo, S. (2007). A choice behavior for morphine reveals experience-dependent drug preference and underlying neural substrates in developing larval zebrafish. *Neuroscience* **146**, 1109–1116.

Bylund, M., Andersson, E., Novitch, B. G., and Muhr, J. (2003). Vertebrate neurogenesis is counteracted by Sox1-3 activity. *Nat. Neurosci.* **6**, 1162–1168.

Byrd, C. A., and Brunjes, P. C. (1998). Addition of new cells to the olfactory bulb of adult zebrafish. *Ann. N.Y. Acad. Sci.* **855**, 274–276.

Byrd, C. A., and Brunjes, P. C. (2001). Neurogenesis in the olfactory bulb of adult zebrafish. *Neuroscience* **105**, 793–801.

Callier, S., Snapyan, M., Le Crom, S., Prou, D., Vincent, J. D., and Vernier, P. (2003). Evolution and cell biology of dopamine receptors in vertebrates. *Biol. Cell* **95**, 489–502.

Candy, J., and Collet, C. (2005). Two tyrosine hydroxylase genes in teleosts. *Biochim. Biophys. Acta* **1727**, 35–44.

Carney, T. J., Dutton, K. A., Greenhill, E., Delfino-Machin, M., Dufourcq, P., Blader, P., and Kelsh, R. N. (2006). A direct role for Sox10 in specification of neural crest-derived sensory neurons. *Development* **133**, 4619–4630.

Caron, S. J., Prober, D., Choy, M., and Schier, A. F. (2008). In vivo birthdating by BAPTISM reveals that trigeminal sensory neuron diversity depends on early neurogenesis. *Development* **135**, 3259–3269.

Castro, A., Becerra, M., Manso, M. J., Tello, J., Sherwood, N. M., and Anadon, R. (2009). Distribution of growth hormone-releasing hormone-like peptide: Immunoreactivity in the central nervous system of the adult zebrafish (Danio rerio). *J. Comp. Neurol.* **513**, 685–701.

Challet, E., Micelli, D., Pierre, J., Reperant, J., Masicotte, G., Herbin, M., and Vesselkin, N. P. (1996). Distribution of serotonin-immunoreactivity in the brain of the pigeon (Columbia livia). *Anat. Embryol. (Berl.)* **193**, 209–227.

Chandrasekhar, A. (2004). Turning heads: Development of vertebrate branchiomotor neurons. *Dev. Dyn.* **229**, 143–161.

Chapouton, P., Adolf, B., Leucht, C., Ryu, S., Driever, W., and Bally-Cuif, L. (2006). Her5 expression reveals a pool of neural stem cells in the adult zebrafish midbrain. *Development* **133**, 4293–4303.

Ciruna, B., Jenny, A., Lee, D., Mlodzik, M., and Schier, A. F. (2006). Planar cell polarity signalling couples cell division and morphogenesis during neurulation. *Nature* **439**, 220–224.

Clarke, J. (2009). Role of polarized cell divisions in zebrafish neural tube formation. *Curr. Opin. Neurobiol.* **19**, 134–138.

Colman, J. R., Baldwin, D., Johnson, L. L., and Scholz, N. L. (2009). Effects of the synthetic estrogen, 17alpha-ethinylestradiol, on aggression and courtship behavior in male zebrafish (Danio rerio). *Aquat. Toxicol.* **91**, 346–354.

Colwill, R. M., Raymond, M. P., Ferreira, L., and Escudero, H. (2005). Visual discrimination learning in zebrafish (Danio rerio). *Behav. Processes* **70**, 19–31.

Cornell, R. A., and Eisen, J. S. (2000). Delta signaling mediates segregation of neural crest and spinal sensory neurons from zebrafish lateral neural plate. *Development* **127**, 2873–2882.

Cornell, R. A., and Eisen, J. S. (2002). Delta/Notch signaling promotes formation of zebrafish neural crest by repressing Neurogenin1 function. *Development* **129**, 2639–2648.

Cui, W. W., Low, S. E., Hirata, H., Saint-Amant, L., Geisler, R., Hume, R. I., and Kuwada, J. Y. (2005). The zebrafish shocked gene encodes a glycine transporter and is essential for the function of early neural circuits in the CNS. *J. Neurosci.* **25**, 6610–6620.

Cui, W. W., Saint-Amant, L., and Kuwada, J. Y. (2004). Shocked Gene is required for the function of a premotor network in the zebrafish CNS. *J. Neurophysiol.* **92**, 2898–2908.

Darland, T., and Dowling, J. E. (2001). Behavioral screening for cocaine sensitivity in mutagenized zebrafish. *Proc. Natl. Acad. Sci. USA.* **98**, 11691–11696.

Davis, R. E., and Klinger, P. D. (1994). NMDA receptor antagonist MK 801 blocks learning of conditioned stimulus-unconditioned stimulus contiguity but not fear of conditioned stimulus in goldfish (Carassius auratus L.). *Behav. Neurosci.* **108**, 935–940.

Dee, C. T., Hirst, C. S., Shih, Y. H., Tripathi, V. B., Patient, R. K., and Scotting, P. J. (2008). Sox3 regulates both neural fate and differentiation in the zebrafish ectoderm. *Dev. Biol.* **320**, 289–301.

Dee, C. T., Gibson, A., Rengifo, A., Sun, S. K., Patient, R. K., and Scotting, P. J. (2007). A change in response to Bmp signalling precedes ectodermal fate choice. *Int. J. Dev. Biol.* **51**, 79–84.

Dicke, U., Wallstein, M., and Roth, G. (1997). 5-HT-like immunoreactivity in the brains of plethodontid and salamandrid salamanders (Hydromantes italicus, Hydromantes genei, Plethodon jordani, Desmognathus ochrophaeus, Pleurodeles waltl): an immunohistochemical and biocytin double-labelling study. *Cell Tissue Res.* **287**, 513–523.

Dornseifer, P., Takke, C., and Campos-Ortega, J. A. (1997). Overexpression of a zebrafish homologue of the Drosophila neurogenic gene Delta perturbs differentiation of primary neurons and somite development. *Mech. Dev.* **63**, 159–171.

Douglass, A. D., Kraves, S., Deisseroth, K., Schier, A. F., and Engert, F. (2008). Escape behavior elicited by single, channelrhodopsin-2-evoked spikes in zebrafish somatosensory neurons. *Curr. Biol.* **18**, 1133–1137.

Duboule, D. (1994). Temporal colinearity and the phylotypic progression: A basis for the stability of a vertebrate Bauplan and the evolution of morphologies through heterochrony. *Dev. Suppl.* 135–142.

Dufour, S., Weltzien, F. A., Sebert, M. E., Le Belle, N., Vidal, B., Vernier, P., and Pasqualini, C. (2005). Dopaminergic inhibition of reproduction in teleost fishes: Ecophysiological and evolutionary implications. *Ann. N.Y. Acad. Sci.* **1040**, 9–21.

Eddins, D., Petro, A., Williams, P., Cerutti, D. T., and Levin, E. D. (2009). Nicotine effects on learning in zebrafish: The role of dopaminergic systems. *Psychopharmacology (Berl.)* **202**, 103–109.

Egan, R. J., Bergner, C. L., Hart, P. C., Cachat, J. M., Canavello, P. R., Elegante, M. F., Elkhayat, S. I., Bartels, B. K., Tien, A. K., Tien, D. H., et al. (2009). Understanding behavioral and physiological phenotypes of stress and anxiety in zebrafish. *Behav. Brain Res.* **205**, 38–44.

Ekstrom, P. (1987). Distribution of choline acetyltransferase-immunoreactive neurons in the brain of a cyprinid teleost (Phoxinus phoxinus L.). *J. Comp. Neurol.* **256**, 494–515.

Elworthy, S., Pinto, J. P., Pettifer, A., Cancela, M. L., and Kelsch, R. N. (2005). Phox2b function in the enteric nervous system is conserved in zebrafish and is sox10-dependent. *Mech. Dev.* **122**, 659–669.

Engeszer, R. E., Patterson, L. B., Rao, A. A., and Parichy, D. M. (2007). Zebrafish in the wild: A review of natural history and new notes from the field. *Zebrafish* **4**, 21–40.

Eriksson, K. S., Peitsaro, N., Karlstedt, K., Kaslin, J., and Panula, P. (1998). Development of the histaminergic neurons and expression of histidine decarboxylase mRNA in the zebrafish brain in the absence of all peripheral histaminergic systems. *Eur. J. Neurosci.* **10**, 3799–3812.

Facchin, L., Burgess, H. A., Siddiqi, M., Granato, M., and Halpern, M. E. (2009). Determining the function of zebrafish epithalamic asymmetry. *Philos. Trans. R. Soc. Lond. B Biol. Sci.* **364**, 1021–1032.

Fame, R. M., Brajon, C., and Ghysen, A. (2006). Second-order projection from the posterior lateral line in the early zebrafish brain. *Neural Dev.* **1**, 4.

Fausett, B. V., and Goldman, D. (2006). A role for alpha1 tubulin-expressing Muller glia in regeneration of the injured zebrafish retina. *J. Neurosci.* **26**, 6303–6313.

Fetcho, J. R., and Faber, D. S. (1988). Identification of motoneurons and interneurons in the spinal network for escapes initiated by the mauthner cell in goldfish. *J. Neurosci.* **8**, 4192–4213.

Filippi, A., Tiso, N., Deflorian, G., Zecchin, E., Bortolussi, M., and Argenton, F. (2005). The basic helix-loop-helix olig3 establishes the neural plate boundary of the trunk and is necessary for development of the dorsal spinal cord. *Proc. Natl. Acad. Sci. USA* **102**, 4377–4382.

Gahtan, E., and Baier, H. (2004). Of lasers, mutants, and see-through brains: Functional neuroanatomy in zebrafish. *J. Neurobiol.* **59**, 147–161.

Gahtan, E, Tanger, P, and Baier, H. (2005). Visual prey capture in larval zebrafish is controlled by identified reticulospinal neurons downstream of the tectum. *J. Neurosci.* **25**, 9294–9303.

Gato, A., and Desmond, M. E. (2009). Why the embryo still matters: CSF and the neuroepithelium as interdependent regulators of embryonic brain growth, morphogenesis and histiogenesis.. *Dev. Biol.* **327**, 263–272.

Geldmacher-Voss, B., Reugels, A. M., Pauls, S., and Campos-Ortega, J. A. (2003). A 90-degree rotation of the mitotic spindle changes the orientation of mitoses of zebrafish neuroepithelial cells. *Development* **130**, 3767–3780.

Gerlai, R., Lahav, M., Guo, S., and Rosenthal, A. (2000). Drinks like a fish: Zebra fish (Danio rerio) as a behavior genetic model to study alcohol effects. *Pharmacol. Biochem. Behav.* **67**, 773–782.

Ghysen, A, and Dambly-Chaudière, C. (2004). Development of the zebrafish lateral line. *Curr. Opin. Neurobiol.* **14**, 67–73.

Ghysen, A., and Dambly-Chaudiere, C. (2007). The lateral line microcosmos. *Genes Dev.* **21**, 2118–2130.

Gnuegge, L., Schmid, S., and Neuhauss, S. C. (2001). Analysis of the activity-deprived zebrafish mutant macho reveals an essential requirement of neuronal activity for the development of a fine-grained visuotopic map. *J. Neurosci.* **21**, 3542–3548.

Godinho, L., Mumm, J. S., Williams, P. R., Schroeter, E. H., Koerber, A., Park, S. W., Leach, S. D., and Wong, R. O. (2005). Targeting of amacrine cell neurites to appropriate synaptic laminae in the developing zebrafish retina. *Development* **132**, 5069–5079.

Grandel, H., Kaslin, J., Ganz, J., Wenzel, I., and Brand, M. (2006). Neural stem cells and neurogenesis in the adult zebrafish brain: Origin, proliferation dynamics, migration and cell fate. *Dev. Biol.* **295**, 263–277.

Gray, M., Moens, C. B., Amacher, S. L., Eisen, J. S., and Beattie, C. E. (2001). Zebrafish deadly seven functions in neurogenesis. *Dev. Biol.* **237**, 306–323.

Grunwald, D. J., Kimmel, C. B., Westerfield, M., Walker, C., and Streisinger, G. (1988). A neural degeneration mutation that spares primary neurons in the zebrafish. *Dev. Biol.* **126**, 115–128.

Haddon, C., and Lewis, J. (1996). Early ear development in the embryo of the zebrafish, Danio rerio. *J. Comp. Neurol.* **365**, 113–128.

Haddon, C., Smithers, L., Schneider-Maunoury, S., Coche, T., Henrique, D., and Lewis, J. (1998). Multiple delta genes and lateral inhibition in zebrafish primary neurogenesis. *Development* **125**, 359–370.

Heanue, T. A., and Pachnis, V. (2008). Ret isoform function and marker gene expression in the enteric nervous system is conserved across diverse vertebrate species. *Mech. Dev.* **125**, 687–699.

Hernandez-Lagunas, L., Choi, I. F., Kaji, T., Simpson, P., Hershey, C., Zhou, Y., Zon, L., Mercola, M., and Artinger, K. B. (2005). Zebrafish narrowminded disrupts the transcription factor prdm1 and is required for neural crest and sensory neuron specification. *Dev. Biol.* **278**, 347–357.

Higashijima, S., Mandel, G., and Fetcho, J. R. (2004a). Distribution of prospective glutamatergic, glycinergic, and GABAergic neurons in embryonic and larval zebrafish. *J. Comp. Neurol.* **480**, 1–18.

Higashijima, S., Schaefer, M., and Fetcho, J. R. (2004b). Neurotransmitter properties of spinal interneurons in embryonic and larval zebrafish. *J. Comp. Neurol.* **480**, 19–37.

Holzschuh, J., Wada, N., Wada, C., Schaffer, A., Javidan, Y., Tallafuss, A., Bally-Cuif, L., and Schilling, T. F. (2005). Requirements for endoderm and BMP signaling in sensory neurogenesis in zebrafish. *Development* **132**, 3731–3742.

Hong, E., and Brewster, R. (2006). N-cadherin is required for the polarized cell behaviors that drive neurulation in the zebrafish. *Development* **133**, 3895–3905.

Honjo, Y., Kniss, J., and Eisen, J. S. (2008). Neuregulin-mediated ErbB3 signaling is required for formation of zebrafish dorsal root ganglion neurons. *Development* **135**, 2615–2625.

Houart, C., Caneparo, L., Heisenberg, C., Barth, K., Take-Uchi, M., and Wilson, S. (2002). Establishment of the telencephalon during gastrulation by local antagonism of Wnt signaling. *Neuron* **35**, 255–265.

Huang, X., and Saint-Jeannet, J. P. (2004). Induction of the neural crest and the opportunities of life on the edge. *Dev. Biol.* **275**, 1–11.

Itoh, M., Kim, C. H., Palardy, G., Oda, T., Jiang, Y. J., Maust, D., Yeo, S. Y., Lorick, K., Wright, G. J., Ariza-McNaughton, L., et al. (2003). Mind bomb is a ubiquitin ligase that is essential for efficient activation of Notch signaling by Delta. *Dev. Cell* **4**, 67–82.

Jessen, J. R., Topczewski, J., Bingham, S., Sepich, D. S., Marlow, F., Chandrasekhar, A., and Solnica-Krezel, L. (2002). Zebrafish trilobite identifies new roles for Strabismus in gastrulation and neuronal movements. *Nat. Cell Biol.* **4**, 610–615.

Jia, S., Wu, D., Xing, C., and Meng, A. (2009). Smad2/3 activities are required for induction and patterning of the neuroectoderm in zebrafish. *Dev. Biol.* **333**, 273–284.

Jiang, Y. J., Brand, M., Heisenberg, C. P., Beuchle, D., Furutani-Seiki, M., Kelsh, R. N., Warga, R. M., Granato, M., Haffter, P., Hammerschmidt, M., et al. (1996). Mutations affecting neurogenesis and brain morphology in the zebrafish, Danio rerio. *Development* **123**, 205–216.

Kapsimali, M., Le Crom, S., and Vernier, P. (2003). A natural history of vertebrate dopamine receptors. In: *Dopamine receptors and transporters* (M. Laruelle, A. Sidhu and P. Vernier, eds), pp. 1–43. Marcel Dekker Inc., New York.

Kapsimali, M., Vidal, B., Gonzalez, A., Dufour, S., and Vernier, P. (2000). Distribution of the mRNA encoding the four dopamine $D_1$-receptor subtypes in the brain of the European eel (Anguilla anguilla): Comparative approach to the functions of $D_1$-receptors in vertebrates. *J. Comp. Neurol.* **419**, 320–343.

Kaslin, J., Ganz, J., Geffarth, M., Grandel, H., Hans, S., and Brand, M. (2009). Stem cells in the adult zebrafish cerebellum: Initiation and maintenance of a novel stem cell niche. *J. Neurosci.* **29**, 6142–6153.

Kaslin, J., and Panula, P. (2001). Comparative anatomy of the histaminergic and other aminergic systems in zebrafish (Danio rerio). *J. Comp. Neurol.* **440**, 342–377.

Kily, L. J., Cowe, Y. C., Hussain, O., Patel, S., McElwaine, S., Cotter, F. E., and Brennan, C. H. (2008). Gene expression changes in a zebrafish model of drug dependency suggest conservation of neuro-adaptation pathways. *J. Exp. Biol.* **211**, 1623–1634.

Kim, D. J., Seok, S. H., Baek, M. W., Lee, H. Y., Na, Y. R., Park, S. H., Lee, H. K., Dutta, N. K., Kawakami, K., and Park, J. H. (2009). Estrogen-responsive transient expression assay using a brain aromatase-based reporter gene in zebrafish (Danio rerio). *Comp. Med.* **59**, 416–423.

Kim, H., Kim, S., Chung, A. Y., Bae, Y. K., Hibi, M., Lim, C. S., and Park, H. C. (2008). Notch-regulated perineurium development from zebrafish spinal cord. *Neurosci. Lett.* **448**, 240–244.

Knaut, H., Blader, P., Strahle, U., and Schier, A. F. (2005). Assembly of trigeminal sensory ganglia by chemokine signaling. *Neuron* **47**, 653–666.

Kobayashi, M., Osanai, H., Kawakami, K., and Yamamoto, M. (2000). Expression of three zebrafish Six4 genes in the cranial sensory placodes and the developing somites. *Mech. Dev.* **98**, 151–155.

Kohashi, T., and Oda, Y. (2008). Initiation of Mauthner- or non-Mauthner-mediated fast escape evoked by different modes of sensory input. *J. Neurosci.* **28**, 10641–10653.

Korn, H., and Faber, D. S. (2005). The Mauthner cell half a century later: A neurobiological model for decision-making? *Neuron* **47**, 13–28.

Kuhlman, J., and Eisen, J. S. (2007). Genetic screen for mutations affecting development and function of the enteric nervous system. *Dev. Dyn.* **236**, 118–127.

Lam, C. S., Marz, M., and Strahle, U. (2009). Gfap and nestin reporter lines reveal characteristics of neural progenitors in the adult zebrafish brain. *Dev. Dyn.* **238**, 475–486.

LaMora, A., and Voigt, M. M. (2009). Cranial sensory ganglia neurons require intrinsic N-cadherin function for guidance of afferent fibers to their final targets. *Neuroscience* **159**, 1175–1184.

Larson, E. T., O'Malley, D. M., and Melloni, R. H., Jr. (2006). Aggression and vasotocin are associated with dominant-subordinate relationships in zebrafish. *Behav. Brain Res.* **167**, 94–102.

Lau, B., Bretaud, S., Huang, Y., Lin, E., and Guo, S. (2006). Dissociation of food and opiate preference by a genetic mutation in zebrafish. *Genes Brain Behav.* **5**, 497–505.

Le Douarin, N. M., and Dupin, E. (2003). Multipotentiality of the neural crest. *Curr. Opin. Genet. Dev.* **13**, 529–536.

Lele, Z., Folchert, A., Concha, M., Rauch, G. J., Geisler, R., Rosa, F., Wilson, S. W., Hammerschmidt, M., and Bally-Cuif, L. (2002). parachute/n-cadherin is required for morphogenesis and maintained integrity of the zebrafish neural tube. *Development* **129**, 3281–3294.

Levin, E. D., Bencan, Z., and Cerutti, D. T. (2007). Anxiolytic effects of nicotine in zebrafish. *Physiol. Behav.* **90**, 54–58.

Levin, E. D., and Chen, E. (2004). Nicotinic involvement in memory function in zebrafish. *Neurotoxicol. Teratol.* **26**, 731–735.

Levin, E. D., Limpuangthip, J., Rachakonda, T., and Peterson, M. (2006). Timing of nicotine effects on learning in zebrafish. *Psychopharmacology (Berl.)* **184**, 547–552.

Levkowitz, G., Zeller, J., Sirotkin, H. I., French, D., Schilbach, S., Hashimoto, H., Hibi, M., Talbot, W. S., and Rosenthal, A. (2003). Zinc finger protein too few controls the development of monoaminergic neurons. *Nat. Neurosci.* **6**, 28–33.

Lewis, J. L., Bonner, J., Modrell, M., Ragland, J. W., Moon, R. T., Dorsky, R. I., and Raible, D. W. (2004). Reiterated Wnt signaling during zebrafish neural crest development. *Development* **131**, 1299–1308.

Li, P., Shah, S., Huang, L., Carr, A. L., Gao, Y., Thisse, C., Thisse, B., and Li, L. (2007). Cloning and spatial and temporal expression of the zebrafish dopamine D1 receptor. *Dev. Dyn.* **236**, 1339–1346.

Liao, J. C., and Fetcho, J. R. (2008). Shared versus specialized glycinergic spinal interneurons in axial motor circuits of larval zebrafish. *J. Neurosci.* **28**, 12982–12992.

Liedtke, D., and Winkler, C. (2008). Midkine-b regulates cell specification at the neural plate border in zebrafish. *Dev. Dyn.* **237**, 62–74.

Lillesaar, C., Stigloher, C., Tannhaeuser, B., Wullimann, M. F., and Bally-Cuif, L. (2009). Axonal projections originating from raphe serotonergic neurons in the developing and adult zebrafish, Danio rerio, using transgenics to visualize raphe-specific pet1 expression. *J. Comp. Neurol.* **512**, 158–182.

Lillesaar, C., Tannhauser, B., Stigloher, C., Kremmer, E., and Bally-Cuif, L. (2007). The serotonergic phenotype is acquired by converging genetic mechanisms within the zebrafish central nervous system. *Dev. Dyn.* **236**, 1072–1084.

Lister, J. A., Cooper, C., Nguyen, K., Modrell, M., Grant, K., and Raible, D. W. (2006). Zebrafish Foxd3 is required for development of a subset of neural crest derivatives. *Dev. Biol.* **290**, 92–104.
Liu, D. W., and Westerfield, M. (1990). The formation of terminal fields in the absence of competitive interactions among primary motoneurons in the zebrafish. *J. Neurosci.* **10**, 3947–3959.
Liu, K. S., and Fetcho, J. R. (1999). Laser ablations reveal functional relationships of segmental hindbrain neurons in zebrafish. *Neuron* **23**, 325–335.
Liu, K. S., Gray, M., Otto, S. J., Fetcho, J. R., and Beattie, C. E. (2003). Mutations in deadly seven/notch1a reveal developmental plasticity in the escape response circuit. *J. Neurosci.* **23**, 8159–8166.
Lopez-Patino, M. A., Yu, L., Cabral, H., and Zhdanova, I. V. (2008). Anxiogenic effects of cocaine withdrawal in zebrafish. *Physiol. Behav.* **93**, 160–171.
Lowery, L. A., and Sive, H. (2004). Strategies of vertebrate neurulation and a re-evaluation of teleost neural tube formation. *Mech. Dev.* **121**, 1189–1197.
Lowery, L. A., and Sive, H. (2005). Initial formation of zebrafish brain ventricles occurs independently of circulation and requires the nagie oko and snakehead/atp1a1a.1 gene products. *Development* **132**, 2057–2067.
Lowery, L. A., and Sive, H. (2009). Totally tubular: The mystery behind function and origin of the brain ventricular system. *Bioessays* **31**, 446–458.
Ma, P. M. (1994). Catecholaminergic systems in the zebrafish. I. Number, morphology, and histochemical characteristics of neurons in the locus coeruleus. *J. Comp. Neurol.* **344**, 242–255.
Ma, P. M. (1997). Catecholaminergic systems in the zebrafish. III. Organization and projection pattern of medullary dopaminergic and noradrenergic neurons. *J. Comp. Neurol.* **381**, 411–427.
Mahler, J., and Driever, W. (2007). Expression of the zebrafish intermediate neurofilament Nestin in the developing nervous system and in neural proliferation zones at postembryonic stages. *BMC Dev. Biol.* **7**, 89.
Maves, L., Jackman, W., and Kimmel, C. B. (2002). FGF3 and FGF8 mediate a rhombomere 4 signaling activity in the zebrafish hindbrain. *Development* **129**, 3825–3837.
McDowell, A. L., Dixon, L. J., Houchins, J. D., and Bilotta, J. (2004). Visual processing of the zebrafish optic tectum before and after optic nerve damage. *Vis. Neurosci.* **21**, 97–106.
McGraw, H. F., Nechiporuk, A., and Raible, D. W. (2008). Zebrafish dorsal root ganglia neural precursor cells adopt a glial fate in the absence of neurogenin1. *J. Neurosci.* **28**, 12558–12569.
McHenry, M. J., Feitl, K. E., Strother, J. A., and Van Trump, W. J. (2009). Larval zebrafish rapidly sense the water flow of a predator's strike. *Biol. Lett.* **5**, 477–479.
McLean, D. L., and Fetcho, J. R. (2004a). Ontogeny and innervation patterns of dopaminergic, noradrenergic, and serotonergic neurons in larval zebrafish. *J. Comp. Neurol.* **480**, 38–56.
McLean, D. L., and Fetcho, J. R. (2004b). Relationship of tyrosine hydroxylase and serotonin immunoreactivity to sensorimotor circuitry in larval zebrafish. *J. Comp. Neurol.* **480**, 57–71.
McLean, D. L., and Fetcho, J. R. (2008). Using imaging and genetics in zebrafish to study developing spinal circuits in vivo. *Dev. Neurobiol.* **68**, 817–834.
McLean, D. L., Fan, J., Higashijima, S., Hale, M. E., and Fetcho, J. R. (2007). A topographic map of recruitment in spinal cord. *Nature* **446**, 71–75.
McLean, D. L., Masino, M. A., Koh, I. Y., Lindquist, W. B., and Fetcho, J. R. (2008). Continuous shifts in the active set of spinal interneurons during changes in locomotor speed. *Nat. Neurosci.* **11**, 1419–1429.

Metscher, B. D., and Ahlberg, P. E. (1999). Zebrafish in context: Uses of a laboratory model in comparative studies. *Dev. Biol.* **210**, 1–14.

Miklosi, A., and Andrew, R. J. (1999). Right eye use associated with decision to bite in zebrafish. *Behav. Brain Res.* **105**, 199–205.

Miyasaka, N., Morimoto, K., Tsubokawa, T., Higashijima, S., Okamoto, H., and Yoshihara, Y. (2009). From the olfactory bulb to higher brain centers: Genetic visualization of secondary olfactory pathways in zebrafish. *J. Neurosci.* **29**, 4756–4767.

Mongeon, R., Gleason, M. R., Masino, M. A., Fetcho, J. R., Mandel, G., Brehm, P., and Dallman, J. E. (2008). Synaptic homeostasis in a zebrafish glial glycine transporter mutant. *J. Neurophysiol.* **100**, 1716–1723.

Mueller, T., and Guo, S. (2009). The distribution of GAD67-mRNA in the adult zebrafish (teleost) forebrain reveals a prosomeric pattern and suggests previously unidentified homologies to tetrapods. *J. Comp. Neurol.* **516**, 553–568.

Mueller, T., Vernier, P., and Wullimann, M. F. (2004). The adult central nervous cholinergic system of a neurogenetic model animal, the zebrafish Danio rerio. *Brain Res.* **1011**, 156–169.

Mueller, T., Vernier, P., and Wullimann, M. F. (2006). A phylotypic stage in vertebrate brain development: GABA cell patterns in zebrafish compared with mouse. *J. Comp. Neurol.* **494**, 620–634.

Mueller, T., and Wullimann, M. F. (2003). Anatomy of neurogenesis in the early zebrafish brain. *Brain Res. Dev. Brain Res.* **140**, 137–155.

Mumm, J. S., Williams, P. R., Godinho, L., Koerber, A., Pittman, A. J., Roeser, T., Chien, C. B., Baier, H., and Wong, R. O. (2006). In vivo imaging reveals dendritic targeting of laminated afferents by zebrafish retinal ganglion cells. *Neuron* **52**, 609–621.

Munderloh, C., Solis, G. P., Bodrikov, V., Jaeger, F. A., Wiechers, M., Malaga-Trillo, E., and Stuermer, C. A. (2009). Reggies/flotillins regulate retinal axon regeneration in the zebrafish optic nerve and differentiation of hippocampal and N2a neurons. *J. Neurosci.* **29**, 6607–6615.

Munson, C., Huisken, J., Bit-Avragim, N., Kuo, T., Dong, P. D., Ober, E. A., Verkade, H., Abdelilah-Seyfried, S., and Stainier, D. Y. (2008). Regulation of neurocoel morphogenesis by Pard6 gamma b. *Dev. Biol.* **324**, 41–54.

Murakami, T., Morita, Y., and Ito., H. (1983). Extrinsic and intrinsic fiber connections of the telencephalon in a teleost, Sebasticus marmoratus. *J. Comp. Neurol.* **216**, 115–131.

Nam, R. H., Kim, W., and Lee, C. J. (2004). NMDA receptor-dependent long-term potentiation in the telencephalon of the zebrafish. *Neurosci. Lett.* **370**, 248–251.

Nechiporuk, A., Linbo, T., and Raible, D. W. (2005). Endoderm-derived Fgf3 is necessary and sufficient for inducing neurogenesis in the epibranchial placodes in zebrafish. *Development* **132**, 3717–3730.

Nelson, J. S. (2006). Fishes of the World, 4th edn. J. Wiley & Sons.

Nevin, L. M., Taylor, M. R., and Baier, H. (2008). Hardwiring of fine synaptic layers in the zebrafish visual pathway. *Neural Dev.* **3**, 36.

Nicolson, T. (2005). The genetics of hearing and balance in zebrafish. *Annu. Rev. Genet.* **39**, 9–22.

Ninkovic, J., and Bally-Cuif, L. (2006). The zebrafish as a model system for assessing the reinforcing properties of drugs of abuse. *Methods* **39**, 262–274.

Ninkovic, J., Folchert, A., Makhankov, Y. V., Neuhauss, S. C., Sillaber, I., Straehle, U., and Bally-Cuif, L. (2006). Genetic identification of AChE as a positive modulator of addiction to the psychostimulant D-amphetamine in zebrafish. *J. Neurobiol.* **66**, 463–475.

Nyholm, M. K., Abdelilah-Seyfried, S., and Grinblat, Y. (2009). A novel genetic mechanism regulates dorsolateral hinge-point formation during zebrafish cranial neurulation. *J. Cell Sci.* **122**, 2137–2148.

Okuda, Y., Yoda, H., Uchikawa, M., Furutani-Seiki, M., Takeda, H., Kondoh, H., and Kamachi, Y. (2006). Comparative genomic and expression analysis of group B1 sox genes in zebrafish indicates their diversification during vertebrate evolution. *Dev. Dyn.* **235**, 811–825.

Orger, M. B., Kampff, A. R., Severi, K. E., Bollmann, J. H., and Engert, F. (2008). Control of visually guided behavior by distinct populations of spinal projection neurons. *Nat. Neurosci.* **11**, 327–333.

Park, H. C., Boyce, J., Shin, J., and Appel, B. (2005). Oligodendrocyte specification in zebrafish requires notch-regulated cyclin-dependent kinase inhibitor function. *J. Neurosci.* **25**, 6836–6844.

Park, H. C., and Appel, B. (2003). Delta-Notch signaling regulates oligodendrocyte specification. *Development* **130**, 3747–3755.

Parmentier, C., Taxi, J., Balment, R., Nicolas, G., and Calas, A. (2006). Caudal neurosecretory system of the zebrafish: Ultrastructural organization and immunocytochemical detection of urotensins. *Cell Tissue Res.* **325**, 111–124.

Pather, S., and Gerlai, R. (2009). Shuttle box learning in zebrafish (Danio rerio). *Behav. Brain Res.* **196**, 323–327.

Peitsaro, N., Kaslin, J., Anichtchik, O. V., and Panula, P. (2003). Modulation of the histaminergic system and behaviour by alpha-fluoromethylhistidine in zebrafish. *J. Neurochem.* **86**, 432–441.

Pellegrini, E., Mouriec, K., Anglade, I., Menuet, A., Le Page, Y., Gueguen, M. M., Marmignon, M. H., Brion, F., Pakdel, F., and Kah, O. (2007). Identification of aromatase-positive radial glial cells as progenitor cells in the ventricular layer of the forebrain in zebrafish. *J. Comp. Neurol.* **501**, 150–167.

Perez, S. E., Yanez, J., Marin, O., Anadon, R., Gonzalez, A., and Rodriguez-Moldes, I. (2000). Distribution of choline acetyltransferase (ChAT) immunoreactivity in the brain of the adult trout and tract-tracing observations on the connections of the nuclei of the isthmus. *J. Comp. Neurol.* **428**, 450–474.

Pietsch, J., Delalande, J. M., Jakaitis, B., Stensby, J. D., Dohle, S., Talbot, W. S., Raible, D. W., and Shepherd, I. T. (2006). lessen encodes a zebrafish trap100 required for enteric nervous system development. *Development* **133**, 395–406.

Piront, M. L., and Schmidt, R. (1988). Inhibition of long-term memory formation by anti-ependymin antisera after active shock-avoidance learning in goldfish. *Brain Res.* **442**, 53–62.

Portavella, M., Salas, C., Vargas, J. P., and Papini, M. R. (2003). Involvement of the telencephalon in spaced-trial avoidance learning in the goldfish (Carassius auratus). *Physiol. Behav.* **80**, 49–56.

Portavella, M., Torres, B., Salas, C., and Papini, M. R. (2004). Lesions of the medial pallium, but not the lateral pallium, disrupt spaced-trial avoidance learning in goldfish (Carassius auratus). *Neurosci. Lett.* **362**, 75–78.

Pradel, G., Schachner, M., and Schmidt, R. (1999). Inhibition of memory consolidation by antibodies against cell adhesion molecules after active avoidance conditioning in zebrafish. *J. Neurobiol.* **39**, 197–206.

Pradel, G., Schmidt, R., and Schachner, M. (2000). Involvement of L1.1 in memory consolidation after active avoidance conditioning in zebrafish. *J. Neurobiol.* **43**, 389–403.

Puelles, L., and Rubenstein, J. L. (2003). Forebrain gene expression domains and the evolving prosomeric model. *Trends Neurosci.* **26**, 469–476.

Ragland, J. W., and Raible, D. W. (2004). Signals derived from the underlying mesoderm are dispensable for zebrafish neural crest induction. *Dev. Biol.* **276**, 16–30.

Raible, D. W. (2006). Development of the neural crest: Achieving specificity in regulatory pathways. *Curr. Opin. Cell Biol.* **18**, 698–703.

Raible, D. W., and Kruse, G. J. (2000). Organization of the lateral line system in embryonic zebrafish. *J. Comp. Neurol.* **421**, 189–198.

Raible, D. W., Wood, A., Hodsdon, W., Henion, P. D., Weston, J. A., and Eisen, J. S. (1992). Segregation and early dispersal of neural crest cells in the embryonic zebrafish. *Dev. Dyn.* **195**, 29–42.

Ramdya, P. and Engert, F. (2008). Emergence of binocular functional properties in a monocular neural circuit. *Nat. Neurosci.* **11**(9), 1083–1090.

Ravi, V., and Venkatesh, B. (2008). Rapidly evolving fish genomes and teleost diversity. *Curr. Opin. Genet. Dev.* **18**, 544–550.

Reichenbach, B., Delalande, J. M., Kolmogorova, E., Prier, A., Nguyen, T., Smith, C. M., Holzschuh, J., and Shepherd, I. T. (2008). Endoderm-derived Sonic hedgehog and mesoderm Hand2 expression are required for enteric nervous system development in zebrafish. *Dev. Biol.* **318**, 52–64.

Reimer, M. M., Sorensen, I., Kuscha, V., Frank, R. E., Liu, C., Becker, C. G., and Becker, T. (2008). Motor neuron regeneration in adult zebrafish. *J. Neurosci.* **28**, 8510–8516.

Reiner, A., and Northcutt, R. G. (1992). An immunohistochemical study of the telencephalon of the senegal bichir (Polypterus senegalus). *J. Comp. Neurol.* **319**, 359–386.

Reyes, R., Haendel, M., Grant, D., Melancon, E., and Eisen, J. S. (2004). Slow degeneration of zebrafish Rohon-Beard neurons during programmed cell death. *Dev. Dyn.* **229**, 30–41.

Rink, E., and Wullimann, M. F. (2002). Connections of the ventral telencephalon and tyrosine hydroxylase distribution in the zebrafish brain (Danio rerio) lead to identification of an ascending dopaminergic system in a teleost. *Brain Res. Bull.* **57**, 385–387.

Rink, E., and Wullimann, M. F. (2004). Connections of the ventral telencephalon (subpallium) in the zebrafish (Danio rerio). *Brain Res.* **1011**, 206–220.

Rink, E., and Wullimann, M. F. (2001). The teleostean (zebrafish) dopaminergic system ascending to the subpallium (striatum) is located in the basal diencephalon (posterior tuberculum). *Brain Res.* **889**, 316–330.

Ritter, D. A., Bhatt, D. H., and Fetcho, J. R. (2001). In vivo imaging of zebrafish reveals differences in the spinal networks for escape and swimming movements. *J. Neurosc.* **21**, 8956–8965.

Robertson, G. N., McGee, C. A., Dumbarton, T. C., Croll, R. P., and Smith, F. M. (2007). Development of the swimbladder and its innervation in the zebrafish, Danio rerio. *J. Morphol.* **268**, 967–985.

Rodriguez, F., Duran, E., Gomez, A., Ocana, F. M., Alvarez, E., Jimenez-Moya, F., Broglio, C., and Salas, C. (2005). Cognitive and emotional functions of the teleost fish cerebellum. *Brain Res. Bull.* **66**, 365–370.

Roeser, T., and Baier, H. (2003). Visuomotor behaviors in larval zebrafish after GFP-guided laser ablation of the optic tectum. *J. Neurosci.* **23**, 3726–3734.

Rogers, C. D., Moody, S. A., and Casey, E. S. (2009). Neural induction and factors that stabilize a neural fate. *Birth Defects Res. C Embryo Today* **87**, 249–262.

Saint-Amant, L., and Drapeau, P. (1998). Time course of the development of motor behaviors in the zebrafish embryo. *J. Neurobiol.* **37**, 622–632.

Salas, C., Broglio, C., Duran, E., Gomez, A., Ocana, F. M., Jimenez-Moya, F., and Rodriguez, F. (2006). Neuropsychology of learning and memory in teleost fish. *Zebrafish* **3**, 157–171.

Salas, C., Broglio, C., and Rodriguez, F. (2003). Evolution of forebrain and spatial cognition in vertebrates: conservation across diversity. *Brain Behav. Evol.* **62**, 72–82.

Sassa, T., Aizawa, H., and Okamoto, H. (2007). Visualization of two distinct classes of neurons by gad2 and zic1 promoter/enhancer elements in the dorsal hindbrain of developing zebrafish reveals neuronal connectivity related to the auditory and lateral line systems. *Dev. Dyn.* **236**, 706–718.

Satou, C., Kimura, Y., Kohashi, T., Horikawa, K., Takeda, H., Oda, Y., and Higashijima, S. (2009). Functional role of a specialized class of spinal commissural inhibitory neurons during fast escapes in zebrafish. *J. Neurosci.* **29**, 6780–6793.

Saverino, C., and Gerlai, R. (2008). The social zebrafish: Behavioral responses to conspecific, heterospecific, and computer animated fish. *Behav. Brain Res.* **191**, 77–87.

Schier, A. F., Neuhauss, S. C., Harvey, M., Malicki, J., Solnica-Krezel, L., Stainier, D. Y., Zwartkruis, F., Abdelilah, S., Stemple, D. L., Rangini, Z., et al. (1996). Mutations affecting the development of the embryonic zebrafish brain. *Development* **123**, 165–178.

Schier, A. F., and Talbot, W. S. (2005). Molecular genetics of axis formation in zebrafish. *Annu. Rev. Genet.* **39**, 561–613.

Schmidt, J. T. (1994). C-kinase manipulations disrupt activity-driven retinotopic sharpening in regenerating goldfish retinotectal projection. *J. Neurobiol.* **25**, 555–570.

Schmidt, J. T., Fleming, M. R., and Leu, B. (2004). Presynaptic protein kinase C controls maturation and branch dynamics of developing retinotectal arbors: Possible role in activity-driven sharpening. *J. Neurobiol.* **58**, 328–340.

Schmidt, J. T., Schmidt, R., Lin, W. C., Jian, X. Y., and Stuermer, C. A. (1991). Ependymin as a substrate for outgrowth of axons from cultured explants of goldfish retina. *J. Neurobiol.* **22**, 40–54.

Scholpp, S., Delogu, A., Gilthorpe, J., Peukert, D., Schindler, S. and Lumsden, A. (2009). her6 regulates the neurogenetic gradient and neuronal identity in the thalamus. *Proc. Natl Acad. Sci. USA* **106**(47), 19895–19900.

Scholpp, S., Wolf, O., Brand, M., and Lumsden, A. (2006). Hedgehog signalling from the zona limitans intrathalamica orchestrates patterning of the zebrafish diencephalon. *Development* **133**, 855–864.

Schwerte, T., Prem, C., Mairosl, A., and Pelster, B. (2006). Development of the sympatho-vagal balance in the cardiovascular system in zebrafish (Danio rerio) characterized by power spectrum and classical signal analysis. *J. Exp. Biol.* **209**, 1093–1100.

Serra, E. L., Medalha, C. C., and Mattioli, R. (1999). Natural preference of zebrafish (Danio rerio) for a dark environment. *Braz. J. Med. Biol. Res.* **32**, 1551–1553.

Shepherd, I. T., Beattie, C. E., and Raible, D. W. (2001). Functional analysis of zebrafish GDNF. *Dev. Biol.* **231**, 420–435.

Shepherd, I. T., Pietsch, J., Elworthy, S., Kelsh, R. N., and Raible, D. W. (2004). Roles for GFRalpha1 receptors in zebrafish enteric nervous system development. *Development* **131**, 241–249.

Sherpa, T., Fimbel, S. M., Mallory, D. E., Maaswinkel, H., Spritzer, S. D., Sand, J. A., Li, L., Hyde, D. R., and Stenkamp, D. L. (2008). Ganglion cell regeneration following whole-retina destruction in zebrafish. *Dev. Neurobiol.* **68**, 166–181.

Shin, J., Poling, J., Park, H. C., and Appel, B. (2007). Notch signaling regulates neural precursor allocation and binary neuronal fate decisions in zebrafish. *Development* **134**, 1911–1920.

Smeets, W. J., and Gonzalez, A. (2000). Catecholamine systems in the brain of vertebrates: New perspectives through a comparative approach. *Brain Res. Rev.* **33**, 308–379.

Smeets, W. J., and Steinbusch, H. W. (1988). Distribution of serotonin immunoreactivity in the forebrain and midbrain of the lizard GEkko gecko. *J. Comp. Neurol.* **271**, 419–434.

Stern, C. D. (2006). Evolution of the mechanisms that establish the embryonic axes. *Curr. Opin. Genet. Dev.* **16**, 413–418.

Stigloher, C., Chapouton, P., Adolf, B., and Bally-Cuif, L. (2008). Identification of neural progenitor pools by E(Spl) factors in the embryonic and adult brain. *Brain Res. Bull.* **75**, 266–273.

Stuermer, C. A., Rohrer, B., and Munz, H. (1990). Development of the retinotectal projection in zebrafish embryos under TTX-induced neural-impulse blockade. *J. Neurosci.* **10**, 3615–3626.

Sumbre, G, Muto, A, Baier, H, and Poo, MM. (2008). Entrained rhythmic activities of neuronal ensembles as perceptual memory of time interval. *Nature* **456**, 102–106.

Swain, H. A., Sigstad, C., and Scalzo, F. M. (2004). Effects of dizocilpine (MK-801) on circling behavior, swimming activity, and place preference in zebrafish (Danio rerio). *Neurotoxicol. Teratol.* **26**, 725–729.

Takei, Y., and Loretz, C. A. (2006). Endocrinology. In: *The physiology of fishes* (D.H. Evans and J.B. Claiborne, eds), 3rd Edition, pp. 271–318. CRC Press, Boca Raton.

Takke, C., Dornseifer, P., v Weizsäcker, E., and Campos-Ortega, J. A. (1999). her4, a zebrafish homologue of the Drosophila neurogenic gene E(spl), is a target of NOTCH signalling. *Development* **126**, 1811–1821.

Tawk, M., Araya, C., Lyons, D. A., Reugels, A. M., Girdler, G. C., Bayley, P. R., Hyde, D. R., Tada, M., and Clarke, J. D. (2007). A mirror-symmetric cell division that orchestrates neuroepithelial morphogenesis. *Nature* **446**, 797–800.

Thummel, R., Kassen, S. C., Enright, J. M., Nelson, C. M., Montgomery, J. E., and Hyde, D. R. (2008). Characterization of Müller glia and neuronal progenitors during adult zebrafish retinal regeneration. *Exp. Eye Res.* **87**, 433–444.

Topp, S., Stigloher, C., Komisarczuk, A. Z., Adolf, B., Becker, T. S., and Bally-Cuif, L. (2008). Fgf signaling in the zebrafish adult brain: Association of Fgf activity with ventricular zones but not cell proliferation. *J. Comp. Neurol.* **510**, 422–439.

Vargas, J. P., Rodriguez, F., Lopez, J. C., Arias, J. L., and Salas, C. (2000). Spatial learning-induced increase in the argyrophilic nucleolar organizer region of dorsolateral telencephalic neurons in goldfish. *Brain Res.* **865**, 77–84.

Vecino, E., Perez, M. T., and Ekstrom, P. (1995). Localization of enkephalinergic neurons in the central nervous system of the salmon (Salmo salar L.) by in situ hybridization and immunocytochemistry. *J. Chem. Neuroanat.* **9**, 81–97.

Veldman, M. B., Bemben, M. A., Thompson, R. C., and Goldman, D. (2007). Gene expression analysis of zebrafish retinal ganglion cells during optic nerve regeneration identifies KLF6a and KLF7a as important regulators of axon regeneration. *Dev. Biol.* **312**, 596–612.

Vernier, P., Moret, F., Callier, S., Snapyan, M., Wersinger, C., and Sidhu, A. (2004). The degeneration of dopamine neurons in Parkinson's disease: Insights from embryology and evolution of the mesostriatocortical system. *Ann. N.Y. Acad. Sci.* **1035**, 231–249.

Vernier, P., and Wullimann, M. (2009). Evolution of the posterior tuberculam and preglomerular nuclear complex. In: *Springer Encyclopedia of Neuroscience* (M.D. Binder, N. Hirokawa and U. Windhorst, eds), Part 5, pp. 1318–1326. Springer Verlag, Berlin, Heidelberg.

Volkoff, H., and Peter, R. E. (2006). Feeding behavior of fish and its control. *Zebrafish* **3**, 131–140.

Wada, H., and Okamoto, H. (2009). Roles of planar cell polarity pathway genes for neural migration and differentiation. *Dev. Growth Differ.* **51**, 233–240.

Wai, M. S., Lorke, D. E., Webb, S. E., and Yew, D. T. (2006). The pattern of c-fos activation in the CNS is related to behavior in the mudskipper, Periophthalmus cantonensis. *Behav. Brain Res.* **167**, 318–327.

Webb, K. J., Norton, W. H., Trumbach, D., Meijer, A. H., Ninkovic, J., Topp, S., Heck, D., Marr, C., Wurst, W., Theis, F. J., et al. (2009). Zebrafish reward mutants reveal novel transcripts mediating the behavioral effects of amphetamine. *Genome Biol.* **10**, R81.

Williams, F. E., White, D., and Messer, W. S. J. (2002). A simple spatial alternation task for assessing memory function in zebrafish. *Behav. Processes* **2002**, 125–132.

Wilson, S. W., Brand, M., and Eisen, J. S. (2002). Patterning the zebrafish central nervous system. *Results Probl. Cell Differ.* **40**, 181–215.

Wilson, S. W., and Houart, C. (2004). Early steps in the development of the forebrain. *Dev. Cell* **6**, 167–181.

Wulliman, M. F. (1998). The central nervous system. In: *Physiology of Fishes* (D.H. Evans, ed.), p. 245. CRC Pres, Boca Raton.

Wullimann, M., and Vernier, P. (2006). Evolution of the nervous system in fishes. In: *Evolution of Nervous Systems: A Comprehensive Reference* (J.H. Kaas, ed.), . Elsevier Inc.

Wullimann, M. F., and Rink, E. (2001). Detailed immunohistology of Pax6 protein and tyrosine hydroxylase in the early zebrafish brain suggests role of Pax6 gene in development of dopaminergic diencephalic neurons. *Dev. Brain Res.* **131**, 173–191.

Wullimann, M. F., Meyer, D. L., and Northcutt, R. G. (1991). The visually related posterior pretectal nucleus in the non-percomorph teleost Osteoglossum bicirrhosum projects to the hypothalamus: A DiI study. *J. Comp. Neurol.* **312**, 415–435.

Wullimann, M. F., and Mueller, T. (2004). Teleostean and mammalian forebrains contrasted: Evidence from genes to behavior. *J. Comp. Neurol.* **475**, 143–162.

Wullimann, M. F., and Rink, E. (2002). The teleostean forebrain: A comparative and developmental view based on early proliferation, Pax6 activity and catecholaminergic organization. *Brain Res. Bull.* **57**, 363–370.

Xu, X., Bazner, J., Qi, M., Johnson, E., and Freidhoff, R. (2003). The role of telencephalic NMDA receptors in avoidance learning in goldfish (Carassius auratus). *Behav. Neurosci.* **117**, 548–554.

Yaksi, E., von Saint Paul, F., Niessing, J., Bundschuh, S. T., and Friedrich, R. W. (2009). Transformation of odor representations in target areas of the olfactory bulb. *Nat. Neurosci.* **12**, 474–482.

Yamamoto, K., Ruuskanen, J., Wulliman, M., and Vernier, P. (2010). Two tyrosine hydroxylase genes in vertebrates: Comparative distribution with other monoaminergic markers in the zebrafish brain. *Mol. Cell. Neurosci.* **43**, 394–402.

Yeo, S. Y., Kim, M., Kim, H. S., Huh, T. L., and Chitnis, A. B. (2007). Fluorescent protein expression driven by her4 regulatory elements reveals the spatiotemporal pattern of Notch signaling in the nervous system of zebrafish embryos. *Dev. Biol.* **301**, 555–567.

Yoshida, M., Okamura, I., and Uematsu, K. (2004). Involvement of the cerebellum in classical fear conditioning in goldfish. *Behav. Brain Res.* **153**, 143–148.

Yu, L., Tucci, V., Kishi, S., and Zhdanova, I. V. (2006). Cognitive aging in zebrafish. *PLoS ONE* **1**, e14.

Yurco, P., and Cameron, D. A. (2005). Responses of Muller glia to retinal injury in adult zebrafish. *Vision Res.* **45**, 991–1002.

Zikopoulos, B., and Dermon, C. R. (2005). Comparative anatomy of alpha(2) and beta adrenoceptors in the adult and developing brain of the marine teleost the red porgy (Pagrus pagrus, Sparidae): [(3)H]clonidine and [(3)H]dihydroalprenolol quantitative autoradiography and receptor subtypes immunohistochemistry. *J. Comp. Neurol.* **489**, 217–240.

Zupanc, G. K., Hinsch, K., and Gage, F. H. (2005). Proliferation, migration, neuronal differentiation, and long-term survival of new cells in the adult zebrafish brain. *J. Comp. Neurol.* **488**, 290–319.

# 3

# ZEBRAFISH VISION: STRUCTURE AND FUNCTION OF THE ZEBRAFISH VISUAL SYSTEM

*STEPHAN C.F. NEUHAUSS*



The zebrafish (*Danio rerio*) has been one of the foremost model organisms to study vertebrate development. The ability to combine superb embryology, due to its rapid extracorporal development in combination with transparent embryos, with genetics has made it one of the most studied organisms worldwide. While the strongest contribution of the zebrafish model is still in the genetic analysis of development, there is growing interest in using this animal model for research in integrative physiology, as evidenced by this book devoted to zebrafish. The visual system of the zebrafish lends itself ideally to such an integrative approach, by combining cross-disciplinary technologies. The following chapter describes the properties of vision in zebrafish, by discussing the anatomical structure of the visual system and the visual behaviors that it supports. I will detail the advances that have recently been made, point out gaps in our knowledge and discuss future directions.



*Zebrafish: Volume 29*
FISH PHYSIOLOGY


DOI: 10.1016/S1546-5098(10)02903-1

## 1. VISUAL ECOLOGY

The development, structure and function of the zebrafish visual system are shaped by its ecology. The zebrafish is a shoaling cyprinid native to the flood planes of the Indian sub-continent. It is a diurnal (active during the day) species that inhabits shallow, slow-moving waters, including rice paddy fields. The zebrafish is omnivorous with its natural diet mainly consisting of zooplankton, insects and plant material (reviewed in Spence et al. 2008). A recent expedition found zebrafish mainly on silt-covered bottoms of rivers with overhanging or submerged vegetation, resulting in a biotope with a large range of luminance levels (Engeszer et al. 2007b).

Overall our knowledge about its wild ecology is scanty and awaits more thorough investigations which will be essential to interpret behavioral experiments in the laboratory (McClure et al. 2006; Engeszer et al. 2007b; Spence et al. 2008).

This little sketch of the ecology of the zebrafish already allows an assessment of the likely challenges posed to the zebrafish visual system. The diurnal life style in complex structured surface water suggests the need for well-developed photopic vision, with the ability to distinguish colors. Additionally, ultraviolet-sensitive vision and maybe polarized light sensitivity may be useful to better detect (for human eyes) translucent prey.

The breeding behavior of the zebrafish with hundred of eggs spawned and the absence of paternal care illustrates a strong ecologic pressure on the larvae to rapidly develop a nervous system in order to be able to respond to environmental stimuli and capture prey. In the laboratory care must be taken to prevent the parental fish from feasting on their freshly laid eggs. Hence the rapid early development of the zebrafish can be rationalized as an ecological adaptation to orient, escape predators and catch prey as early as possible. In the following pages it will become clear that one of the endearing features of the zebrafish to vision scientists is the rapid maturation of the visual system supporting a number of visual behaviors already in young larvae. The initiation of prey capture movements and food-seeking behavior coincides with the depletion of the yolk supplied by the mother.

In the following chapter I will discuss the structure and function of the zebrafish visual system under the tacit assumption that these features are adaptive traits shaped by selection.

## 2. STRUCTURE OF THE VISUAL SYSTEM

The visual system of the zebrafish can be loosely grouped into three parts, namely the eye (including the retina), retinofugal projections connecting the retina to visual centers in the rest of the brain, and visual centers in the brain receiving and processing visual information.

The anterior segment of the eye contains, among other structures, the cornea and the lens, which function in projecting a focused image onto the retina.

The retina, as an outpocketing of the diencephalon, has long enjoyed the status of one of the most intensely studied parts of the central nervous system. Due to its position in the body it is an approachable part of the brain with clearly demarcated layers and a clear association between stimulus input, structure and behavioral and physiological output (Dowling 1987). Therefore it is probably the part of the zebrafish nervous system where we currently have the most detailed structural and functional information, albeit – as will be apparent in the following paragraphs – by no means approaching a complete picture.

The main retinofugal projections terminate in the optic tecta of the dorsal midbrain, apart from a number of additional retinofugal areas. The retinotectal projection is one of the most popular systems to study axon guidance and map formation, where studies of the zebrafish model have been a significant contributor, especially concerning the genetic analysis of axon pathfinding and mapping (Culverwell and Karlstrom 2002; Hutson et al. 2004).

How visual information is processed in the brain is still largely unknown in the zebrafish. Only recently attempts have been undertaken to understand the physiology of the zebrafish optical tectum, mainly relying on optical recordings aided by its transparency and accessibility.

### 2.1. Anterior Segment of the Eye

The ocular anterior segment of the vertebrate eye has the function to focus incoming light onto the photoreceptors of the retina and to regulate intraocular pressure. It is comprised of the cornea, lens, iris, ciliary body, and specialized tissue at the iridocorneal angle. The anatomy of the anterior segment in zebrafish is comparable to other vertebrates'. The transparent zebrafish embryo in combination with transgenic technologies makes a three-dimensional time lapse analysis feasible, allowing for a stunning visualization of the dynamics of lens formation (Greiling and Clark 2009). The zebrafish lens is spherical with a high refractive index, accounting for practically all the refractive power of the eye. This contrasts with mammals and birds, where the lens is concave and light is primarily refracted by the cornea at the phase transition between air and aqueous medium. The development of the lens is different from the well-studied mammalian situation in that it forms by delamination rather than invagination. Hence a spherical cell mass arises rather than a hollow lens vesicle, as in mammals and birds. Another difference to mammals appears at the iridocorneal angle. Instead of a trabecular meshwork, zebrafish have an annular ligament (Soules and Link 2005; Dahm et al. 2007). Nevertheless, there are functional analogous structures for production and transport of the aqueous humor (Gray et al. 2009).

There is no anatomical evidence for accommodation in zebrafish (Easter and Nicola 1996). Accommodation, the process whereby the optics of the eyes are changed to put an object into focus, is achieved in some teleosts by altering the position of the lens relative to the retina, by virtue of muscle fibers attached to the lens (Ott 2006). The ventral lentis retractor muscle is attached to the lens. This small muscle is probably vestigial and there is no evidence for lens motility with respect to the retina (Gray et al. 2009). In the absence of any functional or anatomical evidence, it seems likely that zebrafish vision operates with a fixed depth of field corrected to infinity.

## 2.2. Retina

The zebrafish possesses a canonical vertebrate retina with the neural retina consisting of three nuclear layers, separated by two plexiform (synaptic) layers (Figs 3.1 and 3.2). The outer retina, which is in close contact with the retinal pigment epithelium (RPE), contains rod photoreceptors and four types of cone photoreceptor cells, including an ultraviolet-sensitive cone type. Hence zebrafish are tetrachromats that extend their visual range into the ultraviolet.

The inner nuclear layer constitutes the cell bodies of bipolar, horizontal, and amacrine interneurons, as well as the cell somata of Muller glia cells. Synaptic contacts between photoreceptors and the inner retina are formed in the outer plexiform layer. Closest to the lens is the ganglion cell layer, containing displaced amacrine cells and ganglion cells, whose axons form the optic nerve and, after entering the central brain, the optic tract. Synaptic contacts between ganglion cells and the inner nuclear cells are formed in the inner plexiform layer and its relative thickness denotes the great complexity of these connections in the zebrafish retina.

### 2.2.1. Development of the Retina

Zebrafish retina development is extraordinarily rapid, starting with an evagination of the diencephalon, that subsequently forms the optic lobes at around 10 hpf (hours post fertilization). Development proceeds in an inside-out fashion with ganglion cells differentiating before inner nuclear cells, followed by cells of the outer retina (see Table 3.1).

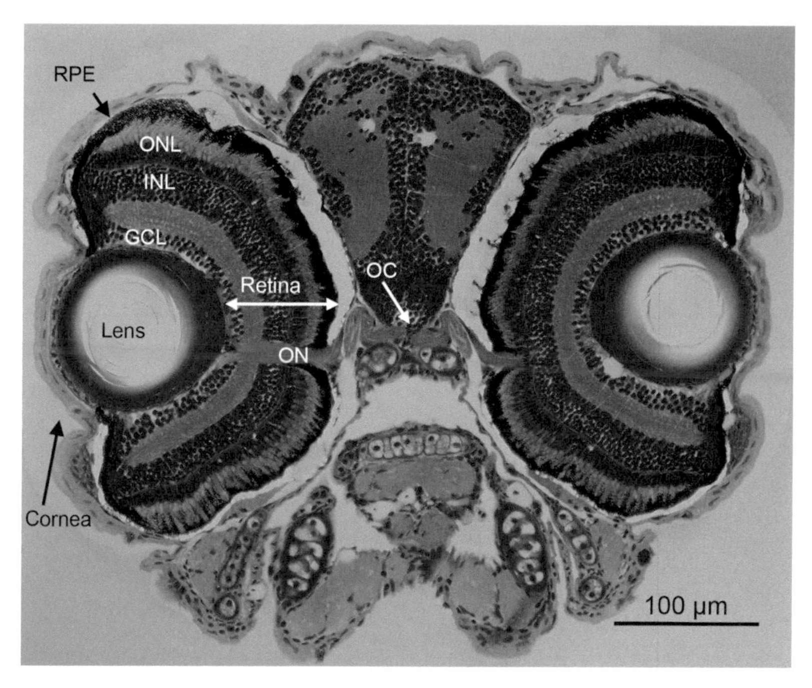


**Fig. 3.1.** The visual system morphology of a 6-day-old larva. Histological section through the eyes of a zebrafish larva. The vertebrate typical layering of the retina, the optic nerve and chiasm are visible. ONL, outer nuclear layer; INL, inner nuclear layer; GCL, ganglion cell layer; ON, optic nerve; RPE, retinal pigment epithelium; OC, optic chiasm. See color plate section

The ganglion cell differentiation occurs around 32 hpf, and is followed by the formation of cells of the inner nuclear layer (Schmitt and Dowling 1994, 1999; Hu and Easter 1999). By 55 hpf the appearance of rod and cone outer segments as well as rod and cone synaptic terminals can be detected (Schmitt and Dowling 1999). Synaptic structures indicative of functional maturation (ribbon triads) arise within photoreceptor synaptic terminals at about 65 hpf. Bipolar cell ribbon synapses form at about 70 hpf. Signal transmission from photoreceptors to second-order neurons starts around 84 hpf and becomes fully functional at 5 days post fertilization (Biehlmaier et al. 2003). This rapid maturation of the visual system is also apparent by electrophysiological properties and the wealth of visually guided behaviors that can be evoked at this early larval stage.

In a sense the zebrafish retina never stops developing, since even the mature adult retina is still proliferating. The main proliferative compartment is the circumferential germinal zone at the ciliary margin, where all cell

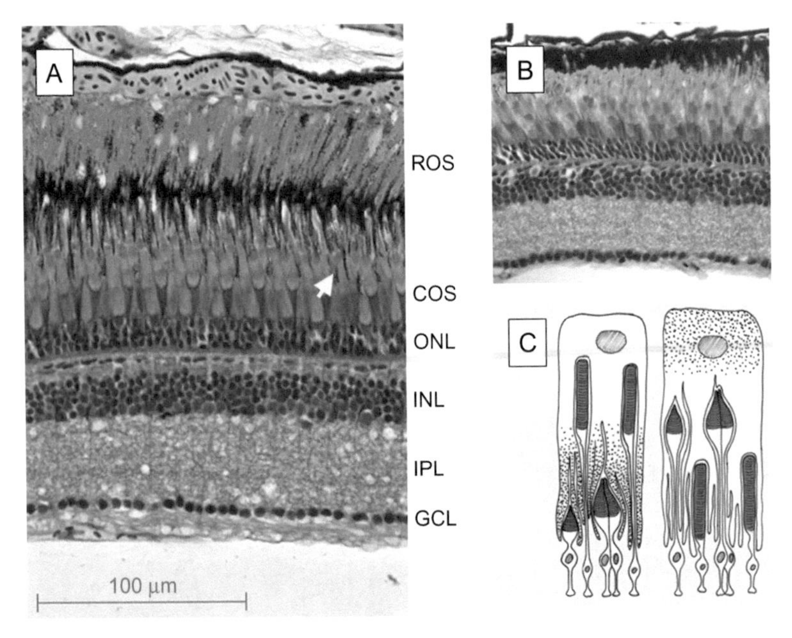


**Fig. 3.2.** Radial section of the adult zebrafish retina. (A) Light-adapted retina. (B) Dark-adapted retina. In the light-adapted retina pigment granules reach to the outer tips of the RPE microvilli (arrow), while in the dark-adapted retina almost all pigment granules are located in the basal site of the RPE. (C) Schematic drawing of photoreceptor cells and RPE in the dark-adapted (left) and light-adapted (right) state. Note the shift in the relative position of the rod and cone outer segments in the different adaptation states. ONL, outer nuclear layer; INL, inner nuclear layer; GCL, ganglion cell layer; ROS, rod outer segments; COS, cone outer segments. Adapted from Hodel et al. (2006) with permission of Wiley Inc. See color plate section

types of the retina are constantly generated. Additionally, rod photoreceptors originating from rod precursor cells of the inner retina are added throughout life (Raymond et al. 2006).

### 2.2.2. Photoreceptors

The zebrafish outer retina has five different photoreceptor types, one rod and four cone types, that arise early in development. These photoreceptors are readily distinguishable by morphology and are characterized by the peak sensitivity of their respective visual pigments.

While there is only one anatomical rod type using rhodopsin as its chromophore, cones can be subdivided into four different types with distinct cellular morphology and opsin content.

**Table 3.1**
Development of Zebrafish Vision

| | Retina morphology | Extra-retinal structures | Visual behavior |
|---|---|---|---|
| 11 hpf | optic vesicle formation | | |
| 26 hpf | retinal pigment epithelium pigmented | lens vesicle detachment, anterior chamber formation | |
| 32 hpf | retinal pigment epithelium pigmented | retinal ganglion cell axons exit the retina | |
| 36 hpf | | optic axons reach optic chiasm | |
| 40 hpf | ganglion cells differentiation starts | optic axons reach the tectum | |
| 50 hpf | opsin expression initiated | | |
| 55 hpf | photoreceptor outer segments appear | | |
| 60 hpf | ribbon synaptic terminals | | |
| 65 hpf | pedicles and spherules form | axonal arborization on the tectum | |
| 70 hpf | | extraocular eye muscles functional | visual startle response |
| 72 hpf | | eye and lens emmetropic | |
| 74 hpf | | | optokinetic response |
| 84 hpf | pedicules formed | | |
| 4 dpf | spinule formation | | |
| 5 dpf | synaptic plasticity; pigment granule migration | | robust optokinetic response |
| 6 dpf | | | robust optomotor response |

Ultraviolet-sensitive cones are short single cones (SSC), with short outer segments. Long single cones (LSC) contain blue-sensitive opsin. Red- and green-sensitive cones have even longer outer segments and are fused to form double cones (DC).

In teleost fishes cone photoreceptors are typically arranged in a regular mosaic pattern, that may differ among examined teleost species (Lyall 1957; Ali et al. 1978; Stenkamp et al. 2001).

The zebrafish retina is characterized by a row mosaic, aptly named since it is composed of rows of alternating blue- and UV-sensitive single cones that in turn alternate with double cones. These parallel rows are aligned in such a way that the green-sensitive members of the double cones flank the short single cones, and the long single cones are nearer to the red-sensitive member of the double cone.

Rod photoreceptors have been thought to be randomly packed in the teleost retina. However, a recent study using transgenically labeled rod photoreceptors demonstrated that rod photoreceptors are also arranged in a mosaic pattern in the zebrafish (Fadool 2003).

The functional significance of these mosaics is still debated. A regular pattern could be beneficial for optimal packing density, but the fact that different teleosts use different mosaic arrangements is still in search of a satisfying explanation. One possibility is that the crystalline arrangement of the photoreceptor could play a role for polarized-light vision (Hawryshyn 2000; Flamarique and Harosi 2002); however, currently there is no evidence that the zebrafish is capable to determine the e-vector direction of polarized light (the direction of the electric field component of light in electromagnetic light theory).

Glutamate, the neurotransmitter, released by photoreceptors, occurs at specialized synaptic structures. Such ribbon synapses have evolved for tonic glutamate release and are unique to photoreceptor, bipolar and inner hair cell synapses of vertebrates. The synaptic terminals associated with these ribbon synapses of the photoreceptors are intricately built structures, termed pedicles for cones and spherules for rods, representing contact points of horizontal and bipolar cell dendrites (Haverkamp et al. 2000). These structures have important signal integration properties and their early development in the zebrafish emphasizes the early functional maturation of the visual system (Biehlmaier et al. 2003).

Zebrafish mutants disrupting the architecture of the photoreceptor synapse have been identified, offering the opportunity to use genetics to study the development and maintenance of these baroque synapses (Allwardt et al. 2001; Van Epps et al. 2004; Biehlmaier et al. 2007).

### 2.2.3. Visual Pigments

The different photoreceptor types are characterized by the peak sensitivity of their respective pigments. There are four cone classes in the zebrafish retina, namely ultraviolet-, short- (S, blue), medium- (M, green), and long-wavelength (L, red) sensitive cones, which express SWS1, SWS2, RH2 and LWS opsin, respectively. Rod photoreceptors contain rhodopsin, encoded by the RH1 gene (Raymond et al. 1993; Robinson et al. 1993; Vihtelic et al. 1999; Hamaoka et al. 2002; Chinen et al. 2003).

Interestingly there are multiple gene duplications among the opsin genes, so that the zebrafish genome harbors two functional red opsin genes (LWS-1 and LWS-2), four green opsin genes (RH2-1, RH2-2, RH2-3, and RH2-4), while there are only single copies of the SWS1 and SWS2 genes (Vihtelic et al. 1999; Chinen et al. 2003). These genes likely arise via tandem duplications, since they can be found in distinct clusters in the genome. The relative expression levels of these paralogous genes in the zebrafish eye as measured by real-time RT-PCR are very unequal, with RH2-2 and LWS-1 accounting for most of the total expression of the green and red opsins, respectively (Chinen et al. 2003).

The peak sensitivities of the different opsins have been measured by microspectrophotometry on isolated photoreceptors and spectrophotometry of reconstituted photopigments from cDNAs. Such measurements have been done for both rods (501–503 nm) and cones (UV cone, 360–361 nm; S cone, 407–417 nm; M cone 473–480 nm; L cone, 556–564 nm) (Nawrocki et al. 1985; Robinson et al. 1993; Cameron 2002; Allison et al. 2004).

The zebrafish likely uses A1 (retinol)-based visual pigments, as concluded by extraction and bleaching studies. Additionally, the in vitro measurements of the spectral sensitivity of 11-*cis* retinal reconstituted opsins correlates well with microspectrophotometrical measurements in intact photoreceptors (Nawrocki et al. 1985; Raymond et al. 1993; Robinson et al. 1993; Vihtelic et al. 1999; Hamaoka et al. 2002; Chinen et al. 2003). However, many cyprinids possess an A1–A2 interchange system, where the A1 pigment can be replaced by an A2 (2,3-dehydroretinol) based chromophore (Kusmic and Gualtieri 2000). Exchange of the visual pigment can modulate the peak sensitivities in response to changes in the environment and maturation stage, with A2-based chromophores typically having absorption maxima shifted to longer wavelengths. A recent microspectrophotometrical study obtained evidence that A2-visual-based pigments are also present in the zebrafish retina. Thyroid hormone treatment can significantly shift the peak and half-band width of opsin absorption, consistent with A1–A2 interchange (Allison et al. 2004). Further studies will have to determine if this interchange system is used by the zebrafish to tune its vision to different environments.

#### 2.2.4. Non-Visual Pigments

Apart from visual pigment, a number of additional opsins have been identified outside of photoreceptors. Little is known regarding their function, but since some are expressed in cells of the visual system, it is conceivable that they play a modulatory role in vision, for instance by influencing circadian rhythmicity.

One of the non-visual pigments that likely plays a role in setting the circadian clock is melanopsin. At least two melanopsin genes have been reported in the zebrafish genome. In contrast to mammals, where it is expressed in intrinsically photosensitive retinal ganglion cells, this gene is expressed in a subset of horizontal cells in the zebrafish retina (Bellingham et al. 2002). Recently another paralog has been isolated, that is also expressed in the eye (Bellingham et al. 2006). If and what role this opsin plays in setting the circadian clock is currently unknown.

Exo-rhodopsin is an interesting case of subfunctionalization, where the duplication of a gene leads to the functional divergence of the two initial redundant paralogs. This gene originated as a duplication of the rhodopsin gene by the teleost-specific whole genome duplication. It has lost its expression in the retina and is exclusively expressed in pinealocytes of the pineal organ in extant teleosts (Mano et al. 1999).

The zebrafish also harbors two paralogs of VAL-opsin. This opsin was originally isolated from the salmon eye as vertebrate ancient opsin (Soni and Foster 1997). In the zebrafish one paralog (VAL-opsin A) is expressed in deep cells of the brain and the eye, including the iris and non-GABA-ergic horizontal cells (Kojima et al. 2000). The second paralog (VAL-opsin B) is likewise expressed in the brain and in a subset of amacrine cells in the eye (Kojima et al. 2008).

Finally, a novel opsin family, termed tmt-opsin (teleost multiple tissue) has been identified. As the name implies, it is expressed in multiple neuronal and non-neuronal tissues and is a prospective pigment for the light-sensitive component of peripheral clocks (Moutsaki et al. 2003).

All of these opsins are candidates for non-photoreceptor-mediated vision that may play a role in setting the circadian clock and may contribute to background adaptation by body color changes.

#### 2.2.5. Inner Retina

The inner retina contains the interneurons that mediate the vertical signal flow (bipolar cells) and the horizontal signal flow (horizontal and amacrine cells) of visual information. There are multiple subtypes of horizontal, bipolar, and amacrine cells that differ in their morphology, neurochemistry, and physiology. The characterization of inner retinal cells of the zebrafish is certainly not complete yet (as is true for most other organisms as well). Similar to other cyprinid fish, the zebrafish contains four different horizontal cell types, as concluded from the application of various staining methods (Connaughton and Dowling 1998; Yazulla and Studholme 2001; Connaughton et al. 2004; Song et al. 2008; Li et al. 2009). The two small field horizontal cells (H1, H2) connect exclusively to cones. This is also true for one large field horizontal cell (H3), which responds best to UV light (Connaughton and Nelson 2007; Li et al. 2009). The other large field horizontal cell is the rod-specific rod horizontal cell.

By morphological criteria at least 17 bipolar cell types have been identified in the zebrafish retina. These cells can be divided by virtue of the position of their dendritic terminals into three groups (ON-type, OFF-type, multi-stratified ON- and OFF-type) that suggests to reflect a functional division as well (Connaughton and Nelson 2000; Connaughton et al. 2004). While there is ample evidence that depolarizing (OFF-type) bipolar cells are depolarized via AMPA and kainate glutamate receptor-mediated mechanisms in both teleosts and mammals (Dowling 1987), there is a potential difference in the mechanisms by which hyperpolarizing (ON-type) bipolar cells respond to glutamate released by photoreceptors. In the mammalian retina glutamate depolarizes ON-bipolar cells by activating a metabotropic glutamate receptor (mGluR6) that ultimately leads to the closure of a cation channel (Masu et al. 1995). Recently this channel has been identified as a transient receptor potential like (TRP) channel in mouse bipolar cells (Shen et al. 2009). Although there is firm evidence that such a mechanism also plays a role in the rod ON-pathway of the teleost retina, there is mounting evidence for an alternative mechanism taking part in the cone ON-response. Studies in the bass and white perch retina established that the ON-response can be mediated by a glutamate-activated chloride channel with transporter-like pharmacology (Grant and Dowling 1995, 1996). Since excitatory amino acid transporters (EAAT) are both present in the retina and exhibit a large chloride conductance concomitant to glutamate transport, these molecules are excellent candidates for mediating the cone ON-response (Grant and Dowling 1995). Consistent with this hypothesis, pharmacological blocking of glutamate transporters abolishes the cone ON-response. More recent studies in larval zebrafish and the closely related giant danio (*Danio aequipinnatus*) confirmed these results (Wong et al. 2004, 2005a, 2005b; Wong and Dowling 2005). The best candidate protein for mediating this response is EAAT5, which has been located on bipolar dendrites in the outer retina of goldfish (Kamermans et al. 2004). In the zebrafish there are bipolar cells that make use of both mechanisms, namely glutamate binding to mGluR6 and glutamate transport-dependent chloride influx, for glutamate-dependent hyperpolarization (Connaughton and Nelson 2000).

The jury is still out as to whether this mechanism is specific for the teleost retina or may also play a role in the mammalian retina as well (Joselevitch and Kamermans 2008).

The complexity found in the bipolar cells is probably more than matched with the amacrine cells. The teleost retina may contain up to 70 different amacrine cell types, distinguished by morphology and neurochemical properties (Wagner and Wagner 1988).

In the zebrafish seven amacrine cell types have been identified on the basis of morphology alone (Connaughton et al. 2004). Future studies using molecular and physiological methods likely will be able to subdivide further these cell types into additional specific classes (Marc and Cameron 2001). An unusual cell type, the interplexiform cell, extends processes into both plexiform layers. Since their cell bodies are situated in the amacrine cell layer, they may be regarded as a specialized amacrine cell type, despite them carrying information from the inner to the outer retina. These cells are dopaminergic and likely play a crucial role in network adaptation of the retina to light (Witkovsky 2004). A zebrafish mutant screen for dominant mutations affecting vision yielded the *night blindness b* (*nbb*) mutant. Dark-adapted heterozygous mutants show abnormal visual threshold fluctuations arising in the inner retina, which can be linked to a defect in the dopaminergic interplexiform cells. These cells receive input from the olfactory epithelium via the olfactoretinal centrifugal pathway, the only retinopedal projection. Excision of the olfactory epithelium or depletion of dopaminergic interplexiform cells can phenocopy the mutant defect. This defect is likely mediated via the rod pathway and also affects circadian rhythmicity of visual sensitivity (Li and Dowling 2000a, 2000b). In this context it is interesting to note that olfactory input is involved in retinal ganglion cell activity (Huang et al. 2005) and in the modulation of visual sensitivity (Maaswinkel and Li 2003).

Finally, the ganglion cell layer of zebrafish contains at least 11 ganglion cells types; these can be distinguished by their morphology (Mangrum et al. 2002; Ott et al. 2007).

The advance of transgenic techniques making it feasible to label specific cell types including their arborization pattern by virtue of cell-specific marker gene expression will complement the currently used dye-filling techniques and serial reconstruction techniques. Such transgenic labeling techniques can be coupled with patch-clamp recordings, allowing a detailed morphological and physiological characterization of the labeled cell type. This should yield to much more refined and complicated wiring diagrams in the near future.

### 2.2.6. Morphological Adaptation of the Retina

Vision functions over a remarkable range of up to 10 log units of light intensity, calling for exquisite gain control mechanisms to ensure proper visual function. Additionally, light and concomitant generation of reactive oxygen species are potentially harmful for photoreceptors, attested by photoreceptor damage following excessive illumination. In order to ensure vision under a variety of light conditions, multiple mechanisms have evolved to adapt the retina to varying illumination levels. Particularly striking are retinomotor movements in response to light that have been described in teleosts and amphibia (Ali 1971). These changes in the outer retina are immediately apparent on histological sections (Fig. 3.2). They can be divided into two components, namely the migration of melanosomes, melanin-containing particles, within the microvilli of the retinal pigment epithelium (RPE) and positional changes of photoreceptor cells. During the light-adapted state cones contract, rods elongate, and pigment granules are found at the apical tips of the RPE. Thus rods are shaded by pigments, presumably to avoid excessive bleaching or damage, while cone outer segments are optimally exposed to light (Ali 1971). The situation is reversed in the dark-adapted state, where rods are contracted and closer to the inner limiting membrane and cones are elongated. Pigment granules of the RPE are condensed basally, allowing maximal access of light to rod outer segments. Interestingly the short single (UV) cones do not exhibit retinomotor movements and have a fixed position in relation to the other photoreceptors.

Retinomotor movements are established early in development, with pigment granule migration apparent already at early larval stages. Changes in the relative position of photoreceptors become apparent at around 15 dpf, when the rod system starts to become fully functional (Menger et al. 2005; Hodel et al. 2006). The kinetics of the two components are quite different with the rearrangement of the pigment granules being slower than the rearrangement of the photoreceptors (Hodel et al. 2006). Finally, retinomotor movements are not only regulated directly by light, but also by endogenous oscillators (Menger et al. 2005).

Another form of plasticity takes place at photoreceptor synapses of teleosts, where spinules (protrusions of horizontal cells) are formed in the light-adapted retina. These are transient structures that retract in the dark-adapted retina (Wagner 1980). Their formation is triggered by decreased glutamate release from photoreceptors and increased dopamine release from interplexiform cells (Kohler et al. 1990; Wagner and Djamgoz 1993). This process is well documented in the zebrafish retina and can be observed as early as 5 days post fertilization, although the number and complexity of spinules increases with ensuing maturation of the retina (Biehlmaier et al. 2003).

### 2.2.7. Biochemical Adaptation of the Retina

Much is known about biochemical adaptation mechanisms acting in photoreceptors. Biochemical adaptation mechanisms in photoreceptors mainly act on the visual transduction cascade, mostly mediated by changes of intracellular calcium levels (reviewed in Koch 1992; Pugh et al. 1999). The zebrafish with its cone-dominant retina has particularly contributed to our understanding of cone photoreceptor adaptation. Cones carry the main burden of light adaptation, since they function over approximately 7 log units of light intensity and can practically not be saturated.

One central biochemical step in adaptation is visual pigment phosphorylation. Studies in the zebrafish have identified a cone-specific kinase (Grk7) that is distinct from its rod counterpart (rhodopsin kinase or Grk1) (Rinner et al. 2005a; Wada et al. 2006). Downregulation of this kinase leads to significantly prolonged recovery time of cone visual pigment and accordingly to reduced temporal resolution in behavioral experiments (Rinner et al. 2005a).

Another study made use of a mutant (*no optokinetic response f, nof*) carrying a mutation in the alpha subunit of cone transducin, affecting exclusively cone photoreceptors. Interestingly the cones in this mutant still respond to bright light, indicative of a transducin-independent light-induced cytoplasmic calcium increase. The source of this calcium is currently unknown (Brockerhoff et al. 2003). An elegant biochemical study using zebrafish cones found that cytoplasmic calcium levels not only regulate the site of light-stimulated opsin phosphorylation, but also the extent of phosphorylation of unbleached pigments (Kennedy et al. 2004). These studies effectively combine the favorable genetics with the cone-dominant nature of the zebrafish retina.

### 2.2.8. Influence of Circadian Rhythms on Vision

Circadian regulation of visual sensitivity can be thought of as yet another mechanism of light adaptation. Behavioral assessment of visual performance during different times of the day has uncovered a surprisingly high difference in visual sensitivity between day and night time. The threshold light intensity for detection of visual stimuli in the adult escape response paradigm is about 2 log units lower in the late afternoon as compared to early mornings (Li and Dowling 1998). This strong circadian influence on visual function is also reflected by studies of the electroretinogram, directly measuring sum field potentials of the retina in response to light (Li and Dowling 1998; Allison et al. 2004). This circadian rhythm of visual performance persists when fish are kept in constant light, arguing for an endogenous circadian clock, working to decrease visual sensitivity to light onset during the subjective night.

A number of different mechanisms of the zebrafish visual system are regulated by an endogenous circadian clock. Examples are melatonin synthesis, which is elevated at night and downregulated during the day (Cahill 1996), and increased expression of IRBP (interphotoreceptor retinoid-binding protein) mRNA during the day compared to the night (Rajendran et al. 1996). Dopamine produced by interplexiform neurons likely plays major roles in regulating the light sensitivity of the retina during different parts of the day. Dopamine release is influenced by lighting conditions and it has a variety of neuromodulatory functions, including a role in retinomotor movements, gap junctions connections between photoreceptors and horizontal cells, and regulating excitability of bipolar cells (reviewed in Witkovsky 2004).

### 2.2.9. Regenerative Properties

The regenerative potential of lower vertebrates is immense and the retina is no exception. The adult zebrafish retina has recently become a well-investigated preparation to study the regeneration of photoreceptors following pharmacological and light-induced degeneration (reviewed in Hitchcock and Raymond 2004). Teleost retinas contain a proliferative stem cell zone at the ciliary margin, the circumferential germinal zone. This zone becomes active early on in development and adds new neurons to the retina throughout the lifetime of the fish (Marcus et al. 1999). All retinal cell types originate from this stem cell niche, with the exception of rod photoreceptors. These arise from rod precursor cells which reside in the inner nuclear layer of the retina and differentiate while migrating to the outer retina and thereby interstitially adding rods to the outer retina (Hitchcock and Raymond 2004; Raymond et al. 2006).

The zebrafish retina can respond to lesions (pharmacological, physical or light-induced) by rapid regeneration. Interestingly this injury-induced mitotic activity is initiated in the inner nuclear layer. In the case of light-induced photoreceptor lesions, these inner retinal cells migrate to the outer nuclear layer, where they keep on dividing and differentiate into cone photoreceptors (Vihtelic et al. 2006). Pharmacological damage to the inner retina, without affecting the outer retina, also led to rapid regeneration of the damaged cell types (Fimbel et al. 2007). In both cases Müller glia cells of the retina are the source of regeneration (Yurco and Cameron 2005; Fausett and Goldman 2006; Bernardos et al. 2007). Müller glia cells even proliferate at a low frequency in the uninjured retina, where they likely function as the rod precursor cells (Raymond et al. 2006; Bernardos et al. 2007). Upon damage they re-enter the cell cycle and are able to produce all cell types of the retina. Hence they are multipotent stem cells that can be activated upon damage of other retinal cells (Fimbel et al. 2007; Bernardos et al. 2007) Therefore the zebrafish holds great promise as a research preparation to decipher the molecular orchestration of this neuronal regeneration (Cameron et al. 2005; Wehman et al. 2005; Kassen et al. 2007; Fausett et al. 2008; Qin et al. 2009).

The overall hope is that information gained from these studies will one day be used to coax human retinal cells to regenerate with obvious medical applications.

## 3. RETINAL PROJECTIONS

The projection of the vertebrate retina to its main target, the optic tectum or its mammalian homolog the superior colliculus, has long enjoyed the status of one of the most commonly used preparations to study neuronal map formation (Ruthazer and Cline 2004; Lemke and Reber 2005). The ease of labeling the projection by injecting lipophilic tracer dyes into the eyes of (fixed) animals and the quasi two-dimensional target region are major advantages of this preparation (Fig. 3.3). Early studies have established the topographic mapping of the zebrafish projection (Stuermer 1988), where all axons grow across the midline, hence terminating on the contralateral tectum forming a complete optic chiasm. Terminal arborization in the tectum is topographic, with axons from neighboring retinal ganglion cells terminating at neighboring positions on the tectum. The projection pattern is inversed with nasal axons terminating in the posterior tectum and dorsal axons terminating on the ventral tectum (Fig. 3.3) (Stuermer 1988).

Apart from the main projection to the optic tectum, there are nine additional arborization fields of the retinofugal projection. With a sole exception, they are all contralateral (Burrill and Easter 1994).

In the zebrafish it is possible to inject a large number of larvae with tracer dyes and subsequently analyze their projection pattern. This has enabled researchers to perform a large-scale genetic screen successfully isolating mutants with projection and mapping defects (Baier et al. 1996). In this screen using an ingenious high-throughput labeling method 114 mutants in about 35 genes were identified (Baier et al. 1996; Culverwell and Karlstrom

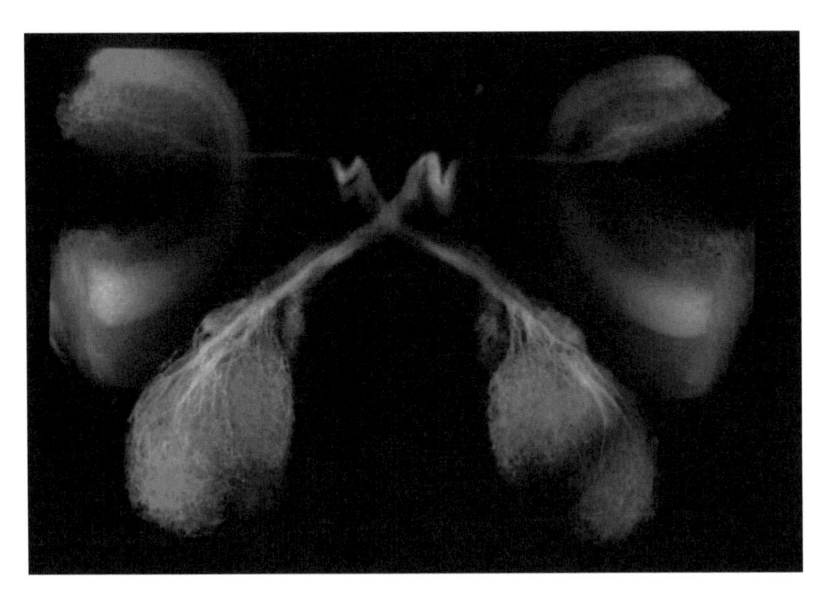

**Fig. 3.3.** Retinotectal projection. The eyes of a 5-day-old zebrafish larva have been injected with two lipophilic tracer dyes (DiO, DiI). The reversed retinotectal projection is visible with anterior cells of the eye projecting to the posterior ipsilateral tectum. Image courtesy of Chi-Bin Chien, reprinted with permission of Elsevier. See color plate section

2002). Recently another screen using a transgenic line with labeled retinal ganglion cell axons isolated 13 additional mutants (Xiao et al. 2005). These mutants turned out to be more specific for the retinotectal projection featuring less pleiotropic phenotypes. Interestingly there is no overlap of mutants found in these two screens, indicating that many additional genes are awaiting isolation in future screens.

The analysis of such mutants shows that pathfinding (finding the tectum) is at least partially under separate genetic control as mapping (finding the correct position on the tectum), since mutant classes exist that affect one but not the other process.

The role of neuronal activity and axonal competition in shaping the retinotectal projection has been investigated in some detail in the zebrafish. Earlier studies argue for no role of neuronal activity in setting up the projection, since blocking activity by tetradotoxin (TTX) during early larval development did not affect the projection pattern (Stuermer et al. 1990). A subsequent genetic analysis found that the initial projection is not changed by either pharmacological or genetic neuronal inactivation, but subsequent shifting of the axon termination due to disparate growth of retina and tectum may be affected by lack of neuronal activity (Gnuegge et al. 2001). An elegant in vivo imaging study showed that growth and branching is influenced by activity-based competition. Axonal arbors from transgenic activity-suppressed neurons show inhibition of growth and branch formation, which is relieved when nearby (competing) axonal arbors are suppressed as well (Hua et al. 2005).

Axonal competition does not seem to play a role in finding the appropriate target on the tectum, but is likely involved in restricting arbor size and branch refinement, since solitary axons are targeted correctly to their tectal position but display larger and more complex arborizations (Gosse et al. 2008). Interestingly synaptogenesis plays a direct role in determining axonal arbor properties by promoting initial branch extension and selective branch stabilization (Meyer and Smith 2006).

## 4. OPTIC TECTUM

The optic tectum, constituting the dorsal-most aspect of the midbrain, is the main projection area of retinal ganglion axons. Apart from visual input, it also receives other sensory inputs. Hence it is likely involved in processing and integrating multiple sensory inputs. The homologous structure in mammals is the superior colliculus, which plays a prominent role in the control of eye movements (Hall and Moschovakis 2003).

The (optic) tectum in the zebrafish is a multilayered structure, with four layers receiving input from retinal ganglion cell axons. The most superficial of these is the stratum opticum, right above the stratum fibrosum et griseum superficiale (SFGS), which receives most of the retinal ganglion cell axons and can be further divided into at least three sublaminae. There are two additional deeper layers, the stratum griseum centrale (SGC) and a projection zone between the stratum album centrale and the stratum periventriculare (SAC/SPV) (Huber and Crosby 1933; von Bartheld and Meyer 1987; Xiao et al. 2005; Xiao and Baier 2007). Retinal axons are specified to terminate exclusively in one of these four layers. The analysis of a mutant (*dragnet*) isolated in the aforementioned screen using transgenically labeled retinal ganglion cell axons implied a role for collagen IV in anchoring secreted guidance cues to the surface of the tectum (Xiao and Baier 2007). Similar studies, using the power of zebrafish genetics, will likely further our understanding of tectal development in the near future.

Although the substantial innervation of the tectum by the optic fibers suggests a prominent role in vision, the specific role of the tectum not only in visual processing, but also in sensory integration is still largely unknown. Experiments in the goldfish have established that ablation of the tectum affects the optomotor response but not the optokinetic response (see below for a description of these behaviors) (Springer et al. 1977). This question was revisited by modern techniques using laser ablation of the larval tectum. Interestingly both optokinetic (OKR) and optomotor (OMR) responses were unaffected and visual acuity remained intact. Only the frequency of saccades was diminished in the OKR paradigm without affecting tracking movements. Therefore the optic tectum is probably not involved in motion detection, including second-order motion (Roeser and Baier 2003). Taking into account data from other teleosts, it seems likely that assays testing for visual grasp behavior, visual orientation, or visual discrimination would reveal a role of the tectum (Akert 1949; Yager et al. 1977; Davis and Klinger 1987; Salas et al. 1997).

The interest in the electrophysiological properties of the zebrafish optic tectum has only recently been rekindled by the advances in optical activity imaging technologies. Earlier studies described four tectal cell types, which differ in spontaneous activity responses to stationary spots, and a number of additional stimulation routines, including diverse motion stimulations. These studies failed to identify tectal cells with orientation specificity or spatially separated ON and OFF areas (Sajovic and Levinthal 1982a,b). Further pharmacological manipulations revealed evidence for intratectal delayed inhibitory input, shaping the properties of the visual responses (Sajovic and Levinthal 1983).

It took more than two decades and new optical recording technology until renewed interest in zebrafish tectal physiology started. Larvae can be immobilized in low-melting agarose, allowing simultaneous stimulation, e.g. by using miniature LCD screens or projecting stimuli via lenses onto the eyes, and measuring activity of individual neurons by two-photon calcium imaging (Niell and Smith 2005; Ramdya and Engert 2008; Sumbre et al. 2008). In such experiments additional response types of tectal neurons could be identified, including direction-selective cells (Niell and Smith 2005). This new technology enabled a fascinating study showing that by forcing retinal ganglion cell axons onto a single tectum by removing one tectum, binocular function emerges in the normally monocular zebrafish larva. The inputs from the two eyes are functionally integrated in the rewired tectum, which is particularly surprising for direction-selective neurons. This direction selectivity likely arises through temporally asymmetric inhibitory input, which is probably established independently of visual input (Ramdya and Engert 2008).

Finally, ablation studies in combination with high-speed locomotion analysis found a role for the optic tectum in prey capture of larval zebrafish. The tectum and identified reticulospinal neurons are part of a neural circuit coordinating prey capture movements (Gahtan et al. 2005).

## 5. VISUAL BEHAVIOR

Any analysis of vision is incomplete without an assessment of visually mediated behavior. The zebrafish larva expresses a number of visually mediated behaviors that made it possible to screen for mutant strains deficient in these behaviors (Brockerhoff et al. 1995, 1998; Neuhauss et al. 1999; Gross et al. 2005; Muto et al. 2005). Such a forward genetic approach has been recently complemented by the morpholino technology, where the translation of any protein of choice can be downregulated and subsequently assayed for any resulting defect (Nasevicius and Ekker 2000). Hence the zebrafish offers the unique opportunity to combine powerful genetics with a robust behavioral read-out to study visual system function.

The visual system is particularly well suited for studying behavior in the laboratory since the relevant stimulus properties can be well controlled (brightness, spectral content, contrast, duration). Furthermore the retina is the only relevant light-receiving structure with well-defined properties. Hence there are a number of sophisticated visual assays that can be used for rapidly screening through a large number of specimens or more elaborate assays able to define even subtle changes in visual performance.

Much less is known about the visual performance of zebrafish in its natural environment. Visual behavior in the laboratory certainly serves as a guide to what the zebrafish is capable of in the wild, but our knowledge about the natural visual behavior of zebrafish is still in its infancy.

## 5.1. Visual Behavior in the Laboratory

Most of the visual behavior assays have been developed with the aim to screen for zebrafish mutants affected in vision. The zebrafish has been the first vertebrate where large-scale forward genetic screens have been performed, opening the possibility to also conduct behavioral screens instead of being restricted to morphological screens.

Such screens have been pioneered by John Clark, a student of George Streisinger at the University of Oregon, whose seminal work established the zebrafish as a genetic model organism. He described many of the behavioral assays currently in use in his unpublished doctoral thesis (Clark 1981). As part of his thesis project, he also initiated a screen to isolate mutant strains with heritable defects in visual behavior.

### 5.1.1. Development of Visual Behavior

The rapid morphological maturation of the visual system during the course of development already suggests that this rapid maturation will also be reflected in the rapid maturation of behavioral responses to light.

The first emerging visually mediated behavioral response is the visual startle response, where larvae respond with a rapid increase in body movements to a sudden decrease in brightness (Kimmel et al. 1974). This simple behavior emerges just when the synaptic ribbons of the retina are formed at around the third day of development (Easter and Nicola 1996). Slightly later, around 80 hpf, the optokinetic response emerges. This is the earliest emerging visual behavior requiring form vision (Clark 1981; Easter and Nicola 1996, 1997).

Slightly later the optomotor response (OMR) and positive phototaxis develops (Brockerhoff et al. 1995). It should be noted that the preference for a lit compartment changes during development. Lightly pigmented larvae display positive phototaxis, while this changes to a preference for dark compartments in older larvae and adults, correlated with the increase in body pigmentation of the animals (Watkins et al. 2004). Although some studies found such an innate preference for dark compartments (Serra et al. 1999), other studies found no preference for dark or light environments in habituated adult zebrafish (Gerlai et al. 2000).

Some visual behaviors are not developed in the larva and can only be evoked in the adult, such as the dorsal light reflex and the escape response.

Larval behavior is strongly dominated by cone vision. Although rod photoreceptors with short outer segments and synaptic vesicle-containing synapses are present in the larval zebrafish retina, a number of studies concluded from electroretinograms (ERG) that rods do not contribute to vision at early larval stages (Branchek 1984; Bilotta et al. 2001). A contribution of rods to the electroretinogram becomes apparent at 15 dpf, when the rods are also morphologically more mature (Bilotta et al. 2001). However, a recent study found small rod responses in aforementioned *nof* mutant larvae which lack phototransduction in cones. This abstract reports that dark-adapted 6-day-old mutant larvae display OKR and ERG responses, which are attributed to rod function (Mills et al. 2009).

#### 5.1.2. Visual Background Adaptation (VBA)

A very simple reflex behavior is visual background adaptation, where the zebrafish adjusts the distribution of melanin pigment-containing vesicles (melanosomes) in pigment cells in response to changes in ambient light. The skin of the zebrafish contains large star-shaped dark pigment cells called melanophores (or melanocytes), which are filled with melanosomes, their eponymous melanin-containing organelles.

At low light levels, melanosomes are widely distributed throughout the melanophore's processes, giving the larva a dark appearance, presumably aiding the larva to be camouflaged against a dark background. At brighter ambient light, the melanosomes concentrate at the center of the melanocyte, giving the larva a light appearance. These responses are controlled by a fast (neural) and a slower hormonal system (Jain and Bhargava 1978; Fujii 2000).

These changes of body coloration are achieved by molecular motors, that carry melansomes either out to the periphery of the cell or back to the center of the melanophore on microtubule tracts (Nascimento et al. 2003).

Subcellular shift of melanin distribution is triggered by light perception of the retina and signaled via a direct retino-hypothalamic projection to the hypothalamus, which in turn activates the secretion of two hormones from the pituitary, which act on the melanophores (Balm and Groneveld 1998).

Any defect in sensing ambient light levels would therefore be expected to result in a lack of background adaptation, evidenced by permanently dispersed melanosomes, independent of ambient light levels. Indeed, a number of mutants lacking this background adaptation (termed "expanded melanophore phenotype" in the original reports (Haffter et al. 1996; Kelsh et al. 1996), turned out to be blind in behavioral assays (Neuhauss et al. 1999) (Fig. 3.4), Therefore lack of background adaptation in a mutant is often correlated with behavioral blindness and can serve as a first indication of a visual defect (Figs 3.4 and 3.5).

**Fig. 3.4.** Visual background adaptation in adult zebrafish. An adult blind fish (top) is markedly darker in appearance than an unaffected wild-type fish with normal light perception (below). See color plate section

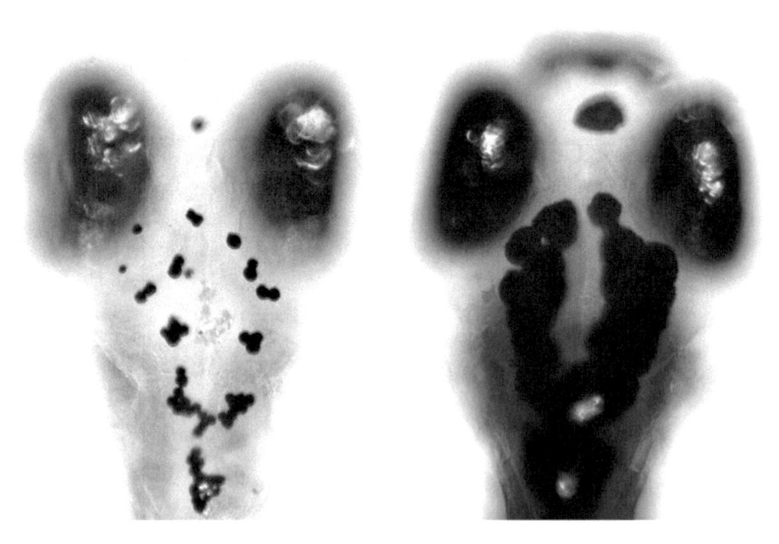

**Fig. 3.5.** Visual background adaptation in larval zebrafish. Close-up of the head region showing expanded melanocytes of a behaviorally blind mutant larva (right) in comparison to a wild-type sibling (left) under bright illumination. See color plate section

### 5.1.3. Optokinetic Response (OKR)

Probably the most robust behavior is the optokinetic response (OKR), which is triggered by large field movements in the visual field (Fig. 3.6A). Zebrafish larvae respond with stereotyped tracking eye movements that are comparable to eye movements in other vertebrates (Huang and Neuhauss 2008). These eye movements consist of two components: a compensatory smooth pursuit movement tracking the moving object followed by a resetting fast phase in the opposite direction, also referred to as a saccade.

The optokinetic eye movements in larval zebrafish are coupled, with no overall direction selectivity (Beck et al. 2004; Rinner et al. 2005b). At low spatial frequencies the efficacy of the eye movements is greater in response to temporal-to-nasal stimulation. Intriguingly, this directional asymmetry is reversed at higher spatial frequency, which could be construed as ganglion cells of the temporal and nasal retina responding differently to whole-field motion depending on spatial frequency (Qian et al. 2005).

In the larval zebrafish the OKR is easily evoked by immobilizing the larvae in a viscous medium, such as methylcellulose solution or low melting agarose. Since eye movements are suppressed by whole body movements, it is important to restrain the larvae with minimal interference of eye movements. The stimulus can be presented by an actual drum fitted with moving stripes, by temporally controlled light-emitting diodes (LED), or by

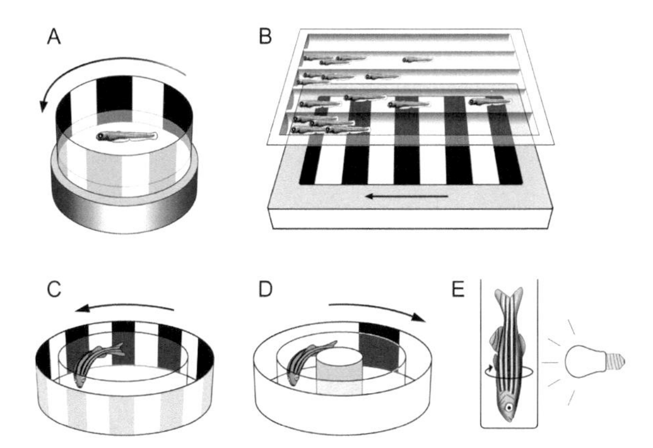


**Fig. 3.6.** Visual behavior assays in larval (A,B) and adult (C–E) zebrafish. (A) Optokinetic response (OKR), (B) population screening for larval optomotor response (OMR), (C) adult OMR, (D) adult escape response, and (E) dorsal light response (DLR). Arrows indicate direction of moving grating (A–D), and movement of fish (E). Actual size of larvae and adult fish is proportionally much smaller than depicted in the diagram. For details see text. Adapted from Neuhauss (2003), reprinted with permission of Blackwill, Inc.

projecting or reflecting computer-generated images simulating a rotating drum. Various quantification methods are employed from simple counting saccades to sophisticated computer-based kinetic analysis (reviewed in Huang and Neuhauss 2008).

Since this behavior is easily scored and robustly displayed in larvae, it has been successfully employed in a number of genetic screens, yielding a treasure trove of mutant strain defective in vision (reviewed in Gross and Perkins 2008; Neuhauss 2003).

Due to the need to immobilize the animal, it has been technically more challenging to measure the OKR in the adult zebrafish. Embedding in a viscous medium is not feasible, since the gills of the fish will be clogged and in contrast to the larva, oxygen uptake through the skin is not sufficient. Hence special holding devices are needed (Beck et al. 2004), preferably with water flushing over the gills (Mueller and Neuhauss, in press).

The neuronal substrate of the OKR is not known in zebrafish. Analogous to birds and mammals, the OKR is likely mediated by pretectal nuclei receiving direct innervation from the retina. This notion is supported by ablation studies, showing that ablation of the optic tectum did not abolish the OKR (Springer et al. 1977; Roeser and Baier 2003).

A recent study addressed the question, which of the two main pathways through the retina, the ON- or the OFF-pathway or both, is needed to drive the OKR. A pharmacological block of the ON-pathway abolished the OKR, while another behavioral assay provided evidence for an intact OFF-pathway in these OKR-deficient larvae. Additional support for the conclusion that the OFF-pathway cannot drive the OKR came from *nrc* (*no optokinetic response c*) mutant larvae, which show no OKR, but have an intact OFF-pathway (Emran et al. 2007). From this study it is clear that the ON pathway is necessary to drive the OKR. Future studies will determine if it is also sufficient.

The OKR can also be adapted to screen for mutations affecting exclusively red cone photoreceptors by using illumination, only stimulating one cone type. Given the overlap of the cone pigment absorption spectra, only the red cone can be separately stimulated by far red light. A genetic screen using far red illumination was performed resulting in the isolation of the *partial optokinetic response b* (*pob*) mutant (Brockerhoff et al. 1997). Homozygous *pob* larva lacked an OKR under red illumination but responded normally under white light. Consistent with red light insensitivity, mutants show a sharp drop in relative sensitivity at wavelengths longer than 550 nm in the spectral electroretinogram. Subsequent histological analysis revealed a specific loss of red cones in the mutant. The mutated locus encodes a protein of unknown function that is localized to the inner segment and synapse of all photoreceptor cells. Furthermore, it colocalizes with markers of the trans-Golgi and the endoplasmic reticulum, suggesting a function in protein trafficking and/or sorting (Taylor et al. 2005).

#### 5.1.4. Optomotor Response (OMR)

Larval and adult zebrafish have a tendency to follow moving visual objects in the surround. This optomotor response (OMR) can be evoked in the larva by placing free-swimming larvae in a container surrounded by a moving drum fitted with stripes (Clark 1981; Bilotta 2000). Alternatively apparent moving stripes can be presented to the larvae, for instance by placing them in a Plexiglas container on top of a flipped computer monitor (Neuhauss et al. 1999; Orger and Baier 2005) (Fig. 3.6B). The majority of 6 dpf wild-type larvae will swim in the direction of the perceived motion, following a leading stripe. This is a population assay in that several families can be tested at once. After the trial period the response can be evaluated by assessing the distribution of the larvae. Randomly distributed larvae did not perceive the stripes, while the majority of seeing larvae cluster around the end of the container where the apparent movements of the stimulus were heading to. This assay works for larvae, since they have no shoaling behavior yet, so that each larva determines its swimming direction individually. Adult shoaling zebrafish have to be tested individually for the OMR, by placing them inside a round testing chamber surrounded by a rotating drum containing alternating black and white stripes (Fig. 3.6C). Both assays can be used to define spatial and temporal frequencies of motion detection (Maaswinkel and Li 2003) or by using different spectral illumination to probe for chromatic inputs to motion detection (Krauss and Neumeyer 2003).

#### 5.1.5. Visual Motor Response (VMR)

The two most commonly used assays (OKR and OMR) to assess visual performance in zebrafish are both based on motion detection. Recently a motion independent visual assay based on the visual startle reflex has been described (Prober et al. 2006; Emran et al. 2007, 2008). This assay, termed the visual motor behavioral assay (VMR), uses an automated tracking system that simultaneously monitors the movements of single larvae individually placed in a 96-well plate. The lights are then turned on for 30 minutes followed by a 30-minute dark period, while recording and quantifying locomotor responses of the larvae. Larvae show the startle response, and a brief spike of activity upon light being turned on, followed by a drop to sub-baseline levels of activity. Similarly larvae increase their locomotor activity over the course of a couple of minutes (Emran et al. 2008). Apart from testing vision independent of motion detection, this assay additionally allows testing separately for ON and OFF responses and enables testing spectral sensitivities independent of motion detection.

### 5.1.6. Adult Escape Response

The only behavioral assay that has been used so far to isolate adult mutant fish with vision defects is the escape response. This response is triggered when the fish is threatened by a potential predator, mimicked in the experiment by a single black stripe. In the experimental set-up the fish can hide behind an opaque object when threatened. By moving the threatening stripe, the fish will move concomitantly aiming to be at the opposite site of the opaque object, hidden from the stripe (Fig. 3.6D). This behavior obviously depends on the fish's ability to see and can be used to assess visual performance (Li and Dowling 1997, 2000b). By varying light intensity, absolute rod and cone threshold levels as well as the course of dark adaptation after light adaptation can be reliably measured (Li and Dowling 1997). Similarly the strong influence of the circadian clock on visual sensitivity of the zebrafish was established with this method (Li and Dowling 1998).

This assay was also used for the first systematic screen for dominant mutations affecting vision in the adult, identifying a number of mutant strains with progressive dominant retinal degeneration (Li and Dowling 1997, 2000a; Maaswinkel et al. 2003, 2005).

Such mutations are of particular biomedical interest, since a high proportion of human outer retinal dystrophies are progressive and affect older patients.

### 5.1.7. Additional Assays

There are several additional assays that are based on vision that have not been used much in zebrafish. An interesting assay has been described by Clark based on the ability of zebrafish larvae to prey on paramecia (Clark 1981). Larvae that successfully preyed on green paramecia can be easily scored by their bellies turning green due to ingested paramecia. Preying success is dependent on brightness and blind larvae failed to ingest larvae. Recent study confirmed that prey capture is mainly visually mediated (Gahtan et al. 2005; McElligott and O'Malley 2005).

The pilot screen using this feeding assay remains the only reported use of this assay in the literature.

Zebrafish larvae are phototactic and prefer the illuminated part of a holding container over the dark part. Since this behavioral preference is time consuming to test and not very strong, it has not been used to isolate larvae defective in vision (Brockerhoff et al. 1995).

Most fishes (and some marine invertebrates as well) have a tendency to turn their back towards a light source. This dorsal light response (DLR) was first described by von Holst, who showed that some fish tilt their dorsal side towards a light source placed at the side of the tank (Holst 1935). Animals use this dorsal light response in conjunction with their sense of balance to hold their posture in the water. In the goldfish this behavior is robust enough that it can be used to determine spectral and visual sensitivity of vision (Yager 1968; Powers 1978). In zebrafish the DLR is only obvious in mutants with defects in their vestibular system (Nicolson et al. 1998). In wild-type fish the response can be evoked by placing the fish upside-down in a tube to perplex the sense of gravity (Fig. 3.6E). Illumination from the side will cause the fish to turn its back to the light source and it will also follow it when the light is slowly moved around the tube.

#### 5.1.8. Motion Detection

Motion is for most animals a most relevant visual cue, so it is not surprising that already the immature larval visual system of the zebrafish supports the detection of motion. In the natural environment, motion detection is dominated by luminance-defined (first-order or Fourier) motion cues. In higher vertebrates there is much evidence that second-order cues, such as movement of contrast, flicker or texture, can be used to perceive complex movements (Ramachandran et al. 1973). The processing of such second-order motion cues, also referred to as non-Fourier motion, was long assumed to necessitate a cerebral cortex (Chubb and Sperling 1988; Cavanagh and Mather 1989). Hence it comes as a surprise that already zebrafish larvae can utilize second-order motion cues, as has been shown by the elegant use of the optomotor response (Orger et al. 2000). Second-order motion cues are extracted earlier in the visual pathway as has been assumed and clearly no visual cortex is needed for this task.

An interesting question is to what extent color information is used for motion detection. In humans, as for a number of other animals, motion perception receives stronger inputs from L and M cones as compared to S cones (Wandell et al. 1999). The predominant use of longer-wavelength cones for motion detection is a common trait found in most investigated organisms. The relative contribution of the different cone types has also been studied in the adult goldfish, employing the optomotor response (Schaerer and Neumeyer 1996). In both the dark- and light-adapted state there was a single maximum in the action spectrum. While the dark-adapted spectrum coincided well with the absorption maximum of rod photoreceptors, the light-adapted action spectrum was found to be located in the long-wavelength range (620–660 nm). Using stripes of different color and intensity, the optomotor response was found to be dependent on modulation of the L-cone type. Hence motion detection can be assumed to be "color blind" (Schaerer and Neumeyer 1996). This result was confirmed by similar experiments in adult zebrafish, where also the optomotor response was at a minimum whenever the L-cone type was not modulated. The single maximum for the light-adapted zebrafish action spectrum is at lower wavelength (550–600 nm), likely reflecting differences in cone type absorption spectra and relative sensitivities (Krauss and Neumeyer 2003).

The influence of chromatic input to motion detection was also addressed in the larval zebrafish, using the OMR in motion-nulling experiments (Orger and Baier 2005). Intriguingly the situation appears to be slightly different in larval zebrafish as compared to adults. In larvae the obtained results are closer to the situation in higher vertebrates, with motion vision deriving predominantly not only from red, but also green cones. The luminance channels dominating motion detection likely pool red and green cones signals, after being spatially filtered (Orger and Baier 2005). Whether there is a real difference in the chromatic influence on motion detection between larval and adult zebrafish is still debatable. One explanation may be found in a difference in threshold sensitivities of the cones. It may very well be that the overproportional increase in red cone sensitivity may come to dominate the action spectrum of the adult (Saszik et al. 1999; Orger and Baier 2005).

### 5.1.9. Color Vision

Color vision is prevalent in multiple fish species (Douglas and Hawryshyn 1990). The presence of four cone types with different spectral sensitivities and morphological evidence for the appropriate neural circuitry, argues strongly – although indirectly – for color vision in zebrafish.

The cone contribution to the electroretinogram has been assessed and evidence for color opponent mechanisms has been found (Hughes et al. 1998; Patterson et al. 2002). Similar cone mechanisms have been found in recordings of the massed electrical potential of the adult optic tectum (McDowell et al. 2004).

We have surprisingly little information about color vision from behavioral experiments. Psychophysical experiments using the OMR found the response to be "color blind", as described in the previous paragraph (Krauss and Neumeyer 2003). This result can be construed as an argument that motion-based assays may not be adequate tools to investigate color vision in fish.

In order to address spectral sensitivity in motion-independent psychophysical experiments, Risner et al. used an appetitive instrumental conditioning paradigm to train zebrafish to monochromatic stimuli (Risner et al. 2006). Visual thresholds as determined by varying stimulus irradiance were used to derive a spectral sensitivity function, with several peaks near the absorption maxima of the various cone pigments. This function also found evidence for opponent mechanisms, these being comparable to mechanisms proposed for the goldfish (Neumeyer 1984).

Visual discrimination learning for different colors is possible in zebrafish (Colwill et al. 2005). However, since the color stimuli were not controlled for isoluminance, it is not possible yet to conclude that the trained fish could distinguish color as opposed to differences in luminance.

More sophisticated behavioral experiments on zebrafish color vision, for instance addressing color constancy, have not been reported.

## 6. COMPLEX VISUAL BEHAVIOR

Most of the literature on zebrafish vision is concerned with reflexive responses under highly controlled laboratory conditions. Such visually mediated responses are well suited for screens aiming to identify mutants affecting vision. Among the prerequisites of utilizing behavioral assays for large-scale screen is their robustness and the more directly they rely on vision the better (Fleisch and Neuhauss 2006). Hence such reflexive behaviors are very well suited to analyze physiological properties of the visual system, but tell us little about the natural use of vision. In the life of the fish, vision is embedded into complex behaviors that often rely on more than one sensory modality. In the following I will briefly discuss some more complex behaviors that have a large contribution of vision.

### 6.1. Visual Lateralization

Behavioral lateralization is a common feature of all vertebrates and the zebrafish is no exception. The study of these behavioral asymmetries is particularly attractive in the framework of linking behavioral asymmetries to anatomical asymmetries. Evolution of cerebral lateralization, especially in the context of emergence of language, has generated considerable interest and brain lateralization of lower vertebrates is of obvious evolutionary relevance to these questions.

The best way to study behavioral asymmetry in the zebrafish is preferential eye use that becomes already apparent at larval stages. Teleosts show visual lateralization, revealed by preferential eye use depending on the type of visual stimulus presented. Novel objects are typically viewed with the right eye, while familiar objects are preferentially viewed with the left eye (Miklosi et al. 1997). The right eye is used when the situation calls for a decision based on the visual input. Possibly the neural circuit served by the right eye allows for the inhibition of an immediate response to enable careful inspection of the visual scene. This is apparent in a number of behaviors, e.g. in the decision to approach an object (Miklosi et al. 2001) or in the decision to bite (Miklosi and Andrew 1999). Lateralized behavioral responses can already be observed at larval stages, making genetic approaches possible (Watkins et al. 2004; Sovrano and Andrew 2006). Analysis of the *frequent-situ-inversus* (*fsi*) line in zebrafish shows that some, but intriguingly not all, behavioral asymmetries correlate with neuroanatomical asymmetries of the diencephalon (Barth et al. 2005). Hence the exact anatomical basis of behavioral lateralization is currently unknown. Neuroanatomical asymmetries and their development can be analyzed in the zebrafish brain and are amenable for a genetic dissection (reviewed in Bianco and Wilson 2009).

### 6.2. Social Preference

Zebrafish have a tendency to associate with conspecifics, probably a precondition for shoaling and of importance for mate choice. This behavioral tendency is largely mediated by vision. Fish prefer to associate with fish sporting horizontal stripes over vertical stripes, even when presented by a video monitor (Turnell et al. 2003). Interestingly, this preference turned out to be learned as has been elucidated by cross-rearing experiments employing adult pigment mutant strains (McCann and Carlson 1982; Engeszer et al. 2004; Spence and Smith 2007).

Early experience determines adult shoaling preference, so that wild-type embryos raised with pigment-deficient fish prefer to shoal with these fish rather than with wild-type fish. This behavior is clearly learned since isolated fish showed no preference associated with pigment patterns. Shoaling preference turned out to be stable and difficult to change after it has been established (Engeszer et al. 2007a). There is a marked gender difference with males responding to species identity and stripe pattern, while no such preference has been discovered for females (Engeszer et al. 2008). This is an unexpected result since the selection pressure to join a shoaling group of conspecifics should be the same for males and females. Another interesting point of this study is that the differences perceived by the males were not the same as those perceived by human observers, stressing the point that the visual world may look very different to a fish than to a human observer.

Reassuringly these observations did not differ between inbred laboratory strains and wild zebrafish (Engeszer et al. 2008).

It is clear from the studies discussed above that visual cues play a major role in shoaling preference. The role of visual cues and learning in mating preference is less clear. Olfactory cues play an important role for mate choice that may dominate visual contributions (Gerlach 2006; Gerlach and Lysiak 2006).

## 7. CONCLUDING REMARKS

Much of the motivation to study vision in zebrafish originates from its status as one of the best-studied vertebrate model organisms. The rapid maturation of visual system development allows for behavioral genetic screens that will be increasingly refined in the future with advances in transgenic technology (Fadool and Dowling 2008). The ease of rapid and transient inactivation of any gene of choice by current and yet-to-be-developed technologies, such as morpholino based downregulation or siRNA approaches, allows for the rapid assessment of gene function in the context of vision. Currently these techniques are mainly applicable to the early larval stage, where the visual system is mature enough to enable study of most aspects of vision. An additional bonus is that at this stage vision is mainly cone driven, which ideally complements the extensive literature on rod vision and its genetic analysis of the nocturnal mouse retina.

Recently technologies have been developed to inactivate any gene of choice, allowing for heritable disruption for any gene of choice in the near future (Doyon et al. 2008; Meng et al. 2008; Moens et al. 2008). This will lift the limitation to study vision at the larval stage, when most isolated mutants are affected and where antisense technologies are applicable. In order to fully benefit from these advances in zebrafish gene technology, a more thorough understanding of the adult retina including robust visual assays needs to be developed.

Our physiological understanding of the retina lags far behind our genetic knowledge. The transparent larva will lend itself ideally for the application of optic imaging technologies to study the neural circuit in the behaving larvae. First steps have been undertaken in this direction and will likely revolutionize our understanding of tectal function. Similarly the explanatory power of cellular physiology has only been sporadically employed in the zebrafish. This technology in combination with live labeling techniques of identified cell types in the retina should deepen our understanding of retinal function.

Much research, not to speak of financial research support, is motivated by biomedical considerations (Gross and Perkins 2008). There are many heritable eye diseases, many of them progressive and surprisingly common that lack any therapeutic potential so far. In particular, age-related retinal dystrophies, foremost being age-related macular degeneration, will be of increasing medial concern for the aging societies in the developed world. The well-developed genetic technology of the zebrafish allows for the generation of disease models that can be studied in unprecedented detail. A particular advantage is the fast development of the zebrafish, which facilitates the study of progressive diseases.

In order to take full advantage of the zebrafish model system, a much more thorough understanding of its natural ecology is needed. Such an understanding is indispensable in guiding appropriate behavioral studies in the laboratory, clearly not restricted to visual behavior (Miklosi and Andrew 2006; Spence et al. 2008). Finally, the visual system lends itself ideally to study evolution of vision. With the enormous genomic resources available for many species across the tree of life, we can now begin to apply a comparative analysis taking genomic and functional information into account. In this context comparative approaches using the Medaka fish (*Oryzias latipes*), another genetically well-described species, will prove to be useful (Mitani et al. 2006).

Impressive progress in understanding vision has been achieved by using the zebrafish as a vision model. Now the foundation has been laid to take full advantage of multi-disciplinary approaches to further deepen our knowledge about the workings of vertebrate vision, including our own.

## ACKNOWLEDGMENTS

I would like to thank Jim Fadool, Matthias Gesemann, Edda Kastenhuber, and Corinne Hodel for helpful comments on the manuscript. Images were kindly provided by Chi-Bin Chien (Fig. 3.3), Corinne Hodel (Figs 3.2 and 3.5), and Colette Maurer (Fig. 3.4). Work in the author's laboratory is supported by the Swiss National Science Foundation and the European Commission (Integrated 6th framework project ZF-MODELS, 7th framework project RETICIRC).

## REFERENCES

Akert, K. (1949). Der visuelle Greifreflex. *Helv. Physiol. Pharmacol. Acta* **7**, 112–134.

Ali, M. A. (1971). Retinomotor response: Characteristics and mechanisms. *Vis. Res.* **11**, 1225–1288.

Ali, M. A., Harosi, F. I., and Wagner, H. J. (1978). Photoreceptors and visual pigments in a cichlid fish, Nannacara anomala. *Sens. Processes* **2**, 130–145.

Allison, W. T., Haimberger, T. J., Hawryshyn, C. W., and Temple, S. E. (2004). Visual pigment composition in zebrafish: Evidence for a rhodopsin-porphyropsin interchange system. *Vis. Neurosci.* **21**, 945–952.

Allwardt, B. A., Lall, A. B., Brockerhoff, S. E., and Dowling, J. E. (2001). Synapse formation is arrested in retinal photoreceptors of the zebrafish nrc mutant. *J. Neurosci.* **21**, 2330–2342.

Baier, H., Klostermann, S., Trowe, T., Karlstrom, R. O., Nusslein-Volhard, C., and Bonhoeffer, F. (1996). Genetic dissection of the retinotectal projection. *Development* **123**, 415–425.

Balm, P. H., and Groneveld, D. (1998). The melanin-concentrating hormone system in fish. *Ann. N.Y. Acad. Sci.* **839**, 205–209.

Barth, K. A., Miklosi, A., Watkins, J., Bianco, I. H., Wilson, S. W., and Andrew, R. J. (2005). fsi zebrafish show concordant reversal of laterality of viscera, neuroanatomy, and a subset of behavioral responses. *Curr. Biol.* **15**, 844–850.

Beck, J. C., Gilland, E., Tank, D. W., and Baker, R. (2004). Quantifying the ontogeny of optokinetic and vestibuloocular behaviors in zebrafish, medaka, and goldfish. *J. Neurophysiol.* **92**, 3546–3561.

Bellingham, J., Chaurasia, S. S., Melyan, Z., Liu, C., Cameron, M. A., Tarttelin, E. E., Iuvone, P. M., Hankins, M. W., Tosini, G., and Lucas, R. J. (2006). Evolution of melanopsin photoreceptors: Discovery and characterization of a new melanopsin in nonmammalian vertebrates. *PLoS Biol.* **4**, e254.

Bellingham, J., Whitmore, D., Philp, A. R., Wells, D. J., and Foster, R. G. (2002). Zebrafish melanopsin: Isolation, tissue localisation and phylogenetic position. *Brain Res. Mol. Brain Res.* **107**, 128–136.

Bernardos, R. L., Barthel, L. K., Meyers, J. R., and Raymond, P. A. (2007). Late-stage neuronal progenitors in the retina are radial Muller glia that function as retinal stem cells. *J. Neurosci.* **27**, 7028–7040.

Bianco, I. H., and Wilson, S. W. (2009). The habenular nuclei: A conserved asymmetric relay station in the vertebrate brain. *Philos. Trans. R. Soc. Lond. B Biol. Sci.* **364**, 1005–1020.

Biehlmaier, O., Makhankov, Y., and Neuhauss, S. C. (2007). Impaired retinal differentiation and maintenance in zebrafish laminin mutants. *Invest. Ophthalmol. Vis. Sci.* **48**, 2887–2894.

Biehlmaier, O., Neuhauss, S. C., and Kohler, K. (2003). Synaptic plasticity and functionality at the cone terminal of the developing zebrafish retina. *J. Neurobiol.* **56**, 222–236.

Bilotta, J. (2000). Effects of abnormal lighting on the development of zebrafish visual behavior. *Behav. Brain Res.* **116**, 81–87.

Bilotta, J., Saszik, S., and Sutherland, S. E. (2001). Rod contributions to the electroretinogram of the dark-adapted developing zebrafish. *Dev. Dyn.* **222**, 564–570.

Branchek, T. (1984). The development of photoreceptors in the zebrafish, brachydanio rerio. II. Function. *J. Comp. Neurol.* **224**, 116–122.

Brockerhoff, S. E., Dowling, J. E., and Hurley, J. B. (1998). Zebrafish retinal mutants. *Vision Res.* **38**, 1335–1339.

Brockerhoff, S. E., Hurley, J. B., Janssen-Bienhold, U., Neuhauss, S. C., Driever, W., and Dowling, J. E. (1995). A behavioral screen for isolating zebrafish mutants with visual system defects. *Proc. Natl Acad. Sci. USA* **92**, 10545–10549.

Brockerhoff, S. E., Hurley, J. B., Niemi, G. A., and Dowling, J. E. (1997). A new form of inherited red-blindness identified in zebrafish. *J. Neurosci.* **17**, 4236–4242.

Brockerhoff, S. E., Rieke, F., Matthews, H. R., Taylor, M. R., Kennedy, B., Ankoudinova, I., Niemi, G. A., Tucker, C. L., Xiao, M., Cilluffo, M. C., et al. (2003). Light stimulates a transducin-independent increase of cytoplasmic $Ca^{2+}$ and suppression of current in cones from the zebrafish mutant nof. *J. Neurosci.* **23**, 470–480.

Burrill, J. D., and Easter, S. S., Jr. (1994). Development of the retinofugal projections in the embryonic and larval zebrafish (Brachydanio rerio). *J. Comp. Neurol.* **346**, 583–600.

Cahill, G. M. (1996). Circadian regulation of melatonin production in cultured zebrafish pineal and retina. *Brain Res.* **708**, 177–181.

Cameron, D. A. (2002). Mapping absorbance spectra, cone fractions, and neuronal mechanisms to photopic spectral sensitivity in the zebrafish. *Vis. Neurosci.* **19**, 365–372.

Cameron, D. A., Gentile, K. L., Middleton, F. A., and Yurco, P. (2005). Gene expression profiles of intact and regenerating zebrafish retina. *Mol. Vis.* **11**, 775–791.

Cavanagh, P., and Mather, G. (1989). Motion: The long and short of it. *Spat. Vis.* **4**, 103–129.

Chinen, A., Hamaoka, T., Yamada, Y., and Kawamura, S. (2003). Gene duplication and spectral diversification of cone visual pigments of zebrafish. *Genetics* **163**, 663–675.

Chubb, C., and Sperling, G. (1988). Drift-balanced random stimuli: A general basis for studying non-Fourier motion perception. *J. Opt. Soc. Am. A* **5**, 1986–2007.

Clark, D. T. (1981). Visual responses in the developing zebrafish (Brachydanio rerio). Ph.D.: University of Oregon Press, Eugene.

Colwill, R. M., Raymond, M. P., Ferreira, L., and Escudero, H. (2005). Visual discrimination learning in zebrafish (Danio rerio). *Behav. Processes* **70**, 19–31.

Connaughton, V., and Nelson, R. (2007). Light responses from presumed horizontal and amacrine cells in zebrafish retina. *Invest. Ophthalmol. Vis. Sci.* **48**, 5957. E-abstract.

Connaughton, V. P., and Dowling, J. E. (1998). Comparative morphology of distal neurons in larval and adult zebrafish retinas. *Vis. Res.* **38**, 13–18.

Connaughton, V. P., Graham, D., and Nelson, R. (2004). Identification and morphological classification of horizontal, bipolar, and amacrine cells within the zebrafish retina. *J. Comp. Neurol.* **477**, 371–385.

Connaughton, V. P., and Nelson, R. (2000). Axonal stratification patterns and glutamate-gated conductance mechanisms in zebrafish retinal bipolar cells. *J. Physiol.* **524**(Pt 1), 135–146.

Culverwell, J., and Karlstrom, R. O. (2002). Making the connection: Retinal axon guidance in the zebrafish. *Semin. Cell Dev. Biol.* **13**, 497–506.

Dahm, R., Schonthaler, H. B., Soehn, A. S., van Marle, J., and Vrensen, G. F. (2007). Development and adult morphology of the eye lens in the zebrafish. *Exp. Eye Res.* **85**, 74–89.

Davis, R. E., and Klinger, P. D. (1987). Spatial discrimination in goldfish following bilateral tectal ablation. *Behav. Brain Res.* **25**, 255–260.

Douglas, R. H., and Hawryshyn, C. W. (1990). Behavioural studies of fish vision: An analysis of visual capabilities. In: *The Visual System of Fish* (R.H. Douglas and B.A. Djamgoz, eds), pp. 373–418. Chapman & Hall, London.

Dowling, J. E. (1987). *The Retina: An approachable Part of the Brain.* Belknap Press, Cambridge.

Doyon, Y., McCammon, J. M., Miller, J. C., Faraji, F., Ngo, C., Katibah, G. E., Amora, R., Hocking, T. D., Zhang, L., Rebar, E. J., et al. (2008). Heritable targeted gene disruption in zebrafish using designed zinc-finger nucleases. *Nat. Biotechnol.* **26**, 702–708.

Easter, S. S., Jr., and Nicola, G. N. (1996). The development of vision in the zebrafish (Danio rerio). *Dev. Biol.* **180**, 646–663.

Easter, S. S., Jr., and Nicola, G. N. (1997). The development of eye movements in the zebrafish (Danio rerio). *Dev. Psychobiol.* **31**, 267–276.

Emran, F., Rihel, J., Adolph, A. R., Wong, K. Y., Kraves, S., and Dowling, J. E. (2007). OFF ganglion cells cannot drive the optokinetic reflex in zebrafish. *Proc. Natl Acad. Sci. USA* **104**, 19126–19131.

Emran, F., Rihel, J. and Dowling, J. E. (2008). A behavioral assay to measure responsiveness of zebrafish to changes in light intensities. *J. Vis. Exp* (20), pii, 923.

Engeszer, R. E., Barbiano, L. A., Ryan, M. J., and Parichy, D. M. (2007a). Timing and plasticity of shoaling behaviour in the zebrafish, Danio rerio. *Anim. Behav.* **74**, 1269–1275.

Engeszer, R. E., Patterson, L. B., Rao, A. A., and Parichy, D. M. (2007b). Zebrafish in the wild: A review of natural history and new notes from the field. *Zebrafish* **4**, 21–40.

Engeszer, R. E., Ryan, M. J., and Parichy, D. M. (2004). Learned social preference in zebrafish. *Curr. Biol.* **14**, 881–884.

Engeszer, R. E., Wang, G., Ryan, M. J., and Parichy, D. M. (2008). Sex-specific perceptual spaces for a vertebrate basal social aggregative behavior. *Proc. Natl Acad. Sci. USA* **105**, 929–933.

Fadool, J. M. (2003). Development of a rod photoreceptor mosaic revealed in transgenic zebrafish. *Dev. Biol.* **258**, 277–290.

Fadool, J. M., and Dowling, J. E. (2008). Zebrafish: A model system for the study of eye genetics. *Prog. Retin. Eye Res.* **27**, 89–110.

Fausett, B. V., and Goldman, D. (2006). A role for alpha1 tubulin-expressing Muller glia in regeneration of the injured zebrafish retina. *J. Neurosci.* **26**, 6303–6313.

Fausett, B. V., Gumerson, J. D., and Goldman, D. (2008). The proneural basic helix-loop-helix gene ascl1a is required for retina regeneration. *J. Neurosci.* **28**, 1109–1117.

Fimbel, S. M., Montgomery, J. E., Burket, C. T., and Hyde, D. R. (2007). Regeneration of inner retinal neurons after intravitreal injection of ouabain in zebrafish. *J. Neurosci.* **27**, 1712–1724.

Flamarique, I. N., and Harosi, F. I. (2002). Visual pigments and dichroism of anchovy cones: A model system for polarization detection. *Vis. Neurosci.* **19**, 467–473.

Fleisch, V. C., and Neuhauss, S. C. (2006). Visual behavior in zebrafish. *Zebrafish* **3**, 191–201.

Fujii, R. (2000). The regulation of motile activity in fish chromatophores. *Pigment Cell Res.* **13**, 300–319.

Gahtan, E., Tanger, P., and Baier, H. (2005). Visual prey capture in larval zebrafish is controlled by identified reticulospinal neurons downstream of the tectum. *J. Neurosci.* **25**, 9294–9303.

Gerlach, G. (2006). Pheromonal regulation of reproductive success in female zebrafish: Female suppression and male enhancement. *Anim. Behav.* **72**, 1119–1124.

Gerlach, G., and Lysiak, N. (2006). Kin recognition and inbreeding avoidance in zebrafish, Danio rerio, is based on phenotype matching. *Anim. Behav.* **71**, 1371–1377.

Gerlai, R., Lahav, M., Guo, S., and Rosenthal, A. (2000). Drinks like a fish: Zebra fish (Danio rerio) as a behavior genetic model to study alcohol effects. *Pharmacol. Biochem. Behav.* **67**, 773–782.

Gnuegge, L., Schmid, S., and Neuhauss, S. C. (2001). Analysis of the activity-deprived zebrafish mutant macho reveals an essential requirement of neuronal activity for the development of a fine-grained visuotopic map. *J. Neurosci.* **21**, 3542–3548.

Gosse, N. J., Nevin, L. M., and Baier, H. (2008). Retinotopic order in the absence of axon competition. *Nature* **452**, 892–895.

Grant, G. B., and Dowling, J. E. (1995). A glutamate-activated chloride current in cone-driven ON bipolar cells of the white perch retina. *J. Neurosci.* **15**, 3852–3862.

Grant, G. B., and Dowling, J. E. (1996). On bipolar cell responses in the teleost retina are generated by two distinct mechanisms. *J. Neurophysiol.* **76**, 3842–3849.

Gray, M. P., Smith, R. S., Soules, K. A., John, S. W., and Link, B. A. (2009). The aqueous humor outflow pathway of zebrafish. *Invest. Ophthalmol. Vis. Sci.* **50**, 1515–1521.

Greiling, T. M. and Clark, J. I. (2009). Early lens development in the zebrafish: A three-dimensional time-lapse analysis. *Dev. Dyn* **238**, 2254–2265.

Gross, J. M., and Perkins, B. D. (2008). Zebrafish mutants as models for congenital ocular disorders in humans. *Mol. Reprod. Dev.* **75**, 547–555.

Gross, J. M., Perkins, B. D., Amsterdam, A., Egana, A., Darland, T., Matsui, J. I., Sciascia, S., Hopkins, N., and Dowling, J. E. (2005). Identification of zebrafish insertional mutants with defects in visual system development and function. *Genetics* **170**, 245–261.

Haffter, P., Granato, M., Brand, M., Mullins, M. C., Hammerschmidt, M., Kane, D. A., Odenthal, J., van Eeden, F. J., Jiang, Y. J., Heisenberg, C. P., et al. (1996). The identification of genes with unique and essential functions in the development of the zebrafish, Danio rerio. *Development* **123**, 1–36.

Hall, W. C., and Moschovakis, A. (2003). *The Superior Colliculus: New Approaches for Studying Sensorimotor Integration.* CRC Press.

Hamaoka, T., Takechi, M., Chinen, A., Nishiwaki, Y., and Kawamura, S. (2002). Visualization of rod photoreceptor development using GFP-transgenic zebrafish. *Genesis* **34**, 215–220.

Haverkamp, S., Grunert, U., and Wassle, H. (2000). The cone pedicle, a complex synapse in the retina. *Neuron* **27**, 85–95.

Hawryshyn, C. W. (2000). Ultraviolet polarization vision in fishes: Possible mechanisms for coding e-vector. *Philos. Trans. R. Soc. Lond. B Biol. Sci.* **355**, 1187–1190.

Hitchcock, P. F., and Raymond, P. A. (2004). The teleost retina as a model for developmental and regeneration biology. *Zebrafish* **1**, 257–271.

Hodel, C., Neuhauss, S. C., and Biehlmaier, O. (2006). Time course and development of light adaptation processes in the outer zebrafish retina. *Anat. Rec. A Discov. Mol. Cell Evol. Biol.* **288**, 653–662.

Holst, E. v. (1935). Über den Lichtrückenreflex bei Fischen. *Pubbl. Staz. Zool. Napoli.* **15**, 143–158.

Hu, M., and Easter, S. S. (1999). Retinal neurogenesis: The formation of the initial central patch of postmitotic cells. *Dev. Biol.* **207**, 309–321.

Hua, J. Y., Smear, M. C., Baier, H., and Smith, S. J. (2005). Regulation of axon growth in vivo by activity-based competition. *Nature* **434**, 1022–1026.

Huang, L., Maaswinkel, H., and Li, L. (2005). Olfactoretinal centrifugal input modulates zebrafish retinal ganglion cell activity: A possible role for dopamine-mediated $Ca^{2+}$ signalling pathways. *J. Physiol.* **569**, 939–948.

Huang, Y. Y., and Neuhauss, S. C. (2008). The optokinetic response in zebrafish and its applications. *Front Biosci.* **13**, 1899–1916.

Huber, G. C., and Crosby, E. C. (1933). A Phylogenetic consideration of the optic tectum. *Proc. Natl Acad. Sci. USA* **19**, 15–22.

Hughes, A., Saszik, S., Bilotta, J., Demarco, P. J., Jr., and Patterson, W. F., 2nd (1998). Cone contributions to the photopic spectral sensitivity of the zebrafish ERG. *Vis. Neurosci.* **15**, 1029–1037.

Hutson, L. D., Campbell, D. S., and Chien, C. B. (2004). Analyzing axon guidance in the zebrafish retinotectal system. *Methods Cell Biol.* **76**, 13–35.

Jain, A. K., and Bhargava, H. N. (1978). Transitory colour-change mechanism in a fresh-water teleost, Clarias batrachus (L). *Biochem. Exp. Biol.* **14**, 263–269.

Joselevitch, C. and Kamermans, M. (2008). Retinal parallel pathways: Seeing with our inner fish. *Vis. Res* **40**, 943–959.

Kamermans, M., Joselevitch, C., and Klooster, J. (2004). Rod-driven light-responses in mixed-inout bipolar cells. *Invest. Ophthalmol. Vis. Sci.* **45**, 2198. E-abstract.

Kassen, S. C., Ramanan, V., Montgomery, J. E., Burket, C. T., Liu, C. G., Vihtelic, T. S., and Hyde, D. R. (2007). Time course analysis of gene expression during light-induced photoreceptor cell death and regeneration in albino zebrafish. *Dev. Neurobiol.* **67**, 1009–1031.

Kelsh, R. N., Brand, M., Jiang, Y. J., Heisenberg, C. P., Lin, S., Haffter, P., Odenthal, J., Mullins, M. C., van Eeden, F. J., Furutani-Seiki, M., et al. (1996). Zebrafish pigmentation mutations and the processes of neural crest development. *Development* **123**, 369–389.

Kennedy, M. J., Dunn, F. A., and Hurley, J. B. (2004). Visual pigment phosphorylation but not transducin translocation can contribute to light adaptation in zebrafish cones. *Neuron* **41**, 915–928.

Kimmel, C., Patterson, J., and Kimmel, R. (1974). The development and behavioral characteristics of the startle response in the zebra fish. *Dev. Psychobiol.* **7**, 47–60.

Koch, K. W. (1992). Biochemical mechanism of light adaptation in vertebrate photoreceptors. *Trends Biochem. Sci.* **17**, 307–311.

Kohler, K., Kolbinger, W., Kurz-Isler, G., and Weiler, R. (1990). Endogenous dopamine and cyclic events in the fish retina, II: Correlation of retinomotor movement, spinule formation, and connexon density of gap junctions with dopamine activity during light/dark cycles. *Vis. Neurosci.* **5**, 417–428.

Kojima, D., Mano, H., and Fukada, Y. (2000). Vertebrate ancient-long opsin: A green-sensitive photoreceptive molecule present in zebrafish deep brain and retinal horizontal cells. *J. Neurosci.* **20**, 2845–2851.

Kojima, D., Torii, M., Fukada, Y., and Dowling, J. E. (2008). Differential expression of duplicated VAL-opsin genes in the developing zebrafish. *J. Neurochem.* **104**, 1364–1371.

Krauss, A., and Neumeyer, C. (2003). Wavelength dependence of the optomotor response in zebrafish (Danio rerio). *Vis. Res.* **43**, 1273–1282.

Kusmic, C., and Gualtieri, P. (2000). Morphology and spectral sensitivities of retinal and extraretinal photoreceptors in freshwater teleosts. *Micron* **31**, 183–200.

Lemke, G., and Reber, M. (2005). Retinotectal mapping: New insights from molecular genetics. *Annu. Rev. Cell Dev. Biol.* **21**, 551–580.

Li, L., and Dowling, J. E. (1997). A dominant form of inherited retinal degeneration caused by a non-photoreceptor cell-specific mutation. *Proc. Natl. Acad. Sci. USA.* **94**, 11645–11650.

Li, L., and Dowling, J. E. (1998). Zebrafish visual sensitivity is regulated by a circadian clock. *Vis. Neurosci.* **15**, 851–857.

Li, L., and Dowling, J. E. (2000a). Disruption of the olfactoretinal centrifugal pathway may relate to the visual system defect in night blindness b mutant zebrafish. *J. Neurosci.* **20**, 1883–1892.

Li, L., and Dowling, J. E. (2000b). Effects of dopamine depletion on visual sensitivity of zebrafish. *J. Neurosci.* **20**, 1893–1903.

Li, Y. N., Matsui, J. I., and Dowling, J. E. (2009). Selectivity in the horizontal cell-photoreceptor connections in the zebrafish retina. *Invest. Ophthalmol. Vis. Sci.* **50**, 1042. E-abstract.

Lyall, A. H. (1957). Cone arrangements in teleost retinae. *Q. J. Mic. Sci.* **98**, 189–201.

Maaswinkel, H., and Li, L. (2003). Spatio-temporal frequency characteristics of the optomotor response in zebrafish. *Vis. Res.* **43**, 21–30.

Maaswinkel, H., Ren, J. Q., and Li, L. (2003). Slow-progressing photoreceptor cell degeneration in night blindness c mutant zebrafish. *J. Neurocytol.* **32**, 1107–1116.

Maaswinkel, H., Riesbeck, L. E., Riley, M. E., Carr, A. L., Mullin, J. P., Nakamoto, A. T., and Li, L. (2005). Behavioral screening for nightblindness mutants in zebrafish reveals three new loci that cause dominant photoreceptor cell degeneration. *Mech. Ageing Dev.* **126**, 1079–1089.

Mangrum, W. I., Dowling, J. E., and Cohen, E. D. (2002). A morphological classification of ganglion cells in the zebrafish retina. *Vis. Neurosci.* **19**, 767–779.

Mano, H., Kojima, D., and Fukada, Y. (1999). Exo-rhodopsin: A novel rhodopsin expressed in the zebrafish pineal gland. *Brain Res. Mol. Brain Res.* **73**, 110–118.

Marc, R. E., and Cameron, D. (2001). A molecular phenotype atlas of the zebrafish retina. *J. Neurocytol.* **30**, 593–654.

Marcus, R. C., Delaney, C. L., and Easter, S. S., Jr. (1999). Neurogenesis in the visual system of embryonic and adult zebrafish (Danio rerio). off. *Vis. Neurosci.* **16**, 417–424.

Masu, M., Iwakabe, H., Tagawa, Y., Miyoshi, T., Yamashita, M., Fukuda, Y., Sasaki, H., Hiroi, K., Nakamura, Y., Shigemoto, R., et al. (1995). Specific deficit of the ON response in visual transmission by targeted disruption of the mGluR6 gene. *Cell* **80**, 757–765.

McCann, L. I., and Carlson, C. C. (1982). Effect of cross-rearing on species identification in zebra fish and pearl danios. *Dev. Psychobiol.* **15**, 71–74.

McClure, M. M., McIntyre, P. B., and McCune, A. R. (2006). Notes on the natural diet and habitat of eight danionin fishes, including the zebrafish, Danio rerio. *J. Fish Biol.* **69**, 553–570.

McDowell, A. L., Dixon, L. J., Houchins, J. D., and Bilotta, J. (2004). Visual processing of the zebrafish optic tectum before and after optic nerve damage. *Vis. Neurosci.* **21**, 97–106.

McElligott, M. B., and O'Malley, D. M. (2005). Prey tracking by larval zebrafish: Axial kinematics and visual control. *Brain Behav. Evol.* **66**, 177–196.

Meng, X., Noyes, M. B., Zhu, L. J., Lawson, N. D., and Wolfe, S. A. (2008). Targeted gene inactivation in zebrafish using engineered zinc-finger nucleases. *Nat. Biotechnol.* **26**, 695–701.

Menger, G. J., Koke, J. R., and Cahill, G. M. (2005). Diurnal and circadian retinomotor movements in zebrafish. *Vis. Neurosci.* **22**, 203–209.

Meyer, M. P., and Smith, S. J. (2006). Evidence from in vivo imaging that synaptogenesis guides the growth and branching of axonal arbors by two distinct mechanisms. *J. Neurosci.* **26**, 3604–3614.

Miklosi, A., and Andrew, R. J. (1999). Right eye use associated with decision to bite in zebrafish. *Behav. Brain Res.* **105**, 199–205.

Miklosi, A., and Andrew, R. J. (2006). The zebrafish as a model for behavioral studies. *Zebrafish* 3, 227–231.

Miklosi, A., Andrew, R. J., and Gasparini, S. (2001). Role of right hemifield in visual control of approach to target in zebrafish. *Behav. Brain Res.* **122**, 57–65.

Miklosi, A., Andrew, R. J., and Savage, H. (1997). Behavioural lateralisation of the tetrapod type in the zebrafish (Brachydanio rerio). *Physiol. Behav.* **63**, 127–135.

Mills, I. A., Adolph, A. R., and Dowling, J. E. (2009). Rods are Functional in 5-6 Day Old Larval Zebrafish. *Invest. Ophthalmol. Vis. Sci.* **50**, 4566. E-abstract.

Mitani, H., Kamei, Y., Fukamachi, S., Oda, S., Sasaki, T., Asakawa, S., Todo, T., and Shimizu, N. (2006). The medaka genome: Why we need multiple fish models in vertebrate functional genomics. *Genome Dyn.* **2**, 165–182.

Moens, C. B., Donn, T. M., Wolf-Saxon, E. R., and Ma, T. P. (2008). Reverse genetics in zebrafish by TILLING. *Brief. Funct. Genomic. Proteomic.* **7**, 454–459.

Moutsaki, P., Whitmore, D., Bellingham, J., Sakamoto, K., David-Gray, Z. K., and Foster, R. G. (2003). Teleost multiple tissue (tmt) opsin: A candidate photopigment regulating the peripheral clocks of zebrafish? *Brain Res. Mol. Brain Res.* **112**, 135–145.

Muto, A., Orger, M. B., Wehman, A. M., Smear, M. C., Kay, J. N., Page-McCaw, P. S., Gahtan, E., Xiao, T., Nevin, L. M., Gosse, N. J., et al. (2005). Forward genetic analysis of visual behavior in zebrafish. *PLoS Genet.* **1**, e66.

Nascimento, A. A., Roland, J. T., and Gelfand, V. I. (2003). Pigment cells: A model for the study of organelle transport. *Annu. Rev. Cell Dev. Biol.* **19**, 469–491.

Nasevicius, A., and Ekker, S. C. (2000). Effective targeted gene 'knockdown' in zebrafish. *Nat. Genet.* **26**, 216–220.

Nawrocki, L., BreMiller, R., Streisinger, G., and Kaplan, M. (1985). Larval and adult visual pigments of the zebrafish, Brachydanio rerio. *Vis. Res.* **25**, 1569–1576.

Neuhauss, S. C. (2003). Behavioral genetic approaches to visual system development and function in zebrafish. *J. Neurobiol.* **54**, 148–160.

Neuhauss, S. C., Biehlmaier, O., Seeliger, M. W., Das, T., Kohler, K., Harris, W. A., and Baier, H. (1999). Genetic disorders of vision revealed by a behavioral screen of 400 essential loci in zebrafish. *J. Neurosci.* **19**, 8603–8615.

Neumeyer, C. (1984). On spectral sensitivity in the goldfish. Evidence for neural interactions between different "cone mechanisms". *Vis. Res.* **24**, 1223–1231.

Nicolson, T., Rusch, A., Friedrich, R. W., Granato, M., Ruppersberg, J. P., and Nusslein-Volhard, C. (1998). Genetic analysis of vertebrate sensory hair cell mechanosensation: The zebrafish circler mutants. *Neuron* **20**, 271–283.

Niell, C. M., and Smith, S. J. (2005). Functional imaging reveals rapid development of visual response properties in the zebrafish tectum. *Neuron* **45**, 941–951.

Orger, M. B., and Baier, H. (2005). Channeling of red and green cone inputs to the zebrafish optomotor response. *Vis. Neurosci.* **22**, 275–281.

Orger, M. B., Smear, M. C., Anstis, S. M., and Baier, H. (2000). Perception of Fourier and non-Fourier motion by larval zebrafish. *Nat. Neurosci.* **3**, 1128–1133.

Ott, M. (2006). Visual accommodation in vertebrates: Mechanisms, physiological response and stimuli. *J. Comp. Physiol. A Neuroethol. Sens. Neural. Behav. Physiol.* **192**, 97–111.

Ott, M., Walz, B. C., Paulsen, U. J., Mack, A. F., and Wagner, H. J. (2007). Retinotectal ganglion cells in the zebrafish, Danio rerio. *J. Comp. Neurol.* **501**, 647–658.

Patterson, W. F., 2nd, McDowell, A. L., Hughes, A., and Bilotta, J. (2002). Opponent and nonopponent contributions to the zebrafish electroretinogram using heterochromatic flicker photometry. *J. Comp. Physiol. A Neuroethol. Sens. Neural Behav. Physiol.* **188**, 283–293.

Powers, M. K. (1978). Light-adapted spectral sensitivity of the goldfish: A reflex measure. *Vis. Res.* **18**, 1131–1136.

Prober, D. A., Rihel, J., Onah, A. A., Sung, R. J., and Schier, A. F. (2006). Hypocretin/orexin overexpression induces an insomnia-like phenotype in zebrafish. *J. Neurosci.* **26**, 13400–13410.

Pugh, E. N., Jr., Nikonov, S., and Lamb, T. D. (1999). Molecular mechanisms of vertebrate photoreceptor light adaptation. *Curr. Opin. Neurobiol.* **9**, 410–418.

Qian, H., Zhu, Y., Ramsey, D. J., Chappell, R. L., Dowling, J. E., and Ripps, H. (2005). Directional asymmetries in the optokinetic response of larval zebrafish (Danio rerio). *Zebrafish* **2**, 189–196.

Qin, Z., Barthel, L. K. and Raymond, P. A. (2009). Genetic evidence for shared mechanisms of epimorphic regeneration in zebrafish. *Proc. Natl Acad. Sci. USA* **106**, 9310–9315.

Rajendran, R. R., Van Niel, E. E., Stenkamp, D. L., Cunningham, L. L., Raymond, P. A., and Gonzalez-Fernandez, F. (1996). Zebrafish interphotoreceptor retinoid-binding protein: Differential circadian expression among cone subtypes. *J. Exp. Biol.* **199**, 2775–2787.

Ramachandran, V. S., Rao, V. M., and Vidyasagar, T. R. (1973). Apparent movement with subjective contours. *Vis. Res.* **13**, 1399–1401.

Ramdya, P., and Engert, F. (2008). Emergence of binocular functional properties in a monocular neural circuit. *Nat. Neurosci.* **11**, 1083–1090.

Raymond, P. A., Barthel, L. K., Bernardos, R. L., and Perkowski, J. J. (2006). Molecular characterization of retinal stem cells and their niches in adult zebrafish. *BMC Dev. Biol.* **6**, 36.

Raymond, P. A., Barthel, L. K., Rounsifer, M. E., Sullivan, S. A., and Knight, J. K. (1993). Expression of rod and cone visual pigments in goldfish and zebrafish: A rhodopsin-like gene is expressed in cones. *Neuron* **10**, 1161–1174.

Rinner, O., Makhankov, Y. V., Biehlmaier, O., and Neuhauss, S. C. (2005a). Knockdown of cone-specific kinase GRK7 in larval zebrafish leads to impaired cone response recovery and delayed dark adaptation. *Neuron* **47**, 231–242.

Rinner, O., Rick, J. M., and Neuhauss, S. C. (2005b). Contrast sensitivity, spatial and temporal tuning of the larval zebrafish optokinetic response. *Invest. Ophthalmol. Vis. Sci.* **46**, 137–142.

Risner, M. L., Lemerise, E., Vukmanic, E. V., and Moore, A. (2006). Behavioral spectral sensitivity of the zebrafish (Danio rerio). *Vis. Res.* **46**, 2625–2635.

Robinson, J., Schmitt, E. A., Harosi, F. I., Reece, R. J., and Dowling, J. E. (1993). Zebrafish ultraviolet visual pigment: Absorption spectrum, sequence, and localization. *Proc. Natl Acad. Sci. USA* **90**, 6009–6012.

Roeser, T., and Baier, H. (2003). Visuomotor behaviors in larval zebrafish after GFP-guided laser ablation of the optic tectum. *J. Neurosci.* **23**, 3726–3734.

Ruthazer, E. S., and Cline, H. T. (2004). Insights into activity-dependent map formation from the retinotectal system: A middle-of-the-brain perspective. *J. Neurobiol.* **59**, 134–146.

Sajovic, P., and Levinthal, C. (1982a). Visual cells of zebrafish optic tectum: Mapping with small spots. *Neuroscience* **7**, 2407–2426.

Sajovic, P., and Levinthal, C. (1982b). Visual response properties of zebrafish tectal cells. *Neuroscience* **7**, 2427–2440.

Sajovic, P., and Levinthal, C. (1983). Inhibitory mechanism in zebrafish optic tectum: Visual response properties of tectal cells altered by picrotoxin and bicuculline. *Brain Res.* **271**, 227–240.

Salas, C., Herrero, L., Rodriguez, F., and Torres, B. (1997). Tectal codification of eye movements in goldfish studied by electrical microstimulation. *Neuroscience* **78**, 271–288.

Saszik, S., Bilotta, J., and Givin, C. M. (1999). ERG assessment of zebrafish retinal development. *Vis. Neurosci.* **16**, 881–888.

Schaerer, S., and Neumeyer, C. (1996). Motion detection in goldfish investigated with the optomotor response is "color blind". *Vis. Res.* **36**, 4025–4034.

Schmitt, E. A., and Dowling, J. E. (1994). Early eye morphogenesis in the zebrafish, Brachydanio rerio. *J. Comp. Neurol.* **344**, 532–542.

Schmitt, E. A., and Dowling, J. E. (1999). Early retinal development in the zebrafish, Danio rerio: Light and electron microscopic analyses. *J. Comp. Neurol.* **404**, 515–536.

Serra, E. L., Medalha, C. C., and Mattioli, R. (1999). Natural preference of zebrafish (Danio rerio) for a dark environment. *Braz. J. Med. Biol. Res.* **32**, 1551–1553.

Shen, Y., Heimel, J. A., Kamermans, M., Peachey, N. S., Gregg, R. G., and Nawy, S. (2009). A transient receptor potential-like channel mediates synaptic transmission in rod bipolar cells. *J. Neurosci.* **29**, 6088–6093.

Song, P. I., Matsui, J. I., and Dowling, J. E. (2008). Morphological types and connectivity of horizontal cells found in the adult zebrafish (Danio rerio) retina. *J. Comp. Neurol.* **506**, 328–338.

Soni, B. G., and Foster, R. G. (1997). A novel and ancient vertebrate opsin. *FEBS Lett.* **406**, 279–283.

Soules, K. A., and Link, B. A. (2005). Morphogenesis of the anterior segment in the zebrafish eye. *BMC Dev. Biol.* **5**, 12.

Sovrano, V. A., and Andrew, R. J. (2006). Eye use during viewing a reflection: Behavioural lateralisation in zebrafish larvae. *Behav. Brain Res.* **167**, 226–231.

Spence, R., Gerlach, G., Lawrence, C., and Smith, C. (2008). The behaviour and ecology of the zebrafish, Danio rerio. *Biol. Rev. Camb. Philos. Soc.* **83**, 13–34.

Spence, R., and Smith, C. (2007). The role of early learning in determining shoaling preferences based on visual cues in the zebrafish, Danio rerio. *Ethology* **113**, 62–67.

Springer, A. D., Easter, S. S., Jr., and Agranoff, B. W. (1977). The role of the optic tectum in various visually mediated behaviors of goldfish. *Brain Res.* **128**, 393–404.

Stenkamp, D. L., Powers, M. K., Carney, L. H., and Cameron, D. A. (2001). Evidence for two distinct mechanisms of neurogenesis and cellular pattern formation in regenerated goldfish retinas. *J. Comp. Neurol.* **431**, 363–381.

Stuermer, C. A. (1988). Retinotopic organization of the developing retinotectal projection in the zebrafish embryo. *J. Neurosci.* **8**, 4513–4530.

Stuermer, C. A., Rohrer, B., and Munz, H. (1990). Development of the retinotectal projection in zebrafish embryos under TTX-induced neural-impulse blockade. *J. Neurosci.* **10**, 3615–3626.

Sumbre, G., Muto, A., Baier, H., and Poo, M. M. (2008). Entrained rhythmic activities of neuronal ensembles as perceptual memory of time interval. *Nature* **456**, 102–106.

Taylor, M. R., Kikkawa, S., Diez-Juan, A., Ramamurthy, V., Kawakami, K., Carmeliet, P., and Brockerhoff, S. E. (2005). The zebrafish pob gene encodes a novel protein required for survival of red cone photoreceptor cells. *Genetics* **170**, 263–273.

Turnell, E. R., Mann, K. D., Rosenthal, G. G., and Gerlach, G. (2003). Mate choice in zebrafish (Danio rerio) analyzed with video-stimulus techniques. *Biol. Bull.* **205**, 225–226.

Van Epps, H. A., Hayashi, M., Lucast, L., Stearns, G. W., Hurley, J. B., De Camilli, P., and Brockerhoff, S. E. (2004). The zebrafish nrc mutant reveals a role for the polyphosphoinositide phosphatase synaptojanin 1 in cone photoreceptor ribbon anchoring. *J. Neurosci.* **24**, 8641–8650.

Vihtelic, T. S., Doro, C. J., and Hyde, D. R. (1999). Cloning and characterization of six zebrafish photoreceptor opsin cDNAs and immunolocalization of their corresponding proteins. *Vis. Neurosci.* **16**, 571–585.

Vihtelic, T. S., Soverly, J. E., Kassen, S. C., and Hyde, D. R. (2006). Retinal regional differences in photoreceptor cell death and regeneration in light-lesioned albino zebrafish. *Exp. Eye Res.* **82**, 558–575.

von Bartheld, C. S., and Meyer, D. L. (1987). Comparative neurology of the optic tectum in ray-finned fishes: Patterns of lamination formed by retinotectal projections. *Brain Res.* **420**, 277–288.

Wada, Y., Sugiyama, J., Okano, T., and Fukada, Y. (2006). GRK1 and GRK7: Unique cellular distribution and widely different activities of opsin phosphorylation in the zebrafish rods and cones. *J. Neurochem.* **98**, 824–837.

Wagner, H. J. (1980). Light-dependent plasticity of the morphology of horizontal cell terminals in cone pedicles of fish retinas. *J. Neurocytol.* **9**, 573–590.

Wagner, H. J., and Djamgoz, M. B. (1993). Spinules: A case for retinal synaptic plasticity. *Trends Neurosci.* **16**, 201–206.

Wagner, H. J., and Wagner, E. (1988). Amacrine cells in the retina of a teleost fish, the roach (Rutilus rutilus): A Golgi study on differentiation and layering. *Philos. Trans. R. Soc. Lond. B Biol. Sci.* **321**, 263–324.

Wandell, B. A., Poirson, A. B., Newsome, W. T., Baseler, H. A., Boynton, G. M., Huk, A., Gandhi, S., and Sharpe, L. T. (1999). Color signals in human motion-selective cortex. *Neuron* **24**, 901–909.

Watkins, J., Miklosi, A., and Andrew, R. J. (2004). Early asymmetries in the behaviour of zebrafish larvae. *Behav. Brain Res.* **151**, 177–183.

Wehman, A. M., Staub, W., Meyers, J. R., Raymond, P. A., and Baier, H. (2005). Genetic dissection of the zebrafish retinal stem-cell compartment. *Dev. Biol.* **281**, 53–65.

Witkovsky, P. (2004). Dopamine and retinal function. *Doc. Ophthalmol.* **108**, 17–40.

Wong, K. Y., Adolph, A. R., and Dowling, J. E. (2005a). Retinal bipolar cell input mechanisms in giant danio. I. Electroretinographic analysis. *J. Neurophysiol.* **93**, 84–93.

Wong, K. Y., Cohen, E. D., and Dowling, J. E. (2005b). Retinal bipolar cell input mechanisms in giant danio. II. Patch-clamp analysis of on bipolar cells. *J. Neurophysiol.* **93**, 94–107.

Wong, K. Y., and Dowling, J. E. (2005). Retinal bipolar cell input mechanisms in giant danio. III. ON-OFF bipolar cells and their color-opponent mechanisms. *J. Neurophysiol.* **94**, 265–272.

Wong, K. Y., Gray, J., Hayward, C. J., Adolph, A. R., and Dowling, J. E. (2004). Glutamatergic mechanisms in the outer retina of larval zebrafish: Analysis of electroretinogram b- and d-waves using a novel preparation. *Zebrafish* **1**, 121–131.

Xiao, T., and Baier, H. (2007). Lamina-specific axonal projections in the zebrafish tectum require the type IV collagen Dragnet. *Nat. Neurosci.* **10**, 1529–1537.

Xiao, T., Roeser, T., Staub, W., and Baier, H. (2005). A GFP-based genetic screen reveals mutations that disrupt the architecture of the zebrafish retinotectal projection. *Development* **132**, 2955–2967.

Yager, D. (1968). Behavioural measures of the spectral sensitivity of the dark-adapted goldfish. *Nature* **220**, 1052–1053.

Yager, D., Sharma, S. C., and Grover, B. G. (1977). Visual function in goldfish with unilateral and bilateral tectal ablation. *Brain Res.* **137**, 267–275.

Yazulla, S., and Studholme, K. M. (2001). Neurochemical anatomy of the zebrafish retina as determined by immunocytochemistry. *J. Neurocytol.* **30**, 551–592.

Yurco, P., and Cameron, D. A. (2005). Responses of Muller glia to retinal injury in adult zebrafish. *Vis. Res.* **45**, 991–1002.

# 4

# THE ZEBRAFISH INNER EAR

*LEILA ABBAS*
*TANYA T. WHITFIELD*





*Zebrafish: Volume 29*
FISH PHYSIOLOGY

DOI: 10.1016/S1546-5098(10)02904-3

The zebrafish is recognized as an excellent model system for the study of the development of the vertebrate inner ear and its functions of balance and hearing. In the zebrafish embryo, the ear and lateral line have been used to study fundamental developmental processes such as induction, axial patterning, neurogenesis, cell migration and morphogenesis. With the increased interest in zebrafish as a disease model and for evolutionary comparisons, attention is now turning to larval, juvenile and adult stages, addressing the physiological issues of maturation, innervation, behavior, senescence and regeneration. In this review, we aim to describe both the early and later events during zebrafish ear development that lead to the fully functioning organ system found in the adult animal.

## 1. INTRODUCTION

### 1.1. Anatomy of the Adult Zebrafish Ear

Recent developments in optical microscopy (Santi et al. 2009), together with the use of a transparent adult zebrafish line (White et al. 2008), give us a beautifully clear picture of the three-dimensional arrangement of the two ears in situ in a mature zebrafish (Fig. 4.1). The adult ear consists of three orthogonally arranged semicircular canals ("anterior", "posterior" and "horizontal/lateral"), and three otolithic organs, the utricle, saccule and lagena, each containing a sensory patch or macula. The semicircular canals and utricle, together with the endolymphatic duct, form the "pars superior", the evolutionarily more ancient vestibular system of the ear, which is concerned with balance, acceleration and gravity-sensing. The saccule and lagena are the "pars inferior" and have the primary function of hearing. Each semicircular canal terminates in an ampulla, a widening of the canal containing a sensory crista perpendicular to the canal axis; the non-ampullary ends of the anterior and posterior semicircular canals merge medially to form the crus commune. The macula neglecta, a pair of small sensory patches without an otolith, is found medially at the base of the crus within the pars superior. Although fish have no outer or middle ear, and lack a more specialized hearing apparatus such as the basilar papilla in birds or the cochlea in mammals, the organs of balance and gravity-sensing remain functionally equivalent in all vertebrate groups (Fay and Popper 1999a, and references within).

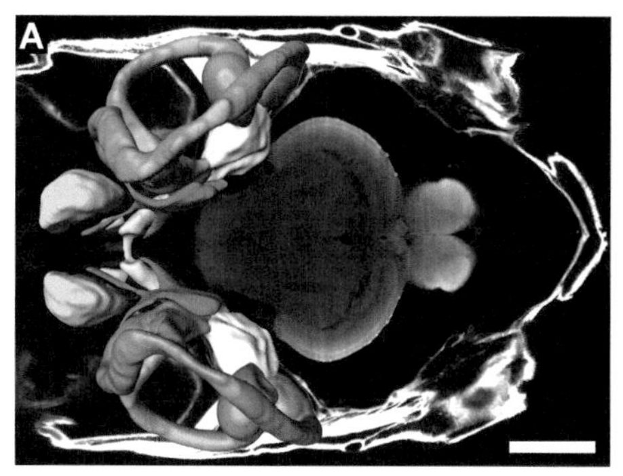

**Fig. 4.1.** Anatomy of the adult zebrafish inner ear. (A) Three-dimensional isosurface renderings of the two inner ears of an adult zebrafish, superimposed on an optical section through a transverse plane of the zebrafish head showing portions of the skull and brain. The images were obtained using a thin-sheet laser imaging optical microscope (TSLIM). Inner ear structures are represented by different colors: semicircular canals (blue), lagena (green), VIIIth nerve (red), saccule and transverse canal (gold), and utricle (yellow). Anterior to the right. Scale bar, 500 μm. © 2009 BioTechniques. Used by Permission (Santi et al. 2009). (B) Three-dimensional reconstruction of the two adult zebrafish inner ears generated using Amira software and conventional histological sections; this corresponds closely to the TSLIM image. Red, anterior semicircular canal; green, lateral/horizontal canal; blue, posterior canal; yellow, sensory chambers and transverse canal. Anterior to the right. Scale bar, 500 μm. Image in panel B courtesy of K. Hammond. See color plate section

The entire labyrinth is continuous and filled with endolymph, a specialized extracellular fluid enriched in $K^+$ ions. The saccule is continuous with the transverse canal, a conduit joining the left and right ear whose posterior end is enclosed blindly within the sinus impar, the main perilymphatic space of the ear (Bang et al. 2001). Dye injection studies have shown that as in other, more complex vertebrate hearing systems, the perilymphatic and endolymphatic compartments are kept separate to preserve the unique ionic composition of each fluid type (Bang et al. 2001). This allows the endolymph to bathe the apical surfaces of the sensory and supporting epithelia, whilst the perilymph, which is more akin to plasma in its ionic composition, bathes the basolateral faces of these cells. The electrochemical gradient between the interior of the cells and these two fluid-filled spaces flanking them is critical for driving ion flow through the hair cells, which allows the signal transduction required for hearing or vestibular sensation to occur.

### 1.2. Hearing in the Zebrafish

The otolithic organs are each overlain by an otolith, a calcified, proteinaceous mass suspended on an underlying sensory macula, cushioned by the gelatinous matrix of the otolithic membrane (see Section 7). The maculae are thickened patches of epithelium containing dense collections of sensory hair cells and supporting cells, arranged in an interspersed manner such that the hair cells are insulated from each other by the surrounding supporting cells. Otoliths have an inertia relative to the rest of the fish, since their higher density and suspension in the otolithic membrane causes them to lag when a sound wave passes through the body. Consequently, the macular epithelium vibrates with respect to the otolith and the sensory hair bundles are deflected: this mechanical stimulation triggers an influx of ions into the hair cell which brings about neurotransmitter release basolaterally, resulting in the relay of this initial signal to the auditory centers of the brain via the VIIIth nerve. This "direct" stimulation of the ear is analogous to the bone conduction found in other vertebrates and represents the only hearing mechanism available in many fish species (Popper and Liu 2000).

The zebrafish is a member of the otophysan group of fishes, known as "hearing specialists" on account of anatomical features allowing efficient sound conduction through the body. The swim bladder, a gas-filled bipartite sac in the abdominal cavity, contributes to "indirect" stimulation of the ear. A sound wave traveling through the water causes alternating rarefaction and compression of the swim bladder wall and this oscillation is functionally analogous to vibration of the tympanic membrane in the tetrapod middle ear. The swim bladder is connected to the saccule by the Weberian ossicles, a series of modified vertebrae that are functionally (although not developmentally) equivalent to the bones of the middle ear (Bird and Mabee 2003; Higgs et al. 2003; Grande and Young 2004) (see Section 8.3 for further details).

The lateral line is a series of sensory organs or neuromasts located superficially within the epidermis of the fish and (at later stages) within grooves or canals. Each neuromast consists of a group of mechanosensitive hair cells that share many anatomical and functional features with hair cells in the inner ear. Lateral line neuromasts act as a local flow detector: when swimming, the fish generates a laminar flow along the body that is detected by its own lateral line. The zebrafish lateral line is an important model system for the study of cell migration, disease modeling, hair cell regeneration and toxicology, and has been the subject of several excellent recent reviews (Ghysen and Dambly-Chaudiere 2007; Aman and Piotrowski 2009; Froehlicher et al. 2009; Ma and Raible 2009); it will not be covered, except in passing, in this chapter. On account of their similar embryological origins, anatomical features and neuronal pathways, the auditory system, the equilibrium-sensing vestibular system and the lateral line are sometimes collectively referred to as the octavolateral system in the fish.

## 2. DEVELOPMENT OF THE INNER EAR

### 2.1. Otic Induction and Placode Formation

The inner ear arises from an ectodermal thickening, the otic placode, which, like other sensory placodes, originates in a common preplacodal region (PPR) surrounding the neural plate at the end of gastrulation (about 10 hours post fertilization (hpf)). Specification of the PPR, which is marked by the expression of *dlx*, *eya*, *six,* and *irx* family genes, is known to be dependent on the correct levels of both bone morphogenetic protein (BMP) and fibroblast growth factor (FGF) signaling (Schimmang 2007, Esterberg and Fritz 2009 and references within). A recent study in zebrafish suggests that the *crossveinless-2* (*cv2*) gene product, acting downstream of *dlx3b/4b*, plays an important role in the modulation of BMP activity during PPR specification. Morpholino knockdown and DNA injection experiments demonstrated that *cv2* and *dlx3b/4b* are both required and sufficient to drive expression of the PPR markers *eya1* and *six4.1* (Esterberg and Fritz 2009).

Induction of the otic placode from the PPR occurs during early somite stages in zebrafish, and has been well studied. For an otic fate to be realized, a combination of competence factors (Foxi1, Dlx3b/4b) expressed within the PPR, together with inducing factors (Fgf3, Fgf8 and BMP antagonists) expressed in adjacent tissues (developing hindbrain and mesendoderm), is required (Mendonsa and Riley 1999; Phillips et al. 2001, 2004; Léger and Brand 2002; Maroon et al. 2002; Liu et al. 2003; Nissen et al. 2003; Solomon et al. 2003, 2004; Mackereth et al. 2005; Hans et al. 2007; Esterberg and Fritz 2009; Kwon and Riley 2009). Loss of function of either competence or inducing factors results in the loss or reduction of otic tissue. The otic placode itself continues to express *dlx*, *eya* and *six* family genes, but is now distinguished from the PPR by an otic-specific combination of additional markers, including those of the *pax* and *soxE* families (Hans et al. 2004; Mackereth et al. 2005; Yan et al. 2005; Dutton et al. 2009).

### 2.2. Cavitation and Early Patterning of the Otic Placode and Vesicle

The otic placode is morphologically evident from 16 hpf, when it is already beginning to cavitate to form an elongated, hollow epithelial ball – the otic vesicle. This differs from the mechanism of otic vesicle formation in birds and mammals, in which the presumptive otic domain invaginates to form an otic cup, which then closes to form a luminated vesicle. The process of patterning the otic vesicle, however, begins well before cavitation of the otic placode. Several genes are expressed in distinctly asymmetric patterns in otic tissue from very early stages (Whitfield and Hammond 2007; Whitfield et al. 2002, and references within). The establishment of this axial pattern within the early otic epithelium depends on a complex interplay between extrinsic signals, such as Fgf, Wnt and Hh, and factors that are intrinsic to the ear, which act to interpret these signals and to reinforce or refine otic patterning (Kwak et al. 2002; Hammond et al. 2003; Lecaudey et al. 2007). Within the otic epithelium, there are also complex regulatory relationships between gene products. The loss of *sox10* function, for example, results in the de-repression of several genes in the otic epithelium, most notably *fgf8*. This has severe patterning consequences: in the *colourless* (*sox10*) mutant ear, although most cell types differentiate, there are deficits in patterning of all areas, including the sensory patches, semicircular canal tissue and neurogenic region (Dutton et al. 2009).

## 2.3. Otic Sensory Development and Neurogenesis

Two of the very earliest patterning events in the otic placode and vesicle are the establishment of the prosensory and proneural domains, which will give rise to sensory hair cells and neuroblasts, respectively. In the zebrafish otic vesicle, the *atoh* genes have an early proneural function in establishing the prosensory domain, and are also required later for hair cell differentiation. Interestingly, *atoh1b* is expressed in a contiguous domain in the very early otic placode (10.5 hpf), and is required for differentiation of the very first hair cells to appear, while *atoh1a* is expressed later, in two separate domains, which prefigure the utricular and saccular maculae, at 14.5 hpf (Millimaki et al. 2007). The activity of *atoh* genes, together with Notch-mediated lateral inhibition, leads to the specification of pairs of hair cells at the anterior and posterior of the otic vesicle at around 22 hpf (Section 4). Concomitantly, an anterior ventral region of the otic epithelium gives rise to neuroblasts that will form the statoacoustic (VIIIth) ganglion beneath the ear; at present, it is unclear whether or not this overlaps with the prosensory domain. The neuroblasts express both *ngn1* and *neuroD* as, or just after, they exit the otic epithelium (Korzh et al. 1998). Notch signaling is required in both the prosensory and proneural domains in the ear. In the *mindbomb* (*mib*) (an E3 ubiquitin ligase) mutant, in which Notch signaling is attenuated, lateral inhibition fails in both domains, resulting in a vast excess of both hair cells in the ear and neurons in the statoacoustic ganglion (Haddon et al. 1998). An excellent comparative overview of the establishment of the otic proneural field is given in Abelló and Alsina (2007).

## 2.4. Morphogenesis: Semicircular Canal Development

The semicircular canals begin to form at 45 hpf, as a set of epithelial evaginations into the lumen of the otic vesicle (Fig. 4.2). The projection from the lateral wall of the vesicle trifurcates to connect with the projections from the anterior and posterior poles first, shortly followed by the ventral projection, to form three pillars of tissue around which the lumens of the semicircular canals run (Waterman and Bell 1984).

### 2.4.1. The Role of Extracellular Matrix in Semicircular Canal Morphogenesis

In *Xenopus*, the driving force underlying the initial outgrowth of the semicircular canal projections has been shown to be dependent on the production of hyaluronic acid (HA), which is secreted into the space between the otic epithelium and the underlying mesenchyme by cells of the projection (Haddon and Lewis 1991). In zebrafish, a pathway leading to the production of HA and other extracellular matrix components in the semicircular canals has been proposed through gene knockdown and mutant analysis, in which Dfna5 function is required for the transcription of the *UDP-glucose dehydrogenase* (*ugdh*) gene, which itself codes for an enzyme that is essential for HA production. In both *dfna5* morphants and *jekyll* (*ugdh*) mutants, semicircular canal projections fail to grow out and fuse, and the ear is swollen. Cells within the projections proliferate, but lack the requisite directional force at the leading edge, which causes disorganization of the epithelium (Neuhauss et al. 1996; Walsh and Stainier 2001; Busch-Nentwich et al. 2004). Mutations in the human *DFNA5* gene lead to progressive high-frequency sensorineural hearing loss; why a gene required for hearing in the mammal should have a role in semicircular canal morphogenesis in the zebrafish is currently unclear. Neither the hair cell phenotype in the zebrafish *dfna5* morphants, nor the expression of the *DFNA5* gene in the mammalian vestibular system, has yet been characterized, however.

The matrix glycoprotein Sparc has been shown to function in morphogenesis of the facial skeleton and the inner ear, alongside its additional role in otolith formation (Section 7.3) (Kang et al. 2008; Rotllant et al. 2008). The *sparc* gene is expressed in the otic vesicle from 14 somites until 72 hpf, as well as in the notochord, floorplate, somites and developing facial skeletal elements. Knockdown of this gene with a morpholino causes a severe reduction in the lower jaw similar to that seen in *sox9* mutants, which also lack *sparc* expression, suggesting that the *sox9* genes may lie upstream and contribute to activating the transcription of *sparc*. The *sparc* morphants have normal patterning of the otic vesicle aside from a loss of *otx1b* expression (see below). Formation of the semicircular canal projections is abnormal, and they fail to connect to form the central hub of tissue (Rotllant et al. 2008).

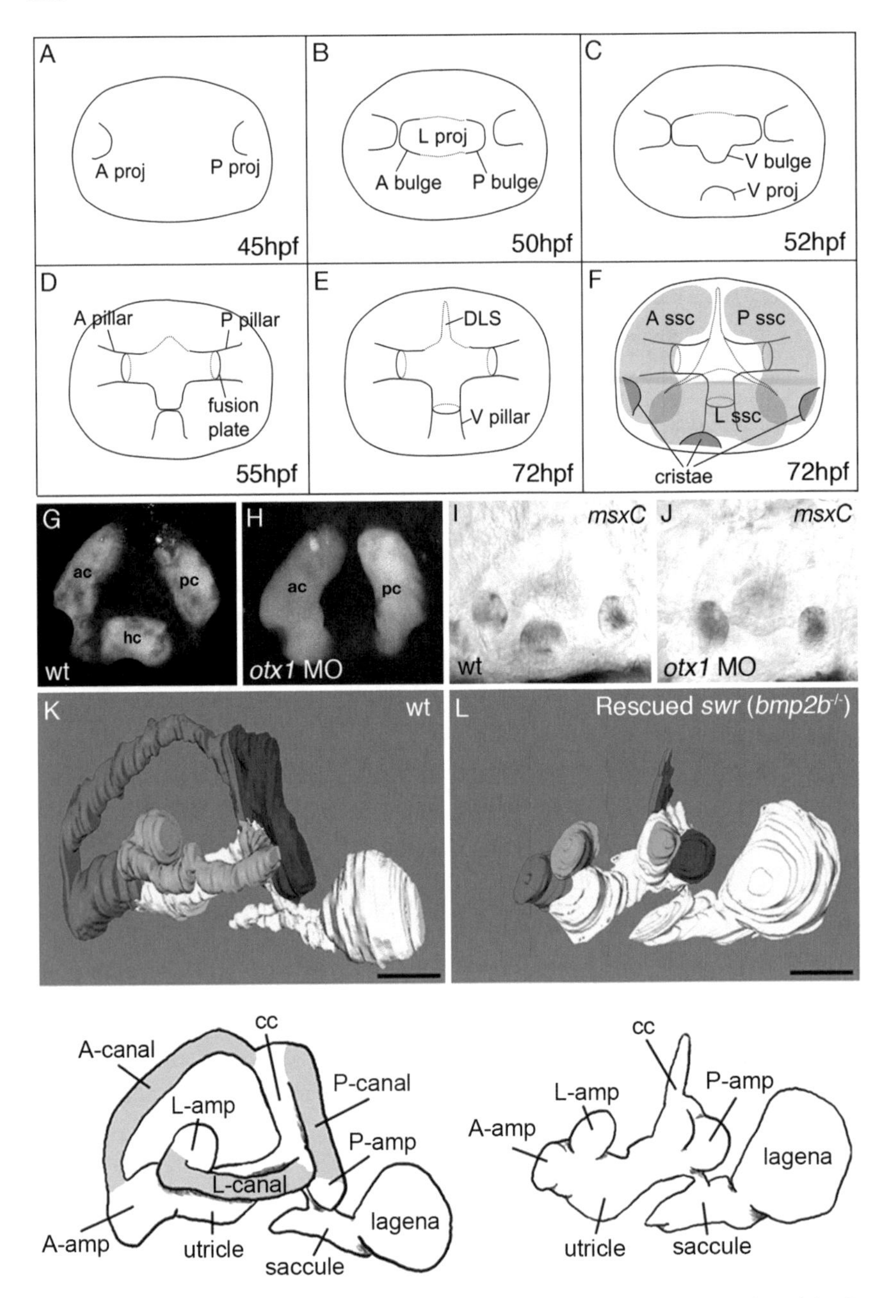


**Fig. 4.2.** The zebrafish semicircular canal system. (A–F) Schematic diagrams of semicircular canal development in the zebrafish otic vesicle. (A) Canal development begins at about 45 hpf with the formation of anterior and then posterior projections of tissue. (B.) At 50 hpf, two

### 2.4.2. The Role of Otx1b in Semicircular Canal Morphogenesis

The loss of *otx1b* expression in the otic vesicle of *sparc* morphants is likely to account for the failure of the lateral (horizontal) canal to develop. *otx1b* (formerly *otx1*) codes for a homeodomain transcription factor that is normally expressed in a ventral domain in the zebrafish otic vesicle (Li et al. 1994). In mice mutant for *otx1* and in zebrafish *otx1b* morphants, the lateral (horizontal) canal and crista fail to develop, and the utricular and saccular maculae are incompletely separated (Morsli et al. 1999; Fritzsch et al. 2001; Hammond and Whitfield 2006). This is very similar to the phenotype seen in the *sparc* morphants (Rotllant et al. 2008). In the *van gogh/tbx1* mutant, the otic vesicle is greatly reduced in size, with a striking loss of semicircular canal tissue; *otx1b* expression is also missing in the otic vesicle of this mutant (Whitfield et al. 1996). While the loss of otic *otx1b* expression may account for the lack of the lateral canal in the *tbx1* mutant, the phenotype is more severe than in the *otx1b* morphant, with a near-complete absence of semicircular canal tissue. Consistent with this, *tbx1* expression is found in all wild-type canal projections until 72 hpf and transplantation experiments suggest that *tbx1* function is needed cell-autonomously for canal initiation to occur (Piotrowski et al. 2003).

Otx1 is thought to have been a key player in the evolution of complexity in the inner ear; its expression is absent in the inner ear of the lamprey, which lacks a lateral canal and crista and has a single undivided macula, elements of which correspond to each of the two separate maculae in zebrafish. The similarity of the zebrafish *otx1b* morphant ear and mouse *Otx1* mutant ear to the lamprey ear suggests that a new domain of *otx1*

---

bulges of a lateral projection form, which grow anteriorly and posteriorly to meet the growing anterior and posterior projections, respectively. (C, D) At 52–55 hpf, the anterior and posterior projections touch and then fuse at a fusion plate, making anterior and posterior pillars. A ventral projection now forms in the vesicle, and grows towards a ventral bulge in the lateral projection. (E) By 72 hpf, the ventral projection and bulge have fused, forming the ventral pillar. The dorsolateral septum separates the anterior and posterior canal lumens in the dorsal part of the ear. (F) The lumens of the canals themselves (gray shading) run around the pillars. (G) Visualization of the three canal lumens at 4 dpf by injection of the ear with a fluorescent dye. (H) In an *otx1* morphant, the ventral pillar and lateral (horizontal) canal fail to form. (I) The three cristae are marked by *msxC* expression at 3 dpf. (J) The lateral (horizontal) crista fails to form in an *otx1* morphant. Panels G–J are reproduced from *Development* (Hammond and Whitfield 2006). (K, L) Three-dimensional reconstructions of a wild-type adult ear (K) and that of a rescued *swr* (*bmp2b*) homozygous mutant, which lacks formation of the semicircular canal ducts (gray shading in sketches below panels). Panels K, L are reproduced from *PLoS ONE* (Hammond et al. 2009). A, anterior; ac, anterior canal; amp, ampulla; cc, crus commune; DLS, dorsolateral septum; hc, horizontal (lateral) canal; L, lateral; MO, morphant; P, posterior; pc, posterior canal; proj, projection; ssc, semicircular canal; V, ventral; wt, wild-type. Color coding for K and L as for Fig. 4.1B. Scale bar (K, L), 500 μm. See color plate section expression in the vertebrate inner ear is likely to have been important in the acquisition of the lateral canal during evolution (Morsli et al. 1999; Fritzsch et al. 2001; Hammond and Whitfield 2006).

### 2.4.3. The Potential Role of Calcium Ions in Semicircular Canal Morphogenesis

The *Drosophila frequenin* gene and its vertebrate homolog, *neuronal calcium sensor-1* (*ncs1*) have been shown to have an important role in the spatiotemporal mediation of calcium signaling during neuronal development and synaptic transmission (Pongs et al. 1993; Hilfiker 2003). NCS proteins have a high affinity for intracellular calcium and are potent transducers of calcium ion flux, but little is known about their role in vertebrate development. Blasiole and co-workers (Blasiole et al. 2005) cloned the zebrafish orthologs of human NCS-1, *ncs1a* and *ncs1b* (now renamed *frequenin homolog* (*freq*) *a* and *b*). Expression of *ncs1a* is found in the developing ear amongst other anatomical structures; morpholino knockdown resulted in a small head phenotype, with delayed and dysmorphic semicircular canal projections and pillars in the ear. Otic expression of *dfna5* or *ugdh* was still present, however, suggesting that Ncs1a does not act upstream of this pathway. Expression of *ncs1a* in *jekyll* mutants or *dfna5* morphants, however, has not yet been examined.

So what is the function of Ncs1a during ear development? A proteomic screen for Ncs1a binding proteins (Petko et al. 2009) found seven putative partners that were expressed in the inner ear; based on these findings, a model was proposed in which Ncs1a might be functioning in intracellular trafficking of secreted components such as HA. Further work will be required to test this interesting proposal. The calcium channel Trpc1 was also identified as a potential binding partner and found to be expressed in the developing semicircular canals; however, the significance of this interaction is currently unclear. The developing semicircular canal projections also express the plasma membrane $Ca^{2+}$-ATPase isoform 2, *atp2b1a*, morpholino knockdown of which affects canal development and other otic structures (see Section 7.4 on otolith development) (Cruz et al. 2009). Thus, semicircular canal formation may depend on active calcium transport, possibly detected by Ncs1a, for the correct protrusion and outgrowth of canal tissue.

### 2.4.4. The Role of BMP Signaling in Semicircular Canal Morphogenesis

In addition to their role in preplacodal and placodal stages of otic development, BMPs have special and necessary roles during morphogenesis of the semicircular canal system. In zebrafish, *Xenopus*, chick and mouse, the expression of various BMP genes is conserved in developing sensory patches in the ear (Mowbray et al. 2001; Whitfield 2002; and references within). In the zebrafish, the earliest BMP family member to be expressed in the ear is *bmp2b*, expression of which by 24 hpf has concentrated, along with *bmp4*, at the anterior and posterior ends of the otic vesicle (Mowbray et al. 2001); inhibitors of the pathway such as *smad6* are also found to be expressed in the otic vesicle (Mowbray et al. 2001). Although there is no direct test for the role of BMPs in the early stages of zebrafish semicircular development, conditional disruption of BMP signaling in the ears of mouse and chick embryos demonstrates a critical role for this pathway in formation of both the semicircular canals and the cristae (Chang et al. 2008). The loss of canal tissue may be mediated by a concomitant loss of *Bmp2* expression from the "canal genesis zone" adjacent to the cristae, suggesting that in order for the non-sensory canal epithelium to grow out, the sensory cristae must be specified first (Chang et al. 2004).

The zebrafish *gallery* mutant only forms lateral and delayed, anterior canal protrusions; the posterior and ventral bulges fail to form (Omata et al. 2007). Implantation of a bead soaked in the BMP inhibitor Noggin gave a partial rescue of the phenotype, such that the anterior and lateral semicircular canal projections were able to fuse, lending weight to the idea that the BMP pathway is critical in some way for initiating the correct morphology of the semicircular canal system in the zebrafish.

After fusion of the epithelial projections to form the semicircular pillars in the zebrafish ear, the lumen of each of the three canals is now defined, and can be visualized by filling with paint or a fluorescent dye (Bever and Fekete 2002; Hammond and Whitfield 2006) (Fig. 4.2). Although the topological arrangement of the canals is now complete, much further outgrowth is required for the canals to acquire their final shape. This takes place during larval and juvenile stages, during which many body systems of the fish undergo considerable morphogenetic and physiological change, corresponding to a juvenile metamorphosis (Budi et al. 2008, and references within).

BMP signaling also appears to be required for these later stages of semicircular canal growth and remodeling in the zebrafish ear (Hammond et al. 2009). As *swr* (*bmp2b*) mutants are early embryonic lethal, preventing an analysis of ear development, an experimental approach was taken in which RNA encoding *bmp2b* or its downstream effector *smad5* was injected into embryos homozygous for the *swirl* (*bmp2b*) mutation. The RNA was sufficient to rescue the early dorsoventral patterning defects, but later developmental stages were completed in the absence of *bmp2b* function. These "rescued" *bmp2b* mutant fish reached adulthood, but displayed a vestibular behavioral defect (a loss of postural control while turning), appearing to result from a failure to detect angular motion stimuli. Detection of gravitational stimuli appeared to be intact, however. Although the early development of the otic vesicle appeared normal (possibly due to the continued presence of rescuing RNA), the ears of adult fish showed a complete lack of semicircular canal duct tissue. Nevertheless, all sensory end-organs – including the semicircular canal ampullae and their cristae – were preserved. This explains the specific behavioral phenotype – the gravitoinertial stimuli-sensing apparatus (maculae and otoliths) was present, but the semicircular canal ducts were missing, preventing any input to the sensory cristae. This is the first time a role in post-embryonic otic development has been ascribed to a zebrafish gene.

## 3. GENERATION AND HOMEOSTASIS OF ENDOLYMPH IN THE ZEBRAFISH INNER EAR

Concomitant with the morphogenetic events described above, the volume of the otic vesicle increases as it fills with a specialized extracellular fluid, the endolymph. In amniotes, endolymph is high in $K^+$ and low in $Na^+$ (Wangemann 1995; Lang et al. 2007); in teleost fish, the $Na^+$ concentration is higher but the increased $K^+$ concentration is still seen (Bernard et al. 1986; Ghanem et al. 2008). Various ion channels, pumps and transporters are known to be critical for the tightly regulated ionic balance of mammalian endolymph: loss or disruption of a single key player in the regulatory system can lead to deafness and balance disturbances in human patients and mouse models (Leibovici et al. 2008). In the zebrafish, the mechanisms underlying the generation and maintenance of endolymph are still relatively poorly understood, but recent research is gradually illuminating our knowledge of this physiology. Initial data point toward conservation of mechanisms, despite the anatomical differences between the fish and mammalian ear.

### 3.1. The Role of Ion Channels, Pumps and Transporters

The $Na^+$-$K^+$-ATPase pump is required for the regulation of sodium and potassium homeostasis by mediating the active transport of three sodium ions in exchange for two potassium ions to establish a chemical and electrical gradient across the basolateral membrane of the cell (Wangemann 2002). The role of the $Na^+$-$K^+$-ATPase pump has been well studied by several groups. This sodium pump consists of an $\alpha$ and a $\beta$ subunit complex in the plasma membrane – the expression of different $\alpha$ and $\beta$ subunits in different cell types allows a broadening of pump specificity and function. The expression of six of these subunits has been described in the developing wild-type zebrafish ear

(Rajarao et al. 2001; Blasiole et al. 2002, 2003). Mutations in the *atp1a1a.1* gene underlie the *snakehead/small heart* phenotype (Yuan and Joseph 2004; Lowery and Sive 2005), characterized by a failure to inflate the hindbrain ventricles, together with heart defects, delayed pigmentation and a defective touch response (Schier et al. 1996); the ear in these mutants is smaller in size, with rudimentary otoliths (Lowery and Sive 2005). A morpholino directed specifically against the *atp1a1a.1* subunit was found to block otolith formation in a dose-dependent manner, with only half the usual number of hair cells forming in the anterior macula at 72hpf; a complement of other defects such as a lack of circulation entailed that these embryos failed to survive past 4 days post fertilization (dpf) (Blasiole et al. 2006). However, a knockdown of *atp1a1a.2* gave a milder phenotype, perhaps reflecting its more restricted expression pattern, or functional redundancy with other Atp1a1a subunits. In *atp1a1a.2* morphants, the semicircular canals failed to form correctly, although expression of *ncs1a* and *ugdh* was still present in the disorganized epithelial projections, implying that *atp1a1a.2* function is not required for the expression of either of these genes. Morpholino knockdown experiments indicated that the Atp1b2b and Atp1a1a.1 subunits associate in the ear to give a functional pump that is required to regulate otolith formation (Blasiole et al. 2006). The expression of the *b2b* subunit is thought to be a downstream target of the transcription factor Eya4: the expression of *atp1b2b* mirrors that of *eya4,* and *eya4* morphants have a similar otic phenotype to the *atp1b2b* knockdown and can be rescued by the injection of full-length *eya4* RNA (Wang et al. 2008). Again, formation of the semicircular canal system was disrupted in the morphants, suggesting that the ionic milieu may be critical in regulating canal tissue outgrowth and fusion.

In the *little ears* (*lte*) mutant strain, the otic vesicle is patterned and grows normally until 72 hpf, when it collapses due to a loss of endolymph from the otic lumen (Whitfield et al. 1996; Abbas and Whitfield 2009). Cloning of this mutant revealed the lesion to be in the sodium-potassium-chloride cotransporter gene *nkcc1* (*slc12a2*): the phenotype it causes is markedly similar to the collapse of the semicircular canals and the endolymphatic compartment in the cochlea of the *Nkcc1* mutant mouse (Delpire et al. 1999; Dixon et al. 1999; Flagella et al. 1999). The ears of *lte* mutant embryos had lost the expression of *nkcc1* itself, and also had reduced expression levels of the potassium channel *kcnq1* and subunits of the sodium pump, implying that there is complex interplay at the transcriptional level between the different proteins regulating the ion balance within the endolymph (Abbas and Whitfield 2009). These data demonstrate that fluid regulation in the zebrafish ear is controlled by similar mechanisms to those in the mammal and that the zebrafish should prove to be a valid model for the investigation of human endolymphatic disorders.

In the embryo, the main site of expression of endolymph-producing genes appears to be the semicircular canal projections and pillars (Abbas and Whitfield 2009). However, in the adult zebrafish inner ear, rings of ionocytes surround the sensory maculae (Fig. 4.3). These express $Na^{+}$-$K^{+}$-ATPase and, together with carbonic anhydrase-expressing transitional cells, are thought to regulate the pH of the endolymph bathing the hair cells, neutralizing the $H^{+}$ ions released by the $H^{+}$-ATPase-expressing hair cells by secreting $HCO_3^{-}$ ions (Shiao et al. 2005) (see Section 7).

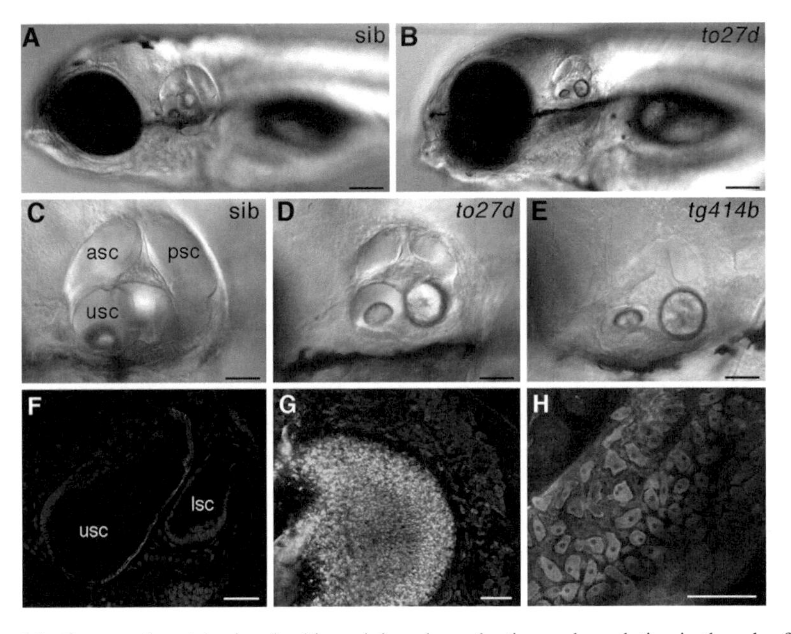


**Fig. 4.3.** Genes and proteins involved in endolymph production and regulation in the zebrafish ear. (A–E) Phenotype of the *little ears* (*lte*) mutant, which carries a mutation in the *nkcc1* (*slc12a2*) gene. At 5 dpf, the ear in the mutant (B, D, E) has collapsed due to a loss of endolymphatic fluid. Two different alleles (D, E) illustrate the variation in degree of collapse. (F) Expression of Nkcc1 protein in the zebrafish ear at 6 dpf is confined to epithelial cells lining the utriculosaccular cavity (usc). Scale bar, 50 μm. Panels A–G are reproduced from *Development* (Abbas and Whitfield 2009). (G) Expression of $H^{+}$-ATPase (green) in hair cells of the adult lagenar macula, surrounded by several rows of ionocytes, which express $Na^{+}$-$K^{+}$-ATPase (red). Scale bar, 50 μm. H. Higher magnification of the ionocyte region. Ionocytes express $Na^{+}$-$K^{+}$-ATPase (red) and are surrounded by transitional cells that express Carbonic Anhydrase (Cahz; green). Scale bar, 30 μm. Panels G and H are reproduced, with permission, from Shiao et al. 2005. asc, anterior semicircular canal; lsc, lateral semicircular canal; psc, posterior semicircular canal; sib, phenotypically wild-type sibling; usc, utriculosaccular cavity 2. See color plate section

## 4. ZEBRAFISH SENSORY HAIR CELLS

### 4.1. Hair Cell Morphology

Auditory and vestibular stimuli are detected by sensory hair cells in the inner ear. The hair cell is a specialized sensory epithelial cell, with an apical hair bundle consisting of a single kinocilium and numerous stereocilia. Stereocilia are modified microvilli filled with cross-linked F-actin filaments, which are rooted at the base in the cuticular plate, a dense mesh of actin and associated cytoskeletal proteins in the hair cell apical cytoplasm. The structure, development and function of the hair cell are well conserved between fish and mammals (Nayak et al. 2007). Zebrafish hair cells, like mammalian hair cells, are susceptible to injury by a range of ototoxic agents (Harris et al. 2003; Ou et al. 2007), but unlike mammalian auditory hair cells, are able to regenerate after such ototoxic damage (Section 4.5).

### 4.2. Hair Cell Development

Hair cells are initially specified at the anterior and posterior ends of the otic vesicle. The very first hair cells to differentiate have been termed "tether cells", as they appear to tether the otoliths, which nucleate at the tips of the kinocilia of these early hair cells (Riley et al. 1997). At all developmental stages, including in the adult, the actin-rich hair cell stereociliary bundles can be visualized by staining with fluorescently coupled phalloidin (Fig. 4.4; and see, for example, Haddon and Lewis 1996; Liang and Burgess 2009). Hair cells can also be visualized in the zebrafish ear and lateral line by staining with HCS-1, Myo6, Myo7a, acetylated tubulin or Pax2 antibodies, or with vital dyes such as DASPEI, Yo-Pro-1, FM1-43 and FM4-64 (Fig. 4.4; and see, for example, Whitfield et al. 1996; Seiler and Nicolson 1999; Santos et al. 2006; Coffin et al. 2007; Millimaki et al. 2007; Cruz et al. 2009; Gleason et al. 2009; Jing and Malicki 2009). In addition, transgenic lines expressing green fluorescent protein (GFP) in lateral line neuromasts and maculae of the ear have been generated, which allow spectacular imaging of these sensory organs in the live embryo, and are facilitating the analysis of mutant phenotypes (Fig. 4.4; see, for example, Xiao et al. 2005; Lecaudey et al. 2008; Behra et al. 2009; Tanimoto et al. 2009). The reader is also referred to other recent and comprehensive reviews of hair cell development and function (Nicolson 2005a,b).

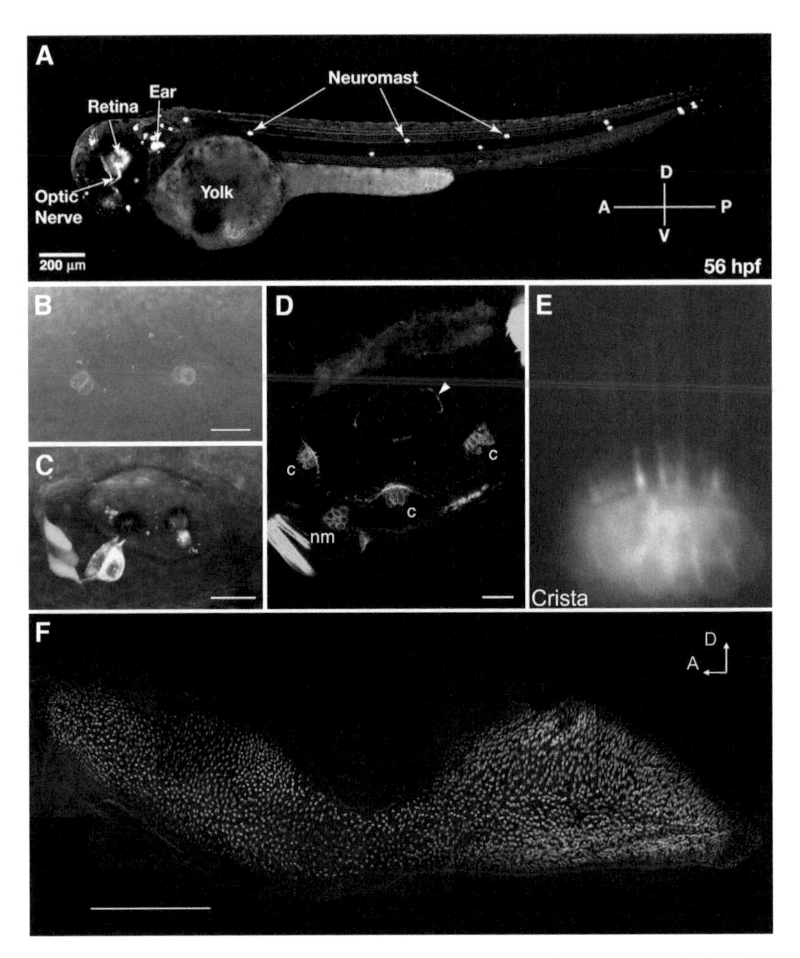


**Fig. 4.4.** Sensory hair cells in the zebrafish inner ear and lateral line. (A) Lateral view of 56 hpf live Tg(pou4f3::gap43-GFP) embryo showing GFP expression in mechanosensory hair cells (neuromasts) of the lateral line and inner ear. Retinal ganglion cells and the optic nerve are also visible. Scale bar, 200 µm. Reproduced, with permission, from Xiao et al. (2005). (B) Pairs of tether cells at either end of the wild-type otic vesicle, revealed by HCS-1 antibody staining at 35 hpf. Reproduced, with permission, from Cruz et al. (2009). (C) Early projections of the VIIIth nerves to the macular hair cells. Stacked image of the lateral part of the otic vesicle in a Tg(zCREST2-isl1:GFP) embryo (Uemura et al. 2005) at 27 hpf. Anterior macular hair cells are labeled with FM1–43 (orange) and are contacted at the base by peripheral axons of the VIIIth ganglion neurons (green). Stereociliary bundles (orange spikes) can be seen on the apical surface of the two hair cells, contacting the anterior otolith. Scale bar, 20 µm. Reproduced, with permission, from Tanimoto et al. (2009). (D) FITC-phalloidin (green) and HCS-1 antibody (red) staining of the hair cells in zebrafish inner ear at 4 dpf. Rings of actin (of which one is marked by an arrowhead) mark the position of the fusion plates for the three semicircular canals. c, cristae; nm, neuromast. Scale bar, 25 µm. Reproduced, with permission, from Cruz et al. (2009). (E) High-magnification view of a crista of a Tg(pou4f3::gap43-GFP) embryo.

## 4.3. Hair Cell Differentiation and Function

Recent advances in genomic technology have permitted the identification of new genes involved in hair cell differentiation and maintenance. McDermott and coworkers (McDermott et al. 2007) screened microarrays with RNA extracted from a pure population of adult, lagenar hair cells. Over a thousand genes deemed to be expressed in the hair cells were identified that fell into several broad categories: membrane transport, synaptic transmission, transcriptional control, cell adhesion and signal transduction and cytoskeletal organization. In the membrane transport category, five genes were found encoding ion transporters, pumps or channels, including the voltage-gated potassium channel *kcnd3*, a candidate for the A-type $K^+$ conductance, which is needed to maintain the hair cell resting potential (Sugihara and Furukawa 1995); in situ hybridization showed expression of this gene in anterior maculae of 4 dpf larvae, suggesting that it has a role at both embryonic and adult stages. Seven genes for synaptic transmission were identified, for example *synaptojanin* (Section 6.1), required for endocytosis and cytoskeletal dynamics. The transcription factor category included an ortholog of the human *SALL4* gene, mutations in which are causative of Okihiro-Duane syndrome (Hayes et al. 1985), in which patients often present with sensorineural deafness. In terms of cytoskeletal dynamics, 19 genes were isolated, of which nine had not previously been reported. Alongside genes with putative roles in stereociliary and cuticular plate morphogenesis through their association with cortical actin, for example a protein similar to human Coronin1b, six genes were found encoding ciliary or microtubule associated proteins. *ift172* encodes a protein required for ciliary turnover and has been shown to be mutated in the zebrafish *moe* mutant (Sun et al. 2004), which was initially characterized by its cystic kidney phenotype, a symptom often associated with ciliary defects. McDermott et al. (2007) examined the hair cells in the cristae of *moe* mutants at 4 dpf and found them to be lacking kinocilia. Likewise, their examination of the *seahorse/leucine-rich repeat-containing 6-like* (*lrrc6l*) polycystic kidney mutant (Sun et al. 2004; Serluca et al. 2009) showed defects in hair cell kinocilia – at 4 dpf, kinocilia were enlarged with bloated bases filled with α-tubulin. These results demonstrate a necessity for a functional ciliary turnover system in order for the hair cells to be properly maintained.

Reproduced, with permission, from Behra et al. (2009). (F) Hair cells in the complete adult saccular sensory epithelium, dissected and stained with Alexa Fluor 488 Phalloidin (green). Several pictures were taken at different positions of the sample and tiled together with Adobe Photoshop 7.0. Scale bar, 150 μm. Reproduced, with permission, from Liang and Burgess (2009). See color plate section

## 4.4. miRNAs in Sensory Hair Cell Development

The importance of microRNAs (miRNAs) in modulating gene expression has come to prominence in recent years. Noncoding miRNAs are 17–25 nucleotides in length and are produced from stem-loop precursors, encoded by miR genes. They function by translational suppression or mRNA cleavage and often have exquisitely tissue-specific expression patterns. Three miRNAs – miR-183, miR-96 and miR-182 – have been found to be expressed specifically in vertebrate hair cells (Wienholds et al. 2005; Weston et al. 2006). In addition, miR-200a, miR-15a-1 and miR-18a are expressed in sensory epithelia of the developing zebrafish inner ear or lateral line (Wienholds et al. 2005; Friedman et al. 2009). Morpholino-based knockdown of miR-15a-1 or miR-18a has a robust effect on hair cell development, reducing the number of cells in the ear and in the lateral line neuromasts (Friedman et al. 2009). Targets for inner ear miRNAs have not yet been identified in the zebrafish; however, it is tempting to extrapolate based on comparable data from the mouse, in which ion transporters (*nkcc1/slc12a2*), tight junction components (*claudin12*) and growth factors (*bdnf*) have all been proposed as targets based on in silico sequence analysis (Friedman et al. 2009).

## 4.5. Hair Cell Regeneration

In the mammal, hair cell regeneration does not apparently take place in the sensory epithelium of the postnatal cochlea, leading to permanent deafness after damage. In the fish, however, hair cells continue to develop throughout life. The zebrafish lateral line system has been used extensively as a model for hair cell regeneration and has been the subject of recent excellent studies and reviews (Collado et al. 2008; Behra et al. 2009; Brignull et al. 2009; Ma and Raible 2009, and references within). Alongside birds and amphibia, the adult zebrafish has been shown to exhibit regeneration of hair cells in the inner ear. Schuck and Smith 2009 exposed adult fish to damaging noise levels (100 Hz at 179 dB) for 36 hours and examined the condition of the saccular epithelium immediately and at various time points afterwards. They observed an immediate decrease in hair bundle density, ragged and splayed stereocilia and a prevalence of lesions and scars formed by support cells, presumably expanding to fill in the 'holes' left by the dying hair cells – BrdU staining marks an increase in proliferating cells at 2 days post sound exposure (dpse). By 7 and 14 dpse, the affected sites have equivalent levels of hair cell recovery, displaying some cells with bundle-less cuticular plates and others with characteristically short, immature hair cell bundles. The caudal region of the saccule was more severely damaged than other areas initially but recovered to comparable levels in terms of hair bundle density at these later stages; this differential damage may well reflect frequency-dependent functions within the hair cells but as yet, tonotopic mapping has not been determined in the zebrafish ear.

## 5. MECHANOTRANSDUCTION

Mechanotransduction is the conversion of stereociliary deflections into electrical activity. When a vibratory stimulus is received, the inertia of the otolith relative to the hair bundles results in shearing forces that deflect extracellular filaments between the stereocilia, the tip links. This movement opens mechanosensitive cation channels; the inward transduction current elicited depolarizes the hair cell membrane and neurotransmitter release is triggered at the afferent synapse.

### 5.1. Interstereociliary Links

Each mature hair bundle has up to 300 stereocilia arranged in rows that are connected to each other by very fine extracellular lateral and oblique filaments (Nicolson et al. 1998). Four different link types are thought to connect adjacent stereocilia – tip links, horizontal top connectors, shaft connectors and ankle links (Nayak et al. 2007). The "gating spring" hypothesis proposes that the tip links are directly connected to the mechanotransduction channels (see Section 5) and act like springs to pull them open during the mechanical shearing of the hair bundle, leading to an influx of cations (Nicolson et al. 1998). Cadherin23 has been shown to be a key component of the tip link in zebrafish (Söllner et al. 2004a): the *sputnik* mutant, originally isolated on the basis of its balance defect (Granato et al. 1996), has been found to carry a mutation in *cadherin23* and displays splayed hair bundles in which the tip links are lacking, but the other connections, for example the ankle links, remain intact (Söllner et al. 2004a). An antibody raised to Cadherin23 shows that the protein is located at the distal tips of the stereociliary bundle; immunoreactivity is absent from the mutants. In other vertebrates, however, there may be additional components forming the tip link (Rzadzinska and Steel 2009) – There are null alleles for *the waltzer/Cdh23* mouse strain, in which tip link-like interstereociliary connections are still present in the hair cells. Fish rely on a combination of visual and gravitational/static input to maintain postural control, and *sputnik* (*cdh23*) fish respond to this challenge by orienting their dorsal side towards the light to swim on their side when illuminated from the side, demonstrating that they have lost their gravity-sensing capability.

The unconventional, non-filament-forming myosins also have a key role in hair bundle integrity. They are known to be critical motor proteins in the vertebrate inner ear, with mutations in MYOVI and MYOVIIA leading to deafness in humans (Weil et al. 1995), mice (Avraham et al. 1995; Gibson et al. 1995) and zebrafish (Ernest et al. 2000). Hair cells in zebrafish *mariner/myoVIIa* mutants have short, splayed hair bundles in which the kinocilia and stereocilia become detached and larvae show a deafness and balance phenotype (Ernest et al. 2000). MyosinVIIa localizes to the cytoplasm of the hair cell and to the hair bundle itself and is thought to have a functional role, operating in series with the mechanotransduction channel; it has also been proposed to be associated with the lateral links between stereocilia. MyosinVI is also in the cytoplasm and cuticular plate of hair cells where it is thought to have a role in endocytosis (see Section 5.3); it too is found in the stereocilia – however, this is thought to be a novel role in anamniote hair cells (Coffin et al. 2007).

### 5.2. TRP Channels

The transient receptor potential (TRP) channels have been proposed as candidates for the mechanosensitive cation channel. In the mouse, Mucolipin3 is a six-transmembrane domain protein with similarities to the TRP family; mutations in *Mcoln3* are causative in the *varitint-waddler* strain, in which mice display hearing loss, balance defects and altered pigmentation. The search for a cation channel with a similar function in the zebrafish has identified NompC/Trpn1 (Sidi et al. 2004), a TRP channel expressed selectively in the hair cells of the inner ear and lateral line organs in the larval zebrafish. Abrogation of NompC/Trpn1 function with a morpholino gave embryos with no response to acoustic stimuli and a swimming defect characteristic of a vestibular or balance defect. In the morphants, a loss of transduction-dependent endocytosis, demonstrated by a failure in uptake of the fluorescent styryl dye FM1-43, and a loss of evoked microphonic potentials – a readout of the sum of stimulus-evoked transepithelial return currents resulting from asymmetric ionic currents across the hair cell membrane (Corey and Hudspeth 1983) – suggest that NompC/Trpn1 has a critical role in hair cell mechanotransduction. However, NompC/Trpn1 may not be the sole constituent of the vertebrate mechanotransduction channel – it is localized to the kinocilium in *Xenopus* hair cells (Shin et al. 2005), a structure that is lacking in some vertebrate hair cells, for example those in the cochlea. Perhaps more pertinently, *NompC/Trpn1* genes have been found in zebrafish and *Xenopus,* but as yet, no mammalian *Trpn1* gene has been identified, implying that there has either been a divergence in hair cell mechanotransduction during evolution, or that there is an additional channel or channels not yet found in the zebrafish.

## 5.3. The Role of Endocytosis

Hair cells stain intensely with the vital styryl dye FM1-43, a marker for apical endocytosis (Fig. 4.4) (Seiler and Nicolson 1999). There is a "pericuticular" zone lying between the actin-rich cuticular plate and the heavily invaginated apical cell surface, which is filled with small vesicles, 30–200 nm in diameter, indicating that this is a region of high membrane turnover and trafficking. When hair cells are labeled with FM1-43, "threads" of fluorescence can be seen spanning the apico-basal extent of the cell and it is thought that this transport may be occurring on a microtubule-based network for transcytosis (Seiler and Nicolson 1999). This endocytic pathway has been shown to be calmodulin-dependent; inhibitors of $Ca^{2+}$ transport, $Ca^{2+}$ channels, or Calmodulin itself all block FM1-43 uptake (Seiler and Nicolson 1999).

Hair cell apical endocytosis is also dependent on the unconventional myosins (Section 5.1). Mutations in *myosinVI* underlie the *satellite* mutant, in which hair cells have shortened and thinner, split hair bundles, bulging of the cuticular plate and spreading of the apical membrane up the bases of the stereocilia, implying a role for MyoVI in tethering the apical membrane to the cuticular plate (Seiler et al. 2004). Vesicles were found to accumulate within the cuticular plate, suggesting a role for MyosinVI in trafficking.

Both MyosinVI and MyosinVIIa localize to pericuticular vesicles and both the *satellite/myo6b* and *mariner/myo7a* mutants are defective in FM1-43 uptake (Seiler and Nicolson 1999; Seiler et al. 2004). Inhibitors of mechanotransduction such as aminoglycoside antibiotics are thought to act through myosin-based endocytic mechanisms – neither the *sputnik* nor the *mariner* mutants are sensitive to streptomycin-induced hair cell death (Seiler and Nicolson 1999). Thus, apical endocytosis might have several roles in hair cell function: a housekeeping role in the maintenance of the structural integrity or the extracellular links between the stereocilia, shown by the stereociliary disorganization in the *satellite* mutant; a means of communication between the apical and basolateral domains of the hair cell, shown by the spreading of FM1-43 across the cell; and a role in the functional mechanics of the hair bundle in $Ca^{2+}$- and Calmodulin-dependent processes (Seiler and Nicolson 1999; Seiler et al. 2004).

## 5.4. Stereociliary Maintenance

Other human deafness gene homologs have been identified in the fish as being involved with the maintenance of stereocilia. *Anosmin-1a* and *anosmin-1b* are the zebrafish duplicates of the human *KAL1* gene, mutations in which cause X-linked Kallmann syndrome (Legouis et al. 1991; Hardelin et al. 1992), a developmental disorder with defects in cell migration and axon targeting; vestibular defects have also been reported to be linked with this syndrome (Hill et al. 1992). Anosmin1 is an extracellular matrix protein expressed in the lateral line neuromasts and cristae of larval zebrafish (Ernest et al. 2007). Interestingly, it is localized to the surfaces of the kinocilia and the stereocilia (and to a lesser extent, within the cell bodies) of hair cells at the distal extremities of the cristae, but not within the medial regions, which may suggest that the crista hair cells are a heterogeneous population. Mammalian cristae have functionally and molecularly distinct regions (Lindeman 1966; Fritzsch 2003); in zebrafish cristae, there is also heterogeneity in hair cell morphology (Bang et al. 2001), and at least one marker – *aldh1a3* – is expressed in a regionalized pattern (Pittlik et al. 2008). However, no functional partition has been reported as yet within cristae in the zebrafish.

The *transmembrane inner ear* (*tmie*) gene has also recently been reported to have a role in hair cell maturation and function in the zebrafish (Shen et al. 2008; Gleason et al. 2009). *Tmie* codes for a transmembrane protein and the human ortholog lies in the critical region for the deafness locus *DFNB6* (Naz et al. 2002); mutations in the mouse *Tmie* gene are causative for the *Spinner* phenotype, in which mice display hearing loss and vestibular dysfunction with stereociliary defects (Mitchem et al. 2002). In *tmie* mutants, hair cells were present, but failed to take up the fluorophore FM4-64 and showed a complete lack of electrical activity. Ultrastructural analysis revealed several structural defects in the hair cells, including shortened stereocilia, and a lack of tip links and insertional plaques (Gleason et al. 2009). Morphants for *tmie* (Shen et al. 2008) showed similar hair cell functional deficits, but also had additional structural defects in the ear, which may result from non-specific effects of the morpholino. These data suggest that, as in the mouse, *Tmie* might function in hair cell maturation and stereociliary maintenance; it is also possible that it is a component of the transduction apparatus (Gleason et al. 2009).

## 6. NEURAL ACTIVITY

### 6.1. Synaptic Transmission

In response to the mechanical stimulus of the otolith shearing the hair bundle, a graded receptor potential is triggered in the hair cell, resulting in neurotransmitter release at the afferent synapse with high temporal accuracy and fidelity. A specialized structure, the ribbon synapse, is found at the interface of the hair cells and the acousticolateralis afferent neurons, which permits very high rates of exocytosis. A large pool of reserve vesicles is apposed to the ribbon synapse, which aid in reducing signal latency. The molecular basis of synaptic vesicle release and recycling is not currently well known, but a trio of zebrafish mutants with defects in various aspects of this process have helped elucidate some of the underlying mechanisms (Sidi et al. 2004; Obholzer et al. 2008; Trapani et al. 2009).

The *comet* mutant was isolated on the basis of its vestibular defects, and corresponds to a mutation in the gene encoding the lipid phosphatase *synaptojanin1* (*synj1*) (Trapani et al. 2009). Mutants have large basal blebs close to the ribbon synapses of their hair cells, caused by an imbalance of endo- and exocytosis. Measurement of post-synaptic electrical activity in the posterior lateral line ganglion neurons in *comet* mutants indicated a frequency-dependent fatigue. The evoked microphonic potentials remained constant, however, suggesting that the initial process of mechanotransduction was normal in these animals. Observation of the ribbon synapse in over-stimulated mutant hair cells showed an increase in the number of large (70 nm, compared to a "normal" synaptic vesicle size of 40 nm), coated vesicles but fewer releasable synaptic vesicles and fewer reserve, pool vesicles, indicating a defect in synaptic vesicle recycling (Trapani et al. 2009). Hair cells have a tendency to fire spontaneously in the absence of any mechanical stimulus – *comet* mutant hair cells showed a delay of several minutes in returning to this spontaneous firing condition, compared to a latency of seconds for wild-type cells. This may be indicative of the time taken to replenish the synaptic vesicle pool at the ribbon synapse – the rate-limiting step in the mutants is proposed to be the membrane recycling that allows the formation of new vesicles from endocytic compartments and the refilling of the synapse. The data support the idea that vesicle numbers are critical for the temporal fidelity of vesicle release at ribbon synapses and that Synj1 plays an essential role in facilitating vesicle recycling. The acoustic startle response, a stereotyped escape behavior, remains intact in *comet* mutants; however, the vestibular deficit manifesting as a swimming defect worsens with repeated stimulation, which is again suggestive that the initial synaptic transmission occurs as normal but becomes fatigued with repetition (Trapani et al. 2009).

Double homozygous mutant larvae carrying both the *comet* mutation and *gemini*, a mutation in the *Cav1.3* L-type calcium channel (Sidi et al. 2004) demonstrated that this calcium channel was necessary for the blebbing seen in the *comet* ribbon synapses (Trapani et al. 2009). L-type calcium channels drive the bulk of voltage-gated $Ca^{2+}$ entry into hair cells and thus are essential for many essential steps in auditory processing, including neurotransmitter release onto the postsynaptic afferents (Fuchs et al. 2003). The bulk of calcium ion influx into hair cells occurs basolaterally. Cav1.3 is the predominant avian calcium channel (Kollmar et al. 1997) and mutations in *Cav1.3* in the mouse lead to deafness (Platzer et al. 2000; Namkung et al. 2001). Exocytosis is reduced in isolated sensory epithelia from these mice (Brandt et al. 2003). In zebrafish, the *gemini* mutant was isolated on the basis of its abnormal swimming behavior (Granato et al. 1996) and has a deafness and vestibular phenotype (Nicolson et al. 1998), despite normal hair cell viability, afferent synaptogenesis and peripheral innervation (Sidi et al. 2004). Mutant larvae show a decrease in microphonic potentials but FM1-43 uptake is normal, implying that endocytosis, and therefore probably mechanotransduction, is occurring normally (Sidi et al. 2004). Immunostaining revealed Cav1.3 to be localized in focal spots organized into ring-shaped structures within specific areas of the hair cell basal membrane. In the lateral line neuromasts, these rings are sited in apposition to the afferent terminals of the lateral line nerve, where ribbon synapse-specific $Ca^{2+}$ channels would be predicted to sit; there is also a 1:1 correspondence between the number of Cav1.3-positive clusters and the number of ribbon synapses per cell. The lesion in the *cav1.3* gene in *gemini* leads to a loss of this channel; concomitant loss of the calcium ion influx at the ribbon synapse leads to the deafness and circling phenotype (Sidi et al. 2004).

The importance of glutamate in ribbon synapse transmission has been shown by altering the levels of the vesicular glutamate transporters 1 and 3 (Vglut1, Vglut3) (Obholzer et al. 2008). Both proteins are localized to the basal half of the hair cell – double staining with the Ribeye antibody demonstrates that Vglut3 is present in all ribbon synapses. Morpholino knockdown of the *vglut1* gene gives embryos with deafness and vestibular phenotypes; they also have smaller eyes and are less sensitive to touch stimuli, suggesting an additional role for Vglut1 in the CNS. Mutations in the *vglut3* gene underlie the *asteroid* mutant, in which larvae lack action currents in the acousticolateralis neurons – there is a decrease in the number of ribbon synapse-associated vesicles and each vesicle is predicted to contain too little glutamate to trigger action potentials in the downstream glutamatergic afferent neurons (Obholzer et al. 2008). This loss of synaptic transmission from the hair cells to the first-order neurons suggests a role for Vglut3 in vesicle recruitment, recycling or indeed, both.

### 6.2. Signal Transduction from Ear to Brain

The VIIIth or octaval nerve transmits the neural signals induced by mechanical stimulation of the hair cells to the brain. The lateral dendrites of the Mauthner neuron in the reticulospinal system receive inputs directly from this auditory nerve and so the Mauthner neuron is triggered to initiate the acoustic startle response (Burgess and Granato 2007; Zottoli 1977). Tanimoto et al. (2009) looked at the developmental acquisition of this auditory response in embryos by examining the innervation of macular hair cells and its relationship to the startle response circuitry. By taking whole-cell readings from Mauthner cells in the hindbrain of intact embryos, inward currents were generated in response to a 500 Hz tone from the age of 62 hpf onwards. No currents were observed in response to any frequency or volume of stimulus before 40 hpf; in the intervening times, currents were observed with long and unstable onset latencies. Interestingly, all components of the requisite neural circuitry – connections between neural processes of the VIIIth ganglion and the macular hair cells, and the Ribeye and Vglut3 components of the hair cell presynaptic bodies – are present by 27 hpf. The central projection from the VIIIth nerve to the Mauthner neuron has also been shown to be present from 23 hpf (Kimmel et al. 1990). It may be that the hair cells themselves are immature: microphonic potentials measured in response to an acoustic stimulus are not observed before 40 hpf and again, appear to require a "maturation" phase over the subsequent few hours, in which the response amplitude increases and the latency decreases (Tanimoto et al. 2009). This is in contrast to the response of hair cells to a purely mechanical stimulus, such as displacement of the otolith with a jet of water, which can evoke both microphonic potentials in the hair cells and stimulus-evoked currents in the Mauthner cell in pre-40 hpf embryos (Tanimoto et al. 2009).

These results suggest that the afferents of the VIIIth nerve are functional before the sound-evoked response develops in the Mauthner cell, which may in turn rely upon the functional maturation of the inner ear including the development of sound sensitivity within the hair cell population. A candidate for this step-change in sensitivity is the NompC/Trpn1 channel (Section 5.2), which is expressed from 48 hpf in the maculae (Sidi et al. 2003), and may be critical for induction of the acoustically invoked escape response.

## 7. OTOLITHS

The otoliths, or "ear stones", are composite bodies consisting largely of calcium carbonate with a small percentage (0.2–10% in different fish species) of organic matter in the form of fibrous, collagen-like proteins such as otolin (Murayama et al. 2002). As mentioned above (Section 1.2), it is the shearing action of the otoliths deflecting the apical hair bundles of the underlying hair cells that triggers the ion influx cascade culminating in auditory nerve firing.

At 20 hpf, the otoliths are first visible as two specks at the anterior and posterior poles, contacting the kinocilia of the tether cells, which are the first hair cells to develop of the presumptive anterior and posterior maculae. The otoliths sit on the gelatinous otolithic membrane, which is connected to the microvilli of the supporting cells (Nicolson 2005b) (Fig. 4.5).

The larger sensory maculae in the adult fish inner ear each have an associated otolith. The utricular (anterior) and saccular (posterior) otoliths develop first (see Fig. 4.4C (27 hpf) and Fig. 4.3C (5 dpf) for illustrations). In the adult, there are three otoliths in each ear: the lapillus overlies the utricular macula, the arrow-headed sagitta sits within the saccule and the crenellated asteriscus (Fig. 4.5B) lies upon the lagenar macula (Platt 1993). The utricle is thought to be the main vestibular organ in the zebrafish and loss of the utricular otolith is incompatible with viability (Riley and Moorman 2000); the saccule is primarly concerned with the sense of hearing (Fay and Popper 1999b) (Section 5.2). The mammalian counterpart of the otoliths, the otoconia, are mostly produced embryonically and must be maintained throughout life – they are subject to demineralization and alterations in structure and composition (Hughes et al. 2006). Exposure to drugs, disease and aging can all lead to otoconial degeneration; increased dizziness and loss of postural control seen in the elderly has been proposed to result at least in part from degradation of the otoconia (Lim 1984).

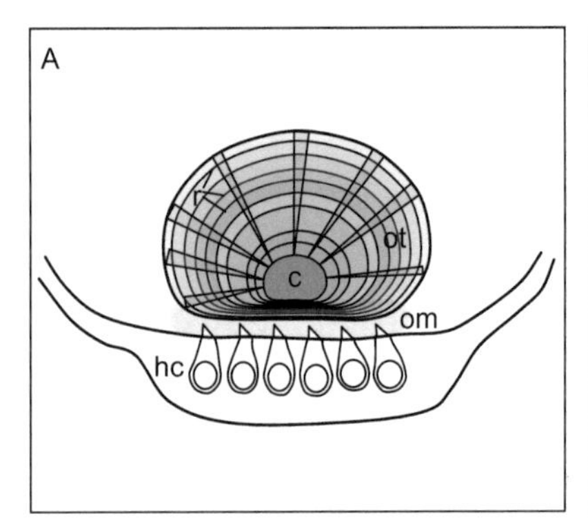


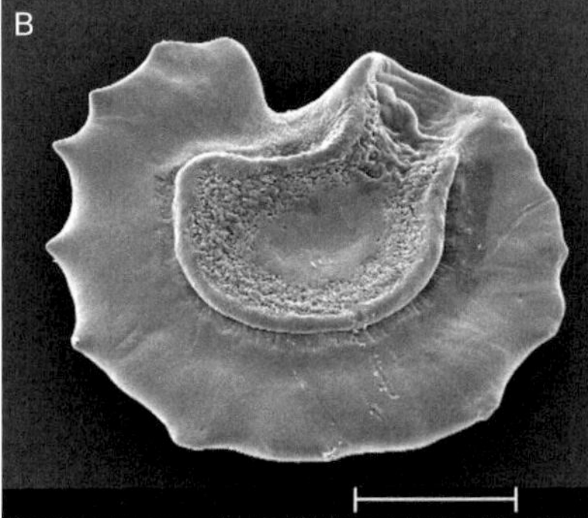


**Fig. 4.5.** Zebrafish otoliths. (A) Schematic diagram of a cross-section through a zebrafish otolith in the second week of development. The otolith (ot) sits over the sensory macula above the stereociliary bundles of the hair cells (hc). These and the otolith are thought to be embedded in an otolithic membrane (om), but this has not been well described in the zebrafish. The otolith contains a central proteinaceous core (c) and shows daily growth rings (r) of proteinaeous matrix and $CaCO_3$ mineralization. (B) Scanning electron micrograph of the lagenar otolith (asteriscus) from an adult wild-type zebrafish. The macular surface is shown; anterior is to the top of the panel. Scale bar, 200 μm.

## 7.1. Initiation of Otolith Formation

Otolith development in the zebrafish begins with the aggregation of free-floating core particles, thought to be largely composed of glycogen (Pisam et al. 2002). These particles traverse into the lumen of the vesicle as spherules of glycogen via an apocrine pathway (involving secretion by budding from the cell membrane) and progressively coagulate into globules 3 μm in diameter by 23 hpf. These globules aggregate directly above the first hair cells in the sensory patches in the ear, where they appear to be tethered by the hair cell kinocilia (Riley et al. 1997; Pisam et al. 2002). By 30 hpf, the glycogen globules have more than doubled in size and form the central core of the developing otolith, becoming coated by two, concentric, fibrillar layers by 50 hpf (Pisam et al. 2002). The glycogen aggregates rapidly become mineralized with a crystalline casing of the aragonitic form of calcium carbonate (Söllner et al. 2003); the growing otolith also becomes anchored to the otolithic membrane. The utricular and saccular otoliths begin developing around 19–22 hpf; the lagenar otolith arises later around 11–12 dpf (Riley and Moorman 2000). The otoliths continue to grow daily throughout the life of the fish – sectioning through a mature otolith reveals rings of accretion, similar to the growth rings of a tree.

The initial seeding of the otolith may be disrupted by several factors. In zebrafish mutant lines where the underlying patterning of the ear is disrupted, there is scope for otolith formation to go awry. For example, in the *acerebellar*/*fgf8* mutant, a single otolith frequently forms, overlying a single sensory patch (Phillips et al. 2001). Otolith aggregation is also inhibited in morphants for the chaperone protein GP96 (Sumanas et al. 2003). GP96 has been implicated in the processing of newly synthesized proteins in the endoplasmic reticulum, but it looks likely here that its role may be to do with its potential extracellular chaperone activity. A low dose of *gp96* morpholino reduces the size of the utricular otolith but increases the size of the saccular; higher doses give a single, enlarged otolith, which may or may not be tethered. GP96 thus appears to be required for the initial seeding of otolith material, which might be mediated by its extracellular chaperone/intracellular processing roles (Sumanas et al. 2003). As development proceeds, seeding particles gradually disappear from the otic vesicle and otoliths continue to grow by the surface deposition of solutes from the endolymph (Riley et al. 1997).

## 7.2. Otolith Growth

For the actual synthesis of the otolith, three processes must occur (Hughes et al. 2006). Both otolithic matrix proteins and the gelatinous membrane in which the otolith will be embedded must be produced. Many proteins are now thought to be required for otolith assembly in the zebrafish. The glycoprotein core must be assembled correctly (Section 7.3) and there must be a local increase in $Ca^{2+}$ and $CO_3^-$ ions – $Ca^{2+}$ transport is mediated by the plasma membrane calcium ATPase (Cruz et al. 2009). The enzyme carbonic anhydrase is required for otolith maintenance in the adult (Shiao et al. 2005) and probably has a role at embryonic stages too, based on its expression pattern within the developing otic vesicle (Peterson et al. 1997) (Sections 3.1, 7.4). These steps must occur in a specific order and within a certain time frame in order for correct otolith formation to be initiated – for example, calcification must only occur overlying the maculae, as multiple, smaller otoliths spread throughout the ear could impair balance and hearing.

Correct otolith formation also relies on correct patterning of the maculae, with the correct arrangement of hair cells and supporting cells. In the *mib* mutant, where hair cells are produced at the expense of supporting cells, otolith seeding occurs but the otoliths fail to enlarge and remain tiny (Haddon et al. 1999). Formation of the maculae relies on boundary formation between presumptive sensory and non-sensory epithelia. At the interface of the two, transitional cells arise which are required for the secretion of the otolithic membrane, a key component of the inner ear architecture (Murayama et al. 2002) (Fig. 4.4). There must also be the establishment of the correct ionic milieu in the endolymph of the otic vesicle – disruption of the ionic composition within the developing ear (Section 3) can lead to a loss of otoliths, for example when subunits of the $Na^+/K^+$-ATPase are knocked out using a morpholino or are inhibited by ouabain, a chemical antagonist of this pump (Blasiole et al. 2006).

### 7.3. Role of Calcium-binding Proteins in Otolith Formation

Zebrafish have at least six major matrix proteins required for otolith initiation and subsquent growth – Otoconin-1 (Otoc1), Otolith Matrix Protein (Omp1), Otolin-1, Otopetrin1 (Otop1), Sparc and Starmaker (Stm) (Söllner et al. 2003, 2004b; Murayama et al. 2005; Kang et al. 2008; Petko et al. 2008). The *starmaker* (*stm*) gene was amongst the first to be identified as being required for the correct morphology of the otolith (Söllner et al. 2003). In the absence of Stm function, the initial steps of seeding and tethering (Section 7.1) occur as normal, but the otolith becomes "star-shaped", with its fibrous matrix being disorganized and excluded from inorganic calcium carbonate crystals within it. X-ray diffraction analysis of the crystals in *stm* morphants shows that they are composed of the more stable, calcite form of calcium carbonate, compared to the smaller, aragonite crystals found in the wild type. The Starmaker protein is thought to be a zebrafish structural homolog of the human dentin sialophosphoprotein (DSPP), mutations in which cause dentinogenesis imperfecta and hearing loss, suggesting a role for the human protein in the inner ear (Xiao et al. 2001). The protein is highly acidic with a repeating DSS motif (Kaplon et al. 2008) and it is the phosphorylation of these serine residues that could create spatial structures capable of binding high numbers of calcium ions, a process essential for bridging inorganic crystals with the protein backbone of the otolith. In the zebrafish, Starmaker protein is concentrated at the surface of the otolith facing the sensory epithelium, and thus may regulate growth at this surface (Söllner et al. 2003). The protein lacks a rigid tertiary structure and exists as a highly dynamic, labile structure with an extended, rod-shaped conformation, which allows its interaction with other components of otolith biogenesis such as Omp-1 and Otolin-1 (Murayama et al. 2005; Kaplon et al. 2008, 2009).

Omp-1 is a glycoprotein thought to be a member of the transferrin family, whereas Otolin-1 is a collagenous protein of the type VIII/X family (Murayama et al. 2005). Results of morpholino knockdown experiments show that neither protein is required for the initial seeding or tethering of the otolith, but that they are required for correct otolith growth – *zomp-1* morphants have smaller, slower-growing otoliths, whereas *zotolin-1* morphants have smaller, fused otoliths. In both cases, hair and supporting cells form normally but there is agenesis of the semicircular canal system. Murayama et al. (2005) speculate as to the roles of these proteins in otolith biogenesis – the lack of Otolin-1 immunoreactivity in *zomp-1* morphants suggests that the latter is required for the deposition of Otolin-1; however, the converse arrangement does not seem to be the case. Otolin-1 may be required for the anchoring of otoliths in the otolithic membrane (the development of which has still not been characterized fully in the zebrafish) and the stability of the organic matrix within the otoliths – *zotolin-1* morphant otoliths appear to swell upon fixation and undergo more rapid decalcification than their untreated siblings.

Whereas other biomineralized structures such as bones or teeth rely upon direct epithelial contact for their calcification, otoliths form within an extracellular space, and so depend upon high levels of matter being secreted into the otic vesicle. The *backstroke* (*bks*) mutant, where otoliths fail to form, was found to be a lesion in the *otopetrin1* gene (Söllner et al. 2004b). Mutations in the murine *Otopetrin* gene cause the mouse *tilted* and *mergulhador* phenotypes – mice have a vestibular disruption caused by a lack of otoconia (Hurle et al. 2003). In zebrafish, *otop1* is expressed in the developing sensory epithelium from 16.5 hpf and becomes progressively restricted to the hair cells of the maculae by 80 hpf (Hughes et al. 2004; Söllner et al. 2004b). In *bks* mutants, the loss of otoliths is thought to be caused by a secretory defect, as Stm protein trafficking from the hair cells is affected: ectopic, high levels of the protein are found in the epithelium and no Stm-positive seeding particles are seen at 24 hpf. At later stages, Stm is found in the otolithic membrane in *bks*, suggesting that its secretion is only being blocked in a subset of cells, presumably the hair cells, but this later appearance of Stm in the lumen of the otic cavity cannot compensate for the earlier defects (Söllner et al. 2004b).

The formation of the underlying matrix on which calcium carbonate deposition occurs is a critical component of otolith genesis. The crystal polymorph of $CaCO_3$ in the otolith is dependent on which matrix proteins are present – for example, birds and mammals have predominantly calcite in their otoconia, embedded in Otoconin90, whereas garfish have vateritic calcium carbonate surrounding a core of Otoconin54 (Pote and Ross 1991; Wang et al. 1998). As discussed above, Omp-1 has been shown to be the most abundant protein in the zebrafish (Hughes et al. 2004) but it has been proposed that it may lie downstream of Otoc1, the zebrafish ortholog of Otoconin90, in the otolith biogenesis pathway (Petko et al. 2008). A morpholino-based knockdown of both genes gives a synergistic phenotype compared with either gene individually, with a full spectrum of otolith phenotypes – size reduction, absence and extra otolithic bodies (Petko et al. 2008). Otoconin1 is another highly acidic protein and its role here may be to provide a rigid scaffold for $CaCO_3$ binding and the recruitment of other proteins into the matrix.

Analysis of the otoliths of other fish species such as Medaka, trout and catfish (Kang et al. 2008; Nemoto et al. 2008) has led to the identification of the other proteins that are likely to be critical for otolith formation and stability. The protein Sparc has been found to have a key role in directing normal otolith growth in zebrafish (Kang et al. 2008). A secreted, acidic, cysteine-rich glycoprotein, it has collagen- and $Ca^{2+}$-binding motifs and is found on the surface of the otoliths and throughout the epithelium at 28 hpf, suggesting that it is incorporated at early stages of biomineralization. *Sparc* morphants have a variety of otolith defects based on a core defect of reduced growth with an increased protein content, suggesting that recruitment of $CaCO_3$ was lacking – this is in comparison with the *stm* morphant phenotype, where mineralization of the otoliths was in fact enhanced. The otolith abnormalities in *sparc* morphants encompass all of the phenotypes found in *otoc-1*, *omp-1* and *otolin-1* morphants, consistent with the role of Sparc in interacting with some or all of the other otolith matrix proteins. In other systems, Sparc has been found in basement membranes, where it modulates the assembly of extracellular matrix but is not involved itself in structural support (Bradshaw and Sage 2001). Its C-terminal extracellular domain can bind collagen, and may act as a chaperone during matrix assembly; its N-terminal region has affinity for $Ca^{2+}$ (Bradshaw and Sage 2001), and thus Sparc may act as a linker between the inorganic and organic phases of the otolith.

### 7.4. Role of Ion Channels and Tight Junctions in Otolith Formation

The ionic composition of the endolymph plays a critical role in otolith formation. Blasiole and colleagues (Blasiole et al. 2006) demonstrated the contribution of certain subunits of the $Na^+$-$K^+$-ATPase pump in directing otolith formation. Of the six pump subunits expressed in the ear, morpholinos directed against *atp1a1a.1* and *atp1b2b* were found to cause abnormalites in otolith formation, with a combination of the two abrogating otolith formation completely. This suggests that the $Na^+$ and $K^+$ concentrations within the endolymph may be crucial for otolith seeding, although the morphants had additional underlying otic defects, which may be clouding the underlying mechanism. Note that the loss of *nkcc1* function in the *little ears* mutant has no apparent effect on otolith formation (Abbas and Whitfield 2009), despite a presumed disruption of $Na^+$ and $K^+$ concentrations leading to endolymph collapse, so perhaps these ions are not as critical as has been proposed.

In order to preserve the exquisitely balanced ionic composition of the endolymph, the otic vesicle must remain a sealed chamber. The cells of the otic epithelium are polarized in an apico-basal manner and form apical tight junctions between cells, components of which are known to be the Claudin family of proteins. A retroviral insertion in *claudinj* described by Hardison et al. (2005) results in embryos with tiny otoliths but no other apparent defects; the fish die from starvation at 12 dpf, underlining the importance of at least the utricular otolith for survival (Riley and Moorman 2000). Sections through the otic vesicle at 31 dpf show that mutants have enlarged seeding particles that are distributed randomly throughout the otic vesicle, rather than being concentrated at the anterior and posterior poles; at 5 dpf, the reduced otolith size is accompanied by an accumulation of unincorporated material within the ear. This even distribution of seeding particles may be reducing the efficiency of otolith seeding – the usual localization of Claudinj protein dorsolaterally in the otic epithelium might create a microenvironment that excludes otolithic matter from accumulating there, forcing it to build up preferentially in ventromedial regions (Hardison et al. 2005).

Since the otolith is mostly composed of calcium carbonate, a mechanism must exist within the otic epithelium to allow calcium carbonate to accrue and precipitate within the endolymph. The plasma membrane calcium ATPase isoform 2 (ATP2B2/PCMA1) is an active $Ca^{2+}$ transporter that can remove calcium ions from cells against steep concentration gradients and it is thought to have specific importance in the physiology of the inner ear. The *atp2b1a* gene is expressed ubiquitously in the embryonic zebrafish ear, with increased levels in the five sensory patches; in the adult, this expression is seen within the macular hair cells and in the ionocytes and transitional cells surrounding them (Cruz et al. 2009). Knockdown of the gene in the embryo has specific effects on the ear – larvae display balance defects with decreased otolith accretion, ectopic otolith positioning and occasional fusion. An interesting aspect of the morphant phenotype is a delay in otolith seeding, suggesting that although some aspects of otolith biogenesis are clearly dependent on the correct spatiotemporal aspect, other features may be more flexible. The number of hair cells is also reduced in the morphants – perhaps a critical concentration of calcium ions must be built up before calcium carbonate can be laid down onto the organic matrix. Meanwhile, the enzyme carbonic anhydrase (*cahz*) is expressed from the 20 somite stage at the anterior and posterior poles of the otic vesicle (Kudoh et al. 2001; Thisse et al. 2001) and is maintained in the ear throughout embryonic, larval and adult stages (Fig. 4.3). As yet, no knockdown of this gene has been published, but it is tempting to speculate that it would have an essential role in otolith formation as it catalyzes the formation of bicarbonate ions. Preliminary data (L. Abbas, unpublished) show that this is indeed the case, since treating embryos with the carbonic anhydrase inhibitor acetazolamide leads to a decrease in otolith size.

The cadherin family of homotypic cell adhesion molecules has also been shown to be involved in otolith formation, indicating a surprising and novel extracellular role for these proteins (Clendenon et al. 2009). Cadherin11 (Cdh11) is expressed only weakly in lateral cell–cell contacts within the otic epithelium; instead, the majority of protein is found within the cell cytoplasm, from where it is transported into the endolymph in vesicles and deposited on the surface of the otoliths. Morphants for *cdh11* still produce these vesicles but have inhibited otolith growth, suggesting that Cdh11 might be required for the transport of packages of material needed for biomineralization. These vesicles are increased in size in the morphants, raising the possibility that they collide and fuse with each other rather than the surface of the otolith, inhibiting the deposition of otolith constituents (Clendenon et al. 2009).

### 7.5. Role of Cilia in Otolith Formation

As described above, the kinocilium present on each tether cell is the site of otolith seeding and particle aggregation from 21.5 hpf onwards (Pisam et al. 2002). Riley et al. (1997) demonstrated that ciliary flow in the otic vesicle acts to move the seeding particles around to prevent them aggregating prematurely. Constraining the seeding particles with laser tweezers could cause large, untethered masses to form, which would fuse with the already-tethered growing otolith. In addition, growth of one otolith could be enhanced at the expense of the other, suggesting that there is a finite amount of otolith-generating material present. While the Riley group proposed that the short cilia present on all otic epithelial cells were responsible for the ciliary flow, a recent study suggests that these short cilia are in fact immotile, and that it is the longer tether kinocilia themselves – beating at a frequency of 34 Hz – that create local vortices to attract otolith precursor particles (Colantonio et al. 2009). The dynein regulatory complex is thought to be critical for tether cell kinocilia motility – the zebrafish homolog of the trypanin subunit of the axonemal regulatory complex is *gas8*, morphants for which display variable otolith abnormalities amongst other developmental defects. Morpholino knockdown of the *lrrc50* and *dnah9* genes, both of which are also required for cilia motility, gave the same phenotype (Colantonio et al. 2009). Several other mutants have been described that show defects in ciliary formation, including in the ear (Omori et al. 2008; Huang and Schier 2009; Wilkinson et al. 2009); these will be useful tools for examining the role of cilia in otolith formation and ear patterning in more detail.

### 7.6. Morphology and Remodeling of the Otoliths in the Adult

In the adult fish, there is a logical conundrum – if the otolith continues to grow daily, how can this occur without the risk of crushing the delicate underlying sensory epithelium? Shiao et al. (2005) examined the morphology of the adult otolith and by studying the diurnal growth rings in longitudinal sections, saw a decreased growth rate along the central sulcus, a groove running along the medial face of the otolith on the surface proximal to the hair cells. They looked at the complement of ion channels expressed within and just peripheral to the maculae and found that the hair cells express $H^+$-ATPase apically, which they suggest will bring about a localized acidification of the endolymph. This in turn will hinder the formation of calcium carbonate, thus slowing the growth of the otolith or perhaps even dissolving it in the region of the sulcus; this would help to protect the hair cells from mechanical damage. Conversely, the transitional cells surrounding the maculae express Cahz (Fig. 4.3), which generates $HCO_3^-$ ions, and produce otolith matrix protein in a milieu of alkaline endolymph, allowing rapid biomineralization in a ridge-like thickened ring on the proximal surface of the otolith. Growth on the distal surface of the otolith is thought to be slower, relying on passive diffusion of otolith components generated by these proximal cells (Shiao et al. 2005).

## 8. ADULT HEARING

### 8.1. Growth of the Sensory Maculae

In fish, unlike most other vertebrates, hair cells are added continually to the sensory maculae in the ear as the fish grows. Between day 7 and day 21, this is thought to be in the region of 10–15 cells per day in the utricle and saccule, dropping to ten per day in the utricle and lagena and five per day in the saccule at one year of age (Bang et al. 2001). Higgs and coworkers counted a total of 2100 hair cells in the mature zebrafish saccule and 3500 in the lagena (Higgs et al. 2003). It has been argued, however, that after 10 months of age, there is no net addition of hair cells but cells are produced to replace those that have been lost (Higgs et al. 2002). The sensory maculae also undergo morphological changes during larval and juvenile development: the posterior saccular patch matures from its characteristic "ping-pong bat" shape at 5 dpf (Haddon and Lewis 1996) into a spindle-shaped area at 29 dpf via a pear-shaped intermediate at 21 dpf (Bang et al. 2001). The lagenar macula begins as an oval patch developing at 21 dpf and elongates over the subsequent week to assume a pear-shaped morphology. The utricular macula, however, keeps its initial rounded shape (Platt 1993; Bang et al. 2001).

The density of hair cells varies along the length of the maculae, with considerably more cells at the caudal than the rostral end having been reported in the saccule (Higgs et al. 2002) and more cells around the periphery than in the center being recorded for the lagena and the saccule (Bang et al. 2001; Schuck and Smith 2009). Each macula has characteristic hair cell polarity patterns (reviewed in Whitfield et al. 2002), and hair bundle morphology varies between hair cell zones within a single patch (Bang et al. 2001).

### 8.2. Development of Hearing Thresholds

The range of tones detected by adult fish ranges from 100 Hz to 4000 Hz with a peak in the 600–1000 Hz frequency band; the mean hearing sensitivity of 127 dB decreases either side of the best frequency of 800 Hz (Higgs et al. 2002). No change in hearing sensitivity or bandwidth is observed as the macula increases in size (Higgs et al. 2002). It has been proposed that more hair cells are needed to maintain the consistency of hearing as the distance between peripheral auditory structures and the ear increases with body length.

As the zebrafish increases in length from juvenile to adult, there is no change in auditory sensitivity, response latency or amplitude using ABR measurements as a readout for stimulation (Higgs et al. 2003). However, there is a gradual expansion of the maximum detectable sound frequency from 200 Hz in a 10 mm fish to 4000 Hz in a 20 mm animal, which is coincident with the development of the Weberian ossicles, a set of modified vertebrae that conduct vibrations from the swim bladder to the inner ear. The ossicles begin to develop when the fish is 7 mm in length; by 20 mm, they are well formed and, together with ligaments, form an unbroken chain with no gaps between elements (Section 8.3; Fig. 4.6). Concomitant with ossicle formation is an expansion in the size of the swim bladder, the anterior chamber becoming more spherical and the posterior more elongate; both increase in length (Higgs et al. 2003; Robertson et al. 2007). It has been posited that the early responses to frequencies of 200 Hz and lower might be mediated by direct stimulation of the ear and lateral line, whereas responses to 400 Hz and above are largely auditory, with the increase in detectable frequency resulting from increased swim bladder and Weberian ossicle size rather than the addition of high-frequency-sensing hair cells in the saccule (Higgs et al. 2003); thus, the developmental changes in hearing ability may be driven by the development of auxiliary structures rather than by the ear itself.

The importance of the swim bladder as an accessory organ in zebrafish hearing has been demonstrated by several groups. At 4 dpf, zebrafish larvae fail to respond to a tonal stimulus, whereas at 5 dpf and beyond, an acoustically evoked behavioral response (AEBR) can be elicited (Zeddies and Fay 2005). This is eliminated in adults, but not in larvae, if the swim bladder is deflated. Although it is likely to be the anterior chamber that is necessary for hearing, both the anterior and posterior chambers must be extinguished for abrogation of the AEBR, as, if the posterior chamber remains intact, it can reinflate the anterior via the ductus communicans. In a high-throughput screen for hearing defects, Bang et al. (2002) assessed adult fish on the basis of their acoustic startle reflex, an unconditioned reflex triggered by acoustic and visual stimuli, modulated by input from the lateral line. Signals from the VIIIth nerve are transmitted to the Mauthner neuron and reticulospinal "escape network" to effect a sharp and rapid contraction of the body musculature allowing a quick getaway. Fish that failed to escape in response to a 400 Hz tone were examined by X-ray and nearly half were found to have abnormal swim bladder morphology or position relative to the Weberian apparatus. A similar number were found to have vertebral abnormalities, which also may have been affecting sound transmission from the swim bladder to the inner ear.

The gas within the swim bladder is compressed by sound pressure waves and consequently, the volume change relayed to the ear is proportional to the loudness of the sound received by the swim bladder: the Weberian ossicles are thought to provide a 40 dB gain in pressure sensitivity. However, despite this greater input to the mature inner ear compared to that

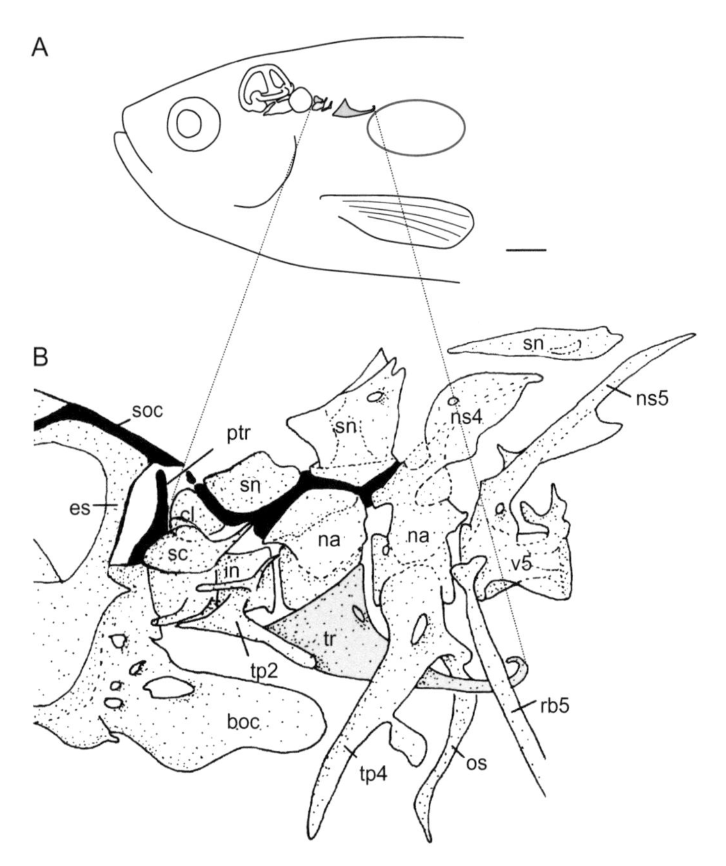


**Fig. 4.6.** Weberian ossicles in the zebrafish. (A) Sketch of an adult zebrafish head showing the position of the Weberian ossicular chain relative to the anterior chamber of the swimbladder (blue) and the inner ear. (B) Drawing of the Weberian apparatus from a 30 mm long adult zebrafish. The four ossicles, the tripus (tr), intercalarium (in), scaphium (sc) and claustrum (cl) are lightly colored for clarity. Stippling indicates bone; black shading indicates cartilage. boc, basioccipital; es, exoccipital strut; na, neural arch; ns4, neural spine of vertebra 4; ns5, neural spine of vertebra 5; os, os suspensorium; ptr, perineural tube remnant; rb5, pleural rib of vertebra 5; soc, supraoccipital; sn, supraneural; tp2, transverse process of vertebra 2; tp4, transverse process of vertebra 4; v5, vertebra 5. Ligaments are not shown. Anterior to the left. Reproduced, with permission, from Grande and Young (2004). See color plate section

perceived by larvae for a given sound pressure level, the magnitude of the acoustic startle response elicited remains the same, leading to the interesting hypothesis that adult fish must somehow temper their response to maintain an appropriate reaction to a given stimulus (Zeddies and Fay 2005).

It may be the case that larval fish are responding to the "nearfield component" of sound wave propagation, a local particle flow close to the sound source which decays rapidly with distance and is acting directly on the otoliths, whereas adults are capable of hearing long-range pressure changes in the "farfield", mediated by the concerted actions of the swim bladder and the Weberian apparatus. Zeddies and Fay (2005) suggest that thresholds might be different in animals tested at further distances from the sound source; this may account for the difference in maximum frequency response observed in their work (1200 Hz) and that seen by Higgs et al. (4000 Hz; Higgs et al. 2002, 2003).

### 8.3. Development and Anatomy of the Weberian Ossicles

The axial skeleton of the zebrafish is derived from somitic mesoderm in the embryo. Vertebrae are derived from the sclerotome, a mesenchymal population in the ventral somite, which migrates to surround axial midline structures and differentiates as cartilage and bone. The precaudal vertebrae form such structures as the neural arches and spines, and the most anterior will develop into the Weberian ossicles, a novel adaptation in the otophysan lineage. Comparison of the ontogeny and homology of the Weberian ossicles in different taxa has shown complex patterns of homology amongst the component parts (Grande and Young 2004). In sister taxa, the Gonorynchiformes and the Clupeomorpha, rudiments of the ossicles are present, suggesting that the Weberian apparatus may have evolved as a series of subunits and only gained its function as a sound-transducing unit once all the elements were present and in contact with each other (Bird and Mabee 2003). In the zebrafish, the ossicles remain unfused throughout the life of the fish. This is assumed to be a secondary condition, since the basal state is thought to involve the fusion of the second and third bony elements (Bird and Mabee 2003).

The Weberian ossicles consist of four, tiny, paired bones, the scaphium, claustrum, intercalarium and tripus, which lie adjacent to the first four vertebrae and connect the swim bladder to the ears via the sinus impar (Bang et al. 2001; Bird and Mabee 2003) (Fig. 4.6). Their development is an example of axial regionalization, where vertebral types become adapted to differing physiological requirements by differential growth rates – for example, the neural arch of the first precaudal vertebra becomes modified into the cup-shaped scaphium (Bird and Hernandez 2009). The most anterior bone, the scaphium, is connected at one end to the posterolateral wall of the sinus impar; its posterior end is connected to the anterolateral process, the manubrium, of the intercalarium by the interossicular ligament. Dorsomedial and connected to the scaphium is the claustrum, a triangular-shaped bone which extends anteroventrally to encircle the neural tube. The claustrum is immobile, being fused to the first vertebra, and its ventrolateral face supports the wall of the atrium of the sinus impar. The interossicular ligament also connects the rod-shaped intercalarium at its posterior extent to the largest of the ossicles, the tripus. Triangular in shape, the tripus is the closest ossicle to the swim bladder and spans most of the distance between the swim bladder and the ear (around 1 mm). Its anterior process connects to the intercalarium; posteriorly, it articulates with the camera aerea Weberiana of the swim bladder and the os suspensorium, a bone curving along the anterior chamber of the swim bladder. This and other accessory support structures, such as the roofing cartilage and the supraneurals, form part of the wider Weberian apparatus. The scaphium and intercalarium both show shifts in patterns of growth during development to allow gradual but precise growth in a specific region of the bone – their eventual small size permits a tight relay of sounds without any loss of signal to the surrounding tissue. Meanwhile, the tripus has regions of differential growth that result in its anterior part being significantly shorter than its posterior part. Consequently, this allows for signal intensification as a result of swim bladder movement (Bird and Hernandez 2009).

## 9. CONCLUDING REMARKS

Despite the obvious anatomical differences of the zebrafish ear compared to a mammalian ear – the lack of a cochlea, and the intricate Weberian apparatus, for example – many of the molecular mechanisms of inner ear development, endolymph generation, and hair cell development and function are conserved with those in the mammal. The zebrafish embryo has great advantages as a disease model, with many potential applications beyond analysis of the developmental mechanisms involved. Zebrafish can be used to examine and clarify the underlying pathology (see, for example, Seiler et al. 2004, 2005; Dutton et al. 2009), identify new genetic causes for human disorders, screen for compounds to be exploited in the drug discovery process (Zon and Peterson 2005), and to identify factors that protect against ototoxicity (Chiu et al. 2008; Coffin et al. 2009; Ou et al. 2009; Owens et al. 2009). Coupled with continued exciting improvements in transgenic and imaging techniques, which allow the analysis of phenotypes in astonishing detail (Huisken and Stainier 2009), these attributes will ensure that the zebrafish remains at the forefront of hearing research for many years to come.

ACKNOWLEDGMENTS

We thank Herwig Baier, Shawn Burgess, Terry Grande, Kate Hammond, Pung-Pung Hwang, Jin Liang, Yoichi Oda, Peter Santi, and Tong Xiao for contributing images, Bill

Chaudhry and Victoria Hildreth for help with generating the image in Fig. 4.1B, Chris Hill for help with generating the image in Fig. 4.5B, and Kate Hammond for proofreading the manuscript. Work in the Whitfield lab was supported by the MRC (G0300196).

## REFERENCES

Abbas, L., and Whitfield, T. T. (2009). Nkcc1 (Slc12a2) is required for the regulation of endolymph volume in the otic vesicle and swim bladder volume in the zebrafish larva. *Development* **136**, 2837–2848.

Abelló, G., and Alsina, B. (2007). Establishment of a proneural field in the inner ear. *Int. J. Dev. Biol.* **51**, 483–493.

Aman, A., and Piotrowski, T. (2009). Multiple signaling interactions coordinate collective cell migration of the posterior lateral line primordium. *Cell. Adh. Migr.* **3**, 365–368.

Avraham, K. B., Hasson, T., Steel, K. P., Kingsley, D. M., Russell, L. B., Mooseker, M. S., Copeland, N. G., and Jenkins, N. A. (1995). The mouse Snell's waltzer deafness gene encodes an unconventional myosin required for the structural integrity of inner ear hair cells. *Nat. Genet.* **11**, 369–375.

Bang, P., Sewell, W., and Malicki, J. (2001). Morphology and cell type heterogeneities of the inner ear epithelia in adult and juvenile zebrafish (*Danio rerio*). *J. Comp. Neurol.* **438**, 173–190.

Bang, P. I., Yelick, P. C., Malicki, J. J., and Sewell, W. F. (2002). High-throughput behavioral screening method for detecting auditory response defects in zebrafish. *J. Neurosci. Meth.* **118**, 177–187.

Behra, M., Bradsher, J., Sougrat, R., Gallardo, V., Allende, M. L., and Burgess, S. M. (2009). *Phoenix* is required for mechanosensory hair cell regeneration in the zebrafish lateral line. *PLoS Genet.* **5**, e1000455.

Bernard, C., Ferrary, E., and Sterkers, O. (1986). Production of endolymph in the semicircular canal of the frog *Rana esculenta*. *J. Physiol.* **371**, 17–28.

Bever, M. M., and Fekete, D. M. (2002). Atlas of the developing inner ear in zebrafish. *Dev. Dyn.* **223**, 536–543.

Bird, N. C., and Hernandez, L. P. (2009). Building an evolutionary innovation: Differential growth in the modified vertebral elements of the zebrafish Weberian apparatus. *Zoology (Jena)* **112**, 97–112.

Bird, N. C., and Mabee, P. M. (2003). Developmental morphology of the axial skeleton of the zebrafish, *Danio rerio* (Ostariophysi: Cyprinidae). *Dev. Dyn.* **228**, 337–357.

Blasiole, B., Canfield, V., Degrave, A., Thisse, C., Thisse, B., Rajarao, J., and Levenson, R. (2002). Cloning, mapping, and developmental expression of a sixth zebrafish Na,K-ATPase alpha1 subunit gene (atp1a1a.5). *Mech. Dev.* **119** (Suppl 1), S211–214.

Blasiole, B., Canfield, V. A., Vollrath, M. A., Huss, D., Mohideen, M. A., Dickman, J. D., Cheng, K. C., Fekete, D. M., and Levenson, R. (2006). Separate Na,K-ATPase genes are required for otolith formation and semicircular canal development in zebrafish. *Dev. Biol.* **294**, 148–160.

Blasiole, B., Degrave, A., Canfield, V., Boehmler, W., Thisse, C., Thisse, B., Mohideen, M. A., and Levenson, R. (2003). Differential expression of Na,K-ATPase alpha and beta subunit genes in the developing zebrafish inner ear. *Dev. Dyn.* **228**, 386–392.

Blasiole, B., Kabbani, N., Boehmler, W., Thisse, B., Thisse, C., Canfield, V., and Levenson, R. (2005). Neuronal calcium sensor-1 gene *ncs-1a* is essential for semicircular canal formation in zebrafish inner ear. *J. Neurobiol.* **64**, 285–297.

Bradshaw, A. D., and Sage, E. H. (2001). SPARC, a matricellular protein that functions in cellular differentiation and tissue response to injury. *J. Clin. Invest.* **107**, 1049–1054.

Brandt, A., Striessnig, J., and Moser, T. (2003). CaV1.3 channels are essential for development and presynaptic activity of cochlear inner hair cells. *J. Neurosci.* **23**, 10832–10840.

Brignull, H. R., Raible, D. W., and Stone, J. S. (2009). Feathers and fins: Non-mammalian models for hair cell regeneration. *Brain Res.* **1277**, 12–23.

Budi, E. H., Patterson, L. B., and Parichy, D. M. (2008). Embryonic requirements for ErbB signaling in neural crest development and adult pigment pattern formation. *Development* **135**, 2603–2614.

Burgess, H. A., and Granato, M. (2007). Modulation of locomotor activity in larval zebrafish during light adaptation. *J. Exp. Biol.* **210**, 2526–2539.

Busch-Nentwich, E., Söllner, C., Roehl, H., and Nicolson, T. (2004). The deafness gene *dfna5* is crucial for *ugdh* expression and HA production in the developing ear in zebrafish. *Development* **131**, 943–951.

Chang, W., Brigande, J. V., Fekete, D. M., and Wu, D. K. (2004). The development of semicircular canals in the inner ear: Role of FGFs in sensory cristae. *Development* **131**, 4201–4211.

Chang, W., Lin, Z., Kulessa, H., Hebert, J., Hogan, B. L., and Wu, D. K. (2008). Bmp4 is essential for the formation of the vestibular apparatus that detects angular head movements. *PLoS Genet.* **4**, e1000050.

Chiu, L. L., Cunningham, L. L., Raible, D. W., Rubel, E. W., and Ou, H. C. (2008). Using the zebrafish lateral line to screen for ototoxicity. *J. Assoc. Res. Otolaryngol.* **9**, 178–190.

Clendenon, S. G., Shah, B., Miller, C. A., Schmeisser, G., Walter, A., Gattone, V. H., 2nd, Barald, K. F., Liu, Q., and Marrs, J. A. (2009). Cadherin-11 controls otolith assembly: Evidence for extracellular cadherin activity. *Dev. Dyn.* **238**, 1909–1922.

Coffin, A. B., Dabdoub, A., Kelley, M. W., and Popper, A. N. (2007). Myosin VI and VIIa distribution among inner ear epithelia in diverse fishes. *Hear. Res.* **224**, 15–26.

Coffin, A. B., Reinhart, K. E., Owens, K. N., Raible, D. W., and Rubel, E. W. (2009). Extracellular divalent cations modulate aminoglycoside-induced hair cell death in the zebrafish lateral line. *Hear. Res.* **253**, 42–51.

Colantonio, J. R., Vermot, J., Wu, D., Langenbacher, A. D., Fraser, S., Chen, J. N., and Hill, K. L. (2009). The dynein regulatory complex is required for ciliary motility and otolith biogenesis in the inner ear. *Nature* **457**, 205–209.

Collado, M. S., Burns, J. C., Hu, Z., and Corwin, J. T. (2008). Recent advances in hair cell regeneration research. *Curr. Opin. Otolarygol. & Head & Neck Surg.* **16**, 465–471.

Corey, D. P., and Hudspeth, A. J. (1983). Analysis of the microphonic potential of the bullfrog's sacculus. *J. Neurosci.* **3**, 942–961.

Cruz, S., Shiao, J. C., Liao, B. K., Huang, C. J., and Hwang, P. P. (2009). Plasma membrane calcium ATPase required for semicircular canal formation and otolith growth in the zebrafish inner ear. *J. Exp. Biol.* **212**, 639–647.

Delpire, E., Lu, J., England, R., Dull, C., and Thorne, T. (1999). Deafness and imbalance associated with inactivation of the secretory Na-K-2Cl co-transporter. *Nat. Genet.* **22**, 192–195.

Dixon, M. J., Gazzard, J., Chaudhry, S. S., Sampson, N., Schulte, B. A., and Steel, K. P. (1999). Mutation of the Na-K-Cl co-transporter gene *Slc12a2* results in deafness in mice. *Hum. Mol. Genet.* **8**, 1579–1584.

Dutton, K., Abbas, L., Spencer, J., Brannon, C., Mowbray, C., Nikaido, M., Kelsh, R. N., and Whitfield, T. T. (2009). A zebrafish model for Waardenburg syndrome type IV reveals diverse roles for Sox10 in the otic vesicle. *Dis. Model. Mech.* **2**, 68–83.

Ernest, S., Guadagnini, S., Prevost, M. C., and Soussi-Yanicostas, N. (2007). Localization of anosmin-1a and anosmin-1b in the inner ear and neuromasts of zebrafish. *Gene Expr. Patterns* **7**, 274–281.

Ernest, S., Rauch, G.-J., Haffter, P., Geisler, R., Petit, C., and Nicolson, T. (2000). *Mariner* is defective in *myosin VIIA*: A zebrafish model for human hereditary deafness. *Hum. Mol. Genet.* **9**, 2189–2196.

Esterberg, R., and Fritz, A. (2009). *dlx3b/4b* are required for the formation of the preplacodal region and otic placode through local modulation of BMP activity. *Dev. Biol.* **325**, 189–199.

Fay, R. R., and Popper, A. N. (1999a). Comparative hearing: Fish and amphibians. In: *Springer Handbook of Auditory Research* (R.R. Fay and A.N. Popper, eds). Vol. 11. Springer-Verlag, New York.

Fay, R. R., and Popper, A. N. (1999b). Hearing in fishes and amphibians: An introduction. In: *Comparative Hearing: Fish and Amphibians* (R.R Fay and A.N Popper, eds), Vol. 11, pp. 1–14. Springer Verlag, New York.

Flagella, M., Clarke, L. L., Miller, M. L., Erway, L. C., Giannella, R. A., Andringa, A., Gawenis, L. R., Kramer, J., Duffy, J. J., Doetschman, T., et al. (1999). Mice lacking the basolateral Na-K-2Cl cotransporter have impaired epithelial chloride secretion and are profoundly deaf. *J. Biol. Chem.* **274**, 26946–26955.

Friedman, L. M., Dror, A. A., Mor, E., Tenne, T., Toren, G., Satoh, T., Biesemeier, D. J., Shomron, N., Fekete, D. M., Hornstein, E., et al. (2009). MicroRNAs are essential for development and function of inner ear hair cells in vertebrates. *Proc. Natl Acad. Sci. USA* **106**, 7915–7920.

Fritzsch, B. (2003). Special issue on "Functional anatomy of ear connections". *Brain Res. Bull.* **60**, 395–396.

Fritzsch, B., Signore, M., and Simeone, A. (2001). *Otx1* null mutant mice show partial segregation of sensory epithelia comparable to lamprey ears. *Dev. Genes. Evol.* **211**, 388–396.

Froehlicher, M., Liedtke, A., Groh, K. J., Neuhauss, S. C., Segner, H., and Eggen, R. I. (2009). Zebrafish (*Danio rerio*) neuromast: Promising biological endpoint linking developmental and toxicological studies. *Aquat. Toxicol.* 307–319.

Fuchs, P. A., Glowatzki, E., and Moser, T. (2003). The afferent synapse of cochlear hair cells. *Curr. Opin. Neurobiol.* **13**, 452–458.

Ghanem, T. A., Breneman, K. D., Rabbitt, R. D., and Brown, H. M. (2008). Ionic composition of endolymph and perilymph in the inner ear of the oyster toadfish, *Opsanus tau*. *Biol. Bull.* **214**, 83–90.

Ghysen, A., and Dambly-Chaudiere, C. (2007). The lateral line microcosmos. *Genes Dev.* **21**, 2118–2130.

Gibson, F., Walsh, J., Mburu, P., Varela, A., Brown, K. A., Antonio, M., Beisel, K. W., Steel, K. P., and Brown, S. D. M. (1995). A type VII myosin encoded by the mouse deafness gene *shaker-1*. *Nature* **374**, 62–64.

Gleason, M. R., Nagiel, A., Jamet, S., Vologodskaia, M., López-Schier, H., and Hudspeth, A. J. (2009). The transmembrane inner ear (Tmie) protein is essential for normal hearing and balance in the zebrafish. *Proc. Natl Acad. Sci. USA* Epub ahead of print.

Granato, M., van Eeden, F. J. M., Schach, U., Trowe, T., Brand, M., Furutani-Seiki, M., Haffter, P., Hammerschmidt, M., Heisenberg, C.-P., Jiang, Y.-J., et al. (1996). Genes controlling and mediating locomotion behavior of the zebrafish embryo and larva. *Development* **123**, 399–413.

Grande, T., and Young, B. (2004). The ontogeny and homology of the Weberian apparatus in the zebrafish *Danio rerio* (Ostariophysi: Cypriniformes). *Zool. J. Linn. Soc.* **140**, 241–254.

Haddon, C., Jiang, Y.-J., Smithers, L., and Lewis, J (1998). Delta-Notch signalling and the patterning of sensory cell differentiation in the zebrafish ear: Evidence from the *mind bomb* mutant. *Development* **125**, 4637–4644.
Haddon, C., and Lewis, J. (1996). Early ear development in the embryo of the zebrafish, *Danio rerio*. *J. Comp. Neurol.* **365**, 113–123.
Haddon, C., Mowbray, C., Whitfield, T., Jones, D., Gschmeissner, S., and Lewis, J. (1999). Hair cells without supporting cells: Further studies in the ear of the zebrafish *mind bomb* mutant. *J. Neurocytol.* **28**, 837–850.
Haddon, C. M., and Lewis, J. H. (1991). Hyaluronan as a propellant for epithelial movement: The development of semicircular canals in the inner ear of *Xenopus*. *Development* **112**, 541–550.
Hammond, K. L., Loynes, H. E., Folarin, A. A., Smith, J., and Whitfield, T. T. (2003). Hedgehog signalling is required for correct anteroposterior patterning of the zebrafish otic vesicle. *Development* **130**, 1403–1417.
Hammond, K. L., Loynes, H. E., Mowbray, C., Runke, G., Hammerschmidt, M., Mullins, M. C., Hildreth, V., Chaudhry, B., and Whitfield, T. T. (2009). A late role for *bmp2b* in the morphogenesis of semicircular canal ducts in the zebrafish inner ear. *PLoS ONE* **4**, e4368.
Hammond, K. L., and Whitfield, T. T. (2006). The developing lamprey ear closely resembles the zebrafish otic vesicle: *otx1* expression can account for all major patterning differences. *Development* **133**, 1347–1357.
Hans, S., Christison, J., Liu, D., and Westerfield, M. (2007). Fgf-dependent otic induction requires competence provided by Foxi1 and Dlx3b. *BMC Dev. Biol.* **7**, 5.
Hans, S., Liu, D., and Westerfield, M. (2004). Pax8 and Pax2a function synergistically in otic specification, downstream of the Foxi1 and Dlx3b transcription factors. *Development* **131**, 5091–5102.
Hardelin, J. P., Levilliers, J., del Castillo, I., Cohen-Salmon, M., Legouis, R., Blanchard, S., Compain, S., Bouloux, P., Kirk, J., Moraine, C., et al. (1992). X chromosome-linked Kallmann syndrome: Stop mutations validate the candidate gene. *Proc. Natl Acad. Sci. USA* **89**, 8190–8194.
Hardison, A. L., Lichten, L., Banerjee-Basu, S., Becker, T. S., and Burgess, S. M. (2005). The zebrafish gene *claudinj* is essential for normal ear function and important for the formation of the otoliths. *Mech. Dev.* **122**, 949–958.
Harris, J. A., Cheng, A. G., Cunningham, L. L., MacDonald, G., Raible, D. W., and Rubel, E. W. (2003). Neomycin-induced hair cell death and rapid regeneration in the lateral line of zebrafish (*Danio rerio*). *J. Assoc. Res. Otolaryngol.* **4**, 219–234.
Hayes, A., Costa, T., and Polomeno, R. C. (1985). The Okihiro syndrome of Duane anomaly, radial ray abnormalities, and deafness. *Am. J. Med. Genet.* **22**, 273–280.
Higgs, D. M., Rollo, A. K., Souza, M. J., and Popper, A. N. (2003). Development of form and function in peripheral auditory structures of the zebrafish (*Danio rerio*). *J. Acoust. Soc. Am.* **113**, 1145–1154.
Higgs, D. M., Souza, M. J., Wilkins, H. R., Presson, J. C., and Popper, A. N. (2002). Age- and size-related changes in the inner ear and hearing ability of the adult zebrafish (*Danio rerio*). *JARO* **3**, 174–184.
Hilfiker, S. (2003). Neuronal calcium sensor-1: A multifunctional regulator of secretion. *Biochem. Soc. Trans.* **31**, 828–832.
Hill, J., Elliott, C., and Colquhoun, I. (1992). Audiological, vestibular and radiological abnormalities in Kallman's syndrome. *J. Laryngol. Otol.* **106**, 530–534.
Huang, P., and Schier, A. F. (2009). Dampened Hedgehog signaling but normal Wnt signaling in zebrafish without cilia. *Development* **136**, 3089–3098.

Hughes, I., Blasiole, B., Huss, D., Warchol, M. E., Rath, N. P., Hurle, B., Ignatova, E., Dickman, J. D., Thalmann, R., Levenson, R., et al. (2004). Otopetrin 1 is required for otolith formation in the zebrafish *Danio rerio*. *Dev. Biol.* **276**, 391–402.
Hughes, I., Thalmann, I., Thalmann, R., and Ornitz, D. M. (2006). Mixing model systems: Using zebrafish and mouse inner ear mutants and other organ systems to unravel the mystery of otoconial development. *Brain Res.* **1091**, 58–74.
Huisken, J., and Stainier, D. Y. (2009). Selective plane illumination microscopy techniques in developmental biology. *Development* **136**, 1963–1975.
Hurle, B., Ignatova, E., Massironi, S. M., Mashimo, T., Rios, X., Thalmann, I., Thalmann, R., and Ornitz, D. M. (2003). Non-syndromic vestibular disorder with otoconial agenesis in *tilted/mergulhador* mice caused by mutations in otopetrin 1. *Hum. Mol. Genet.* **12**, 777–789.
Jing, X., and Malicki, J. (2009). Zebrafish *ale oko*, an essential determinant of sensory neuron survival and the polarity of retinal radial glia, encodes the p50 subunit of dynactin. *Development* **136**, 2955–2964.
Kang, Y. J., Stevenson, A. K., Yau, P. M., and Kollmar, R. (2008). Sparc protein is required for normal growth of zebrafish otoliths. *J. Assoc. Res. Otolaryngol.* **9**, 436–451.
Kaplon, T. M., Michnik, A., Drzazga, Z., Richter, K., Kochman, M., and Ozyhar, A. (2009). The rod-shaped conformation of Starmaker. *Biochim. Biophys. Acta* **1794**, 1616–1624.
Kaplon, T. M., Rymarczyk, G., Nocula-Lugowska, M., Jakob, M., Kochman, M., Lisowski, M., Szewczuk, Z., and Ozyhar, A. (2008). Starmaker exhibits properties of an intrinsically disordered protein. *Biomacromolecules* **9**, 2118–2125.
Kimmel, C. B., Hatta, K., and Metcalfe, W. K. (1990). Early axonal contacts during development of an identified dendrite in the brain of the zebrafish. *Neuron* **4**, 535–545.
Kollmar, R., Fak, J., Montgomery, L. G., and Hudspeth, A. J. (1997). Hair cell-specific splicing of mRNA for the $\alpha_{1D}$ subunit of voltage-gated $Ca^{2+}$ channels in the chicken's cochlea. *Proc. Natl Acad. Sci. USA* **94**, 14889–14893.
Korzh, V., Sleptsova, I., Liao, J., He, J., and Gong, Z. (1998). Expression of zebrafish bHLH genes *ngn1* and *nrd* defines distinct stages of neural differentiation. *Dev. Dyn.* **213**, 92–104.
Kudoh, T., Tsang, M., Hukriede, N. A., Chen, X., Dedekian, M., Clarke, C. J., Kiang, A., Schultz, S., Epstein, J. A., Toyama, R., et al. (2001). A gene expression screen in zebrafish embryogenesis. *Genome Res.* **11**, 1979–1987.
Kwak, S.-J., Phillips, B. T., Heck, R., and Riley, B. B. (2002). An expanded domain of *fgf3* expression in the hindbrain of zebrafish *valentino* mutants results in mis-patterning of the otic vesicle. *Development* **129**, 5279–5287.
Kwon, H. J., and Riley, B. B. (2009). Mesendodermal signals required for otic induction: Bmp-antagonists cooperate with Fgf and can facilitate formation of ectopic otic tissue. *Dev. Dyn.* **238**, 1582–1594.
Lang, F., Vallon, V., Knipper, M., and Wangemann, P. (2007). Functional significance of channels and transporters expressed in the inner ear and kidney. *Am. J. Physiol. Cell Physiol.* **293**, C1187–C1208.
Lecaudey, V., Cakan-Akdogan, G., Norton, W. H., and Gilmour, D. (2008). Dynamic Fgf signaling couples morphogenesis and migration in the zebrafish lateral line primordium. *Development* **135**, 2695–2705.
Lecaudey, V., Ulloa, E., Anselme, I., Stedman, A., Schneider-Maunoury, S., and Pujades, C. (2007). Role of the hindbrain in patterning the otic vesicle: A study of the zebrafish *vhnf1* mutant. *Dev. Biol.* **303**, 134–143.
Léger, S., and Brand, M. (2002). Fgf8 and Fgf3 are required for zebrafish ear placode induction, maintenance and inner ear patterning. *Mech. Develop.* **119**, 91–108.

Legouis, R., Hardelin, J. P., Levilliers, J., Claverie, J. M., Compain, S., Wunderle, V., Millasseau, P., Le Paslier, D., Cohen, D., Caterina, D., et al. (1991). The candidate gene for the X-linked Kallmann syndrome encodes a protein related to adhesion molecules. *Cell* **67**, 423–435.

Leibovici, M., Safieddine, S., and Petit, C. (2008). Mouse models for human hereditary deafness. *Curr. Top. Dev. Biol.* **84**, 385–429.

Li, Y., Allende, M. L., Finkelstein, R., and Weinberg, E. S. (1994). Expression of two zebrafish *orthodenticle*-related genes in the embryonic brain. *Mech. Dev.* **48**, 229–244.

Liang, J. and Burgess, S. M. (2009). Gross and fine dissection of inner ear sensory epithelia in adult zebrafish (*Danio rerio*). *J. Vis. Exp. 27*, http://www.jove.com/index/details.stp?id=1211, doi:10.3791/1211

Lim, D. J. (1984). Otoconia in health and disease. A review. *Ann. Otol. Rhinol. Laryngol. Suppl.* **112**, 17–24.

Lindeman, H. H. (1966). Cellular pattern and nerve supply of the vestibular sensory epithelia. *Acta. Otolaryngol.* (Suppl. 224), 86+.

Liu, D., Chu, H., Maves, L., Yan, Y.-L., Morcos, P. A., Postlethwait, P., and Westerfield, M. (2003). Fgf3 and Fgf8 dependent and independent transcription factors are required for otic placode specification. *Development* **130**, 2213–2224.

Lowery, L. A., and Sive, H. (2005). Initial formation of zebrafish brain ventricles occurs independently of circulation and requires the *nagie oko* and *snakehead/atp1a1a.1* gene products. *Development* **132**, 2057–2067.

Ma, E. Y., and Raible, D. W. (2009). Signaling pathways regulating zebrafish lateral line development. *Curr. Biol.* **19**, R381–386.

Mackereth, M. D., Kwak, S.-J., Fritz, A., and Riley, B. B. (2005). Zebrafish *pax8* is required for otic placode induction and plays a redundant role with Pax2 genes in the maintenance of the otic placode. *Development* **132**, 371–382.

Maroon, H., Walshe, J., Mahmood, R., Kiefer, P., Dickson, C., and Mason, I (2002). Fgf3 and Fgf8 are required together for formation of the otic placode and vesicle. *Development* **129**, 2099–2108.

McDermott, B. M., Jr., Baucom, J. M., and Hudspeth, A. J. (2007). Analysis and functional evaluation of the hair-cell transcriptome. *Proc. Natl Acad. Sci. USA* **104**, 11820–11825.

Mendonsa, E. S., and Riley, B. B. (1999). Genetic analysis of tissue interactions required for otic placode induction in the zebrafish. *Dev. Biol.* **206**, 100–112.

Millimaki, B. B., Sweet, E. M., Dhason, M. S., and Riley, B. B. (2007). Zebrafish *atoh1* genes: Classic proneural activity in the inner ear and regulation by Fgf and Notch. *Development* **134**, 295–305.

Mitchem, K. L., Hibbard, E., Beyer, L. A., Bosom, K., Dootz, G. A., Dolan, D. F., Johnson, K. R., Raphael, Y., and Kohrman, D. C. (2002). Mutation of the novel gene *Tmie* results in sensory cell defects in the inner ear of spinner, a mouse model of human hearing loss DFNB6. *Hum. Mol. Genet.* **11**, 1887–1898.

Morsli, H., Tuorto, F., Choo, D., Postiglione, M. P., Simeone, A., and Wu, D. K. (1999). *Otx1* and *Otx2* activities are required for the normal development of the mouse inner ear. *Development* **126**, 2335–2343.

Mowbray, C., Hammerschmidt, M., and Whitfield, T. T. (2001). Expression of BMP signalling pathway members in the developing zebrafish inner ear and lateral line. *Mech. Dev.* **108**, 179–184.

Murayama, E., Herbomel, P., Kawakami, A., Takeda, H., and Nagasawa, H. (2005). Otolith matrix proteins OMP-1 and Otolin-1 are necessary for normal otolith growth and their correct anchoring onto the sensory maculae. *Mech. Dev.* **122**, 791–803.

Murayama, E., Takagi, Y., Ohira, T., Davis, J. G., Greene, M. I., and Nagasawa, H. (2002). Fish otolith contains a unique structural protein, otolin-1. *Eur. J. Biochem.* **269**, 688–696.

Namkung, Y., Skrypnyk, N., Jeong, M. J., Lee, T., Lee, M. S., Kim, H. L., Chin, H., Suh, P. G., Kim, S. S., and Shin, H. S. (2001). Requirement for the L-type $Ca^{(2+)}$ channel $\alpha_{(1D)}$ subunit in postnatal pancreatic beta cell generation. *J. Clin. Invest.* **108**, 1015–1022.
Nayak, G. D., Ratnayaka, H. S., Goodyear, R. J., and Richardson, G. P. (2007). Development of the hair bundle and mechanotransduction. *Int. J. Dev. Biol.* **51**, 597–608.
Naz, S., Giguere, C. M., Kohrman, D. C., Mitchem, K. L., Riazuddin, S., Morell, R. J., Ramesh, A., Srisailpathy, S., Deshmukh, D., Griffith, A. J., et al. (2002). Mutations in a novel gene, *TMIE*, are associated with hearing loss linked to the *DFNB6* locus. *Am. J. Hum. Genet.* **71**, 632–636.
Nemoto, Y., Chatani, M., Inohaya, K., Hiraki, Y., and Kudo, A. (2008). Expression of marker genes during otolith development in medaka. *Gene Expr. Patterns* **8**, 92–95.
Neuhauss, S. C. F., Solnica-Krezel, L., Schier, A. F., Zwartkruis, F., Stemple, D. L., Malicki, J., Abdelilah, S., Stainier, D. Y. R., and Driever, W. (1996). Mutations affecting craniofacial development in zebrafish. *Development* **123**, 357–367.
Nicolson, T. (2005a). Fishing for key players in mechanotransduction. *Trends. Neurosci.* **28**, 140–144.
Nicolson, T. (2005b). The genetics of hearing and balance in zebrafish. *Annu. Rev. Genet.* **39**, 9–22.
Nicolson, T., Rüsch, A., Friedrich, R. W., Granato, M., Ruppersberg, J. P., and Nüsslein-Volhard, C. (1998). Genetic analysis of vertebrate sensory hair cell mechanosensation: The zebrafish circler mutants. *Neuron* **20**, 271–283.
Nissen, R. M., Yan, J., Amsterdam, A., Hopkins, N., and Burgess, S. M. (2003). Zebrafish *foxi one* modulates cellular responses to Fgf signaling required for the integrity of ear and jaw patterning. *Development* **130**, 2543–2554.
Obholzer, N., Wolfson, S., Trapani, J. G., Mo, W., Nechiporuk, A., Busch-Nentwich, E., Seiler, C., Sidi, S., Söllner, C., Duncan, R. N., et al. (2008). Vesicular glutamate transporter 3 is required for synaptic transmission in zebrafish hair cells. *J. Neurosci.* **28**, 2110–2118.
Omata, Y., Nojima, Y., Nakayama, S., Okamoto, H., Nakamura, H., and Funahashi, J. (2007). Role of Bone morphogenetic protein 4 in zebrafish semicircular canal development. *Dev. Growth Diff.* **49**, 711–719.
Omori, Y., Zhao, C., Saras, A., Mukhopadhyay, S., Kim, W., Furukawa, T., Sengupta, P., Veraksa, A., and Malicki, J. (2008). Elipsa is an early determinant of ciliogenesis that links the IFT particle to membrane-associated small GTPase Rab8. *Nat. Cell Biol.* **10**, 437–444.
Ou, H. C., Cunningham, L. L., Francis, S. P., Brandon, C. S., Simon, J. A., Raible, D. W., and Rubel, E. W. (2009). Identification of FDA-approved drugs and bioactives that protect hair cells in the zebrafish (*Danio rerio*) lateral line and mouse (Mus musculus) utricle. *J. Assoc. Res. Otolaryngol.* **10**, 191–203.
Ou, H. C., Raible, D. W., and Rubel, E. W. (2007). Cisplatin-induced hair cell loss in zebrafish (*Danio rerio*) lateral line. *Hear. Res.* **233**, 46–53.
Owens, K. N., Coffin, A. B., Hong, L. S., Bennett, K. O., Rubel, E. W., and Raible, D. W. (2009). Response of mechanosensory hair cells of the zebrafish lateral line to aminoglycosides reveals distinct cell death pathways. *Hear. Res.* **253**, 32–41.
Peterson, R. E., Tu, C., and Linser, P. J. (1997). Isolation and characterization of a carbonic anhydrase homologue from the zebrafish (*Danio rerio*). *J. Mol. Evol.* **44**, 432–439.
Petko, J. A., Kabbani, N., Frey, C., Woll, M., Hickey, K., Craig, M., Canfield, V. A., and Levenson, R. (2009). Proteomic and functional analysis of NCS-1 binding proteins reveals novel signaling pathways required for inner ear development in zebrafish. *BMC Neurosci.* **10**, 27.
Petko, J. A., Millimaki, B. B., Canfield, V. A., Riley, B. B., and Levenson, R. (2008). Otoc1: A novel otoconin-90 ortholog required for otolith mineralization in zebrafish. *Dev. Neurobiol.* **68**, 209–222.

Phillips, B. T., Bolding, K., and Riley, B. B. (2001). Zebrafish *fgf3* and *fgf8* encode redundant functions required for otic placode induction. *Dev. Biol.* **235**, 351–365.
Phillips, B. T., Storch, E. M., Lekven, A. C., and Riley, B. B. (2004). A direct role for Fgf but not Wnt in otic placode induction. *Development* **131**, 923–931.
Piotrowski, T., Ahn, D.-G., Schilling, T. F., Nair, S., Ruvinsky, I., Geisler, R., Rauch, G.-J., Haffter, P., Zon, L. I., Zhou, Y., et al. (2003). The zebrafish *van gogh* mutation disrupts *tbx1*, which is involved in the DiGeorge deletion syndrome in humans. *Development* **130**, 5043–5052.
Pisam, M., Jammet, C., and Laurent, D. (2002). First steps of otolith formation of the zebrafish: Role of glycogen? *Cell. Tissue Res.* **310**, 163–168.
Pittlik, S., Domingues, S., Meyer, A., and Begemann, G. (2008). Expression of zebrafish *aldh1a3 (raldh3)* and absence of *aldh1a1* in teleosts. *Gene Exp. Patt.* **8**, 141–147.
Platt, C. (1993). Zebrafish inner ear sensory surfaces are similar to those in goldfish. *Hear. Res.* **65**, 133–140.
Platzer, J., Engel, J., Schrott-Fischer, A., Stephan, K., Bova, S., Chen, H., Zheng, H., and Striessnig, J. (2000). Congenital deafness and sinoatrial node dysfunction in mice lacking class D L-type $Ca^{2+}$ channels. *Cell* **102**, 89–97.
Pongs, O., Lindemeier, J., Zhu, X. R., Theil, T., Engelkamp, D., Krah-Jentgens, I., Lambrecht, H. G., Koch, K. W., Schwemer, J., Rivosecchi, R., et al. (1993). Frequenin–a novel calcium-binding protein that modulates synaptic efficacy in the Drosophila nervous system. *Neuron* **11**, 15–28.
Popper, A. N., and Liu, C. F. (2000). Structure–function relationships in fish otolith organs. *Fish. Res.* **46**, 15–25.
Pote, K. G., and Ross, M. D. (1991). Each otoconia polymorph has a protein unique to that polymorph. *Comp. Biochem. Physiol. B.* **98**, 287–295.
Rajarao, S. J., Canfield, V. A., Mohideen, M. A., Yan, Y. L., Postlethwait, J. H., Cheng, K. C., and Levenson, R. (2001). The repertoire of Na,K-ATPase α and β subunit genes expressed in the zebrafish, *Danio rerio*. *Genome Res.* **11**, 1211–1220.
Riley, B. B., and Moorman, S. J. (2000). Development of utricular otoliths, but not saccular otoliths, is necessary for vestibular function and survival in zebrafish. *J. Neurobiol.* **43**, 329–337.
Riley, B. B., Zhu, C., Janetopoulos, C., and Aufderheide, K. J. (1997). A critical period of ear development controlled by distinct populations of ciliated cells in the zebrafish. *Dev. Biol.* **191**, 191–201.
Robertson, G. N., McGee, C. A., Dumbarton, T. C., Croll, R. P., and Smith, F. M. (2007). Development of the swimbladder and its innervation in the zebrafish, *Danio rerio*. *J. Morphol.* **268**, 967–985.
Rotllant, J., Liu, D., Yan, Y. L., Postlethwait, J. H., Westerfield, M., and Du, S. J. (2008). Sparc (Osteonectin) functions in morphogenesis of the pharyngeal skeleton and inner ear. *Matrix Biol.* **27**, 561–572.
Rzadzinska, A. K., and Steel, K. P. (2009). Presence of interstereocilial links in waltzer mutants suggests *Cdh23* is not essential for tip link formation. *Neuroscience* **158**, 365–368.
Santi, P. A., Johnson, S. B., Hillenbrand, M., GrandPre, P. Z., Glass, T. J., and Leger, J. R. (2009). Thin-sheet laser imaging microscopy for optical sectioning of thick tissues. *BioTechniques* **46**, 287–294.
Santos, F., MacDonald, G., Rubel, E. W., and Raible, D. W. (2006). Lateral line hair cell maturation is a determinant of aminoglycoside susceptibility in zebrafish (*Danio rerio*). *Hear. Res.* **213**, 25–33.
Schier, A. F., Neuhauss, S. C. F., Harvey, M., Malicki, J., Solnica-Krezel, L., Stainier, D. Y. R., Zwartkruis, F., Abdelilah, S., Stemple, D. L., Rangini, Z., et al. (1996). Mutations affecting the development of the embryonic zebrafish brain. *Development* **123**, 165–178.

Schimmang, T. (2007). Expression and functions of FGF ligands during early otic development. *Int. J. Dev. Biol.* **51**, 473–481.

Schuck, J. B., and Smith, M. E. (2009). Cell proliferation follows acoustically-induced hair cell bundle loss in the zebrafish saccule. *Hear. Res.* **253**, 67–76.

Seiler, C., Ben-David, O., Sidi, S., Hendrich, O., Rusch, A., Burnside, B., Avraham, K. B., and Nicolson, T. (2004). Myosin VI is required for structural integrity of the apical surface of sensory hair cells in zebrafish. *Dev. Biol.* **272**, 328–338.

Seiler, C., Finger-Baier, K. C., Rinner, O., Makhankov, Y. V., Schwarz, H., Neuhauss, S. C., and Nicolson, T. (2005). Duplicated genes with split functions: Independent roles of *protocadherin15* orthologues in zebrafish hearing and vision. *Development* **132**, 615–623.

Seiler, C., and Nicolson, T. (1999). Defective calmodulin-dependent rapid apical endocytosis in zebrafish sensory hair cell mutants. *J. Neurobiol.* **41**, 424–434.

Serluca, F. C., Xu, B., Okabe, N., Baker, K., Lin, S. Y., Sullivan-Brown, J., Konieczkowski, D. J., Jaffe, K. M., Bradner, J. M., Fishman, M. C., et al. (2009). Mutations in zebrafish leucine-rich repeat-containing six-like affect cilia motility and result in pronephric cysts, but have variable effects on left-right patterning. *Development* **136**, 1621–1631.

Shen, Y. C., Jeyabalan, A. K., Wu, K. L., Hunker, K. L., Kohrman, D. C., Thompson, D. L., Liu, D., and Barald, K. F. (2008). The *transmembrane inner ear* (*tmie*) gene contributes to vestibular and lateral line development and function in the zebrafish (*Danio rerio*). *Dev. Dyn.* **237**, 941–952.

Shiao, J. C., Lin, L. Y., Horng, J. L., Hwang, P. P., and Kaneko, T. (2005). How can teleostean inner ear hair cells maintain the proper association with the accreting otolith? *J. Comp. Neurol.* **488**, 331–341.

Shin, J. B., Adams, D., Paukert, M., Siba, M., Sidi, S., Levin, M., Gillespie, P. G., and Grunder, S. (2005). *Xenopus* TRPN1 (NOMPC) localizes to microtubule-based cilia in epithelial cells, including inner-ear hair cells. *Proc. Natl Acad. Sci. USA* **102**, 12572–12577.

Sidi, S., Busch-Nentwich, E., Friedrich, R., Schoenberger, U., and Nicolson, T. (2004). *gemini* encodes a zebrafish L-type calcium channel that localizes at sensory hair cell ribbon synapses. *J. Neurosci.* **24**, 4213–4223.

Sidi, S., Friedrich, R. W., and Nicolson, T. (2003). NompC TRP channel required for vertebrate sensory hair cell mechanotransduction. *Science* **301**, 96–99.

Söllner, C., Burghammer, M., Busch-Nentwich, E., Berger, J., Schwartz, H., Riekel, C., and Nicolson, T. (2003). Control of crystal size and lattice formation by Starmaker in otolith biomineralization. *Science* **302**, 282–286.

Söllner, C., Rauch, G. J., Siemens, J., Geisler, R., Schuster, S. C., Tübingen 2000 Screen Consortium. Müller, U. and Nicolson, T. (2004a). Mutations in *cadherin 23* affect tip links in zebrafish sensory hair cells. *Nature* **428**, 955–959.

Söllner, C., Schwarz, H., Geisler, R., and Nicolson, T. (2004). Mutated otopetrin 1 affects the genesis of otoliths and the localization of Starmaker in zebrafish. *Dev. Genes Evol.* **214**, 582–590.

Solomon, K. S., Kudoh, T., Dawid, I. B., and Fritz, A. (2003). Zebrafish *foxi1* mediates otic placode formation and jaw development. *Development* **130**, 929–940.

Solomon, K. S., Kwak, S.-J., and Fritz, A. (2004). Genetic interactions underlying otic placode induction and formation. *Dev. Dyn.* **230**, 419–433.

Sugihara, I., and Furukawa, T. (1995). Potassium currents underlying the oscillatory response in hair cells of the goldfish sacculus. *J. Physiol.* **489**(Pt 2), 443–453.

Sumanas, S., Larson, J. D., and Miller Bever, M. (2003). Zebrafish chaperone protein GP96 is required for otolith formation during ear development. *Dev. Biol.* **261**, 443–455.

Sun, Z., Amsterdam, A., Pazour, G. J., Cole, D. G., Miller, M. S., and Hopkins, N. (2004). A genetic screen in zebrafish identifies cilia genes as a principal cause of cystic kidney. *Development* **131**, 4085–4093.

Tanimoto, M., Ota, Y., Horikawa, K., and Oda, Y. (2009). Auditory input to CNS is acquired coincidentally with development of inner ear after formation of functional afferent pathway in zebrafish. *J. Neurosci.* **29**, 2762–2767.

Thisse, B., Pfumio, S., Fürthauer, M., Loppin, B., Heyer, V., Degrave, A., Woehl, R., Lux, A., Steffan, T., Charbonnier, X. Q. et al. (2001). Expression of the zebrafish genome during embryogenesis. ZFIN online publication.

Trapani, J. G., Obholzer, N., Mo, W., Brockerhoff, S. E., and Nicolson, T. (2009). Synaptojanin1 is required for temporal fidelity of synaptic transmission in hair cells. *PLoS Genet.* **5**, e1000480.

Uemura, O., Okada, Y., Ando, H., Guedj, M., Higashijima, S., Shimazaki, T., Chino, N., Okano, H., and Okamoto, H. (2005). Comparative functional genomics revealed conservation and diversification of three enhancers of the *isl1* gene for motor and sensory neuron-specific expression. *Dev. Biol.* **278**, 587–606.

Walsh, E. C., and Stainier, D. Y. R. (2001). UDP-glucose dehydrogenase required for cardiac valve formation in zebrafish. *Science* **293**, 1670–1673.

Wang, L., Sewell, W. F., Kim, S. D., Shin, J. T., MacRae, C. A., Zon, L. I., Seidman, J. G., and Seidman, C. E. (2008). Eya4 regulation of $Na^+/K^+$-ATPase is required for sensory system development in zebrafish. *Development* **135**, 3425–3434.

Wang, Y., Kowalski, P. E., Thalmann, I., Ornitz, D. M., Mager, D. L., and Thalmann, R. (1998). Otoconin-90, the mammalian otoconial matrix protein, contains two domains of homology to secretory phospholipase A2. *Proc. Natl Acad. Sci. USA* **95**, 15345–15350.

Wangemann, P. (1995). Comparison of ion transport mechanisms between vestibular dark cells and strial marginal cells. *Hear. Res.* **90**, 149–157.

Wangemann, P. (2002). $K^+$ cycling and the endocochlear potential. *Hear. Res.* **165**, 1–9.

Waterman, R. E., and Bell, D. H. (1984). Epithelial fusion during early semicircular canal formation in the embryonic zebrafish, *Brachydanio rerio*. *Anat. Rec.* **210**, 101–114.

Weil, D., Blanchard, S., Kaplan, J., Guilford, P., Gibson, F., Walsh, J., Mburu, P., Varela, A., Levilliers, J., Weston, M. D., et al. (1995). Defective myosin VIIA gene responsible for Usher syndrome type 1B. *Nature* **374**, 60–61.

Weston, M. D., Pierce, M. L., Rocha-Sanchez, S., Beisel, K. W., and Soukup, G. A. (2006). MicroRNA gene expression in the mouse inner ear. *Brain Res.* **1111**, 95–104.

White, R. M., Sessa, A., Burke, C., Bowman, T., LeBlanc, J., Ceol, C., Bourque, C., Dovey, M., Goessling, W., Burns, C. E., et al. (2008). Transparent adult zebrafish as a tool for in vivo transplantation analysis. *Cell Stem Cell* **2**, 183–189.

Whitfield, T. T. (2002). Zebrafish as a model for hearing and deafness. *J. Neurobiol.* **53**, 157–171.

Whitfield, T. T., Granato, M., van Eeden, F. J., Schach, U., Brand, M., Furutani-Seiki, M., Haffter, P., Hammerschmidt, M., Heisenberg, C. P., Jiang, Y. J., et al. (1996). Mutations affecting development of the zebrafish inner ear and lateral line. *Development* **123**, 241–254.

Whitfield, T. T., and Hammond, K. L. (2007). Axial patterning in the developing vertebrate inner ear. *Int. J. Dev. Biol.* **51**, 507–520.

Whitfield, T. T., Riley, B. B., Chiang, M. Y., and Phillips, B. (2002). Development of the zebrafish inner ear. *Dev. Dyn.* **223**, 427–458.

Wienholds, E., Kloosterman, W. P., Miska, E., Alvarez-Saavedra, E., Berezikov, E., de Bruijn, E., Horvitz, H. R., Kauppinen, S., and Plasterk, R. H. (2005). MicroRNA expression in zebrafish embryonic development. *Science* **309**, 310–311.

Wilkinson, C. J., Carl, M., and Harris, W. A. (2009). Cep70 and Cep131 contribute to ciliogenesis in zebrafish embryos. *BMC Cell Biol.* **10**, 17.

Xiao, S., Yu, C., Chou, X., Yuan, W., Wang, Y., Bu, L., Fu, G., Qian, M., Yang, J., Shi, Y., et al. (2001). Dentinogenesis imperfecta 1 with or without progressive hearing loss is associated with distinct mutations in DSPP. *Nat. Genet.* **27**, 201–204.

Xiao, T., Roeser, T., Staub, W., and Baier, H. (2005). A GFP-based genetic screen reveals mutations that disrupt the architectrure of the zebrafish retinotectal projection. *Development* **132**, 2955–2967.

Yan, Y.-L., Willoughby, J., Liu, D., Crump, J. C., Wilson, C., Miller, C. T., Singer, A., Kimmel, C., Westerfield, M., and Postlethwait, J. (2005). A pair of Sox: Distinct and overlapping functions of zebrafish *sox9* co-orthologs in craniofacial and pectoral fin development. *Development* **132**, 1069–1083.

Yuan, S., and Joseph, E. M. (2004). The small heart mutation reveals novel roles of $Na^+/K^+$-ATPase in maintaining ventricular cardiomyocyte morphology and viability in zebrafish. *Circ. Res.* **95**, 595–603.

Zeddies, D. G., and Fay, R. R. (2005). Development of the acoustically evoked behavioral response in zebrafish to pure tones. *J. Exp. Biol.* **208**, 1363–1372.

Zon, L. I., and Peterson, R. T. (2005). In vivo drug discovery in the zebrafish. *Nat. Rev. Drug Discov.* **4**, 35–44.

Zottoli, S. J. (1977). Correlation of the startle reflex and Mauthner cell auditory responses in unrestrained goldfish. *J. Exp. Biol.* **66**, 243–254.

# 5

# ENDOCRINOLOGY OF ZEBRAFISH: A SMALL FISH WITH A LARGE GENE POOL

*ELLEN R. BUSBY*
*GRAEME J. ROCH*
*NANCY M. SHERWOOD*





*Zebrafish: Volume 29*
FISH PHYSIOLOGY


DOI: 10.1016/S1546-5098(10)02905-5

Our knowledge of the endocrine system of zebrafish has expanded greatly due to the availability of the genome. Most striking is the obvious homology of the hormones and receptors with their mammalian counterparts. The gene structure, deduced protein, chromosome location, synteny and phylogenetic relationship for a large proportion of the hormones and receptors identified in humans have been found in zebrafish. Zebrafish have been selected as a biomedical model for their advantages in genetic manipulations and their accessible and transparent development. It is now clear that many hormones and receptors are expressed as maternal transcripts or within 24–48 hours after fertilization. These components have been difficult to study in mammals before birth but zebrafish offer a window to re-examine the role of many hormones that act primarily in the prenatal period including insulin-like growth factor 2. The endocrinology of zebrafish reproductive, stress, growth and thyroidal systems is supported by a full cascade of neurohormones, pituitary hormones, and peripheral hormones including steroids comparable to human forms with duplicate genes in some cases. The metabolic hormones including insulin, IGF and glucagon have overlapping functions with their human counterparts. The osmoregulatory system demonstrates that prolactin is an important hormone for adaptation to fresh water and hence survival in zebrafish, whereas in mammals, prolactin is important in milk production. An evolutionary view of the zebrafish endocrine system allows us to understand fish physiology and the roots of mammalian hormones in terms of structure and function.

## 1. INTRODUCTION

Genomics is a relatively new science that has transformed our understanding of the zebrafish endocrine system. The small size of zebrafish has made it difficult to determine the primary structure of hormones by protein methods. Now, the identification of genes encoding hormones, receptors and intracellular signaling molecules in the endocrine subgenome has allowed zebrafish to be used not only as a biomedical model but as a node in the study of biological evolution. The zebrafish genome is important for tracing the evolutionary roots of human endocrine genes and in understanding the biology of fish. The relationship of zebrafish and human genes is often defined in terms of orthologs in which the genes share the same ancestor. By comparing the zebrafish genome with the human, mouse or other characterized genomes, it is possible to identify orthologs, paralogs and indeed, novel genes.

Zebrafish and other teleost fish offer a unique window on evolution as they have undergone complete or large-scale duplication of the genome that is thought to have occurred near the beginning of the teleost lineage, about 350 million years ago (Amores et al. 1998; Taylor et al. 2001). This fish-specific genome doubling meant that zebrafish had roughly twice the number of genes compared with tetrapods, including humans. However, many of the duplicated genes have not been retained, leaving zebrafish with about 25% of their original duplicates, a substantial reduction that still provides zebrafish with a large gene pool compared with humans and most other tetrapods. This unique situation enables researchers to study the effect of gene duplication regarding structure and function during evolution.

The scope of this chapter is limited, as endocrinology can be viewed in a broad sense to include not only endocrine hormones, but also a vast array of growth factors, immunological molecules and developmental factors. We have restricted this review to the classical system of endocrine and paracrine hormones and their receptors but not including full signaling pathways and transcription factors. However, hormones that affect development or growth will be discussed. Our emphasis will be on genes that have been cloned to verify their expression.

In this chapter we have followed the nomenclature suggested by zfin.com in which hormones (e.g., gonadotropin-releasing hormone) are shown as proteins and genes, respectively, for human as GNRH and *GNRH*, for mouse as GNRH and *Gnrh* and for zebrafish as Gnrh and *gnrh*. Genes that encode two or more hormones, unfortunately, sometimes have gene names that do not correspond to the hormone names.

## 2. REPRODUCTION

### 2.1. Neurohormones

Zebrafish reproduction is very distinct from mammalian reproduction as zebrafish have external fertilization without pregnancy, continuous spawning of 100–200 eggs per week throughout the year, and in both genders the development of a juvenile ovary that is transformed to a testis in males during weeks 4–12 after fertilization (Biran et al. 2008; Clelland and Peng 2009). However, despite these differences, zebrafish possess most of the major hormones identified for humans in respect to reproduction (Table 5.1,

**Table 5.1**
Zebrafish hormones and receptors that have been identified by cloning, immunological methods or annotation

| | | | cDNA/gene | Immu. | in situ | PCR | Develop. | Function | Citations |
|---|---|---|---|---|---|---|---|---|---|
| REPRODUCTION: Neurohormones & Receptors | | | | | | | | | |
| Gonadotropin-releasing hormone | Gnrh2 | *gnrh2* | X | X | X | X | X | X | Powell et al. 1996; Steven et al. 2003; Gopinath et al. 2004; Kuo et al. 2005; Sherwood and Wu 2005; Wu et al. 2006b; Palevitch et al. 2007; Lin and Ge 2009 |
| | Gnrh3 | *gnrh3* | X | X | X | X | X | X | *refs. for Gnrh2 plus:* Torgersen et al. 2002; Abraham et al. 2008 |
| **Gnrh receptor** | **Gnrhr1-4** | ***gnrhr1-4*** | X | | | X | | X | **Tello et al. 2008** |
| Kisspeptin | Kiss1 | *kiss1* | X | | X | X | X | X | Biran et al. 2008; van Aerle et al. 2008; Elizur 2009 Kitahashi et al. 2009; Lee et al. 2009 |
| | Kiss2 | *kiss2* | X | | X | X | X | X | Elizur 2009; Kitahashi et al. 2009; Lee et al. 2009 |
| **Kisspeptin receptor** | **Kiss1ra** | ***kiss1ra*** | X | | | X | | X | **van Aerle et al. 2008; Biran et al. 2008; Lee et al. 2009** |
| | **Kiss1rb** | ***kiss1rb*** | X | | | X | | X | **Biran et al. 2008; Lee et al. 2009** |
| Pituitary adenylate cyclase-activating polypeptide | Pacap1 | *adcyap1a* | X | X | X | X | X | X | Fradinger and Sherwood 2000; Krueckl et al. 2003; Holmberg et al. 2004; Mathieu et al. 2004, 2005a; Fradinger et al. 2005; Wu et al. 2006a, 2008; Toro et al. 2009 |
| | Pacap2 | *adcyap1b* | X | X | X | X | X | X | Wang et al. 2003; Mathieu et al. 2004, 2005a; Fradinger et al. 2005; Wu et al. 2006a, 2008; Blechman et al. 2007; Lin and Ge 2009; Toro et al. 2009 |

| | | | | | | | | | |
|---|---|---|---|---|---|---|---|---|---|
| **PACAP receptor** | **Pac1r1** | ***adcyap1r1*** | X | | X | X | X | X | **Fradinger et al. 2005; Wu et al. 2006a, 2008; Blechman et al. 2007** |
| | **Pac1r2** | ***adcyap1r2*** | X | | | | | | **Roch et al. 2009** |
| | **Vpac1r1** | ***vipr1a*** | X | | | X | X | X | **Fradinger et al. 2005; Wu et al. 2006a, 2008; Roch et al. 2009; Lin and Ge 2009** |
| | **Vpac1r2** | ***vipr1b*** | X | | | | | | **Roch et al. 2009** |
| | **Vpac2r/Phir** | ***vipr2*** | X | | | X | | X | **Wu et al. 2008; Roch et al. 2009** |
| REPRODUCTION: Pituitary Hormones & Receptors | | | | | | | | | |
| Follice stimulating hormone | Fshβ | *fshb* | X | | X | X | X | X | So et al. 2005; Pogoda and Hammerschmidt 2007; Clelland and Peng 2009; Lin and Ge 2009; Toro et al. 2009 |
| **Follicle stimulating hormone receptor** | **Fshr** | ***fshr*** | X | | | X | | X | **Laan et al. 2002; Wang and Ge 2003b; Kwok et al. 2005; So et al. 2005** |
| Luteinizing hormone | Lhβ1 | *lhb1* | X | | X | X | X | X | So et al. 2005; Pogoda and Hammerschmidt 2007; Clelland and Peng 2009; Lin and Ge 2009; Toro et al. 2009 |
| | Lhβ2 | *lhb2* | X | | | | | | So et al. 2005 |
| **Luteinizing hormone receptor** | Lhr | ***lhcgr*** | X | | | X | | X | **Wang and Ge 2003b; Kwok et al. 2005; So et al. 2005; Liu and Ge 2007** |
| Glycoprotein subunit α | Gsu-α | *cga* | X | | X | X | X | X | Nica et al. 2004; So et al. 2005; Pogoda and Hammerschmidt 2007, 2009; Guner et al. 2008; Huang et al. 2008; Toro et al. 2009 |
| REPRODUCTION: Pituitary/Gonadal Hormones & Receptors | | | | | | | | | |
| Inhibin A | Inh α-βA (Inh A) | *inha+inhba* | X | X | | X | | | Wu et al. 2000; Wang and Ge 2003a |

(*Continued*)

**Table 5.1** (*continued*)

| | | | cDNA/gene | Immu. | in situ | PCR | Develop. | Function | Citations |
|---|---|---|---|---|---|---|---|---|---|
| Inhibin B | Inh α-βB (Inh B) | *inha+inhbb* | X | | | X | | X | Wang and Ge 2003b<br>Lin and Ge 2009 |
| Activin A | Act βA-βA1 (ActβA1) | *inhbaa* | X | X | | X | | X | Rodaway et al. 1999<br>Wu et al. 2000; Pang and Ge 2002b; Wang and Ge 2003a, 2003b, 2004; DiMuccio et al. 2005; Ge 2005 |
| | Act βA-βA2 (ActβA2) | *inhbab* | X | X | | | | X | DiMuccio et al. 2005 |
| Activin B | Act βB-βB (ActβB) | *inhbb* | X | | | X | | X | Wittbrodt and Rosa 1994<br>Pang and Ge 1999; Rodaway et al. 1999; Wang and Ge 2003a, 2003b, 2004 |
| Activin AB | Act βA-βB | | X | | | | | | *refs. for ActβA, ActβB* |
| **Activin receptor** | **Actr1a** | ***acvr1a*** | | | | X | | | **Yelick et al. 1998; Roman et al. 2002; Wang and Ge 2003b; Lin and Ge 2009** |
| | **Actr1b** | ***acvr1b*** | X | | | X | | | **Renucci et al. 1996; Wang and Ge 2003b; Lin and Ge 2009** |
| | **Actr2a** | ***acvr2a*** | X | | | X | | X | **Nagaso et al. 1999; Pang and Ge 1999, 2002b; Wang and Ge 2003b; Albertson et al. 2005; DiMuccio et al. 2005; Lin and Ge 2009** |
| | **Actr2b** | ***acvr2b*** | X | | X | X | X | X | ***refs. for Actr2a plus:*** **Garg et al. 1999** |

| | | | | | | | | | |
|---|---|---|---|---|---|---|---|---|---|
| Follistatin | Fst1 | *fst1* | X | X | X | X | X | X | Bauer et al. 1998; Pang and Ge 1999; Wu et al. 2000; Pang and Ge 2002a; Wang and Ge 2003b, 2004; Dal-Pra et al. 2006; Macqueen and Johnston 2008; Lin and Ge 2009 |
| | Fst2 | *fst2* | X | | X | X | X | X | Dal-Pra et al. 2006; Macqueen and Johnston 2008 |
| Follistatin-like | Fstl1 | *fstl1* | X | | X | | X | | Dal-Pra et al. 2006 |
| | Fstl2 | *fstl2* | X | | X | | X | X | *refs. for Fstl1* |
| REPRODUCTION: Gonadal Hormones & Receptors | | | | | | | | | |
| Anti Müllerian hormone | Amh | *amh* | X | | X | X | X | X | Rodriguez-Mari et al. 2005; von Hofsten et al. 2005; Wang and Orban 2007; Siegfried and Nusslein-Volhard 2008 |
| Growth differentiation factor 9 | Gdf9 | *gdf9* | X | | X | X | X | X | Liu and Ge 2007 |
| REPRODUCTION: Gonadal Steroids & Receptors | | | | | | | | | |
| Estradiol | E2 | | | X | | | | X | Lister and Van Der Kraak 2008; Levi et al. 2009; Lin and Ge 2009 |
| **Estradiol receptor-nuclear** | **Erα** | ***esr1*** | **X** | | **X** | **X** | **X** | **X** | **Bardet et al. 2002; Menuet et al. 2002, 2004; Bertrand et al. 2007; Lin and Ge 2009** |
| | **Erβ1** | ***esr2a*** | **X** | | **X** | **X** | **X** | **X** | ***refs. for Erα*** |
| | **Erβ2** | ***esr2b*** | **X** | | **X** | **X** | **X** | **X** | ***refs. for Erα*** |
| **Estradiol receptor-membrane (GPCR for E2)** | **Mer** | ***gper*** | **X** | | **X** | **X** | | **X** | **Liu et al. 2009** |

(*Continued*)

**Table 5.1** (*continued*)

| | | | cDNA/gene | Immu. | in situ | PCR | Develop. | Function | Citations |
|---|---|---|---|---|---|---|---|---|---|
| 17α, 20β dihydroxy-4-pregnen-3-one (Maturation-inducing hormone) | DHP *(MIH)* | | | X | | | | X | Wu et al. 2000 Pang and Ge 2002a; Lister and Van Der Kraak 2008; Lessman 2009 |
| **MIH membrane receptor (Progesterone R-membrane)** | **Mprα** | ***paqr7b*** | **X** | **X** | | **X** | | **X** | **Zhu et al. 2003a, 2003b; Kazeto et al. 2005; Kohli et al. 2005; Hanna et al. 2006; Hanna and Zhu 2009** |
| | **Mprβ** | ***paqr8*** | **X** | **X** | | **X** | | **X** | ***refs. for Mprα*** |
| **Progesterone receptor-nuclear** | **Pr** | ***pr*** | **X** | | | | | | **Bertrand et al. 2007; Nagahama and Yamashita 2008** |
| Testosterone | T | | | | | | | X | Ings and Van Der Kraak 2006; de Waal et al. 2008; Lin and Ge 2009 |
| 11 Keto testosterone | KT | | | X | | | | X | Wang and Orban 2007; de Waal et al. 2008 |
| **Androgen receptor** | **Ar** | ***ar*** | **X** | **X** | **X** | **X** | **X** | **X** | **Bertrand et al. 2007; Jorgensen et al. 2007; de Waal et al. 2008; Gorelick et al. 2008** |
| STRESS: Neurohormones & Receptors | | | | | | | | | |
| Corticotropin releasing-hormone | Crh | *crh* | X | | X | X | X | | Alderman and Bernier 2007, 2009; Chandrasekar et al. 2007 |
| **CRH receptor** | **Crhr1** | ***crhr1*** | **X** | | | **X** | **X** | | **Alderman and Bernier 2009** |
| | **Crhr2** | ***crhr2*** | **X** | | | **X** | **X** | | ***refs. for CRHR1*** |
| Urotensin I | Uts1 | *uts1* | X | X | X | X | X | | Parmentier et al. 2006, 2008; Alderman and Bernier 2007, 2009 |

| | | | | | | | | | |
|---|---|---|---|---|---|---|---|---|---|
| STRESS: Pituitary Hormones & Receptors | | | | | | | | | |
| Adrenocorticotropic hormone | Actha | *pomca* | X | | X | X | X | X | Hansen et al. 2003; Herzog et al. 2003, 2004; Gonzalez-Nunez et al. 2003; Nica et al. 2004; Song et al. 2004; Liu et al. 2006; Dickmeis et al. 2007; To et al. 2007; Dutta et al. 2008; Liu et al. 2008 |
| | Acthb | *pomcb* | X | | | | | | Gonzalez-Nunez et al. 2003 |
| **ACTH receptor** | **Mc2r** | ***mc2r*** | X | | X | X | X | | **Logan et al. 2003b; To et al. 2007 Alsop and Vijayan 2008** |
| STRESS: Steroids & Receptors | | | | | | | | | |
| Cortisol | | | | X | | | X | X | Pottinger and Calder 1995; Ramsay et al. 2006; Barcellos et al. 2007; Dickmeis et al. 2007; Alsop and Vijayan 2008; Alderman and Bernier 2009 |
| **Glucocorticoid receptor** | **Grα** | ***nr3c1*** | X | X | X | X | X | X | **Dickmeis et al. 2007; Mathew et al. 2007; Alsop and Vijayan 2008; Schaaf et al. 2008, 2009** |
| | **Grβ** | ***nr3c1*** | X | | X | X | X | X | **Schaaf et al. 2008** |
| **Mineralocorticoid receptor** | **Mr** | ***nr3c2*** | X | | | X | X | | **Alsop and Vijayan 2008** |
| GROWTH: Neurohormones & Receptors | | | | | | | | | |
| Growth hormone-releasing hormone | Ghrh | *ghrh* | X | | | | | X | Lee et al. 2007; Wang et al. 2007; Wu et al. 2008 |
| **GHRH receptor** | **Ghrhr** | ***ghrhr*** | X | | | | | X | **Lee et al. 2007** |
| GHRH-like peptide | Ghrh-lp1 | *adcyap1a* | X | | X | X | X | | Fradinger and Sherwood 2000 |
| | Ghrh-lp2 | *adcyap1b* | X | | | | | | Wang et al. 2003 |

(*Continued*)

**Table 5.1** (*continued*)

| | | | cDNA/gene | Immu. | in situ | PCR | Develop. | Function | Citations |
|---|---|---|---|---|---|---|---|---|---|
| **GHRH-like peptide receptor (PRP receptor)** | **Ghrh-lpr** | ***ghrhr*** | X | | | X | X | X | **Fradinger et al. 2005; Wu et al. 2006a, 2008; Roch et al. 2009** |
| GROWTH: Pituitary Hormones & Receptors | | | | | | | | | |
| Growth hormone | Gh | *gh* | X | | X | X | X | X | Mommsen 2001; Herzog et al. 2003; Figueiredo et al. 2007; Zhu et al. 2007; Rosa et al. 2008; Lin and Ge 2009 |
| **Growth hormone receptor** | **Ghra** | ***ghra*** | X | | | X | X | X | **Mommsen 2001; Figueiredo et al. 2007; Liongue and Ward 2007** |
| | **Ghrb** | ***ghrb*** | X | | | X | X | | **Liongue and Ward 2007** |
| GROWTH: Other Hormones & Receptors | | | | | | | | | |
| Insulin-like growth factor | Igf1 | *igf1* | X | | X | X | X | X | Chen et al. 2001; Mommsen 2001; Maures et al. 2002; Eivers et al. 2004; Figueiredo et al. 2007; Sang et al. 2008; Lin and Ge 2009 |
| | Igf2a | *igf2a* | X | | X | X | X | X | Maures et al. 2002; Eivers et al. 2004; Sang et al. 2008; White et al. 2009 |
| | Igf2b | *igf2b* | | | X | X | X | X | Sang et al. 2008; White et al. 2009 |
| | Igf3 | *igf3* | X | | | X | | | Wang et al. 2008 |
| **IGF1 receptor** | **Igf1ra** | ***igf1ra*** | X | X | X | X | X | X | **Mommsen 2001; Maures et al. 2002; Eivers et al. 2004; Schlueter et al. 2006, 2007a, 2007b; Lin and Ge 2009** |
| | **Igf1rb** | ***igf1rb*** | X | X | X | X | X | X | **Ayaso et al. 2002; Maures et al. 2002; Eivers et al. 2004; Schlueter et al. 2006, 2007a; 2007b; Lin and Ge 2009** |

| | | | | | | | | | |
|---|---|---|---|---|---|---|---|---|---|
| Ghrelin | Ghs | *ghrl* | X | X | X | X | X | X | Olsson et al. 2008a; Amole and Unniappan 2009; Li et al. 2009a |
| Obestatin | Obestatin | *ghrl* | X | X | X | X | X | | Li et al. 2009a |
| **Ghrelin receptor** | **Ghsr** | ***ghsra*** | X | X | | | | X | **Olsson et al. 2008a** |
| | | **ghsrb** | X | | | | | | **Kaiya et al. 2008** |
| THYROIDAL REGULATION | | | | | | | | | |
| Thyrotropin-releasing hormone | Trh | *trh* | X | X | | | | | Diaz et al. 2002; Aoki Y et al. 2005 |
| Thyroid-stimulating hormone | Tsh | *tsh* | X | | X | | X | | Herzog W et al. 2003, 2004; Nica et al. 2004 |
| Thyroid hormone | T3/T4 | | | X | | | | | Simon et al. 2002 |
| **Thyroid hormone receptor** | **Trαa** | ***traa*** | X | X | X | X | X | X | **Essner et al. 1997;** Liu et al. 2000; **Liu and Chan 2002; Marchand et al. 2001; Takayama et al. 2008** |
| | **Trαb** | ***trab*** | X | X | X | X | X | X | ***refs. for Trαa*** |
| | **Trβ** | ***trb*** | X | | | X | X | X | **Liu et al. 2000; Liu and Chan 2002; Marchand et al. 2001** |
| ENERGY HOMEOSTASIS: Appetite Regulation | | | | | | | | | |
| Neuropeptide Y | Npy | *npy* | X | X | X | | X | | Soderberg et al. 2000; Mathieu et al. 2002; Sundstrom et al. 2005, 2008 |
| Peptide YY | Pyya | *pyya* | X | | X | | | | Soderberg et al. 2000, Sundstrom et al. 2005, 2008 |
| | Pyyb | *pyyb* | X | | | | | | Sundstrom et al. 2005, 2008 |
| **Neuropeptide Y receptor** | **Y1** | ***npy1r*** | X | | | | | | **Salaneck et al. 2008** |
| | **Y2** | ***npy2r*** | X | | | | | X | **Fredriksson et al. 2006** |
| | **Y4** | ***npy4r*** | X | | | | | X | **Starback et al. 1999** |
| | **Y7** | ***npy7r*** | X | | | X | | X | **Berglund et al. 2000; Fredriksson et al. 2006** |

(*Continued*)

**Table 5.1** (*continued*)

| | | | cDNA/gene | Immu. | in situ | PCR | Develop. | Function | Citations |
|---|---|---|---|---|---|---|---|---|---|
| | **Y8a** | ***npy8ar*** | X | | X | | | X | **Ringvall et al. 1997Berglund et al. 2000; Mathieu et al. 2005b** |
| | **Y8b** | ***npy8br*** | X | | X | X | | X | **Lundell et al. 1997 Berglund et al. 2000; Mathieu et al. 2005b** |
| Orexins (hypocretin) | Orexin-a | *hcrt* | X | X | X | X | X | X | Kaslin et al. 2004; Faraco et al. 2006; Prober et al. 2006 |
| | Orexin-b | *hcrt* | X | X | X | X | X | X | *refs. for Orexin-a* |
| **Orexin (hypocretin) receptor** | **Ox1** | ***hcrtr1*** | X | X | X | | X | X | **Prober et al. 2006; Yokogawa et al. 2007** |
| Agouti-related protein | Agrp | *agrp* | X | X | X | X | X | | Song et al. 2003; Forlano & Cone 2007; Drew et al. 2008 |
| Agouti-Signaling peptide | Asip1 | *asip1* | X | | | | | | Klovins and Schioth 2005 |
| | Asip2 | *asip2* | X | | | | | | *refs. for Asip1* |
| α-Melanocyte-stimulating hormone | α-Msha | *pomca* | X | X | X | X | X | X | *refs for STRESS:Acth plus* de Souza et al. 2005; Forlano & Cone 2007 |
| | α-Mshb | *pomcb* | X | | | | | | Gonzalez-Nunez et al. 2003; de Souza et al. 2005 |
| **Melanocortin receptor** | **Mc1r** | ***mc1r*** | X | | | X | X | | **Logan et al. 2003a; Selz et al. 2007; Richardson et al. 2008** |
| | **Mc2r** | ***mc2r*** | X | | X | X | X | | **see STRESS** |
| | **Mc3r** | ***mc3r*** | X | | | X | | | **Logan et al. 2003a** |
| | **Mc4r** | ***mc4r*** | X | | | X | | X | **Ringholm et al. 2002; Logan et al. 2003a** |
| | **Mc5ra** | ***mc5ra*** | X | | | X | | X | ***refs. for Mc4r*** |
| | **Mc5rb** | ***mc5rb*** | X | | | X | | X | ***refs. for Mc4r plus:*** **Logan et al. 2003b** |

| | | | | | | | | | |
|---|---|---|---|---|---|---|---|---|---|
| Pro melanin-concentrating hormone-like protein | Mchl | *pmchl* | X | | | | | | Toro et al. 2009 |
| **Melanin-concentrating hormone receptor** | **Mchr1a** | ***mchr1a*** | X | | | X | X | | **Logan et al. 2003a; Kawauchi and Baker 2004** |
| | **Mchr1b** | ***mchr1b*** | X | | | X | X | | ***refs. for Mchr1a*** |
| | **Mchr2** | ***mchr2*** | X | | | X | X | | ***refs. for Mchr1a*** |
| Cocaine- and amphetamine-regulated transcript protein | Cart1 | *cart1* | X | | | | | | Kehoe and Volkoff 2007; Murashita et al. 2009 |
| | Cart2-5 | *cart2-5* | X | | | | | | Murashita et al. 2009 |
| Cholecystokinin | Cck1 | *cck1* | X | | | | | | Kurokawa et al. 2003 |
| | Cck2 | *cck2* | X | | | | | | Kurokawa et al. 2003 |
| Ghrelin | Ghrl | *ghrl* | X | X | X | X | X | X | see GROWTH |
| **Ghrelin receptor** | **Ghsr** | ***ghsra*** | X | X | | | | X | **see GROWTH** |
| | | ***ghsrb*** | X | | | | | | **see GROWTH** |
| Leptin | Lepa | *lepa* | X | | | X | | | Gorissen et al. 2009; Huising et al. 2006 |
| | Lepb | *lepb* | X | | | X | | | *refs. for Lepa* |
| **Leptin receptor** | **Lepr** | ***lepr*** | X | | | | | | **Liongue and Ward 2007; Rekha et al. 2008** |
| Adiponectin | Adipo-a | *adipoql2* | X | | X | X | X | | Nishio et al. 2008 |
| | Adipo-b | *adipoql* | X | | X | X | X | | *refs. for Adipo-a* |
| **Adiponectin receptor** | **Adipor1a** | ***adipor1a*** | X | | X | X | X | | **Nishio et al. 2008** |
| | **Adipor1b** | ***adipor1b*** | X | | X | X | X | | ***refs. for Adipor1a*** |
| | **Adipor2** | ***adipor2*** | X | | X | X | X | | ***refs. for Adipor1a*** |
| GHRH-like peptide | Ghrh-lp1 | *adcyap1a* | X | X | | X | | | Fradinger and Sherwood 2000; Castro et al. 2009 |
| | Ghrh-lp2 | *adcyap1b* | X | X | | X | | | Wang et al. 2003; Castro et al. 2009 |
| **GHRH-like peptide receptor** | **Ghrhr-lpr** | ***ghrhrl*** | X | | | X | | X | **Wu et al. 2008** |

(*Continued*)

**Table 5.1** (*continued*)

| | | | cDNA/gene | Immu. | in situ | PCR | Develop. | Function | Citations |
|---|---|---|---|---|---|---|---|---|---|
| Pituitary adenylate cyclase-activating polypeptide | Pacap1a | *adcyap1a* | X | | | | X | | see REPRODUCTION |
| | Pacap1b | *adcyap1b* | X | | | X | | X | see REPRODUCTION |
| Vasoactive intestinal polypeptide | Vip1 | *adcyap1a* | X | X | | X | X | X | Mathieu et al. 2001; Olsson et al. 2008b; Wu et al. 2008 |
| | Vip2 | *adcyap1b* | X | | | | | | Wu et al. 2008 |
| **PACAP and VIP receptors** | **Pac1r1** | ***adcyap1r1*** | X | | X | X | X | X | **see REPRODUCTION** |
| | **Pac1r2** | ***adcyap1r2*** | X | | | | | | **see REPRODUCTION** |
| | **Vpac1r1** | ***vipr1a*** | X | | | X | X | X | **see REPRODUCTION** |
| | **Vpac1r2** | ***vipr1b*** | X | | | | | | **see REPRODUCTION** |
| **PHIR** | **Vpac2r/Phir** | ***vipr2*** | X | | | X | | X | **see REPRODUCTION** |
| Galanin | Gal1a | *gal1a* | | X | | | | | Castro et al. 2006 |
| PANCREATIC FUNCTION: Peptide Hormones & Receptors | | | | | | | | | |
| Insulin | Insa | *insa* | X | X | X | | X | | Milewski et al. 1998; Argenton et al. 1999; Biemar et al. 2001; Irwin 2004; Papasani et al. 2006; Zecchin et al. 2007 |
| | Insb | *insb* | X | | X | | X | | Papasani et al. 2006; Irwin 2004 |
| **Insulin receptor** | **Insra** | ***insra*** | X | | X | X | X | | **Toyoshima et al. 2008** |
| | **Insrb** | ***insrb*** | X | | X | X | X | | ***refs. for Insra*** |
| Glucagon | Gcga | *gcga* | X | X | X | | X | | Argenton et al. 1999; Biemar et al. 2001; Zhou and Irwin 2004 |
| | Gcgb | *gcgb* | X | | | | | | Zhou and Irwin 2004 |
| **Glucagon receptor** | **Gcgr** | ***gcgr*** | X | | | | | | **Roch et al. 2009** |

| | | | | | | | | | |
|---|---|---|---|---|---|---|---|---|---|
| Glucagon-like peptide-1 | Glp1a | *gcga* | X | | | | | X | Mommsen and Mojsov 1998 |
| | Glp1b | *gcgb* | X | | | | | | Zhou and Irwin 2004 |
| **Glucagon-like peptide-1 receptor** | **Glp1r** | ***glp1r*** | **X** | | | | | **X** | **Mojsov 2000** |
| Somatostatin | Sst1 | *sst1* | X | X | X | | X | | Argenton et al. 1999; Devos et al. 2002; Tostivint et al. 2004 |
| | Sst2 | *sst2* | X | X | X | X | X | | Devos et al. 2002; Tostivint et al. 2004, 2005 |
| | Sst3 | *sst3* | X | X | X | | X | | Devos et al. 2002; Tostivint et al. 2004, 2008 |
| | Sst4 | *sst4* | X | | | | | | Argenton et al. 1999; Tostivint et al. 2008 |
| **Somatostatin receptors** | **Sstr1-5** | ***sstr1-5*** | **X** | | | | | | **Moaeen-ud-Din and Yang 2009** |
| Pancreatic polypeptide | Pp | *ppy* | X | | | | | | Argenton et al. 1999; Li et al. 2009b |
| GASTROINTESTINAL REGULATION | | | | | | | | | |
| Glucagon-like peptide-2 | Glp2 | *glp2* | X | | | | | | Zhou and Irwin 2004; Roch et al. 2009 |
| **Glucagon-like peptide-2 receptor** | **Glp2r** | ***glp2r*** | **X** | | | | | | **Roch et al. 2009** |
| Glucose-dependent insulinotropic peptide | Gip | *gip* | X | | | | | | Irwin and Zhang 2006; Roch et al. 2009 |
| **Glucose-dependent insulinotropic peptide receptor** | **Gipr** | ***gipr*** | **X** | | | | | | **Irwin and Wong 2005** |
| Motilin | Mln | *mln* | | X | | | | | Olsson et al. 2008a |

(*Continued*)

**Table 5.1** (*continued*)

| | | | cDNA/gene | Immu. | in situ | PCR | Develop. | Function | Citations |
|---|---|---|---|---|---|---|---|---|---|
| OSMOREGULATION: Neurohormones & Receptors | | | | | | | | | |
| Isotocin | It | *oxtl* | X | | X | | X | X | Unger and Glasgow 2003; Herzog et al. 2004; Eaton and Glasgow 2006, 2007; Blechman et al. 2007; Chandrasekar et al. 2007; Eaton et al. 2008; Russek-Blum et al. 2008; Sanek and Grinblat 2008 |
| Vasotocin | Avt | *avpl* | X | X | X | | X | X | Larson et al. 2006; Tessmar-Raible et al. 2007; Eaton et al. 2008 |
| Urotensin II | Uts2α, Uts2β | *uts2a, uts2b* | X | X | X | X | | | Parmentier et al. 2006, 2008; Tostivint et al. 2006 |
| OSMOREGULATION: Pituitary Hormones & Receptors | | | | | | | | | |
| Prolactin | Prl | *prl* | X | | X | X | X | X | Herzog et al. 2003, 2004; Nica et al. 2004; Liu et al. 2006; 2008; Lopez et al. 2006; Dickmeis et al. 2007; Hoshijima and Hirose 2007; Zhu et al. 2007; Dutta et al. 2008; Nguyen and Zhu 2009 |
| **Prolactin receptor** | **Prlra** | ***prlra*** | X | | X | | X | X | **Liu et al. 2006** |
| OSMOREGULATION: Calcium and Phophates Regulation | | | | | | | | | |
| Calcitonin | Calc | *calca* | X | | X | | X | | Alt et al. 2006 |
| | Cgrp | *calca* | X | X | | | X | | *refs. for Calc plus:* Olsson et al. 2008b |
| **Calcitonin receptor** | **Crlr** | ***calcrla*** | X | | X | X | X | X | **Nicoli et al. 2008** |
| Parathyroid hormone | Pth1a | *pth1a* | X | X | X | X | X | X | Gensure et al. 2004; Hogan et al. 2005; Hoshijima and Hirose 2007 |
| | Pth2 | *pth2* | X | X | X | X | X | X | Gensure et al. 2004; Hogan et al. 2005 |

| | | | | | | | | | |
|---|---|---|---|---|---|---|---|---|---|
| **Parathyroid hormone receptor** | **Pth1r** | ***pth1ra*** | X | | | | | X | **Rubin and Juppner 1999; Hoare et al. 2000; Gensure et al. 2004** |
| | **Pth2r** | ***pth2r*** | X | | X | | X | X | **Rubin et al. 1999; Hoare et al. 2000; Gensure et al. 2004; Papasani et al. 2004** |
| | **Pth3r** | ***pth3r*** | X | | | | | X | **Rubin and Juppner 1999; Hoare et al. 2000** |
| Stanniocalcin | Stc1 | *stc1* | X | | X | | X | X | Luo et al. 2005; Wingert et al. 2007; Tseng et al. 2009 |
| | Stc2 | *stc2* | X | | | | | X | Luo et al. 2005 |
| OSMOREGULATION: Cardiovascular Regulation | | | | | | | | | |
| Atrial natriuretic peptide | Anp | *nppa* | | | X | X | X | X | Berdougo et al. 2003; Bendig et al. 2006; Hoshijima and Hirose 2007 |
| Adrenomedullin | Adm2 | *adm2* | X | | X | | X | | Toro et al. 2009 |
| **Adrenomedullin receptor** | **Crlr** | ***calcrla*** | X | | X | X | X | X | **Nicoli et al. 2008** |
| Angiotensin | AgtI-IV | *agt* | X | | | | | X | Watanabe et al. 2009 |
| Renin | Ren | *ren* | X | | X | X | X | | Liang et al. 2004; Song et al. 2004; Hoshijima and Hirose 2007 |
| Apelin | Apln | *apln* | X | | X | X | X | X | Zeng et al. 2007; Eyries et al. 2008 |
| **Apelin receptor** | **Agtrl1a** | ***aplnra*** | X | | X | X | X | X | **Patterson et al. 2007; Tucker et al. 2007; Eyries et al. 2008** |
| | **Agtrl1b** | ***aplnrb*** | X | | X | X | X | X | **Scott et al. 2007; Zeng et al. 2007; Eyries et al. 2008** |
| Bradykinin | Bk | *kng1* | X | | | | | X | Bromee et al. 2005 |
| **Bradykinin receptor** | **B1** | ***bdkrb1*** | X | | | | | | **Bromee et al. 2006** |
| | **B2** | ***bdkrb2*** | X | | | X | | X | ***refs. for B1 plus:*** **Duner et al. 2002; Bromee et al. 2005** |

Hormones are listed in regular font and receptors in bold font. Columns indicate methods of characterization: "cDNA/gene" indicates sequencing or annotation; "Immu." refers to immunohistochemistry, western blotting, ELISA or RIA; "in situ" refers to RNA in situ hybridization; "PCR" indicates RT-PCR or qPCR; "Develop." indicates studies involving zebrafish embryos; and "Function" refers to studies analyzing function in vivo or in vitro.

Sections 2.1.1–2.5.3), with the exception that the major androgen is 11 ketotestosterone rather than dihydrotestosterone, as in humans. Thus, the main outline of hormones in the brain–pituitary–gonadal axis of zebrafish and human is very similar.

### 2.1.1. Gonadotropin-Releasing Hormone

It is unusual for a teleost fish like zebrafish to have only two forms of gonadotropin-releasing hormone (Gnrh) compared to most other teleosts with three forms. However, zebrafish have Gnrh2, which is identical to one of the human forms (GNRH2), and Gnrh3, which is unique to teleost fishes, but 80% identical to the other human form (GNRH1). Protein for Gnrh2 (also known as chicken GnRH II) and Gnrh3 (salmon GnRH) was originally identified by HPLC and RIA methods (Powell et al. 1996), followed by sequencing of both the gene and cDNA encoding each zebrafish hormone (Torgersen et al. 2002; Steven et al. 2003). Synteny analysis shows that the two zebrafish genes have the same chromosomal neighbors as expected for orthologous vertebrate *gnrh2* and *gnrh3*, confirming the forms are correctly identified (Kuo et al. 2005). The conservation of the promoter regions for *gnrh* genes is highlighted by a study in which the human *GNRH1* promoter was fused to a reporter gene encoding red fluorescent protein and then microinjected into a one- or two-celled zebrafish embryo. The human *GNRH* promoter directed expression to the correct area of the zebrafish brain (Torgersen et al. 2002).

The expression of Gnrh2 and Gnrh3 proteins in zebrafish is distinct. Gnrh2-expressing neurons first appear in the midbrain and remain there throughout life (Steven et al. 2003). However, the location of Gnrh3-expressing neurons depends on migration into the brain of neurons that are first detected near the olfactory placode outside of the brain. In larvae where the *gnrh3* promoter is fused to a gene encoding green fluorescent protein, the Gnrh3-labeled neurons send out extensive axonal tracts that enter the anterior brain and are widely distributed to the brain, eye, and pituitary. Subsequently the Gnrh3 neurons migrate along their own tracts into the anterior brain leaving a continuum of cells in the olfactory bulb, terminal nerve ganglion, ventral telencephalon, preoptic area and hypothalamus (Palevitch et al. 2007; Abraham et al. 2008). The Gnrh3-GFP neurons first reach the hypothalamus by 12 days after fertilization. In addition, Gnrh3 neurons appear in the trigeminal nucleus sending axons down the spinal cord to the tail region.

Gnrh3 appears to be the main regulator for the release of gonadotropins in zebrafish, as shown by its brain location, abundance of Gnrh3 peptide and axon terminals in the pituitary (Steven et al. 2003) and ability to increase *lhb* (luteinizing hormone) mRNA in cultured primary pituitary cells

(Lin and Ge 2009). Gnrh3 did not stimulate production of *fshb* (follicle-stimulating hormone) transcripts as it does in other species, which may have been due to non-pulsatile administration of Gnrh3. In contrast, humans and tetrapods have a portal blood system that delivers Gnrh from the base of the hypothalamus to the pituitary. Regardless of whether the hormone is released into blood, pituitary, brain, or other organs, the target sites of Gnrh depend on the location of its receptors.

In zebrafish, four distinct Gnrh receptors, each a 7-transmembrane G protein-coupled receptor (GPCR), have been identified on four different chromosomes. The recombinant form of each of the four receptors expressed in COS cells responded to both Gnrh2 and Gnrh3, although two of the receptors were more sensitive to Gnrh2 than Gnrh3. Downstream signaling occurred via $G_{q/11}$, which resulted in the accumulation of inositol phosphate (Tello et al. 2008). It is curious that humans have only one functioning GNRH receptor, as the second receptor has an early stop codon and other mutations (Pawson et al. 2003). Without question, zebrafish have a functioning neuropeptide that transmits a signal to the gonadotropes in the pituitary to initiate reproduction by the release of gonadotropins; all four receptors are expressed in the pituitary (Tello et al. 2008), but their function in multiple target sites in the brain and other organs is not yet clear.

One of the virtues of zebrafish is their usefulness in developmental studies because of the accessibility and transparency of their embryos. Many hormones and receptors are expressed early in development in zebrafish (Table 5.2) and this includes Gnrh (Sherwood and Wu 2005). Developmental times refer to hours (hpf) or days (dpf) post-fertilization. The transcripts for *gnrh2* and *gnrh3* can be detected at 0.5–1.5 hpf using PCR (Wu et al. 2006b), at 22–28 hpf by in situ hybridization (Gopinath et al. 2004; Kuo et al. 2005; Wu et al. 2006b; Palevitch et al. 2007), but not until 30–31 hpf using immunohistochemistry (IHC) (Wu et al. 2006b). The early expression of zebrafish *gnrh* mRNA at 0.5–1.5 hpf suggests that *gnrh* mRNA is a maternal transcript in the eggs, as embryo-generated transcripts are only detected after 3–4 hpf (Kane and Kimmel 1993). Medaka also have maternal *gnrh3* transcripts. A reporter gene encoding green fluorescent protein fused to the medaka *gnrh3* promoter was expressed in oocytes inside the transgenic mother before spawning (Okubo et al. 2006). The early expression of Gnrh peptides in the zebrafish brain long before reproduction begins may be related to their role in modulating the anterior and posterior boundaries of the midbrain, as well as to the development of the eye cup and stalk as shown by knockdown studies of Gnrh2 and Gnrh3 separately and together (Wu et al. 2006b). It is also speculated that Gnrh may act during development to inhibit surrounding cells from differentiating into Gnrh3 neurons or may affect locomotion (Abraham et al. 2008).

**Table 5.2**
Expression of zebrafish peptide hormones and receptors

X: Expression not tested
White: No expression
Gray: Expression

| | 0-3 hpf | 24 hpf | 48 hpf | 72 hpf | 96-120 hpf | Method | |
|---|---|---|---|---|---|---|---|
| **REPRODUCTION** | | | | | | | |
| Gnrh2 | | | | | | PCR, in situ, Immuno. | Gopinath et al. 2004; Kuo et al. 2005; Wu et al. 2006b; Palevitch et al. 2007 |
| Gnrh3 | | | | | | PCR, in situ, Immuno. | Gopinath et al. 2004; Kuo et al. 2005; Wu et al. 2006b; Palevitch et al. 2007 |
| Kiss1 | X | | X | | X | PCR | Kitahashi et al. 2009 |
| Kiss2 | X | | X | | X | PCR | Kitahashi et al. 2009 |
| Pacap1 | | | | | | PCR, in situ, Immuno. | Krueckl et al. 2003; Mathieu et al. 2004, 2005a; Wu et al. 2006a |
| Pacap2 | | | | | | PCR, in situ, Immuno. | Mathieu et al. 2004, 2005a; Wu et al. 2006a; Toro et al. 2009 |
| **Pac1r1** | | | | | | PCR, in situ | **Wu et al. 2006a; Blechman et al. 2007** |
| **Vpac1r1** | | | | | | PCR | **Wu et al. 2006a** |

| | | | | | | |
|---|---|---|---|---|---|---|
| **Vpac2r/Phir** | | | | | PCR | **Wu et al. 2006a** |
| Fshb | | | | | in situ | Pogoda and Hammerschmidt 2009; Toro et al. 2009 |
| Lhb | | | | | in situ | Pogoda and Hammerschmidt 2009; Toro et al. 2009 |
| Gsuα | | | | | in situ | Pogoda and Hammerschmidt 2007, 2009; Toro et al. 2009; Guner et al. 2008 |
| **Acvrla (ALK2)** | | | | | in situ | **Yelick et al. 1998** |
| **Acvr1a (ALK8)** | | | | | PCR, in situ, Immuno., GFP | **Roman et al. 2002; Payne-Ferreira et al. 2004; Albertson et al. 2005** |
| **Acvr1b (ALK4)** | | | X | X | in situ | **Renucci et al. 1996** |
| **Acvr2a** | | | | | PCR, in situ | **Nagaso et al. 1999; Albertson et al. 2005** |
| **Acvr2b** | | | | | PCR, in situ | **Garg et al. 1999; Nagaso et al. 1999; Albertson et al. 2005** |
| Fst1 | | | | | PCR | Bauer et al. 1998 |
| Fst2 | | | X | X | in situ | Dal-Pra et al. 2006 |
| Fstl1 | | | X | X | in situ | Dal-Pra et al. 2006 |
| Fstl2 | | | X | X | in situ | Dal-Pra et al. 2006 |
| Amh | | | | | PCR, in situ | von Hofsten et al. 2005; Wang and Orban 2007; Rodriguez-Mari et al. 2005 |
| Gdf9 | | | X | X | PCR | Liu and Ge 2007 |
| **Erα (nuclear)** | | | | X | RNase p., in situ | **Bardet et al. 2002; Bertrand et al. 2007** |

(*Continued*)

**Table 5.2** (*continued*)

X: Expression not tested
White: No expression
Gray: Expression

| | 0-3 hpf | 24 hpf | 48 hpf | 72 hpf | 96-120 hpf | | |
|---|---|---|---|---|---|---|---|
| **Erβ1 (nuc)** | | | | | X | RNase p., in situ | **Bardet et al. 2002; Bertrand et al. 2007** |
| **Erβ2 (nuc)** | | | | | X | RNase p., in situ | **Bardet et al. 2002; Bertrand et al. 2007** |
| **Pr (nuc)** | | | | | | in situ | **Bertrand et al. 2007** |
| **Ar (nuc)** | | | | | | PCR, in situ | **Hossain et al. 2008; Bertrand et al. 2007** |
| STRESS | | | | | | | |
| Crh | | | | X | X | PCR, in situ | Chandrasekar et al. 2007; Alderman and Bernier 2009 |
| **CrhrI** | | | | X | X | PCR | **Alderman and Bernier 2009** |
| **CrhrII** | | | | X | X | PCR | **Alderman and Bernier 2009** |
| UI | | | | X | X | PCR | Alderman and Bernier 2009 |
| Acth | | | | | | PCR, in situ | Hansen et al. 2003; Herzog et al. 2003; Nica et al. 2004; To et al. 2007 |

| | | | | | | | |
|---|---|---|---|---|---|---|---|
| **Mc2r** | X | | | | | PCR, in situ | **Logan et al. 2003a**; **To et al. 2007; Alsop and Vijayan 2008** |
| **Grα** | | | | | | PCR, in situ | **Alsop and Vijayan 2008; Schaaf et al. 2008** |
| **Grβ** | | | | | X | PCR, in situ | **Schaaf et al. 2008** |
| **Mr** | | | | | | PCR, in situ | **Alsop and Vijayan 2008** |
| **GROWTH** | | | | | | | |
| Ghrh-lp1 | | | | | | PCR, in situ, Immuno. | Krueckl et al. 2003; Mathieu et al. 2004; Wu et al. 2006a |
| Ghrh-lp2 | | | | | | PCR, in situ, Immuno. | Mathieu et al. 2004, 2005a; **Wu et al. 2006a**; Toro et al. 2009 |
| **Ghrh-lpr** | | | X | X | X | PCR | Wu et al. 2006a |
| Gh | | | | | X | PCR, in situ, Immuno., GFP | Herzog et al. 2003; Guner et al. 2008; Toro et al. 2009 |
| **Ghra** | X | | X | | X | PCR | **Liongue and Ward 2007** |
| **Ghrb** | X | | X | | X | PCR | **Liongue and Ward 2007** |
| Igf1 | | | | | | PCR, in situ | Maures et al. 2002; Chen et al. 2001; Eivers et al. 2004; Sang et al. 2008 |
| Igf2a | | | | | X | PCR, in situ | White et al. 2009; Maures et al. 2002; Sang et al. 2008 |
| Igf2b | | | X | X | X | PCR, in situ | White et al. 2009; Sang et al. 2008 |
| **Igf1ra** | | | | | X | PCR, in situ, Immuno. | **Maures et al. 2002; Eivers et al. 2004** |

(Continued)

**Table 5.2** (*continued*)

X: Expression not tested
White: No expression
Gray: Expression

| | 0-3 hpf | 24 hpf | 48 hpf | 72 hpf | 96-120 hpf | | |
|---|---|---|---|---|---|---|---|
| **Igf1rb** | | | | | X | PCR, in situ, Immuno. | **Maures et al. 2002; Eivers et al. 2004** |
| Ghrl | | | | | | PCR | **Li et al. 2009a** |
| **ENERGY HOMEOSTASIS** | | | | | | | |
| Npy | X | | | | | Immuno | Mathieu et al. 2002 |
| Orexin-a/b | X | | | | | in situ | Faraco et al. 2006; Prober et al. 2006 |
| **Oxr** | X | X | | X | | in situ | **Prober et al. 2006** |
| Agrp | X | | | | X | PCR | Song et al. 2003 |
| α-Msha | X | | | | X | PCR, in situ | Hansen et al. 2003; Herzog et al. 2003; Song et al. 2003 |
| **Mc1r** | X | X | | X | | PCR | **Logan et al. 2003a** |
| **Mc3r** | X | X | | X | | PCR | **Logan et al. 2003a** |
| **Mc4r** | X | X | | X | | PCR | **Logan et al. 2003a** |
| **Mc5ra** | X | X | | X | | PCR | **Logan et al. 2003a** |
| **Mc5rb** | X | X | | X | | PCR | **Logan et al. 2003a** |
| **Mchr1a** | X | X | | X | | PCR | **Logan et al. 2003a** |

| | | | | | | | |
|---|---|---|---|---|---|---|---|
| **Mchr1b** | X | X | | X | | PCR | **Logan et al. 2003a** |
| **Mchr2** | X | X | | X | | PCR | **Logan et al. 2003a** |
| Adipo-a | X | | | | | PCR, in situ | Nishio et al. 2008 |
| Adipo-b | X | | | | | PCR, in situ | Nishio et al. 2008 |
| **Adipor1a** | X | | | | | PCR, in situ | **Nishio et al. 2008** |
| **Adipor1b** | X | | | | | PCR, in situ | **Nishio et al. 2008** |
| **Adipor2** | X | | | | | PCR, in situ | **Nishio et al. 2008** |
| Insa | X | | | | | PCR, in situ, GFP | Argenton et al. 1999; Biemar et al. 2001; Papasani et al. 2006; Zecchin et al. 2007 |
| Insb | X | | | | | PCR, in situ | Papasani et al. 2006; Li et al. 2009b |
| **Insra** | | | | | | PCR, in situ, Immuno. | **Toyoshima et al. 2008** |
| **Insrb** | | | | | | PCR, in situ, Immuno. | **Toyoshima et al. 2008** |
| Gcg | X | | | | X | PCR, in situ, Immuno., GFP | Argenton et al. 1999; Field et al. 2003; Zecchin et al. 2007; Li et al. 2009b |
| Sst1 | X | | | | X | PCR, in situ, Immuno., GFP | Argenton et al. 1999; Devos et al. 2002; Zecchin et al. 2007; Li et al. 2009b |
| Sst2 | X | | | | X | PCR, in situ, Immuno., GFP | Argenton et al. 1999; Devos et al. 2002; Zecchin et al. 2007; Li et al. 2009b |
| Vip | X | | | | | Immuno. | Mathieu et al. 2001; Olsson et al. 2008b |

(*Continued*)

**Table 5.2** (*continued*)

X: Expression not tested
White: No expression
Gray: Expression

| | 0-3 hpf | 24 hpf | 48 hpf | 72 hpf | 96-120 hpf | | |
|---|---|---|---|---|---|---|---|
| **THYROID** | | | | | | | |
| Tsh | X | | | | X | PCR, in situ | **Herzog et al. 2003** |
| **Trαa** | | | | X | X | in situ | **Essner et al. 1997; Liu et al. 2000; Liu and Chan 2002; Bertrand et al. 2007** |
| **Trαb** | X | | | X | X | in situ | **Liu and Chan 2002 Bertrand et al. 2007** |
| **Trβ** | | | | X | X | in situ | **Liu et al. 2000; Liu and Chan 2002; Bertrand et al. 2007** |
| **OSMOREGULATION** | | | | | | | |
| It | X | | | X | | in situ | Unger and Glasgow 2003; Eaton et al. 2008 |
| Avt | X | | | | | in situ | Tessmar-Raible et al. 2007; Eaton et al. 2008 |
| Prl | X | | | | | in situ | Herzog et al. 2003, 2004; Nica et al. 2004; Lopez et al. 2006 |

| | | | | | | | |
|---|---|---|---|---|---|---|---|
| Pth1 | X | | | | | PCR, in situ | Hogan et al. 2005 |
| **Pth2** | X | | | X | X | PCR, in situ | Hogan et al. 2005 |
| Pth2r | X | X | | | X | in situ | **Papasani et al. 2004** |
| Stc1 | X | | | X | X | in situ | Wingert et al. 2007 |
| Anp | X | X | | | | in situ | Berdougo et al. 2003; Bendig et al. 2006 |
| Adm2 | X | | | | | in situ | Toro et al. 2009 |
| **Crlr** | X | | | X | X | in situ | **Nicoli et al. 2008** |
| Ren | | | | | | in situ | Liang et al. 2004; Song et al. 2004 |
| Apln | | | | | | PCR, in situ | Zeng et al. 2007 |
| **Agtrl1a** | | | X | X | X | in situ | **Patterson et al. 2007; Tucker et al. 2007** |
| **Agtrl1b** | | | | X | X | PCR, in situ | **Scott et al. 2007; Zeng et al. 2007** |

X: Expression not tested at this time; White: No expression; Gray; Expression.

Hormones are listed in regular font and receptors in bold font. Methods of investigation are listed: "PCR" refers to RT-PCR or qPCR, "in situ" refers to whole-mount RNA in situ hybridization, "Immuno." indicates immunohistochemistry or western blotting, "RNase p." refers to RNase protection assay and "GFP" indicates live-tracking of GFP-fusion proteins.

### 2.1.2. Kisspeptin

An important neurohormone that affects reproduction in zebrafish is kisspeptin (Kiss). Human KISS regulates GNRH neurons by signaling through kisspeptin receptors located on GNRH neuronal membranes. Within a short time after the discovery of kisspeptin and its receptor in humans and mice, both zebrafish kisspeptin hormone and receptor were cloned (van Aerle et al. 2008; Biran et al. 2008). In addition, fish-specific duplicate copies of kisspeptin and its receptor were reported (Biran et al. 2008; Kitahashi et al. 2009; Lee et al. 2009). Kisspeptins are detected by quantitative real-time PCR (qPCR) on all days post fertilization tested (1, 3, 7, 30 and 45) and in adults (Kitahashi et al. 2009). Expression of kisspeptin in adult zebrafish by qPCR showed that the greatest abundance of *kiss1* mRNA was in the diencephalic-midbrain region (Biran et al. 2008) as expected since kisspeptin neurons synapse on Gnrh neurons containing kisspeptin receptors. However, as with many neuropeptides, *kiss1* transcripts were also expressed in the pituitary, pancreas and intestine (Biran et al. 2008). Although Kitahashi and colleagues found *kiss1* mRNA only in brain and testis, they identified a duplicate designated as *kiss2* with transcripts in the brain, gonads, intestine, and kidney (Kitahashi et al. 2009). Within the brain the duplicate hormones are expressed in distinct locations: *kiss1* mRNA is in the habenula and *kiss2* is in the hypothalamus (Kitahashi et al. 2009) suggesting that subfunctionalization has occurred between the two paralogs.

The duplicate Kiss1 receptors on the other hand are expressed as *kiss1ra* transcripts in the brain and testis, and *kiss1rb* transcripts in brain, pituitary, spleen, gills, kidney, intestine, pancreas, and adipose tissue (Biran et al. 2008). This suggests that the receptors probably have distinct functions with possible overlap in the brain.

High conservation of the kisspeptin-10 cleaved peptide sequence and the similar location of the *kiss* genes within vertebrate chromosomes suggest that kisspeptin has similar roles in fish and mammals. Both recombinant zebrafish Kiss receptors are functional as shown by their activation with human or zebrafish kisspeptin decapeptides (Biran et al. 2008; Lee et al. 2009). Interestingly, there is evidence that both kisspeptins and the Kiss1ra receptor are associated with the onset of puberty in zebrafish (Biran et al. 2008; Kitahashi et al. 2009). Finally, in adult females, injection of Kiss2 resulted in increased expression of *fshb* and *lhb* mRNA in the pituitary, which shows that the zebrafish brain–pituitary axis is similar to that of other vertebrates (Kitahashi et al. 2009).

### 2.1.3. Pituitary Adenylate Cyclase-Activating Polypeptide

PACAP is a polypeptide that increases cAMP in many tissues. In mammals, PACAP affects the release of several hormones including insulin and epinephrine and influences carbohydrate and lipid metabolism (Sherwood et al. 2003). The role of pituitary adenylate cyclase-activating polypeptide (Pacap) in zebrafish reproduction is not as well known as that of Gnrh, but Pacap is known to contribute to ovarian maturation and to inhibit *fsh* and *lh* mRNA in the pituitary. Both forms of Pacap were cloned from zebrafish tissues (Fradinger and Sherwood 2000; Wang et al. 2003), and zebrafish Pacap1 was shown to inhibit expression of *fsh* and *lh* mRNA when introduced into cultured zebrafish pituitary cells for 72 hours. Pacap1 and Pacap2 proteins have been detected in various areas of the brain including the hypothalamus (Mathieu et al. 2004). Likewise, they have been detected in the pituitary as protein (Mathieu et al. 2004) or mRNA (Lin and Ge 2009). Only single *pac1r* and *vpac1r* genes have been cloned from zebrafish and experimentally activated with the two forms of Pacap (Fradinger et al. 2005; Wu et al. 2008), but potential duplicate copies of each receptor have been identified in the genome (Roch et al. 2009).

Pacap appears to have a paracrine role in the gonads. In a study on Pacap2, expression occurred in follicles at various stages: previtellogenic, midvitellogenic, full-grown immature, and full-grown mature; suggesting a role throughout folliculogenesis (Wang et al. 2003). The Vpac2 receptor was also expressed in the ovary, although it is now classified as a Phi receptor (Wu et al. 2008), which may explain why *vpac2r* mRNA neither varied during the ovarian stages nor responded to gonadotropin stimulation (Wang et al. 2003). The regulation of Pacap2 in the zebrafish ovary may be controlled in part by gonadotropins, as Pacap2 expression in a primary culture of follicle cells showed a significant increase within 2 hours after treatment with human chorionic gonadotropin (hCG) or goldfish pituitary extract (Wang et al. 2003). Further proof that Pacap2 is involved in reproduction is given by synthetic zebrafish Pacap2 peptide, which significantly stimulated oocyte maturation (measured by germinal vesicle breakdown, GVBD) and follistatin expression in cultured follicle cells (Wang et al. 2003). These experiments strongly suggest that Pacap2 acts as a local factor in zebrafish ovary.

During development Pacap1, Pacap2 and their receptors are expressed early. The mRNA for *pacap1*, *pacap2*, *pac1r1* and *vpac2r* (*phir*) are likely to be maternal transcripts as they are expressed before 1.5 hpf. The *vpac1r* transcripts were detected at 5–6 hpf, which suggests they are early but not maternal transcripts. The protein forms of both hormones were also detected by 24 hpf with immunohistochemistry (Mathieu et al. 2004). The function of early expression may be related to early brain development as knockdown of either Pacap1 or Pacap2 resulted in changes in the expression of early brain markers and regional boundaries (Wu et al. 2006a), and knockdown of either Pacap or Pac1r decreased the number of dopamine- and isotocin-expressing cells in the brain (Blechman et al. 2007).

## 2.2. Pituitary Hormones: FSH, LH and Glycoprotein Subunit α

All of the major hormones and their receptors associated with the pituitary components of reproduction in humans have been cloned in zebrafish including follicle stimulation hormone (Fsh), luteinizing hormone (Lh), prolactin (Prl), inhibins, activins, follistatin, Pacap1 and Pacap2. Also included is glycoprotein subunit α (Gsuα) that is needed to form a dimer with both Fshβ and Lhβ to make them functional. However, prolactin assumes a non-reproductive role in zebrafish (see Osmoregulation, Section 7.2). Only a single copy of the cDNA for zebrafish *fshb* (So et al. 2005), *gsuα* (Nica et al. 2004; So et al. 2005) and the receptors *fshr* and *lhr* (Laan et al. 2002; Kwok et al. 2005; So et al. 2005) have been cloned, although two copies of *lhb* exist (So et al. 2005). One of the most interesting aspects of *lh* and *fsh* in zebrafish is that both hormone transcripts are expressed not only at high levels in the pituitary, but at low levels in other tissues including the brain. Also, unlike humans who express FSH and LH protein in the same pituitary cell, zebrafish express *fshb* and *lhb* mRNA in two distinct cell populations as revealed by in situ hybridization (So et al. 2005).

At 32 hpf in zebrafish, the gonadotropes can be detected by measurement of *gsuα* mRNA (Nica et al. 2004; Pogoda and Hammerschmidt 2007, 2009). The expression of *fshβ* and *lhβ* is so low that transcripts can only be identified by reverse transcriptase-PCR (RT-PCR) until the juvenile stage (Pogoda and Hammerschmidt 2007). The early expression of *gsuα* is curious, but may relate to an independent function for this gene as loss-of-function experiments result in embryos with reduced growth and non-specific defects (Huang et al. 2008). In zebrafish, pituitary cells are arranged in an anterior–posterior pattern as opposed to human cells that invaginate to form Rathke's pouch during development to produce a dorsal–ventral orientation (Pogoda and Hammerschmidt 2007, 2009). Regulators for organogenesis and cell specification of the anterior pituitary in zebrafish have been reviewed, noting the dependence of gonadotrope expression on Eya1 and repression by Pit1 transcription factors (Nica et al. 2004; Pogoda and Hammerschmidt 2009). It has been postulated that gonadotropes and corticotropes may represent the most ancient hormone-expressing lineage of the anterior pituitary, as they have to be actively repressed by Pit1 to allow differentiation of the somatotropes, lactotropes and thyrotropes (Pogoda and Hammerschmidt 2009).

The functions of Fsh and Lh in zebrafish appear to overlap more than in most vertebrates. Recombinant zebrafish Fsh and Lh activated their respective G-protein coupled receptors in vitro. Although Fsh selectively activated its receptor, Lh was less specific; it activated both Fshr and Lhr, suggesting that Lh may have a broader role during follicular development in addition to its ovulatory role (So et al. 2005). During the ovulatory cycle, levels of *fshb*, *lhb* and *gsuα* mRNA peaked in the pituitary at the time when the oocyte germinal vesicles were migrating within the ovary (Clelland and Peng 2009) and levels remained relatively high prior to germinal vesicle breakdown and ovulation (So et al. 2005). This similar expression profile for both hormones may reflect that zebrafish release eggs multiple times during the spawning season and have follicles at many stages. The RT-PCR expression profiles for the receptors may be more informative. In individual follicles, expression of *fshr* peaked first at the midvitellogenic stage, whereas *lhr* peaked later at the full-grown stage, just before oocyte maturation and ovulation (Kwok et al. 2005). Thus, only the prolonged expression of Lh might have an early effect on the Fshr during the cycle in zebrafish. During development, the expression pattern for both receptors shows the mRNA is barely detectable in immature ovaries (60 dpf) but increases rapidly at sexual maturity (Kwok et al. 2005) to be found only in follicular cells and not oocytes in adults (Wang et al. 2003).

### 2.3. Pituitary/Gonadal Hormones: Activins, Inhibins and Follistatins

The activin–inhibin–follistatin system acts in a coordinated way to modulate reproduction. All three protein hormones are found predominantly in the ovary, testes and pituitary, but also in a number of other organs related to their important role in early development. Activin was originally shown to stimulate FSH secretion in cultured mammalian anterior pituitary cells, whereas inhibin and follistatin inhibit FSH. Also, all three hormones are known to be local gonadal hormones with paracrine or autocrine functions.

Activins and inhibins are dimers formed from three gene products, which are combined to make five distinct hormones in vertebrates. Activin is composed of two β-subunits that are combined to form three types of activin (βA-βA, βB-βB and βA-βB) whereas inhibin uses the same β-subunits combined with an α-subunit to form two types of inhibin (α-βA and α-βB). As a result, the nomenclature of the hormones is complex (Table 5.1). Follistatin is a single chain glycoprotein with a structure distinct from activin and inhibin. Unlike humans, zebrafish have two genes for the activin βA subunit and two genes each for follistatin and follistatin-like molecules. The zebrafish subunits and hormones cloned to date include activin βA1 (Rodaway et al. 1999; Wu et al. 2000; Wang and Ge 2003a), activin βA2 (DiMuccio et al. 2005), activin βB (Wittbrodt and Rosa 1994), follistatin 1 (Bauer et al. 1998), follistatin 2 and follistatin-like 1 and 2 (Dal-Pra et al.

2006). Although the inhibin α-subunit has not been cloned, the sequence is annotated at NCBI (NM_001045204).

The receptors and signaling system for activin have been studied in depth. Analogous to TGFβ and BMP signaling, activin first binds to a type-2 receptor, which is followed by binding a type-1 receptor to form a heterodimeric complex. It is the type-1 receptor that initiates the signaling pathway involving a number of SMAD proteins as well as others. The complex then moves into the nucleus to regulate target genes. In contrast, inhibin is thought to compete with activin for type-2 receptors, creating an inhibition of activin activity, as the α-subunit may prevent binding to type-1 receptors (Ge 2005). The mechanism of inhibition by follistatin is quite different as it binds directly to activin with high affinity. In zebrafish, a full array of activin receptors and Smads have been identified and cloned. Zebrafish type-1 and type-2 activin receptors have been cloned and characterized: Actr1a (Yelick et al. 1998; Roman et al. 2002), Actr1b (Renucci et al. 1996), Actr2a (Nagaso et al. 1999; DiMuccio et al. 2005; Albertson et al. 2005) and Actr2b (Garg et al. 1999; Nagaso et al. 1999; Albertson et al. 2005). The receptors in both zebrafish and humans are single transmembrane proteins with a serine/threonine kinase domain for signaling, and are also known as activin receptor-like kinases (ALK).

Activin has a paracrine action in zebrafish pituitary, in addition to a possible endocrine role. In cultured zebrafish pituitary cells, recombinant goldfish activin βB stimulated expression of *fshb* and suppressed *lhb* transcripts. The *fshb* response matches that in mammals, but a change in the level of *lhb* does not (Lin and Ge 2009). To test this further, human follistatin was applied to the zebrafish pituitary, which resulted in decreased *fshb* mRNA and increased *lhb* mRNA. A paracrine effect for these actions is supported by the presence of mRNA in the pituitary for activin βB, follistatin, and the four activin receptors (Lin and Ge 2009). The presence of both a hormone and its receptor in the same tissue suggests a local action, although the proportion of local versus external control of pituitary function by activin or follistatin is not known.

Pituitary gonadotropins stimulate activins in the gonads. In zebrafish ovary, the expression pattern for the activin–inhibin–follistatin system and its receptors has been examined with RT-PCR. Oocytes contain activin βA, follistatin, activin receptors Actr1a, Actr1b, Actr2a, Actr2b and ALK1-like transcripts (Wang and Ge 2003b). The follicle cells contained all of the above with the addition of activin βB. Thus a complete signaling system is present in the follicle cells and in oocytes for reception of activin hormones that are secreted by the follicle cells. This basic molecular network suggests that the activin–inhibin–follistatin system is capable of modulating oocyte maturation. It is possible that activins act directly on oocytes to increase the number of receptors for the maturation-inducing hormone (MIH). Although expression of the inhibin α-subunit in zebrafish has not been studied, addition of human recombinant inhibin affects zebrafish oocyte maturation (Wu et al. 2000; Wang and Ge 2003b).

The activins are regulated by the pituitary gonadotropins, as the addition of human chorionic gonadotropin (hCG) or goldfish pituitary extract to zebrafish ovarian follicle cells increased both oocyte maturational competence and expression of *actβa1*, *actβa2*, *actr2a* and *actr2b* mRNA. This suggests that activin acts downstream from the gonadotropins. Further, recombinant human activin βA, goldfish activin βB and human inhibin A were shown to promote oocyte maturation in incubated full-grown zebrafish follicles, whereas human follistatin inhibited the actions of all three hormones (Pang and Ge 1999, 2002a,b; Wu et al. 2000; Wang and Ge 2003b; DiMuccio et al. 2005). The lack of antagonism between human activin A and inhibin A in the zebrafish ovary appears to be species specific and may be an artifact of using human hormones in fish. Activin βB appears to have a different role in the ovary as compared with activin βA, which increases with follicular growth and oocyte maturation. In zebrafish follicles treated with hCG, *actβB* mRNA is suppressed whereas *actβA* expression increases. During the ovarian cycle, *actβB* remains at a steady level. Only after germinal vesicle breakdown (GVBD) and ovulation does *actβB* alone show a sharp increase, but its function is unclear. The activin family has not been well studied in zebrafish testis although members are identified in several other fish species.

## 2.4. Gonadal Hormones

### 2.4.1. Anti-Mullerian Hormone

Sexual differentiation in zebrafish is distinct from mammals in that all zebrafish have a non-functional juvenile ovary with immature oocytes at 3 or more weeks after fertilization. Zebrafish do not have a sex chromosome or a *sry* gene, so the initial event in sex determination is not known, but downstream genes appear to be conserved in zebrafish compared with mammals (Siegfried and Nusslein-Volhard 2008).

Anti-Mullerian hormone acts downstream in the zebrafish sexual differentiation path. Zebrafish and humans both secrete an anti-Mullerian hormone (AMH), presumably from Sertoli cells, which is crucial for male differentiation. In humans, AMH initiates degeneration of the Mullerian ducts and inhibits the aromatase enzyme CYP19A1A that converts testosterone to estrogen in the gonads. However, there are no Mullerian ducts in zebrafish, so the role of Amh is somewhat different. Anti-Mullerian

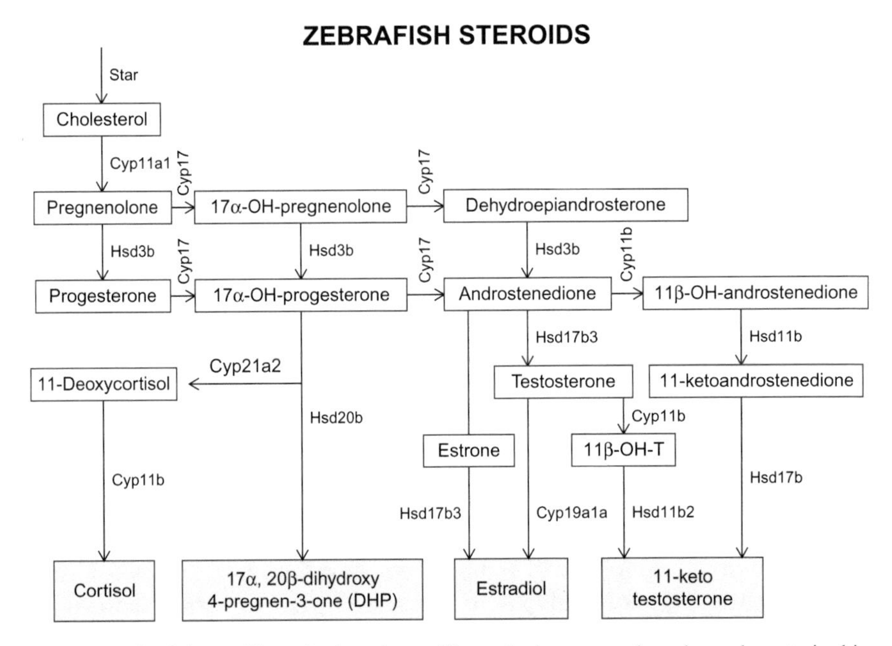


**Fig. 5.1.** Zebrafish steroid synthesis pathway. Biosynthetic enzymes have been characterized in zebrafish except for Cyp21a2. Major circulating hormones have a gray background. Source data from Rodriguez-Mari et al. 2005; Ings and Van Der Kraak 2006; To et al. 2007; Wang and Orban 2007; de Waal et al. 2008; Nagahama and Yamashita 2008; Alsop and Vijayan 2009; Clelland and Peng 2009; Levi et al. 2009.

hormone is a glycoprotein that shares membership with the activins in the transforming growth factor β superfamily (Rodriguez-Mari et al. 2005; von Hofsten et al. 2005).

Anti-Mullerian hormone is expressed very early as detected by qPCR (von Hofsten et al. 2005), but a sizable increase in hormone expression is detected (17–21 dpf) just before the beginning of the transformation of the juvenile ovary into a testis (Rodriguez-Mari et al. 2005; von Hofsten et al. 2005; Wang and Orban 2007). Later, at 30–31 dpf, the differentiated testis expresses high levels of *amh* mRNA but has very low or non-detectable amounts of aromatase. Ovaries at 30–31 dpf show the reverse pattern (Rodriguez-Mari et al. 2005). The differentiation signal in ovaries is not clear, but loss of the germ line results in a male pattern of expression suggesting that a differentiation factor exists in oocytes (Siegfried and Nusslein-Volhard 2008). Surprisingly, loss of the germ line in males does not prevent testes formation. There are other unique features of zebrafish: 11-ketotestosterone is a key differentiation factor in males, but the enzyme (Cyp11b) that converts testosterone to 11-ketotestosterone (11-KT) is expressed later than *amh* mRNA placing Amh earlier in the cascade than 11-KT for male differentiation (Wang and Orban 2007). Nonetheless, the androgen receptor that binds both 11-KT and testosterone is a maternal transcript (Hossain et al. 2008), which allows the possibility that testosterone may be involved in the male differentiation pathway. Also, 11-KT can be formed without Cyp11b, as demonstrated recently, suggesting an alternate pathway for sex determination (de Waal et al. 2008). Anti-Mullerian hormone is expressed in granulosa cells in human and zebrafish females and appears to have a similar role. In zebrafish *amh* mRNA is expressed in granulosa cells surrounding stage II oocytes and then disappears during each cycle, establishing that Amh has roles in addition to male differentiation (Rodriguez-Mari et al. 2005). A receptor for zebrafish Amh has not been reported, but medaka have a type-2 Amh receptor, which is a member of the TGFβ superfamily like activin receptors (Kluver et al. 2007). Large scale analyses of the transcriptome are being used to search for zebrafish genes that are expressed in a dimorphic pattern in the gonads and brain to identify candidate genes that may lead to differentiation (Sreenivasan et al. 2008).

#### 2.4.2. Growth Differentiation Factor 9 and Transforming Growth Factor β1

Growth differentiation factor 9 (GDF9) is structurally similar to the anti-Mullerian hormone and activins, as all are members of the TGFβ superfamily and are expressed in the gonads. Gdf9 is almost exclusively expressed in oocytes in the zebrafish ovary. The highest level of expression is in small primary growth follicles with a gradual decrease in the level during follicular development as measured by qPCR (Liu and Ge 2007). The temporal expression of Gdf9 varies in different species but its function is likely to be important as *Gdf9* gene knockout in mice stops folliculogenesis at the primary follicle stage (Dong et al. 1996).

Another member of the TGFβ superfamily that affects reproduction in zebrafish is Tgfβ1. To date, the main action is inhibition of oocyte maturation via the receptors for Lh and the maturation-inducing factor, DHP (Kohli et al. 2005). Clearly, TGFβ superfamily members have become players in the control of reproduction (Tan et al. 2009).

### 2.5. Gonadal Steroids

The main reproductive steroidal pathway in zebrafish is shown in Fig. 5.1. The steroidal intermediates are similar across many vertebrates, but the end products vary in zebrafish compared with humans; zebrafish have 11-ketotestosterone (11-KT) rather than 5α-dihydroxytestosterone as the final androgen; zebrafish have 17α, 20β-dihydroxy-4-pregnen-3-one (DHP) rather than progesterone as the final progesterone-like product; and zebrafish synthesize only cortisol whereas humans produce cortisol and cortisone. Both species have 17β-estradiol (E2). Zebrafish steroids have been detected by radioimmunoassays (RIA) or enzyme-linked immunosorbent assays (ELISA), which have been used in zebrafish to measure estradiol and testosterone (Ings and Van Der Kraak 2006; Levi et al. 2009), 11-KT (Wang and Orban 2007), DHP and prostaglandins (Lister and Van Der Kraak 2008). Proof of steroidal synthesis also depends on identification of the biosynthetic enzymes involved. Enzymes that have been cloned are shown in Fig. 5.1, whereas the enzyme (Cyp21a2) not yet identified, but postulated to exist in zebrafish, is shown in large font. The final support for the presence of steroids was the cloning of the zebrafish steroid receptors, which are listed in Table 5.1.

#### 2.5.1. Ovarian Steroids: Estradiol

In naturally spawning female zebrafish, estradiol increases within the ovary near the time of ovulation (Lister and Van Der Kraak 2008). Estradiol also increases after the application of hCG to cultured follicles along with several related biosynthetic enzymes (Ings and Van Der Kraak 2006). Unlike mammals, estradiol is an important regulator of genes expressed in the liver during the synthesis of zebrafish vitellogenin, used for the formation of egg yolk; one study revealed 672 hepatic genes that were differentially expressed during vitellogenesis and regulated by estradiol (Levi et al. 2009). In cultured pituitary cells in zebrafish, addition of estradiol increased both *fshb* and *lhb* mRNA suggesting positive feedback, as in other vertebrates. To complete the circuit, pituitary cells express estradiol receptors (*esr1*, *esr2a* and *esr2b* mRNA) (Lin and Ge 2009). These estradiol receptors (ERs) are three of 70 nuclear hormone receptors identified in zebrafish. In comparison, humans have only 48 nuclear hormone receptor genes. In regard to estradiol receptors, zebrafish have a single copy of Erα but have retained a duplicate copy of Erβ (Erβ1 and Erβ2). Only a single copy of each is found in humans (Bertrand et al. 2007).

The three zebrafish estrogen receptors are conventional nuclear hormone receptors (Bardet et al. 2002; Menuet et al. 2002). In addition, zebrafish, like other vertebrates, have an unconventional estrogen receptor that resides in the membrane as a G protein-coupled receptor (GPCR) and mediates rapid non-nuclear functions (Liu et al. 2009). The three nuclear estrogen receptors each bind estradiol (E2) with high affinity, and are activated by E2 in vitro as shown by using an estrogen-regulated reporter (Bardet et al. 2002; Menuet et al. 2002). In contrast to mouse ERs, the zebrafish receptors do not respond to 4OH-E2 and 4OH-tamoxifen (Bardet et al. 2002). The widespread actions of E2 are reflected by the wide and overlapping distribution of all three nuclear ERs in the gonads, pituitary, brain, liver and some non-reproductive organs. Even in the brain there is partial overlap in the location of the three ERs (Menuet et al. 2002). The function of E2 is deduced in part by the location of the receptors in the gonads for reproduction, in the brain and pituitary for regulation of reproduction, and in the liver for vitellogenesis.

It is not clear whether estradiol and its nuclear receptors have an important role in early development. Although RNAse protection assays detected *esr2b* as a maternal transcript at 1–6 hpf (Bardet et al. 2002), expression of the three receptors is not observed until 48–72 hours and is very weak as compared with adults. For instance specific expression of *esr2a1* is seen only at 36 and 48 hpf (Bertrand et al. 2007). The general conclusion is that ER genes are weakly expressed during early development in zebrafish and their importance is questioned (Bardet et al. 2002).

Endogenous E2 is likely present early in development, as the aromatase enzymes (*Cyp19a1a* and *Cyp19a1b* mRNA) are detected in unfertilized eggs, at 1.5 hpf, and at 24–120 hpf (Kishida and Callard 2001). Although fluorescently-tagged E2 localizes along the migration path of primary germ cells (PGCs) in zebrafish embryos at 3 dpf (Costache et al. 2005) suggesting a role in germ cell migration, the PGCs were not unequivocally identified by the presence of vasa, nor was localization of E2 checked in embryos without PGCs (see Gorelick et al. 2008).

Like humans, zebrafish have a plasma membrane bound estrogen receptor. In vitro, the membrane receptor can be activated by E2 resulting in signaling by cAMP or phospholipase C. Like the nuclear ERs, the membrane receptor is found predominantly in testes (early germ and Sertoli cells), ovary and brain but not in liver by RT-PCR (Liu et al. 2009). The overlap in location and signaling paths suggests the possibility of coordinated actions by the membrane and nuclear ERs.

### 2.5.2. Ovarian Steroids: Maturation-Inducing Hormone

Oocyte maturation requires in sequence: luteinizing hormone (Lh), maturation-inducing hormone (Mih) and maturation promoting factor (Mpf) in fish. After exposure to a surge of (Lh), ovarian follicles in zebrafish synthesize and release Mih. Essential for oocyte maturation, Mih in zebrafish is 17α, 20β-*di**h*ydroxy-4-*p*regnen-3-one (DHP) and is the final progesterone-derived product in zebrafish (Fig. 5.1). After the initial growth phase in oocytes in which they acquire yolk protein, and after the oocytes become competent (oocyte size of ~0.7 mm in zebrafish) due to the action of gonadotropins, activins and other factors, they can respond to the steroid DHP. The oocyte response to DHP is to resume meiosis. The downstream factor that directly triggers meiosis is maturation promoting factor (Mpf), which consists of Cdc2 and cyclin B (Nagahama and Yamashita 2008 for review). Cyclin B and Cdc2 have been cloned from zebrafish and are highly conserved. Mpf promotes the M-phase of the cell cycle, which includes nuclear envelope breakdown and chromosome condensation. In zebrafish, final maturation is marked by three periods: migration of the germinal vesicle toward the periphery of the oocyte, breakdown of the germinal vesicle (GVBD), and formation of the first polar body. Finally, ovulation occurs with the transition of the oocyte to a fertilizable egg. The application of this model to zebrafish has been well studied (Pang and Ge 1999, 2002a; Wang and Ge 2004) and reviewed (Clelland and Peng 2009; Lessman 2009).

In naturally spawning groups of female zebrafish, the maturation-inducing hormone (DHP) increased in the ovary near the time of ovulation (Lister and Van Der Kraak 2008). In isolated zebrafish follicles, treatment with DHP combined with either hCG or activins resulted in increased oocyte maturation as measured by the percentage of oocytes with germinal vesicle breakdown (GVBD) (Pang and Ge 2002a). In detail the initial event in oocyte maturation is activated by the binding of DHP (a modified progesterone steroid) to its receptor, which is a steroid receptor embedded in the oocyte outer membrane (Zhu et al. 2003b; Hanna and Zhu 2009). Zebrafish have two receptors that are members of a family of *m*embrane *p*rogestin *r*eceptors (mPRα and mPRβ). Both have high affinity for zebrafish DHP. Recombinant forms of either receptor are activated by DHP, resulting in rapid downstream activation of MAPK but inhibition of cAMP. Progesterone also bound and activated the receptors but with lower potency (Hanna et al. 2006). To show the mPR receptors in zebrafish are necessary for oocyte maturation, translation of mPRα and mPRβ was knocked down. Oocyte maturation stimulated by DHP was subsequently blocked (Zhu et al. 2003b). The location of the mPR receptors is not exclusively in the ovary but also in the testis and pituitary (Hanna and Zhu 2009). Protein levels of mPRα and mPRβ are low in surrounding follicular cells, but increase as oocytes develop. In testis immunostaining of mPRα is found in sperm whereas mPRβ is expressed in spermatogonia and spermatocytes. In the pituitary the receptors are scattered (Hanna and Zhu 2009). Thus DHP is essential for oocyte maturation but its function in testis and pituitary is not clear.

The role of the nuclear progesterone receptor (nPR) is of interest regarding ovulation. Although arachidonic acid and prostaglandins play a role in ovulation (Lister and Van Der Kraak 2008), other factors like progestins may be involved. Mouse *Pr* gene knockout prevented ovulation induced by LH. It has been postulated that DHP in most fish induces oocyte maturation through its membrane receptor and activates ovulation through its nuclear progesterone receptor, possibly by releasing proteinases (Nagahama and Yamashita 2008). This fish model would help to explain coordination between maturation and ovulation.

#### 2.5.3. Testicular Steroids: Testosterone and 11-Ketotestosterone

Zebrafish testes are capable of synthesizing not only testosterone and 11-ketotestosterone, but also three precursor intermediates (see Fig. 5.1) as shown by utilizing thin layer chromatography and [$^3$H]-androstenedione (de Waal et al. 2008). Indeed, testosterone is present in zebrafish testis as shown by RIA, and in the ovarian follicle where it serves as the precursor to estradiol. The level of testosterone can be increased by applying hCG to follicles in vitro (Ings and Van Der Kraak 2006). In turn, testosterone applied to cultured zebrafish pituitary cells results in increased *fshb* and *lhb* mRNA measured with qPCR (Lin and Ge 2009), which implies feedback regulation. The potent 11-ketotestosterone is an end product that can be measured in the testis but not ovary using ELISA (Wang and Orban 2007).

Both testosterone and 11-ketotestosterone have high affinity and potency for the single androgen receptor in zebrafish (Jorgensen et al. 2007; de Waal et al. 2008; Hossain et al. 2008). The zebrafish androgen receptor (Ar) is a nuclear hormone receptor (Jorgensen et al. 2007; de Waal et al. 2008; Hossain et al. 2008). In adults, the androgen receptor is present in the gonads, brain and five non-reproductive tissues with dimorphic expression in the gonads and muscle, as measured by qPCR (Hossain et al. 2008). In the brain, androgen receptor was found in the telencephalon, preoptic and periventricular areas but no dimorphic expression was detected (Gorelick et al. 2008).

In development,the androgen receptor was detected as a maternal transcript by qPCR and expression was maintained until the 5 somite stage (~7–10 hpf). The level of *ar* mRNA increased again at 24 hpf and continued to increase through 14 dpf, the last point of measurement (Hossain et al. 2008). In situ hybridization is not as sensitive as PCR but revealed *ar* mRNA expression in the pronephros at the 14–18 somite stage and at 24–48 hpf but not beyond (Gorelick et al. 2008). Another study did not detect *ar* mRNA by in situ hybridization for the first five days of development (Bertrand et al. 2007).

In summary, the zebrafish reproductive system has a full set of brain, pituitary and gonadal hormones. In some cases zebrafish have duplicates of hormones where only a single copy exists in humans. The similarities between zebrafish and humans are extensive with respect to both hormones and receptors, suggesting that this fish is an excellent model for study of the vertebrate reproductive system. Nonetheless, the zebrafish reproductive system has novelties compared with humans including the early development of a non-functional ovary in both sexes, the lack of a sex chromosome, uncertainty about sex determining factors, and different end products for the testosterone and progesterone pathways.

## 3. STRESS: THE CORTISOL STRESS AXIS

The vertebrate stress axis involves a signaling chain of several hormones and receptors (Table 5.1). The primary stress response is initiated when corticotropin-releasing hormone (CRH) is secreted by hypothalamic neurons to bind its receptor in the anterior pituitary on the surface of corticotrope cells. This is followed by the release of adrenocorticotropic hormone (ACTH), a post-translational product of pro-opiomelanocortin (POMC). ACTH binds its receptor, the type 2 melanocortin receptor (MC2R), in the adrenal cortex of mammals or the interrenal gland in fish. The activation of MC2R triggers a signaling cascade that results in the synthesis of the glucocorticoid (corticosteroid) cortisol, which produces several physiological responses including increased blood pressure and blood sugar, in addition to suppressed immune function.

### 3.1. Neurohormones: Corticotropin-Releasing Hormone and Urotensin I

Zebrafish have genes for all of the major components found in the mammalian stress axis. A single *crh* transcript was confirmed and characterized by two groups (Alderman and Bernier 2007; Chandrasekar et al. 2007) along with the related hormone urotensin-I (UI) (Alderman and Bernier 2007). Expression for *crh* mRNA was detected throughout the zebrafish brain as early as 24 hpf. The single copy of *crh* in zebrafish contrasts that of other closely related cyprinids like carp, which retain a duplicate paralog (Alsop and Vijayan 2009). The cognate receptors for Crh and UI have also been confirmed recently (Alderman and Bernier 2009).

### 3.2. Pituitary Hormone: Adrenocorticotropic Hormone

The Acth sequence has been confirmed within the *pomca* gene (Gonzalez-Nunez et al. 2003; Hansen et al. 2003). A duplicate gene, *pomcb*, was also identified but it lacked the necessary proteolytic cleavage site for Acth synthesis (Gonzalez-Nunez et al. 2003). The zebrafish Acth receptor, Mc2r, has been isolated and characterized; its transcripts are expressed at 25 hpf (Logan et al. 2003a; Alsop and Vijayan 2008). The impact of both Acth and Mc2r on interrenal development has been recently investigated by comparing the phenotypes of various corticotrope null-mutants with a Mc2r morpholino knockdown and dexamethasone treatment, which suppresses *pomc* expression (To et al. 2007). The steroidogenic development of interrenal cells was independent of corticotroph stimulation for the first few days after hatching but then was closely tied to Acth signaling.

### 3.3. Peripheral Steroid: Cortisol

The most well-characterized aspect of the stress axis in zebrafish is the release of cortisol in response to stress and its effect on downstream receptors. The first example of cortisol upregulation in response to stress applied to zebrafish was performed by Pottinger and Calder, by measuring whole-body levels of immunoreactive corticosteroids (Pottinger and Calder 1995). More recently, similar studies have been conducted to determine circulating cortisol levels in zebrafish responding to specific stressors such as crowding (Ramsay et al. 2006) and contact with a predator (Barcellos et al. 2007). The characterization of cortisol and its receptors, the glucocorticoid receptor (GR) and the mineralcorticoid receptor (MR), has also been studied in developing zebrafish. Cortisol levels in the developing embryo were measured immediately after fertilization (representing maternal cortisol in the egg), followed by a gradual decrease in concentration and a resurgence of larval-produced cortisol after hatching at 49 hpf (Alsop and Vijayan 2008). The earliest measured upregulation of cortisol levels in response to stress in zebrafish larvae occurred at 72 hpf after exposure to sea water, or 97 hpf, in response to swirling (Alsop and Vijayan 2008; Alderman and Bernier 2009). The increase in cortisol levels and stress response initiation corresponded well with the expression patterns of two critical glucocorticoid-synthesis factors: steroidogenic acute regulatory protein (Star) and 11 β-hydroxylase.

Cortisol has high affinity for binding both GR and MR. Within mammals, the enzyme 11 β-hydroxysteroid dehydrogenase type 2 (11βHSD2) is expressed in MR-specific tissues, and inactivates cortisol so that aldosterone may bind MR preferentially. Zebrafish lack the necessary steroidogenic enzymes to produce aldosterone, and Alsop and Vijayan found that Mr expression was upregulated in early embryos while Gr expression was downregulated and 11βhsd2 was not expressed, suggesting a possible role for maternal cortisol in the developing embryo involving Mr, before hatching (Alsop and Vijayan 2008). After hatching, both Gr and 11βhsd2 levels were upregulated along with the production of larval cortisol. Zebrafish retain only a single copy of Gr, homologous to the type-2 receptor, unlike the majority of teleosts previously studied that have two genes. This receptor can be spliced into two isoforms, Grα and Grβ, a feature unique to zebrafish and humans (Schaaf et al. 2008). The β-isoform of the zebrafish receptor also performs a homologous function to its human counterpart, acting as an inhibitor of Grα expression, and is found in a similar tissue distribution. Zebrafish therefore present the first non-primate model system for the study of GR, which is speculated to influence the onset of immune-related diseases in humans (Schaaf et al. 2009).

## 4. GROWTH

In most fish, growth continues throughout life making them a good model for understanding the hormonal regulation of growth (Mommsen 2001). Zebrafish are useful in studies on the hormonal control of biochemical aspects of growth. The overall control of growth begins in the brain, although local factors in each tissue are also important. In zebrafish, as in other vertebrates, growth hormone-releasing hormone (GHRH) is synthesized in nerve cells in the brain. After secretion, GHRH binds its receptor on the surface of somatotroph cells in the anterior pituitary to release growth hormone (GH), which in turn acts directly and indirectly through insulin-like growth factor (IGF) to affect growth in many tissues.

### 4.1. Neurohormones

#### 4.1.1. Growth Hormone-Releasing Hormone

The cloned sequence of zebrafish Ghrh shows high conservation with human GHRH (Lee et al. 2007) and functional similarity. Zebrafish Ghrh is identical in sequence to goldfish Ghrh and released Gh from cultured goldfish pituitary cells in a dose-dependent ($10^{-11}$ M to $10^{-6}$ M) manner, within the physiological range (Lee et al. 2007). The zebrafish Ghrh receptor (Ghrhr) is a typical 7-transmembrane receptor that couples with $G_s$ to increase intracellular cAMP. Recombinant zebrafish Ghrhr expressed in vitro is activated by zebrafish Ghrh but not by several other hormones that are structurally related (Lee et al. 2007). These related hormones, which include glucagon and PACAP, are members of the secretin superfamily along with GHRH. This superfamily contains nine hormones in humans and zebrafish (Roch et al. 2009). Most of these superfamily hormones also affect growth and metabolism. The receptors for the secretin superfamily hormones are grouped together in family B of the G protein-coupled receptors (GPCRs).

#### 4.1.2. Growth Hormone-Releasing Hormone-like Peptide

In zebrafish two Ghrh-like peptides (Ghrh-lp) and one specific receptor have been cloned and an additional receptor has been annotated in the genome. Each zebrafish Ghrh-lp gene has synteny with the human GHRH-LP gene (Wang et al. 2007). The function of these peptides still remains elusive. Ghrh-lp has more structural similarity to Ghrh than to Pacap in zebrafish or in human, but does not appear to consistently release growth hormone (Gh).

Ghrh-lp is embedded in the same gene with Pacap so the cloning of the two zebrafish Pacap genes revealed the structure of Ghrh-lps (Fradinger and Sherwood 2000; Wang et al. 2003). Transcripts encoding Ghrh-lps are expressed in the brain, eye, ovary and testis with Ghrh-lp1 also in gastrointestinal tract and Ghrh-lp2 in muscle. The two forms of zebrafish Ghrh-lp, also known as PACAP-related peptide (PRP) in mammals, have not been functionally tested *in vivo* in zebrafish. However, one of the two receptors was cloned and expressed as a recombinant receptor in a cell line (Fradinger et al. 2005; Wu et al. 2008; Roch et al. 2009). Zebrafish Ghrh-lp1, but not Ghrh-lp2 activated zebrafish Ghrh-lp receptor in a dose-dependent manner as measured by an increase in cAMP (Wu et al. 2008). The relatedness of Ghrh with Ghrh-lp1 was examined by cross-activation studies; Ghrh did activate the Ghrh-lp1 receptor using only zebrafish hormones and receptors (Wu et al. 2008), but goldfish Ghrh-lps had little or no effect on the zebrafish Ghrh receptor (Lee et al. 2007). The latter effect needs to be retested in a homologous zebrafish system, as a recent study has indicated there may be cross-talk between both GHRH and GHRH-LP with their respective receptors in chicken (Wang et al. 2007).

The function of Ghrh-lp in zebrafish was addressed by immunohistochemical localization of the hormone in the brain. Castro and colleagues (2009) found Ghrh-lp immunoreactive cells predominated in two discrete areas of the brain: in the lateral tuberal nucleus just above the pituitary with fibers innervating the pituitary, and in the secondary gustatory/visceral nucleus in the hindbrain with fibers innervating the inferior hypothalamus. The former location suggests a neuroendocrine function and the latter a feeding function based on the role of the inferior hypothalamus (Castro et al. 2009). The role of Ghrh-lp in this anatomical connection between taste and feeding centers is worthy of further experiments. There is no consistent proof that Ghrh-lp affects growth by release of Gh, but may affect growth in other ways including feeding. The novelty here is that the Ghrh-lp system appears to be fish-specific, as mammals, including humans, have lost the receptor for GHRH-LP (or PRP) from the genome and the peptide appears to be non-functional (Cardoso et al. 2007; Lee et al. 2007).

#### 4.1.3. Pituitary Adenylate Cyclase-Activating Polypeptide

The details of the Pacap hormones and receptors in zebrafish and humans are discussed in detail in Section 2.1.3. To determine whether Pacap affects growth in zebrafish, cultured pituitaries were treated with zebrafish Pacap1 for 72 h. An increase in *gh* mRNA resulted. Eight different hormones were tested in the system but only Pacap1 increased the *gh* transcripts (Lin and Ge 2009). To show that zebrafish have a Pacap circuit in the pituitary, mRNA for Pacap1 and Pacap2 (*adcyap1a* and *1b* genes) and for the receptors *vpac1r* and *vpac2r* was detected in pituitary cells whereas the receptor *pac1r1* was not detected (Lin and Ge 2009). In development, *adcap1a*, *adcap1b* and *pac1r1* are expressed as maternal transcripts (Wu et al. 2006a) whereas *gh* mRNA is not expressed until 48 hpf (Herzog et al. 2003). This opens the possibility that Pacap might affect Gh in early phases of development.

#### 4.1.4. Somatostatin

In humans growth hormone is stimulated by GHRH and inhibited by somatostatin (SST), which are each released from separate nerve cells in the brain. In addition, somatostatin is widely distributed in the brain beyond the hypothalamus and is one of the major hormones secreted from pancreatic islets along with insulin, glucagon and pancreatic polypeptide. Again, somatostatin has an inhibitory action on the other three peptides. The role of brain somatostatin in control of growth hormone will be considered here but the role of pancreatic somatostatin in metabolism is discussed in Section 6.4.3 below.

Four somatostatin genes (Sst1–4) have been identified (Argenton et al. 1999; Devos et al. 2002; Tostivint et al. 2004, 2008). A number of functions in the central nervous system, including cognitive, sensory and autonomic, are ascribed to somatostatin in addition to its control of the somatotrophs in the pituitary. In zebrafish, Sst1 is expressed in the brain and has been detected as early as 19 hpf in the ventral diencephalon and spinal cord; by 24 hpf it is also expressed in other brain areas and expression in the brain is maintained thereafter (Devos et al. 2002). Five receptor types have been annotated but not cloned. Concerning function, Sst3 is a more potent inhibitor of basal Gh secretion than Sst1 or Sst2 in goldfish (Yunker et al. 2003). Other roles of somatostatin in the pancreas and metabolism are described below (Section 6.4.3).

### 4.2. Pituitary Hormone: Growth Hormone

The actions of growth hormone (Gh) in fish are extensive including both direct and indirect effects. The direct actions in muscle include an increase in DNA and protein synthesis and lipolysis. Indirect actions include an increase in Igf1 transcription and release from liver, which then stimulates lipolysis from adipose tissue, an increase in amino acid transport and intestinal growth in the gut, and in the brain an effect on appetite and physical activity. A more complete list of its functions is available (Mommsen 2001).

The development of the anterior pituitary gland in zebrafish has been analyzed using the expression of the transcription factor *lim3*, which can be used as a marker for the adenohypophysis. To determine when the somatotrophs can first be detected, zebrafish *gh* was cloned and found to be expressed weakly at 42 hpf and strongly at 48 hpf (Herzog et al. 2003; Pogoda and Hammerschmidt 2009). Thus Gh is the last of the six pituitary hormones to be expressed in zebrafish, and the order of expression in the mouse and zebrafish are different, presumably due to development of the zebrafish embryo external to the mother. Although zebrafish have only a single *gh* gene, they have two *gh* receptor genes. These receptors are class 1 cytokine receptors that have only one transmembrane domain, but form homodimers (Liongue and Ward 2007). RT-PCR detected both receptors at 24 and 72 hpf (the only times tested), but the receptors have not been tested for functionality.

The function and regulation of Gh in zebrafish was examined in cultured pituitary cells where *gh* mRNA was detected without any treatment. Eight hormones were individually administered for 72 h on the zebrafish pituitary cells; only zebrafish Pacap1 significantly increased the expression of *gh*. Goldfish activin B and human IGF1 decreased *gh* mRNA presumably as feedback whereas Gnrh3, estradiol, testosterone, human follistatin and human EGF did not change *gh* transcript levels (Lin and Ge 2009). Receptor expression for each of the eight hormones was detected by RT-PCR (Lin and Ge 2009). Two other studies examined loss and gain of growth hormone phenotypes. Overexpression of Gh in a homozygous transgenic zebrafish might be expected to produce a large fish, but instead resulted in a starvation phenotype in which the zebrafish did not grow and were locked in a catabolic state. The reason may be that both growth hormone receptor (*ghr*) mRNA and *igf1* mRNA expression were decreased or downregulated (Figueiredo et al. 2007). This experiment showed that there is an optimal level of Gh needed for induction of growth and that excess hormone could regulate both the Gh receptor and Igf1 transcripts. In contrast, knockdown "did not lead to any discernable morphological changes within 10 dpf" (Zhu et al. 2007). This demonstrates that either Gh is not important in early growth or there is another hormone that compensates for loss of Gh.

### 4.3. Liver Hormone: Insulin-like Growth Factor

In vertebrates, insulin-like growth factor (IGF) is the major regulator of growth. IGF1 is stimulated by growth hormone, and in turn, acts on muscle, liver, adipose tissue, intestine, brain and most tissues to enhance growth.

The "indirect" actions of growth hormone are ascribed to IGF1 and broadly include cell proliferation, differentiation, and survival of most cell types. Although liver is the main source of IGF1, most tissues also produce IGF1 for paracrine or autocrine actions. Whereas IGF1 is associated with postnatal growth, IGF2 is a major regulator of prenatal growth in mammals. Igf2 also has an important role in embryo development, even though zebrafish is a non-placental animal. Structurally IGF1 is related to IGF2 and insulin. In zebrafish, there are six genes: Igf1, two forms of Igf2, one form of Igf3 and two forms of insulin.

Zebrafish Igf1 has a deduced mature protein of 70 amino acids and is expressed in all zebrafish tissues tested by RT-PCR (Chen et al. 2001; Maures et al. 2002). Transcripts of *igf2a* and *igf2b* have distinct expression by in situ hybridization (White et al. 2009). Interestingly, zebrafish have a novel Igf3 that is expressed only in fish gonads (Maures et al. 2002; Sang et al. 2008; Wang et al. 2008). The four zebrafish Igf molecules are encoded on separate genes and different chromosomes.

Functionally, there is a working relationship between zebrafish Igf and Gh. In vivo, the level of *igf1* mRNA was increased 15 h after Gh was injected whereas *igf2* and *igf* receptor transcript levels were not regulated by Gh (Maures et al. 2002). In general Gh regulates *igf1* mRNA. However, overexpression of Gh in homozygous transgenic zebrafish resulted in a significant increase in *Gh* mRNA, but no change in *igf1* expression in liver probably due to downregulation of the Ghr (Figueiredo et al. 2007). In cultured zebrafish pituitary cells, *igf1*, *igf1ra* and *igf1rb* transcripts were expressed, and addition of recombinant human IGF1 resulted in a significant decrease in the level of pituitary growth hormone by negative feedback (Lin and Ge 2009).

With respect to proliferation, zebrafish embryonic cells responded equally to fish, chicken or human IGFs (Maures et al. 2002). To further determine if Igf1 stimulates early growth and development in the zebrafish embryo as in mammals, a loss-of-function experiment was done utilizing a truncated Igf receptor (igf1ra). This resulted in a small embryo, disrupted central nervous systems (CNS) and malformed somites. In severe cases the head and eyes did not form (Eivers et al. 2004). Overexpression of *igf1* mRNA in a normal embryo resulted in development of the anterior region of the embryo at the expense of the trunk and tail. In comparison with the *Igf1* null mouse, both loss-of-function embryos were small but the zebrafish embryo CNS was more severely affected (Eivers et al. 2004). This may occur because the mouse has more redundancy for the IGF system, or because the dominant negative receptor in zebrafish blocked not only Igf1, but also the Igf2. Knockdown of the receptors (Igf1ra and Igf1rb) disrupted the eye and inner ear, but not the patterning of the nervous system (Schlueter et al.

2007a). Also, Igf2b and the Igf1rb have a role in the migration of primordial germ cells (PGC) toward the gonadal ridge (Sang et al. 2008). Clearly the IGF system is important to early development in vertebrates.

Zebrafish may be a very good model for examining the role of Igf2 in prenatal development because the embryo is external to the mother and transparent. In developing embryos, *igf1*, *igf2*, *igf1ra* and *igf1rb* mRNAs were detected by in situ hybridization as ubiquitous maternal transcripts (Table 5.2). Early expression was confirmed by RT-PCR and immunohistochemistry (Maures et al. 2002). In contrast, *igf3* mRNA is not detected in the gonads until about 8 days after fertilization (Wang et al. 2008). To understand the role of Igf2 in the zebrafish embryo, it is important to consider both forms of Igf2 (Igf2a and Igf2b). Recently, the Igf2 hormones were shown to be expressed in distinct locations: Igf2a was expressed in the notochord whereas Igf2b was expressed in nearby midline tissues, forebrain and nephron primordia (White et al. 2009). The expression of both started in the zygote and continued over time. To examine specific functions, Igf2a and Igf2b were knocked down separately with morpholinos. The notochord was noticeably disrupted and segmentation was delayed in 24 hpf embryos after knockdown of either gene. Knockdown of both Igf2 hormones at the same time produced a further reduction in the number of somite pairs (White et al. 2009). In conclusion, both Igf2 molecules were involved in dorsal midline development, but loss of Igf2b had additional effects on the ventral telencephalon and embryonic kidney. A similar phenotype occurred with knockdown of either receptor Igf1ra or Igf1rb. The role of Igf2 ligands and receptors in zebrafish development may open a window on understanding the distinct roles of Igf1 and Igf2 in vertebrates.

The two distinct Igf1/Igf2 receptors (Igf1ra and Igf1rb) are homologous to the single human IGF1R and are ubiquitously expressed in embryos (Ayaso et al. 2002; Maures et al. 2002; Schlueter et al. 2006, 2007a). A specific receptor for Igf3 has been suggested but not identified (Wang et al. 2008). Antibodies raised against either zebrafish Igf1ra or Igf1rb precipitated the receptors after they bound [$^{125}$I]IGF1. These receptors can bind Igf1 or Igf2, but not insulin, suggesting that Igf1 receptors are specific Igf receptors (Maures et al. 2002). Chemical reduction of the receptors followed by immunoprecipitation as smaller molecules, demonstrated that the $\alpha$ and $\beta$ subunits of the receptors were disulfide linked, as they are in other vertebrates. Thus, both Igf1 receptors are expressed and functional and have an adult distribution that is widespread and mostly overlapping (Maures et al. 2002). For loss-of-function studies with Igf receptors, dominant negative truncated forms of either Igf1ra or Igf1rb that still bound the Igf ligands were prepared. Loss of Igf1ra resulted in small embryos with a disrupted CNS and head (Eivers et al. 2004). In comparison, when morpholinos were used to knock down Igf1ra or Igf1rb, growth retardation of the embryo and developmental arrest in somites, eye, inner ear and heart began at about 16–18 hpf (Schlueter et al. 2006, 2007a). Evidence of a reduction in cell proliferation also was provided. Lack of Igf1rb (but not Igf1ra) delayed spontaneous muscle contractility and motoneuron development (Schlueter et al. 2006, 2007a). The knockdown of both receptors together resulted in the death of all embryos by 30 hpf. Knockdown of Igf1rb alone by either morpholinos or a dominant negative construct resulted in disruption of PGC migration to the gonadal ridge (Schlueter et al. 2007b; Sang et al. 2008). Clearly the Igf system is essential for the development, survival and growth of the zebrafish embryo.

### 4.4. Pancreatic Hormone: Insulin

In zebrafish, there are two insulin genes (*insa* and *insb*) and two insulin receptor genes (*insra* and *insrb*). Insulin, its receptors and its role in metabolism are discussed in Section 6.4.1. Here the specific effects of insulin on zebrafish growth are considered. The expression of insulin (Insb) in blastomeres (2–4 hpf) and head (24 and 48 hpf) provides the basis for considering zebrafish insulin as a possible growth, survival and neurotrophic factor. More concrete evidence comes from knockdown of the Insra receptor in zebrafish, which resulted in overall growth retardation at 24 hpf, underdevelopment of the eye and failure of tail elongation (Toyoshima et al. 2008). Loss-of-function for the Insrb receptor was less severe but resulted in pericardial edema (Toyoshima et al. 2008). The knockdown of both receptors produced a more severe phenotype with defects in both the tail and heart (Toyoshima et al. 2008). However, by day 3 of development the *insra* morphant had more severe reduction of the eye, midbrain and forebrain with distortion of the mid-hindbrain boundary whereas the insrb morphant had the same phenotype but milder, with pericardial edema and a distorted jaw. Thus the insulin receptors are required in embryonic growth and cell differentiation, especially of the brain and heart (Toyoshima et al. 2008). There are some differences between zebrafish Insr knockdown and mouse *Insr* knockout, but these may reflect that the zebrafish knockdown is transient. In general terms, the Igf and insulin systems are critical for zebrafish growth and development.

### 4.5. Other Tissue Hormones: Ghrelin and Obestatin

Ghrelin, synthesized mainly in the stomach, is known as a potent stimulator of growth hormone in mammals, although there is a surprisingly mild phenotype in ghrelin null mice. Similarly, ghrelin-receptor knockout mice have normal growth, development and appetite (Kojima and Kangawa 2005). Nonetheless, ghrelin is an important hormone to stimulate food intake and release of GH, although its loss can be compensated for by other systems. In zebrafish, the mature ghrelin hormone is predicted to be an amidated 12- or 19-amino-acid peptide (Amole and Unniappan 2009; Li et al. 2009a). The corresponding ghrelin receptor (also known as the GH secretagog receptor) has been identified by bioinformatics and phylogenetic analysis for zebrafish as a G protein-coupled receptor (Olsson et al. 2008a). Ghrelin mRNA is first expressed in the embryonic pancreas, but in adults ghrelin is present in the nine tissues tested by RT-PCR (Amole and Unniappan 2009).

Morpholino knockdown of ghrelin resulted in delayed body development, small eyes, defective gas bladders, cardiac edema and death at 4–10 days after fertilization (Li et al. 2009a). Growth hormone mRNA was not detected in 56% of the zebrafish morphants treated with a ghrelin morpholino. Normally *gh* transcripts are detected at 48 hpf but were absent even at 60 hpf in this case (Li et al. 2009a). Compensatory stimulators of Gh were not apparent at this early stage of development in zebrafish suggesting ghrelin has an important role in growth of embryos.

Obestatin is encoded within the ghrelin gene in mammals and possibly in fish, although a conventional cut site is lacking in the zebrafish precursor. An inhibitory function has been tentatively ascribed to obestatin. In the ghrelin zebrafish morphants, obestatin mRNA was microinjected and found to further increase the number of morphants lacking *gh* transcripts (Li et al. 2009a). The zebrafish may prove to be useful in studying ghrelin and obestatin effects during early development, as growth and developmental defects were observed in the zebrafish lacking ghrelin but not in the mouse.

## 5. THYROIDAL REGULATION

### 5.1. Thyrotropin-Releasing Hormone and Thyroid-Stimulating Hormone

Zebrafish, like other teleosts, have a thyroid axis similar to that of mammals and amphibians, with a few unique variations. Thyrotropin-releasing hormone (Trh) is detected throughout the zebrafish brain with a wider distribution than in other teleosts or mammals, suggesting possible additional functions in zebrafish (Diaz et al. 2002). Thyroid-stimulating hormone (Tsh) β-subunit and glycoprotein subunit α (Gsuα), both structurally conserved to mammalian orthologs, are produced in the pituitary at levels lower than typically seen in mammals (MacKenzie et al. 2009). Additionally, unexpected expression of Tsh and its receptor (Tshr) has been reported in the gonad of a teleost species (Kumar and Trant 2001; MacKenzie et al. 2009). As there are significantly higher concentrations of thyroid hormone in adult female zebrafish, this suggests a possible role for Tsh in reproduction, although this has not been characterized further (Simon et al. 2002).

### 5.2. Thyroid Hormones: T3, T4

Thyroid hormone plays a significant role in the development of mammals, amphibians and fish and therefore is thought to be conserved throughout evolution (Brown 1997). Due to external fertilization and development, transparent embryos, and complete genetic characterization, zebrafish have been recognized as an ideal model for studying development and the role of thyroid hormone during development and metamorphosis. In early development, maternal thyroid hormone and thyroid hormone receptor mRNA are provided to the oocyte. At mid-blastula stage (2.5 hpf) embryonic expression begins (Table 5.2) (Essner et al. 1997; Liu and Chan 2002). These receptors are available to respond to maternal thyroid hormone as well as potentially down-regulate the thyroid axis until the embryo is developed enough to synthesize embryonic thyroid hormone (Nowell et al. 2001). Additionally, Walpita et al. (2007) showed that the embryo responds to exogenous treatment with triiodothyronine (T3), demonstrating maternal thyroid hormone serves a function in early stages of development. Embryonic thyroid follicle development begins early in zebrafish, compared with mammals, at 60 hpf. This is indicated by production of the thyroid hormone precursor (thyroglobulin) and tetra-iodothyronine (T4) in a functional thyroid gland as early as 72 hpf (Alt et al. 2006; Brown 1997). Similarly to mice, zebrafish do not require Tsh for early follicular development, but basal levels of Tshr may play a role in follicle growth when thyrotropin is absent (Alt et al. 2006).

Thyroid hormone concentrations accumulate in fish oocytes during ovarian maturation, presumably to assist in facilitation of hatching. Exogenous thyroid hormone treatment leads to premature pectoral fin development and accelerated pelvic fin growth (Brown 1997). Thyroid hormone receptor antagonism results in developmental retardation of maturation of gastrointestinal tract, swim bladder, and lower jaw cartilage, and reabsorption of the yolk sac (Liu and Chan 2002). Taken together, it is clear that thyroid hormone signaling is required for the transition from embryo to larval form in zebrafish. A unique characteristic of the mature zebrafish thyroid tissue is its diffuse nature; a loose row of thyroid follicles form in the adult as opposed to the thyroid gland that develops in mammals (MacKenzie et al. 2009; Wendl et al. 2002).

## 6. ENERGY HOMEOSTASIS: FOOD INTAKE AND METABOLISM

Energy homeostasis can be considered a comprehensive approach to metabolism, including central regulation of appetite (energy input), indicators of energy resources from peripheral tissues (gastrointestinal tract and adipose), and energy expenditure and storage regulated by the pancreas. Due to a rise in the incidence of diabetes and obesity, there has been an increased interest in studying all metabolic hormones, including those involved in regulation of appetite and energy homeostasis. As research progresses, it is becoming clear that these regulatory pathways are conserved throughout vertebrates and in general, appetite regulatory mechanisms are conserved between zebrafish and humans. In addition, zebrafish has been demonstrated as an excellent model for studying genetics, regulation and development of systems involved in energy homeostasis.

### 6.1. Neurohormones

The key players in central appetite regulation in the hypothalamus of fish appear to be similar to those of mammals, with the same hormones taking primary roles. Specifically, appetite stimulating hormones like neuropeptide Y (NPY), agouti-related protein (AgRP), orexins, and melanin-concentrating hormone (MCH), as well as appetite inhibiting hormones such as α-melanocyte stimulating hormone (α-MSH) and cocaine- and amphetamine-regulated transcript protein (CART) play a role in central (brain) regulation of food intake in teleosts (Volkoff 2006) (see Table 5.1).

#### 6.1.1. Orexigenic Peptides: NPY, MCH, AgRP, Orexins

NPY, the most powerful appetite stimulant produced in the mammalian brain, has been shown to regulate food intake in teleosts, goldfish and salmon (Silverstein et al. 1998; Narnaware et al. 2000) and is expressed in zebrafish neurons, both during development and in the adult fish (Soderberg et al. 2000; Mathieu et al. 2005b). Clearly, NPY plays an important role in appetite regulation in fish as well as mammals. Less clear is the type of NPY receptor (also known as Y receptor) involved in this function. The Y receptors are a family of G protein-coupled receptors with different members varying in the complement of mammalian and teleost genomes. The mammalian Y receptors include 5 members, Y1, Y2, Y4, Y5, and Y6. Y1 and Y5 are responsible for signaling effects on appetite (Fredriksson et al. 2006). Zebrafish express Y1, Y2, and Y4 orthologs and the unique types Y7, Y8a and b (Fredriksson et al. 2006; Salaneck et al. 2008). Salaneck's recent report of a Y1 receptor in zebrafish is intriguing, as even though they identified the receptor in zebrafish, they were not able to find it in other teleosts, such as medaka or pufferfish. This poses the question of which receptors in these fish are involved in regulating Npy's signal.

Other hormones in the Npy family, the peptide YY's (Pyya and Pyyb), are expressed in gut and brain of zebrafish and have been shown to stimulate appetite, most likely through activation of the same receptor as Npy (Soderberg et al. 2000; Murashita et al. 2009). Until recently, Pyyb had been known as a unique fish PYY called PY, but has now been identified as a fish-specific gene duplication of Pyya (Sundstrom et al. 2008).

Melanin-concentrating hormone (Mch), first discovered in fish, is a 17–19 amino-acid hypothalamic peptide named after its role regulating skin color. Mch stimulates aggregation of pigment in melanocytes, resulting in a lightening of skin color (Takahashi et al. 2004). MCH has a regulatory role in food intake as shown by increased food intake after ICV injections, and elevated pro-melanin concentrating hormone (*Pmch*) mRNA levels in starving mice (Shi 2004). Interestingly, barfin flounder that are reared on a white background have higher plasma Mch levels and greater body lengths and weights. This suggests higher Mch levels also stimulate the hypothalamus to increase food intake, leading to larger size (Takahashi et al. 2004). Contradicting the demonstrated orexigenic role in mammals, and the suggested orexigenic role in flounder, Shimakura et al. (2008) showed a decrease in feeding after ICV injection of Mch in goldfish brain, as well as a decrease in transcripts for the orexigenic peptides Npy and ghrelin. Pro-melanin-concentrating hormone-like (*pmchl*) transcripts are expressed in zebrafish hypothalamus but they have not been well characterized (Toro et al. 2009). Some of the varying MCH activities between animals may be due to the variation in MCH receptors. Humans encode two, rodents encode one, pufferfish encode two and zebrafish encode three receptors (Logan et al. 2003a). The role of MCH may be unclear in these animals; however, it is likely to be involved in appetite regulation.

Other central orexigenic regulators of appetite in mammals include Agouti-related protein and the orexins, A and B (also known as hypocretin), which are expressed in zebrafish hypothalamus (Song et al. 2003; Kaslin et al. 2004). The functions of these peptides in energy homeostatsis have been studied using unique zebrafish models. Overexpression of hypocretin (*hcrt*), a single gene encoding both orexin A and orexin B, leads to zebrafish with insomnia-like behavior, demonstrating that zebrafish orexins also have a role in regulation of the sleep-wakefulness cycle (Prober et al. 2006; Yokogawa et al. 2007). In contrast, overexpression of Agouti-related protein leads to a genetic zebrafish model for obesity (Song and Cone 2007). Zebrafish Agrp is an endogenous antagonist for anorexigenic melanocortin receptors, leading to inhibition of appetite suppression and increased food intake resulting in obese fish.

#### 6.1.2. Anorexigenic Peptides: ACTH, MSH and CART

Differential expression and cleavage of POMC leads to the production of peptide hormones with roles in the stress and pain axes including adrenocorticotropic hormone (ACTH) and β-endorphin, or the production of peptides regulating appetite and pigmentation including α-, β-, and γ-melanocyte-stimulating hormone (MSH). In zebrafish, duplicate Pomc genes are expressed, *pomca* and *pomcb*; the second one lacks processing sites necessary to release mature Acth (Gonzalez-Nunez et al. 2003; de Souza et al. 2005). In teleosts, both Pomc sequences lack a γ-Msh sequence (Gonzalez-Nunez et al. 2003; Hansen et al. 2003). The POMC-derived peptides bind to a family of G protein-coupled receptors, the melanocortin receptor family. In mammals, this family of receptors includes five G protein-coupled receptors, with MC3R and MC4R the subtypes involved in appetite inhibition. In zebrafish, six melanocortin receptors have been identified: Mc1–4r, and duplicate Mc5 receptors (Ringholm et al. 2002). Protein sequence comparison and binding studies have shown that the six zebrafish receptors do not bind the Pomc ligands with the same affinity as the mammalian receptors, implying the fish receptors may have evolved different functions (Logan et al. 2003b; Takahashi and Kawauchi 2006). All six melanocortin receptors are expressed early in embryo development, by 48 hpf, except Mc2r (Logan et al. 2003a). Considering the melanocortin system peptides and receptors, as well as their antagonist, Agrp, have other functions fundamental to development, such as the interrenal actions of Acth and pigment regulation, this early expression is not surprising (Table 5.2).

In zebrafish, the Cart sequence has only been identified as an EST sequence. However, Cart expression in the brain and its role in appetite regulation have been demonstrated in other teleosts (Kehoe and Volkoff 2007).

### 6.2. Peripheral Hormones

Energy homeostasis and the regulation of energy input must have checks and balances with input from peripheral sources. The peripheral regulators bring input from peripheral tissues, such as the gastrointestinal tract and adipose tissue, to the brain, relaying nutritional status either directly or through the afferent portion of the vagus nerve. Peripheral hormones regulating appetite in fish include cholecystokinin (CCK), glucagon-like peptides (GLPs), ghrelin, leptin, insulin and amylin (Volkoff 2006).

#### 6.2.1. Gastrointestinal Hormones: CCK, GLP and Ghrelin

CCK and the GLPs are released from the gastrointestinal tract as indicators of nutrient levels in order to inhibit food intake. These gastrointestinal tract peptides also regulate other physiological functions such as gastric release and gut motility. Sequence information has been identified only for Cck1, Cck2, Glp1 and Glp2, and no functional studies on peripheral appetite regulation have been done in zebrafish (Kurokawa et al. 2003; Roch et al. 2009).

Ghrelin, on the other hand, has been shown to stimulate food intake in goldfish and its expression is up-regulated in the gut and brain of fasting zebrafish (Matsuda et al. 2006; Amole and Unniappan 2009). Recently, the active ghrelin peptide was identified by immunoreactivity in the zebrafish stomach (Olsson et al. 2008a). Activation of ghrelin and the closely related motilin receptors leads to a regulation of intestinal motility in zebrafish (Olsson et al. 2008b).

#### 6.2.2. Adipose Hormones: Leptin

Leptin is recognized in mammals as an adiposity signal and therefore levels of leptin correlate to levels of adipose tissue. Leptin also indicates to the brain that energy stores are high by activating anorexigenic neurons (containing α-MSH and CART) and inhibiting orexigenic neurons (containing NPY and AgRP) to decrease appetite. To construct a zebrafish obesity model, melanocortin receptor antagonist Agrp was overexpressed. This disrupted the balance in central appetite regulation between α-Msh and Agrp, and prevented proper leptin signaling through the α-Msh neurons leading to accumulation of lipids and adipocytes (Song and Cone 2007). The obese zebrafish model revealed that the circuitry of an adiposity signal originated before the evolution of mammals (Song and Cone 2007).

Additionally, a second leptin sequence has been reported for zebrafish (Gorissen et al. 2009). This second leptin shows low sequence conservation compared to the other zebrafish leptin (Lepa) or to mammalian leptin and yet consistent phylogenetic grouping has confirmed this as a leptin sequence (Gorissen et al. 2009). The sequence divergence, taken with differential expression of the two zebrafish transcripts, suggests subfunctionalization of these duplicate genes.

### 6.3. Other Hormones

Insulin and amylin (also known as islet amyloid polypeptide, or IAPP) are both produced in the pancreas in response to energy intake. They signal to the brain the availability of energy resources and when levels are high, lead to a decrease in food intake. Amylin has been shown to regulate food intake in mammals and is highly conserved in vertebrates, with a partial sequence identified in zebrafish (Volkoff 2006).

PACAP and VIP are pleiotropic hormones in mammals and teleosts that may play a role in appetite regulation (Matsuda and Maruyama 2007). Both are expressed in the zebrafish brain (Fradinger and Sherwood 2000; Mathieu et al. 2001; Wu et al. 2008). Although ICV injections in goldfish inhibited food intake (Matsuda et al. 2005), the optimal concentration of peptide is not known. The complete loss of PACAP in a knockout mouse did not alter its food intake or activity (Adams et al. 2008).

### 6.4. Pancreatic Hormones

Pancreas produces the metabolic hormones responsible for regulating energy resources. In zebrafish, the pancreas consists of endocrine tissue, contained in a primary islet with a variable number of differing sized secondary islets, surrounded by exocrine tissue. The islets are composed of insulin producing β-cells, glucagon producing α-cells, somatostatin producing δ-cells, and in adult tissue exclusively, pancreatic polypeptide producing PP cells (Argenton et al. 1999; Li et al. 2009b). Like mammals, the primary islet has a "mantle-core" structure with the β-cells located in the middle of the islet and β-cells and δ-cells around the periphery (Chen et al. 2007; Li et al. 2009b). These strong similarities to human islets have made zebrafish an attractive model for studying mechanisms in regulation, development, and remodeling of endocrine pancreatic tissue.

#### 6.4.1. Insulin

Zebrafish have two insulin genes thought to have originated before the radial speciation of teleosts (Irwin 2004). Both insulins maintain highly conserved cysteine residues indicating that they most likely have both retained functionality (Irwin 2004). Yet, during development, *insa* expression is restricted to the pancreas while *insb* is expressed in both pancreas and brain, indicating the possibility of subfunctionalization (Papasani et al. 2006). The function of these transcripts during development is not clear, but possibilities include roles in selective growth, or differentiation, or as neurotrophic factors (Papasani et al. 2006; Caruso et al. 2008).

The developmental role of insulin in zebrafish is supported by identification of two insulin receptor genes. Toyoshima et al. (2008) have cloned and characterized the gene structure, expression, and potential functions of Insra and Insrb during development. Knockdown of each gene separately and together shows specific individual functions for the insulin receptors with some overlapping functions during early development. Insra appears to play a role in growth and brain development while Insrb regulates heart development (Toyoshima et al. 2008). The earliest expression of the insulin receptor, at 18 hpf, and phenotype of the morphants, precedes pancreatic insulin expression at 24 hpf (Papasani et al. 2006; Toyoshima et al. 2008). Distribution of the two insulin genes and the two receptor genes has not been studied in the adult zebrafish, but insulin expression screening has been done in rainbow trout, demonstrating differential expression throughout adult tissues including brain (Caruso et al. 2008).

### 6.4.2. Glucagon and Glucagon-like Peptide-i

Proglucagon encodes three related peptides: glucagon, glucagon-like peptide-1 (GLP1) and glucagon-like peptide-2 (GLP2). In mammals proglucagon is encoded on a single gene, with expression in pancreatic α-cells producing glucagon while expression in intestinal L-cells produces GLP1 and GLP2. In several teleosts, including zebrafish, the duplicate proglucagon gene lacks a Glp2 sequence (Zhou and Irwin 2004; Roch et al. 2009). Additionally, the Glp2 sequence can be alternatively spliced out of the transcript, allowing the production of glucagon and Glp1 in the pancreas (Lund et al. 1983). This is significant as the role of Glp1 has undergone a functional switch from being an incretin hormone, stimulating the release of insulin in mammals, to being a "better glucagon", stimulating glycogenolysis in teleost liver (Plisetskaya and Mommsen 1996). The zebrafish Glp1 peptide and mammalian GLP1 have equivalent receptor activating potential but zebrafish Glp1 receptor sequence shares high homology to the glucagon receptor. This functional switch is most likely due to fish-specific duplication of the glucagon receptor followed by mutation and subfunctionalization, with Glp1 binding to the duplicate receptor (Mommsen and Mojsov 1998; Mojsov 2000; Irwin and Wong 2005).

### 6.4.3. Somatostatin

Somatostatin is expressed not only in the pancreas, but also in the nervous system, intestine, and stomach, where it functions as a negative regulator of hormone release from endocrine cells. This includes inhibition of growth hormone release from the pituitary and insulin and glucagon from the pancreas. In zebrafish, four genes for somatostatin have been identified and exhibit differential expression. The *Sst1* gene has been identified as an ortholog to the mammalian somatostatin gene, sharing high sequence identity in the C-terminal 14 residues (Devos et al. 2002). This gene is expressed in several regions of the CNS, pancreas and gastrointestinal tract. Differential processing of Sst1 will produce SS-14, found primarily in the brain, stomach and pancreas, and SS-28, the N-terminally extended version, found in intestine (Tostivint et al. 2008). Sst2 is also known as SSIII in non-mammalian vertebrates and is orthologous to cortistatin (CST) in mammals (Tostivint et al. 2005). Sst2 is expressed solely in the pancreas, with brief transient expression in the floor plate during early development (Argenton et al. 1999; Devos et al. 2002). Sst3, also known as SSII, is a divergent somatostatin precursor found in several teleosts, most likely due to duplication of Sst1 during the fish-specific genome duplication (Tostivint et al. 2008). Sst4, previously called the atypical SSII, has only been identified in two teleosts in addition to zebrafish and its origin is currently unknown.

Zebrafish somatostatin receptors are G protein-coupled receptors that fall into five subtypes: Sstr1-Sstr5. Somatostatin receptor type 1 has two variants, and Sstr4 has three variants (Moaeen-ud-Din and Yang 2009). Mammalian data show Sst1 binds all five subtypes with similar potency but varying affinities, but at this point, no studies on expression or binding affinities of the different receptor subtypes have been done in zebrafish.

## 7. OSMOREGULATION

### 7.1. Neurohormones: Isotocin and Vasotocin

Zebrafish possess all of the major hormonal components found in mammals necessary to maintain osmotic balance (Table 5.1). The best characterized zebrafish osmoregulatory neuropeptides are isotocin (It) and arginine vasotocin (Avt), the fish orthologs of mammalian oxytocin and arginine vasopressin, respectively. Unlike oxytocin, best known for its role in uterine contraction, the main function of fish isotocin is similar to that of both arginine vasotocin and vasopressin. Both hormones are released in response to increasing environmental salt concentration for water conservation (Pierson et al. 1995). In zebrafish, both hormones are expressed in magnocellular neurons which extend from the preoptic nucleus into the pituitary's neurohypophysis; *avt* is also expressed in the hypothalamus (Blechman et al. 2007; Eaton et al. 2008). Studies in zebrafish have provided a host of information regarding the development of isotocin and vasotocin-expressing cells critical to forebrain and hypothalamus development (Eaton and Glasgow 2006, 2007; Blechman et al. 2007; Tessmar-Raible et al. 2007; Eaton et al. 2008; Russek-Blum et al. 2008; Sanek and Grinblat 2008). Interestingly, vasotocin expression in zebrafish is also linked to aggression in dominant-subordinate male relationships (Larson et al. 2006).

### 7.2. Pituitary Hormone: Prolactin

Within mammals, the pituitary hormone prolactin (PRL) is involved in milk production during lactation. This is an acquired function of this gene in mammalian lineages, but other functions exist also. Zebrafish *prl* transcripts are expressed in lactotrope cells at 22 hpf, one of the first pituitary cell lineages to differentiate, along with corticotropes, which express *pomc* (Herzog et al. 2003). Similarly to other fish species, zebrafish prolactin plays a crucial function in maintaining salt balance. Examination of the expression of osmoregulatory hormones in response to changing environmental salinity determined both zebrafish prolactin and growth hormone transcripts increased noticeably in response to dilute conditions, but did not change significantly in response to increased salt concentration (Hoshijima and Hirose 2007). This is consistent with other fish studies, indicating zebrafish prolactin is released to reduce water uptake and ion loss from organs including the gills, intestines, kidneys, skin and urinary bladder. Liu et al. determined *prl* expression in embryos is mediated in an autocrine/paracrine manner that can be disrupted by receptor knockdown, a unique example converse to brain regulation in adult fish by dopamine (Liu et al. 2006).

A direct role for prolactin in growth and cell survival has also been examined in zebrafish. Morpholino knockdown for prolactin resulted in larvae with reduced swim bladders, body lengths, eyes and heads – phenotypes that could all be partially rescued (Zhu et al. 2007). The observed dwarfism was similar to a previously characterized *pit1* mutation that also lacked functional lactotropes (Nica et al. 2004). A corresponding study focused on cell survival in *prl* knockdown embryos found an increase of apoptotic cells in the zebrafish morphants, concentrated in the developing eyes and brain (Nguyen and Zhu 2009). Investigations of zebrafish prolactin have yielded several novel functions, further cementing the notion that this is a multifaceted hormone in all vertebrates.

## 7.3. Calcium- and Phosphate-Regulating Hormones

### 7.3.1. Calcitonin and Parathyroid Hormone

Zebrafish also express the major hormones involved in calcium and phosphate homeostasis: calcitonin, parathyroid hormone and stanniocalcin. Calcitonin is secreted to reduce circulating calcium, and is expressed in the ultimobranchial bodies of the zebrafish larva at 60 hpf (Alt et al. 2006). Parathyroid hormone (PTH) exerts an opposing effect to calcitonin, as well as reducing reabsorption of phosphate from the kidneys. Two zebrafish parathyroid hormones have been characterized and found to activate two out of three homologous Pth receptors: Pth1r and Pth3r (Gensure et al. 2004). Pth2r is activated by a related peptide, tuberoinfundibular 39 residue protein precursor (Tip39) which zebrafish also express (Hoare et al. 2000; Papasani et al. 2004).

#### 7.3.2. Stanniocalcin

Stanniocalcin (Stc) is a calcium and phosphate-regulatory hormone that is mainly secreted from a specialized organ in fish known as the corpuscles of Stannius. Its major function in fish is analogous to that of calcitonin, although the primary role of the mammalian STC homolog is less clear. Mammals have two stanniocalcin homologs, STC1 and STC2, and until recently only a single gene was identified in any fish. Luo et al. were able to clone a second isoform of stanniocalcin from zebrafish that was orthologous to mammalian STC2 (Luo et al. 2005). A recent study tested the function of zebrafish Stc1 and found knockdown morphants absorbed more environmental calcium and possessed upregulated expression for the epithelial calcium channel, suggesting a possible regulatory mechanism for Stc1 (Tseng et al. 2009).

### 7.4. Cardiovascular-Regulating Hormones

#### 7.4.1. Atrial Natriuretic Peptide, Adrenomedullin and Angiotensins

Several osmoregulatory hormones involved in cardiovascular regulation have also been characterized in zebrafish. Atrial natriuretic peptide (ANP), a vasodilating hormone secreted from the heart, was initially characterized in zebrafish larvae at 96 hpf (Berdougo et al. 2003). Mutant larvae lacking zebrafish *anp* expression were found to lose ventricular contraction after a short period of time (Bendig et al. 2006). Adrenomedullin, a peptide structurally related to calcitonin that acts as a vasodilator in mammals, was localized to the early larval adenohypophysis but stopped expressing 4 days after hatching (Toro et al. 2009). A cognate receptor for adrenomedullin, also known as calcitonin receptor-like (Crlr), has also been isolated in zebrafish (Nicoli et al. 2008). The angiotensins, a family of vasoconstricting hormones, have not been isolated in zebrafish but their precursor gene has been identified (Watanabe et al. 2009). The angiotensin-activating enzyme renin has been isolated and localized to developing embryos and adult kidneys, similarly to its mammalian homolog (Liang et al. 2004; Song et al. 2004). Bradykinin, a vasodilating peptide hormone from the kinin family, has also been characterized with one of its two cognate receptors in zebrafish (Bromee et al. 2005).

#### 7.4.2. Apelin

The recently discovered peptide hormone apelin has also been characterized in zebrafish. In mammals, apelin is expressed in a variety of tissues and provides a multitude of functions including vasodilation, cardiac contraction, fluid balance and angiogenesis (Chandrasekaran et al. 2008). Both the orthologous apelin and duplicate receptor isoforms were identified in zebrafish (Zeng et al. 2007). A zebrafish mutant known as *grinch* encodes a non-functional apelin receptor (Agtrl1b) and results in a reduction of cardiomyocytes in the developing heart, a trait that was also induced by early receptor activation using ectopic apelin on wild-type fish (Scott et al. 2007). Both apelin and its receptor were also found to impair embryo gastrulation when under- or overexpressed and affect the migration of cells critical for heart formation (Zeng et al. 2007). Zebrafish apelin is upregulated during hypoxia and found to be critical in hypoxia-induced cell proliferation and angiogenic regeneration (Eyries et al. 2008), illustrating one of the novelties uncovered in the course of zebrafish research.

## 8. SUMMARY

A large proportion of the hormones and receptors identified in humans now have been isolated from zebrafish as a result of the availability of the zebrafish genome. These data show that the roots of most of the components in the mammalian endocrine system must be at least as old as the ancestral animals that gave rise to the teleosts and tetrapods. However, information on the function of the zebrafish endocrine components lags behind that for mammals, although the techniques for expressing recombinant proteins have contributed to demonstrating the bioactivity of the zebrafish components. The zebrafish model should advance our understanding of the role of hormones in early development and the role of duplicate forms of hormones and receptors in endocrinology.

## REFERENCES


Abraham, E., Palevitch, O., Ijiri, S., Du, S. J., Gothilf, Y., and Zohar, Y. (2008). Early development of forebrain gonadotrophin-releasing hormone (GnRH) neurones and the role of GnRH as an autocrine migration factor. *J. Neuroendocrinol.* **20**, 394–405.

Adams, B. A., Gray, S. L., Isaac, E. R., Bianco, A. C., Vidal-Puig, A. J., and Sherwood, N. M. (2008). Feeding and metabolism in mice lacking pituitary adenylate cyclase-activating polypeptide. *Endocrinology* **149**, 1571–1580.

Albertson, R. C., Payne-Ferreira, T. L., Postlethwait, J., and Yelick, P. C. (2005). Zebrafish acvr2a and acvr2b exhibit distinct roles in craniofacial development. *Dev. Dyn.* **233**, 1405–1418.

Alderman, S. L., and Bernier, N. J. (2007). Localization of corticotropin-releasing factor, urotensin I, and CRF-binding protein gene expression in the brain of the zebrafish, Danio rerio. *J. Comp. Neurol.* **502**, 783–793.

Alderman, S. L., and Bernier, N. J. (2009). Ontogeny of the corticotropin-releasing factor system in zebrafish. *Gen. Comp. Endocrinol.* **164**, 61–69.

Alsop, D., and Vijayan, M. M. (2008). Development of the corticosteroid stress axis and receptor expression in zebrafish. *Am. J. Physiol. Regul. Integr. Comp. Physiol.* **294**, R711–R719.

Alsop, D., and Vijayan, M. (2009). The zebrafish stress axis: Molecular fallout from the teleost-specific genome duplication event. *Gen. Comp. Endocrinol.* **161**, 62–66.

Alt, B., Reibe, S., Feitosa, N. M., Elsalini, O. A., Wendl, T., and Rohr, K. B. (2006). Analysis of origin and growth of the thyroid gland in zebrafish. *Dev. Dyn.* **235**, 1872–1883.

Amole, N., and Unniappan, S. (2009). Fasting induces preproghrelin mRNA expression in the brain and gut of zebrafish, Danio rerio. *Gen. Comp. Endocrinol.* **161**, 133–137.

Amores, A., Force, A., Yan, Y. L., Joly, L., Amemiya, C., Fritz, A., Ho, R. K., Langeland, J., Prince, V., Wang, Y. L., et al. (1998). Zebrafish hox clusters and vertebrate genome evolution. *Science* **282**, 1711–1714.

Aoki, Y., Takahashi, M., Masuda, T., Tsukamoto, T., Iigo, M., and Yanagisawa, T. (2005). Molecular cloning of prepro-thyrotropin-releasing hormone cDNAs from the common carp Cyprinus carpio and goldfish Carassius auratus. *Gen. Comp. Endocrinol.* **141**, 84–92.

Argenton, F., Zecchin, E., and Bortolussi, M. (1999). Early appearance of pancreatic hormone-expressing cells in the zebrafish embryo. *Mech. Dev.* **87**, 217–221.

Ayaso, E., Nolan, C. M., and Byrnes, L. (2002). Zebrafish insulin-like growth factor-I receptor: Molecular cloning and developmental expression. *Mol. Cell. Endocrinol.* **191**, 137–148.

Barcellos, L. J. G., Ritter, F., Kreutz, L. C., Quevedo, R. M., da Silva, L. B., Bedin, A. C., Finco, J., and Cericato, L. (2007). Whole-body cortisol increases after direct and visual contact with a predator in zebrafish, Danio rerio. *Aquaculture* **272**, 774–778.

Bardet, P. L., Horard, B., Robinson-Rechavi, M., Laudet, V., and Vanacker, J. M. (2002). Characterization of oestrogen receptors in zebrafish (Danio rerio). *J. Mol. Endocrinol.* **28**, 153–163.

Bauer, H., Meier, A., Hild, M., Stachel, S., Economides, A., Hazelett, D., Harland, R. M., and Hammerschmidt, M. (1998). Follistatin and noggin are excluded from the zebrafish organizer. *Dev. Biol.* **204**, 488–507.

Bendig, G., Grimmler, M., Huttner, I. G., Wessels, G., Dahme, T., Just, S., Trano, N., Katus, H. A., Fishman, M. C., and Rottbauer, W. (2006). Integrin-linked kinase, a novel component of the cardiac mechanical stretch sensor, controls contractility in the zebrafish heart. *Genes Dev.* **20**, 2361–2372.

Berdougo, E., Coleman, H., Lee, D. H., Stainier, D. Y., and Yelon, D. (2003). Mutation of weak atrium/atrial myosin heavy chain disrupts atrial function and influences ventricular morphogenesis in zebrafish. *Development* **130**, 6121–6129.

Berglund, M. M., Lundell, I., Cabrele, C., Serra Le Gal, C., Beck-Sickinger, A. G., and Larhammar, D. (2000). Binding properties of three neuropeptide Y-receptor subtypes from zebrafish: Comparison with mammalian Y1 receptors. *Biochem. Pharmacol.* **60**, 1815–1822.

Bertrand, S., Thisse, B., Tavares, R., Sachs, L., Chaumot, A., Bardet, P. L., Escriva, H., Duffraisse, M., Marchand, O., Safi, R., et al. (2007). Unexpected novel relational links uncovered by extensive developmental profiling of nuclear receptor expression. *PLoS Genet.* **3**, e188.

Biemar, F., Argenton, F., Schmidtke, R., Epperlein, S., Peers, B., and Driever, W. (2001). Pancreas development in zebrafish: Early dispersed appearance of endocrine hormone expressing cells and their convergence to form the definitive islet. *Dev. Biol.* **230**, 189–203.

Biran, J., Ben-Dor, S., and Levavi-Sivan, B. (2008). Molecular identification and functional characterization of the kisspeptin/kisspeptin receptor system in lower vertebrates. *Biol. Reprod.* **79**, 776–786.

Blechman, J., Borodovsky, N., Eisenberg, M., Nabel-Rosen, H., Grimm, J., and Levkowitz, G. (2007). Specification of hypothalamic neurons by dual regulation of the homeodomain protein orthopedia. *Development* **134**, 4417–4426.

Bromee, T., Kukkonen, J. P., Andersson, P., Conlon, J. M., and Larhammar, D. (2005). Pharmacological characterization of ligand-receptor interactions at the zebrafish bradykinin receptor. *Br. J. Pharmacol.* **144**, 11–16.

Bromee, T., Venkatesh, B., Brenner, S., Postlethwait, J. H., Yan, Y. L., and Larhammar, D. (2006). Uneven evolutionary rates of bradykinin B1 and B2 receptors in vertebrate lineages. *Gene* **373**, 100–108.

Brown, D. D. (1997). The role of thyroid hormone in zebrafish and axolotl development. *Proc. Natl Acad. Sci. USA* **94**, 13011–13016.

Cardoso, J. C., Vieira, F. A., Gomes, A. S., and Power, D. M. (2007). PACAP, VIP and their receptors in the metazoa: Insights about the origin and evolution of the ligand-receptor pair. *Peptides* **28**, 1902–1919.

Caruso, M. A., Kittilson, J. D., Raine, J., and Sheridan, M. A. (2008). Rainbow trout (Oncorhynchus mykiss) possess two insulin-encoding mRNAs that are differentially expressed. *Gen. Comp. Endocrinol.* **155**, 695–704.

Castro, A., Becerra, M., Manso, M. J., and Anadon, R. (2006). Calretinin immunoreactivity in the brain of the zebrafish, Danio rerio: Distribution and comparison with some neuropeptides and neurotransmitter-synthesizing enzymes. I. Olfactory organ and forebrain. *J. Comp. Neurol.* **494**, 435–459.

Castro, A., Becerra, M., Manso, M. J., Tello, J., Sherwood, N. M., and Anadon, R. (2009). Distribution of growth hormone-releasing hormone-like peptide: Immunoreactivity in the central nervous system of the adult zebrafish (Danio rerio). *J. Comp. Neurol* **513**, 685–701.

Chandrasekar, G., Lauter, G., and Hauptmann, G. (2007). Distribution of corticotropin-releasing hormone in the developing zebrafish brain. *J. Comp. Neurol.* **505**, 337–351.

Chandrasekaran, B., Dar, O., and McDonagh, T. (2008). The role of apelin in cardiovascular function and heart failure. *Eur. J. Heart Fail* **10**, 725–732.

Chen, M. H., Lin, G., Gong, H., Weng, C., Chang, C., and Wu, J. (2001). The characterization of prepro-insulin-like growth factor-1 Ea-2 expression and insulin-like growth factor-1 genes (devoid 81 bp) in the zebrafish (Danio rerio). *Gene* **268**, 67–75.

Chen, S., Li, C., Yuan, G., and Xie, F. (2007). Anatomical and histological observation on the pancreas in adult zebrafish. *Pancreas* **34**, 120–125.

Clelland, E. and Peng, C. (2009). Endocrine/paracrine control of zebrafish ovarian development. *Mol. Cell. Endocrinol* **312**, 42–52.

Costache, A. D., Pullela, P. K., Kasha, P., Tomasiewicz, H., and Sem, D. S. (2005). Homology-modeled ligand-binding domains of zebrafish estrogen receptors alpha, beta1, and beta2: From in silico to in vivo studies of estrogen interactions in Danio rerio as a model system. *Mol. Endocrinol.* **19**, 2979–2990.

Dal-Pra, S., Furthauer, M., Van-Celst, J., Thisse, B., and Thisse, C. (2006). Noggin1 and Follistatin-like2 function redundantly to Chordin to antagonize BMP activity. *Dev. Biol.* **298**, 514–526.

de Souza, F. S., Bumaschny, V. F., Low, M. J., and Rubinstein, M. (2005). Subfunctionalization of expression and peptide domains following the ancient duplication of the proopiomelanocortin gene in teleost fishes. *Mol. Biol. Evol.* **22**, 2417–2427.

de Waal, P. P., Wang, D. S., Nijenhuis, W. A., Schulz, R. W., and Bogerd, J. (2008). Functional characterization and expression analysis of the androgen receptor in zebrafish (Danio rerio) testis. *Reproduction* **136**, 225–234.

Devos, N., Deflorian, G., Biemar, F., Bortolussi, M., Martial, J. A., Peers, B., and Argenton, F. (2002). Differential expression of two somatostatin genes during zebrafish embryonic development. *Mech. Dev.* **115**, 133–137.

Diaz, M. L., Becerra, M., Manso, M. J., and Anadón, R. (2002). Distribution of thyrotropin-releasing hormone (TRH) immunoreactivity in the brain of the zebrafish (*Danio rerio*). *J. Comp. Neurol.* **450**, 45–60.

Dickmeis, T., Lahiri, K., Nica, G., Vallone, D., Santoriello, C., Neumann, C. J., Hammerschmidt, M., and Foulkes, N. S. (2007). Glucocorticoids play a key role in circadian cell cycle rhythms. *PLoS Biol.* **5**, e78.

DiMuccio, T., Mukai, S. T., Clelland, E., Kohli, G., Cuartero, M., Wu, T., and Peng, C. (2005). Cloning of a second form of activin-betaA cDNA and regulation of activin-betaA subunits and activin type II receptor mRNA expression by gonadotropin in the zebrafish ovary. *Gen. Comp. Endocrinol.* **143**, 287–299.

Dong, J., Albertini, D. F., Nishimori, K., Kumar, T. R., Lu, N., and Matzuk, M. M. (1996). Growth differentiation factor-9 is required during early ovarian folliculogenesis. *Nature* **383**, 531–535.

Drew, R. E., Rodnick, K. J., Settles, M., Wacyk, J., Churchill, E., Powell, M. S., Hardy, R. W., Murdoch, G. K., Hill, R. A., and Robison, B. D. (2008). Effect of starvation on transcriptomes of brain and liver in adult female zebrafish (Danio rerio). *Physiol. Genomics* **35**, 283–295.

Duner, T., Conlon, J. M., Kukkonen, J. P., Akerman, K. E., Yan, Y. L., Postlethwait, J. H., and Larhammar, D. (2002). Cloning, structural characterization and functional expression of a zebrafish bradykinin B2-related receptor. *Biochem. J.* **364**, 817–824.

Dutta, S., Dietrich, J. E., Westerfield, M., and Varga, Z. M. (2008). Notch signaling regulates endocrine cell specification in the zebrafish anterior pituitary. *Dev. Biol.* **319**, 248–257.

Eaton, J. L., and Glasgow, E. (2006). The zebrafish bHLH PAS transcriptional regulator, single-minded 1 (sim1), is required for isotocin cell development. *Dev. Dyn.* **235**, 2071–2082.

Eaton, J. L., and Glasgow, E. (2007). Zebrafish orthopedia (otp) is required for isotocin cell development. *Dev. Genes Evol.* **217**, 149–158.

Eaton, J. L., Holmqvist, B., and Glasgow, E. (2008). Ontogeny of vasotocin-expressing cells in zebrafish: Selective requirement for the transcriptional regulators orthopedia and single-minded 1 in the preoptic area. *Dev. Dyn.* **237**, 995–1005.

Eivers, E., McCarthy, K., Glynn, C., Nolan, C. M., and Byrnes, L. (2004). Insulin-like growth factor (IGF) signalling is required for early dorso-anterior development of the zebrafish embryo. *Int. J. Dev. Biol.* **48**, 1131–1140.

Elizur, A. (2009). The KiSS1/GPR54 system in fish. *Peptides* **30**, 164–170.

Essner, J. J., Breuer, J. J., Essner, R. D., Fahrenkrug, S. C., and Hackett, P. B., Jr. (1997). The zebrafish thyroid hormone receptor alpha 1 is expressed during early embryogenesis and can function in transcriptional repression. *Differentiation* **62**, 107–117.

Eyries, M., Siegfried, G., Ciumas, M., Montagne, K., Agrapart, M., Lebrin, F., and Soubrier, F. (2008). Hypoxia-induced apelin expression regulates endothelial cell proliferation and regenerative angiogenesis. *Circ. Res.* **103**, 432–440.

Faraco, J. H., Appelbaum, L., Marin, W., Gaus, S. E., Mourrain, P., and Mignot, E. (2006). Regulation of hypocretin (orexin) expression in embryonic zebrafish. *J. Biol. Chem.* **281**, 29753–29761.

Field, H. A., Si Dong, P. D., Beis, D., and Stainier, D. Y. R. (2003). Formation of the digestive system in zebrafish. II. Pancreas morphogenesis. *Dev. Biol.* **261**, 197–208.

Figueiredo, A. Z., Lanes, C. F. C., Almeida, D. V., Proietti, M. C., and Marins, L. F. (2007). The effect of GH overexpression on GHR and IGF-I gene regulation in different genotypes of GH-transgenic zebrafish. *Comp. Biochem. Physiol. Part D* **2**, 228–233.

Forlano, P. M., and Cone, R. D. (2007). Conserved neurochemical pathways involved in hypothalamic control of energy homeostasis. *J. Comp. Neurol.* **505**, 235–248.

Fradinger, E. A., and Sherwood, N. M. (2000). Characterization of the gene encoding both growth hormone-releasing hormone (GRF) and pituitary adenylate cyclase-activating polypeptide (PACAP) in the zebrafish. *Mol. Cell. Endocrinol.* **165**, 211–219.

Fradinger, E. A., Tello, J. A., Rivier, J. E., and Sherwood, N. M. (2005). Characterization of four receptor cDNAs: PAC1, VPAC1, a novel PAC1 and a partial GHRH in zebrafish. *Mol. Cell. Endocrinol.* **231**, 49–63.

Fredriksson, R., Sjodin, P., Larson, E. T., Conlon, J. M., and Larhammar, D. (2006). Cloning and characterization of a zebrafish Y2 receptor. *Regul. Pept.* **133**, 32–40.

Garg, R. R., Bally-Cuif, L., Lee, S. E., Gong, Z., Ni, X., Hew, C. L., and Peng, C. (1999). Cloning of zebrafish activin type IIB receptor (ActRIIB) cDNA and mRNA expression of ActRIIB in embryos and adult tissues. *Mol. Cell. Endocrinol.* **153**, 169–181.

Ge, W. (2005). Activin and its receptors in fish reproduction. In: *Hormones and their Receptors in Fish Reproduction* (P. Melamed and N. Sherwood, eds), pp. 128–154. World Scientific, Singapore.

Gensure, R. C., Ponugoti, B., Gunes, Y., Papasani, M. R., Lanske, B., Bastepe, M., Rubin, D. A., and Juppner, H. (2004). Identification and characterization of two parathyroid hormone-like molecules in zebrafish. *Endocrinology* **145**, 1634–1639.

Gonzalez-Nunez, V., Gonzalez-Sarmiento, R., and Rodriguez, R. E. (2003). Identification of two proopiomelanocortin genes in zebrafish (Danio rerio). *Brain Res. Mol. Brain Res.* **120**, 1–8.

Gopinath, A., Andrew Tseng, L., and Whitlock, K. E. (2004). Temporal and spatial expression of gonadotropin releasing hormone (GnRH) in the brain of developing zebrafish (Danio rerio). *Gene Expr. Patterns* **4**, 65–70.

Gorelick, D. A., Watson, W., and Halpern, M. E. (2008). Androgen receptor gene expression in the developing and adult zebrafish brain. *Dev. Dyn.* **237**, 2987–2995.

Gorissen, M., Bernier, N. J., Nabuurs, S. B., Flik, G., and Huising, M. O. (2009). Two divergent leptin paralogues in zebrafish (Danio rerio) that originate early in teleostean evolution. *J. Endocrinol.* **201**, 329–339.

Guner, B., Ozacar, A. T., Thomas, J. E., and Karlstrom, R. O. (2008). Graded hedgehog and fibroblast growth factor signaling independently regulate pituitary cell fates and help establish the pars distalis and pars intermedia of the zebrafish adenohypophysis. *Endocrinology* **149**, 4435–4451.

Hanna, R., Pang, Y., Thomas, P., and Zhu, Y. (2006). Cell-surface expression, progestin binding, and rapid nongenomic signaling of zebrafish membrane progestin receptors alpha and beta in transfected cells. *J. Endocrinol.* **190**, 247–260.

Hanna, R. N., and Zhu, Y. (2009). Expression of membrane progestin receptors in zebrafish (Danio rerio) oocytes, testis and pituitary. *Gen. Comp. Endocrinol.* **161**, 153–157.

Hansen, I. A., To, T. T., Wortmann, S., Burmester, T., Winkler, C., Meyer, S. R., Neuner, C., Fassnacht, M., and Allolio, B. (2003). The pro-opiomelanocortin gene of the zebrafish (Danio rerio). *Biochem. Biophys. Res. Commun.* **303**, 1121–1128.

Herzog, W., Zeng, X., Lele, Z., Sonntag, C., Ting, J. W., Chang, C. Y., and Hammerschmidt, M. (2003). Adenohypophysis formation in the zebrafish and its dependence on sonic hedgehog. *Dev. Biol.* **254**, 36–49.

Herzog, W., Sonntag, C., Walderich, B., Odenthal, J., Maischein, H. M., and Hammerschmidt, M. (2004). Genetic analysis of adenohypophysis formation in zebrafish. *Mol. Endocrinol.* **18**, 1185–1195.

Hoare, S. R., Rubin, D. A., Juppner, H., and Usdin, T. B. (2000). Evaluating the ligand specificity of zebrafish parathyroid hormone (PTH) receptors: Comparison of PTH, PTH-related protein, and tuberoinfundibular peptide of 39 residues. *Endocrinology* **141**, 3080–3086.

Hogan, B. M., Danks, J. A., Layton, J. E., Hall, N. E., Heath, J. K., and Lieschke, G. J. (2005). Duplicate zebrafish pth genes are expressed along the lateral line and in the central nervous system during embryogenesis. *Endocrinology* **146**, 547–551.

Holmberg, A., Schwerte, T., Pelster, B., and Holmgren, S. (2004). Ontogeny of the gut motility control system in zebrafish Danio rerio embryos and larvae. *J. Exp. Biol.* **207**, 4085–4094.

Hoshijima, K., and Hirose, S. (2007). Expression of endocrine genes in zebrafish larvae in response to environmental salinity. *J. Endocrinol.* **193**, 481–491.

Hossain, M. S., Larsson, A., Scherbak, N., Olsson, P. E., and Orban, L. (2008). Zebrafish androgen receptor: Isolation, molecular, and biochemical characterization. *Biol. Reprod.* **78**, 361–369.

Huang, W. T., Hsieh, J. C., Chiou, M. J., Chen, J. Y., Wu, J. L., and Kuo, C. M. (2008). Application of RNAi technology to the inhibition of zebrafish GtHalpha, FSHbeta, and LHbeta expression and to functional analyses. *Zoolog. Sci.* **25**, 614–621.

Huising, M. O., Kruiswijk, C. P., and Flik, G. (2006). Phylogeny and evolution of class-I helical cytokines. *J. Endocrinol.* **189**, 1–25.

Ings, J. S., and Van Der Kraak, G. J. (2006). Characterization of the mRNA expression of StAR and steroidogenic enzymes in zebrafish ovarian follicles. *Mol. Reprod. Dev.* **73**, 943–954.

Irwin, D. M. (2004). A second insulin gene in fish genomes. *Gen. Comp. Endocrinol.* **135**, 150–158.

Irwin, D. M., and Wong, K. (2005). Evolution of new hormone function: Loss and gain of a receptor. *J. Hered.* **96**, 205–211.

Irwin, D. M., and Zhang, T. (2006). Evolution of the vertebrate glucose-dependent insulinotropic polypeptide (GIP) gene. *Comp. Biochem. Physiol.* **1**, 385–395.

Jorgensen, A., Andersen, O., Bjerregaard, P., and Rasmussen, L. J. (2007). Identification and characterisation of an androgen receptor from zebrafish Danio rerio. *Comp. Biochem. Physiol. C Toxicol. Pharmacol.* **146**, 561–568.

Kaiya, H., Miyazato, M., Kangawa, K., Peter, R. E., and Unniappan, S. (2008). Ghrelin: A multifunctional hormone in non-mammalian vertebrates. *Comp. Biochem. Physiol. A Mol. Integr. Physiol.* **149**, 109–128.

Kane, D. A., and Kimmel, C. B. (1993). The zebrafish midblastula transition. *Development* **119**, 447–456.

Kaslin, J., Nystedt, J. M., Ostergard, M., Peitsaro, N., and Panula, P. (2004). The orexin/hypocretin system in zebrafish is connected to the aminergic and cholinergic systems. *J. Neurosci.* **24**, 2678–2689.

Kawauchi, H., and Baker, B. I. (2004). Melanin-concentrating hormone signaling systems in fish. *Peptides* **25**, 1577–1584.

Kazeto, Y., Goto-Kazeto, R., and Trant, J. M. (2005). Membrane-bound progestin receptors in channel catfish and zebrafish ovary: Changes in gene expression associated with the reproductive cycles and hormonal reagents. *Gen. Comp. Endocrinol.* **142**, 204–211.

Kehoe, A. S., and Volkoff, H. (2007). Cloning and characterization of neuropeptide Y (NPY) and cocaine and amphetamine regulated transcript (CART) in Atlantic cod (Gadus morhua). *Comp. Biochem. Physiol. A Mol. Integr. Physiol.* **146**, 451–461.

Kishida, M., and Callard, G. V. (2001). Distinct cytochrome P450 aromatase isoforms in zebrafish (Danio rerio) brain and ovary are differentially programmed and estrogen regulated during early development. *Endocrinology* **142**, 740–750.

Kitahashi, T., Ogawa, S., and Parhar, I. S. (2009). Cloning and expression of kiss2 in the zebrafish and medaka. *Endocrinology* **150**, 821–831.

Klovins, J., and Schioth, H. B. (2005). Agouti-related proteins (AGRPs) and agouti-signaling peptide (ASIP) in fish and chicken. *Ann. N. Y. Acad. Sci.* **1040**, 363–367.

Kluver, N., Pfennig, F., Pala, I., Storch, K., Schlieder, M., Froschauer, A., Gutzeit, H. O., and Schartl, M. (2007). Differential expression of anti-Mullerian hormone (amh) and anti-Mullerian hormone receptor type II (amhrII) in the teleost medaka. *Dev. Dyn.* **236**, 271–281.

Kohli, G., Clelland, E., and Peng, C. (2005). Potential targets of transforming growth factor-beta1 during inhibition of oocyte maturation in zebrafish. *Reprod. Biol. Endocrinol.* **3**, 53.

Kojima, M., and Kangawa, K. (2005). Ghrelin: Structure and function. *Physiol. Rev.* **85**, 495–522.

Krueckl, S. L., Fradinger, E. A., and Sherwood, N. M. (2003). Developmental changes in the expression of growth hormone-releasing hormone and pituitary adenylate cyclase-activating polypeptide in zebrafish. *J. Comp. Neurol.* **455**, 396–405.

Kumar, R. S., and Trant, J. M. (2001). Piscine glycoprotein hormone (gonadotropin and thyrotropin) receptors: A review of recent developments. *Comp. Biochem. Physiol. B Biochem. Mol. Biol.* **129**, 347–355.

Kuo, M. W., Lou, S. W., Postlethwait, J., and Chung, B. C. (2005). Chromosomal organization, evolutionary relationship, and expression of zebrafish GnRH family members. *J. Biomed. Sci.* **12**, 629–639.

Kurokawa, T., Suzuki, T., and Hashimoto, H. (2003). Identification of gastrin and multiple cholecystokinin genes in teleost. *Peptides* **24**, 227–235.

Kwok, H. F., So, W. K., Wang, Y., and Ge, W. (2005). Zebrafish gonadotropins and their receptors: I. Cloning and characterization of zebrafish follicle-stimulating hormone and luteinizing hormone receptors – evidence for their distinct functions in follicle development. *Biol. Reprod.* **72**, 1370–1381.

Laan, M., Richmond, H., He, C., and Campbell, R. K. (2002). Zebrafish as a model for vertebrate reproduction: Characterization of the first functional zebrafish (Danio rerio) gonadotropin receptor. *Gen. Comp. Endocrinol.* **125**, 349–364.

Larson, E. T., O'Malley, D. M., and Melloni, R. H., Jr. (2006). Aggression and vasotocin are associated with dominant-subordinate relationships in zebrafish. *Behav. Brain Res.* **167**, 94–102.

Lee, J. Y., Moon, J. S., Eu, Y. J., Lee, C. W., Yang, S. T., Lee, S. K., Jung, H. H., Kim, H. H., Rhim, H., Seong, J. Y., et al. (2009). Molecular interaction between kisspeptin decapeptide analogs and a lipid membrane. *Arch. Biochem. Biophys.* **485**, 109–114.

Lee, L. T., Siu, F. K., Tam, J. K., Lau, I. T., Wong, A. O., Lin, M. C., Vaudry, H., and Chow, B. K. (2007). Discovery of growth hormone-releasing hormones and receptors in nonmammalian vertebrates. *Proc. Natl Acad. Sci. USA* **104**, 2133–2138.

Lessman, C. A. (2009). Oocyte maturation: Converting the zebrafish oocyte to the fertilizable egg. *Gen. Comp. Endocrinol.* **161**, 53–57.

Levi, L., Pekarski, I., Gutman, E., Fortina, P., Hyslop, T., Biran, J., Levavi-Sivan, B., and Lubzens, E. (2009). Revealing genes associated with vitellogenesis in the liver of the zebrafish (Danio rerio) by transcriptome profiling. *BMC Genomics* **10**, 141.

Li, X., He, J., Hu, W., and Yin, Z. (2009a). The essential role of endogenous ghrelin in growth hormone expression during zebrafish adenohypophysis development. *Endocrinology* **150**, 2767–2774.

Li, Z., Wen, C., Peng, J., Korzh, V., and Gong, Z. (2009b). Generation of living color transgenic zebrafish to trace somatostatin-expressing cells and endocrine pancreas organization. *Differentiation* **77**, 128–134.

Liang, P., Jones, C. A., Bisgrove, B. W., Song, L., Glenn, S. T., Yost, H. J., and Gross, K. W. (2004). Genomic characterization and expression analysis of the first nonmammalian renin genes from zebrafish and pufferfish. *Physiol. Genomics* **16**, 314–322.

Lin, S. W., and Ge, W. (2009). Differential regulation of gonadotropins (FSH and LH) and growth hormone (GH) by neuroendocrine, endocrine, and paracrine factors in the zebrafish–an in vitro approach. *Gen. Comp. Endocrinol.* **160**, 183–193.

Liongue, C., and Ward, A. C. (2007). Evolution of Class I cytokine receptors. *BMC Evol. Biol.* **7**, 120.

Lister, A. L., and Van Der Kraak, G. (2008). An investigation into the role of prostaglandins in zebrafish oocyte maturation and ovulation. *Gen. Comp. Endocrinol.* **159**, 46–57.

Liu, L., and Ge, W. (2007). Growth differentiation factor 9 and its spatiotemporal expression and regulation in the zebrafish ovary. *Biol. Reprod.* **76**, 294–302.

Liu, N. A., Liu, Q., Wawrowsky, K., Yang, Z., Lin, S., and Melmed, S. (2006). Prolactin receptor signaling mediates the osmotic response of embryonic zebrafish lactotrophs. *Mol. Endocrinol.* **20**, 871–880.

Liu, N. A., Ren, M., Song, J., Rios, Y., Wawrowsky, K., Ben-Shlomo, A., Lin, S., and Melmed, S. (2008). In vivo time-lapse imaging delineates the zebrafish pituitary proopiomelanocortin lineage boundary regulated by FGF3 signal. *Dev. Biol.* **319**, 192–200.

Liu, X., Zhu, P., Sham, K. W., Yuen, J. M., Xie, C., Zhang, Y., Liu, Y., Li, S., Huang, X., Cheng, C. H., et al. (2009). Identification of a membrane estrogen receptor in zebrafish with homology to mammalian GPER and its high expression in early germ cells of the testis. *Biol. Reprod.* **80**, 1253–1261.

Liu, Y., Lo, L., and Chan, W. (2000). Temporal expression and T3 induction of thyroid hormone receptors $\alpha 1$ and $\beta 1$ during early embryonic and larval development in zebrafish, *Danio rerio*. *Mol. Cell. Endocrinol.* **159**, 187–195.

Liu, Y. W., and Chan, W. K. (2002). Thyroid hormones are important for embryonic to larval transitory phase in zebrafish. *Differentiation* **70**, 36–45.

Logan, D. W., Bryson-Richardson, R. J., Pagan, K. E., Taylor, M. S., Currie, P. D., and Jackson, I. J. (2003a). The structure and evolution of the melanocortin and MCH receptors in fish and mammals. *Genomics* **81**, 184–191.

Logan, D. W., Bryson-Richardson, R. J., Taylor, M. S., Currie, P., and Jackson, I. J. (2003b). Sequence characterization of teleost fish melanocortin receptors. *Ann. N. Y. Acad. Sci.* **994**, 319–330.

Lopez, M., Nica, G., Motte, P., Martial, J. A., Hammerschmidt, M., and Muller, M. (2006). Expression of the somatolactin beta gene during zebrafish embryonic development. *Gene Expr. Patterns* **6**, 156–161.

Lund, P. K., Goodman, R. H., Montminy, M. R., Dee, P. C., and Habener, J. F. (1983). Anglerfish islet pre-proglucagon II. Nucleotide and corresponding amino acid sequence of the cDNA. *J. Biol. Chem.* **258**, 3280–3284.

Lundell, I., Berglund, M. M., Starback, P., Salaneck, E., Gehlert, D. R., and Larhammar, D. (1997). Cloning and characterization of a novel neuropeptide Y receptor subtype in the zebrafish. *DNA Cell Biol.* **16**, 1357–1363.

Luo, C. W., Pisarska, M. D., and Hsueh, A. J. (2005). Identification of a stanniocalcin paralog, stanniocalcin-2, in fish and the paracrine actions of stanniocalcin-2 in the mammalian ovary. *Endocrinology* **146**, 469–476.

MacKenzie, D. S., Jones, R. A., and Miller, T. C. (2009). Thyrotropin in teleost fish. *Gen. Comp. Endocrinol.* **161**, 83–89.

Macqueen, D. J., and Johnston, I. A. (2008). Evolution of follistatin in teleosts revealed through phylogenetic, genomic and expression analyses. *Dev. Genes Evol.* **218**, 1–14.

Marchand, O., Safi, R., Escriva, H., Van Rompaey, E., Prunet, P., and Laudet, V. (2001). Molecular cloning and characterization of thyroid hormone receptors in teleost fish. *J. Mol. Endocrinol.* **26**, 51–65.

Mathew, L. K., Sengupta, S., Kawakami, A., Andreasen, E. A., Lohr, C. V., Loynes, C. A., Renshaw, S. A., Peterson, R. T., and Tanguay, R. L. (2007). Unraveling tissue regeneration pathways using chemical genetics. *J. Biol. Chem.* **282**, 35202–35210.

Mathieu, M., Tagliafierro, G., Angelini, C., and Vallarino, M. (2001). Organization of vasoactive intestinal peptide-like immunoreactive system in the brain, olfactory organ and retina of the zebrafish, Danio rerio, during development. *Brain Res.* **888**, 235–247.

Mathieu, M., Tagliafierro, G., Bruzzone, F., and Vallarino, M. (2002). Neuropeptide tyrosine-like immunoreactive system in the brain, olfactory organ and retina of the zebrafish, Danio rerio, during development. *Brain Res. Dev. Brain Res.* **139**, 255–265.

Mathieu, M., Ciarlo, M., Trucco, N., Griffero, F., Damonte, G., Salis, A., and Vallarino, M. (2004). Pituitary adenylate cyclase-activating polypeptide in the brain, spinal cord and sensory organs of the zebrafish, Danio rerio, during development. *Brain Res. Dev. Brain Res.* **151**, 169–185.

Mathieu, M., Girosi, L., Vallarino, M., and Tagliafierro, G. (2005a). PACAP in developing sensory and peripheral organs of the zebrafish, Danio rerio. *Eur. J. Histochem.* **49**, 167–178.

Mathieu, M., Trombino, S., Argenton, F., Larhammar, D., and Vallarino, M. (2005b). Developmental expression of NPY/PYY receptors zYb and zYc in zebrafish. *Ann. N. Y. Acad. Sci.* **1040**, 399–401.

Matsuda, K., Maruyama, K., Nakamachi, T., Miura, T., Uchiyama, M., and Shioda, S. (2005). Inhibitory effects of pituitary adenylate cyclase-activating polypeptide (PACAP) and vasoactive intestinal peptide (VIP) on food intake in the goldfish, Carassius auratus. *Peptides* **26**, 1611–1616.

Matsuda, K., Shimakura, S., Maruyama, K., Miura, T., Uchiyama, M., Kawauchi, H., Shioda, S., and Takahashi, A. (2006). Central administration of melanin-concentrating hormone (MCH) suppresses food intake, but not locomotor activity, in the goldfish, Carassius auratus. *Neurosci. Lett.* **399**, 259–263.

Matsuda, K., and Maruyama, K. (2007). Regulation of feeding behavior by pituitary adenylate cyclase-activating polypeptide (PACAP) and vasoactive intestinal polypeptide (VIP) in vertebrates. *Peptides* **28**, 1761–1766.

Maures, T., Chan, S. J., Xu, B., Sun, H., Ding, J., and Duan, C. (2002). Structural, biochemical, and expression analysis of two distinct insulin-like growth factor I receptors and their ligands in zebrafish. *Endocrinology* **143**, 1858–1871.

Menuet, A., Pellegrini, E., Anglade, I., Blaise, O., Laudet, V., Kah, O., and Pakdel, F. (2002). Molecular characterization of three estrogen receptor forms in zebrafish: Binding characteristics, transactivation properties, and tissue distributions. *Biol. Reprod.* **66**, 1881–1892.

Menuet, A., Le Page, Y., Torres, O., Kern, L., Kah, O., and Pakdel, F. (2004). Analysis of the estrogen regulation of the zebrafish estrogen receptor (ER) reveals distinct effects of ERalpha, ERbeta1 and ERbeta2. *J. Mol. Endocrinol.* **32**, 975–986.

Milewski, W. M., Duguay, S. J., Chan, S. J., and Steiner, D. F. (1998). Conservation of PDX-1 structure, function, and expression in zebrafish. *Endocrinology* **139**, 1440–1449.

Moaeen-ud-Din, M., and Yang, L. G. (2009). Evolutionary history of the somatostatin and somatostatin receptors. *J. Genet.* **88**, 41–53.

Mojsov, S. (2000). Glucagon-like peptide-1 (GLP-1) and the control of glucose metabolism in mammals and teleost fish. *Am. Zool.* **40**, 246–258.

Mommsen, T. P., and Mojsov, S. (1998). Glucagon-like peptide-1 activates the adenylyl cyclase system in rockfish enterocytes and brain membranes. *Comp. Biochem. Physiol. B Biochem. Mol. Biol.* **121**, 49–56.

Mommsen, T. P. (2001). Paradigms of growth in fish. *Comp. Biochem. Physiol. B Biochem. Mol. Biol.* **129**, 207–219.

Murashita, K., Kurokawa, T., Ebbesson, L. O., Stefansson, S. O., and Ronnestad, I. (2009). Characterization, tissue distribution, and regulation of agouti-related protein (AgRP), cocaine- and amphetamine-regulated transcript (CART) and neuropeptide Y (NPY) in Atlantic salmon (Salmo salar). *Gen. Comp. Endocrinol.* **162**, 160–171.

Nagahama, Y., and Yamashita, M. (2008). Regulation of oocyte maturation in fish. *Dev. Growth Differ.* **50**(Suppl. 1), S195–S219.

Nagaso, H., Suzuki, A., Tada, M., and Ueno, N. (1999). Dual specificity of activin type II receptor ActRIIb in dorso-ventral patterning during zebrafish embryogenesis. *Dev. Growth Differ.* **41**, 119–133.

Narnaware, Y. K., Peyon, P. P., Lin, X., and Peter, R. E. (2000). Regulation of food intake by neuropeptide Y in goldfish. *Am. J. Physiol. Regul. Integr. Comp. Physiol.* **279**, R1025–R1034.

Nguyen, N., and Zhu, Y. (2009). Prolactin functions as a survival factor during zebrafish embryogenesis. *Comp. Biochem. Physiol. A Mol. Integr. Physiol.* **153**, 88–93.

Nica, G., Herzog, W., Sonntag, C., and Hammerschmidt, M. (2004). Zebrafish pit1 mutants lack three pituitary cell types and develop severe dwarfism. *Mol. Endocrinol.* **18**, 1196–1209.

Nicoli, S., Tobia, C., Gualandi, L., De Sena, G., and Presta, M. (2008). Calcitonin receptor-like receptor guides arterial differentiation in zebrafish. *Blood* **111**, 4965–4972.

Nishio, S., Gibert, Y., Bernard, L., Brunet, F., Triqueneaux, G., and Laudet, V. (2008). Adiponectin and adiponectin receptor genes are coexpressed during zebrafish embryogenesis and regulated by food deprivation. *Dev. Dyn.* **237**, 1682–1690.

Nowell, M. A., Power, D. M., Canario, A. V., Llewellyn, L., and Sweeney, G. E. (2001). Characterization of a sea bream (Sparus aurata) thyroid hormone receptor-beta clone expressed during embryonic and larval development. *Gen. Comp. Endocrinol.* **123**, 80–89.

Okubo, K., Sakai, F., Lau, E. L., Yoshizaki, G., Takeuchi, Y., Naruse, K., Aida, K., and Nagahama, Y. (2006). Forebrain gonadotropin-releasing hormone neuronal development: Insights from transgenic medaka and the relevance to X-linked Kallmann syndrome. *Endocrinology* **147**, 1076–1084.

Olsson, C., Holbrook, J. D., Bompadre, G., Jonsson, E., Hoyle, C. H., Sanger, G. J., Holmgren, S., and Andrews, P. L. (2008a). Identification of genes for the ghrelin and motilin receptors and a novel related gene in fish, and stimulation of intestinal motility in zebrafish (Danio rerio) by ghrelin and motilin. *Gen. Comp. Endocrinol.* **155**, 217–226.

Olsson, C., Holmberg, A., and Holmgren, S. (2008b). Development of enteric and vagal innervation of the zebrafish (Danio rerio) gut. *J. Comp. Neurol.* **508**, 756–770.

Palevitch, O., Kight, K., Abraham, E., Wray, S., Zohar, Y., and Gothilf, Y. (2007). Ontogeny of the GnRH systems in zebrafish brain: In situ hybridization and promoter-reporter expression analyses in intact animals. *Cell Tissue Res.* **327**, 313–322.

Pang, Y., and Ge, W. (1999). Activin stimulation of zebrafish oocyte maturation in vitro and its potential role in mediating gonadotropin-induced oocyte maturation. *Biol. Reprod.* **61**, 987–992.

Pang, Y., and Ge, W. (2002a). Gonadotropin regulation of activin betaA and activin type IIA receptor expression in the ovarian follicle cells of the zebrafish, Danio rerio. *Mol. Cell. Endocrinol.* **188**, 195–205.

Pang, Y., and Ge, W. (2002b). Gonadotropin and activin enhance maturational competence of oocytes in the zebrafish (Danio rerio). *Biol. Reprod.* **66**, 259–265.

Papasani, M. R., Gensure, R. C., Yan, Y. L., Gunes, Y., Postlethwait, J. H., Ponugoti, B., John, M. R., Juppner, H., and Rubin, D. A. (2004). Identification and characterization of the zebrafish and fugu genes encoding tuberoinfundibular peptide 39. *Endocrinology* **145**, 5294–5304.

Papasani, M. R., Robison, B. D., Hardy, R. W., and Hill, R. A. (2006). Early developmental expression of two insulins in zebrafish (Danio rerio). *Physiol. Genomics* **27**, 79–85.

Parmentier, C., Taxi, J., Balment, R., Nicolas, G., and Calas, A. (2006). Caudal neurosecretory system of the zebrafish: Ultrastructural organization and immunocytochemical detection of urotensins. *Cell Tissue Res.* **325**, 111–124.

Parmentier, C., Hameury, E., Lihrmann, I., Taxi, J., Hardin-Pouzet, H., Vaudry, H., Calas, A., and Tostivint, H. (2008). Comparative distribution of the mRNAs encoding urotensin I and urotensin II in zebrafish. *Peptides* **29**, 820–829.

Patterson, L. J., Gering, M., Eckfeldt, C. E., Green, A. R., Verfaillie, C. M., Ekker, S. C., and Patient, R. (2007). The transcription factors Scl and Lmo2 act together during development of the hemangioblast in zebrafish. *Blood* **109**, 2389–2398.

Pawson, A. J., Morgan, K., Maudsley, S. R., and Millar, R. P. (2003). Type II gonadotrophin-releasing hormone (GnRH-II) in reproductive biology. *Reproduction* **126**, 271–278.

Payne-Ferreira, T. L., Tong, K. W., and Yelick, P. C. (2004). Maternal alk8 promoter fragment directs expression in early oocytes. *Zebrafish* **1**, 27–39.

Pierson, P. M., Guibbolini, M. E., Mayer-Gostan, N., and Lahlou, B. (1995). ELISA measurements of vasotocin and isotocin in plasma and pituitary of the rainbow trout: Effect of salinity. *Peptides* **16**, 859–865.

Plisetskaya, E. M., and Mommsen, T. P. (1996). Glucagon and glucagon-like peptides in fishes. *Int. Rev. Cytol.* **168**, 187–257.

Pogoda, H. M., and Hammerschmidt, M. (2007). Molecular genetics of pituitary development in zebrafish. *Semin. Cell Dev. Biol.* **18**, 543–558.

Pogoda, H. M., and Hammerschmidt, M. (2009). How to make a teleost adenohypophysis: Molecular pathways of pituitary development in zebrafish. *Mol. Cell. Endocrinol.* **312**(1-2), 2–13.

Pottinger, T. G., and Calder, G. M. (1995). Physiological stress in fish during toxicological procedures: A potentially confounding factor. *Environ. Toxicol. Water Qual.* **10**, 135–146.

Powell, J. F., Krueckl, S. L., Collins, P. M., and Sherwood, N. M. (1996). Molecular forms of GnRH in three model fishes: Rockfish, medaka and zebrafish. *J. Endocrinol.* **150**, 17–23.

Prober, D. A., Rihel, J., Onah, A. A., Sung, R. J., and Schier, A. F. (2006). Hypocretin/orexin overexpression induces an insomnia-like phenotype in zebrafish. *J. Neurosci.* **26**, 13400–13410.

Ramsay, J. M., Feist, G. W., Varga, Z. n. M., Westerfield, M., Kent, M. L., and Schreck, C. B. (2006). Whole-body cortisol is an indicator of crowding stress in adult zebrafish, Danio rerio. *Aquaculture* **258**, 565–574.

Rekha, R. D., Amali, A. A., Her, G. M., Yeh, Y. H., Gong, H. Y., Hu, S. Y., Lin, G. H., and Wu, J. L. (2008). Thioacetamide accelerates steatohepatitis, cirrhosis and HCC by expressing HCV core protein in transgenic zebrafish Danio rerio. *Toxicology* **243**, 11–22.

Renucci, A., Lemarchandel, V., and Rosa, F. (1996). An activated form of type I serine/threonine kinase receptor TARAM-A reveals a specific signalling pathway involved in fish head organiser formation. *Development* **122**, 3735–3743.

Richardson, J., Lundegaard, P. R., Reynolds, N. L., Dorin, J. R., Porteous, D. J., Jackson, I. J., and Patton, E. E. (2008). mc1r Pathway regulation of zebrafish melanosome dispersion. *Zebrafish* **5**, 289–295.

Ringholm, A., Fredriksson, R., Poliakova, N., Yan, Y. L., Postlethwait, J. H., Larhammar, D., and Schioth, H. B. (2002). One melanocortin 4 and two melanocortin 5 receptors from zebrafish show remarkable conservation in structure and pharmacology. *J. Neurochem.* **82**, 6–18.

Ringvall, M., Berglund, M. M., and Larhammar, D. (1997). Multiplicity of neuropeptide Y receptors: Cloning of a third distinct subtype in the zebrafish. *Biochem. Biophys. Res. Commun.* **241**, 749–755.

Roch, G. J., Wu, S., and Sherwood, N. M. (2009). Hormones and receptors in fish: Do duplicates matter? *Gen. Comp. Endocrinol.* **161**, 3–12.

Rodaway, A., Takeda, H., Koshida, S., Broadbent, J., Price, B., Smith, J. C., Patient, R., and Holder, N. (1999). Induction of the mesendoderm in the zebrafish germ ring by yolk cell-derived TGF-beta family signals and discrimination of mesoderm and endoderm by FGF. *Development* **126**, 3067–3078.

Rodriguez-Mari, A., Yan, Y. L., Bremiller, R. A., Wilson, C., Canestro, C., and Postlethwait, J. H. (2005). Characterization and expression pattern of zebrafish Anti-Mullerian hormone (Amh) relative to sox9a, sox9b, and cyp19a1a, during gonad development. *Gene Expr. Patterns* **5**, 655–667.

Roman, B. L., Pham, V. N., Lawson, N. D., Kulik, M., Childs, S., Lekven, A. C., Garrity, D. M., Moon, R. T., Fishman, M. C., Lechleider, R. J., et al. (2002). Disruption of acvrl1 increases endothelial cell number in zebrafish cranial vessels. *Development* **129**, 3009–3019.

Rosa, C. E., Figueiredo, M. A., Lanes, C. F., Almeida, D. V., Monserrat, J. M., and Marins, L. F. (2008). Metabolic rate and reactive oxygen species production in different genotypes of GH-transgenic zebrafish. *Comp. Biochem. Physiol. B Biochem. Mol. Biol.* **149**, 209–214.

Rubin, D. A., Hellman, P., Zon, L. I., Lobb, C. J., Bergwitz, C., and Juppner, H. (1999). A G protein-coupled receptor from zebrafish is activated by human parathyroid hormone and not by human or teleost parathyroid hormone-related peptide. Implications for the evolutionary conservation of calcium-regulating peptide hormones. *J. Biol. Chem.* **274**, 23035–23042.

Rubin, D. A., and Juppner, H. (1999). Zebrafish express the common parathyroid hormone/parathyroid hormone-related peptide receptor (PTH1R) and a novel receptor (PTH3R) that is preferentially activated by mammalian and fugufish parathyroid hormone-related peptide. *J. Biol. Chem.* **274**, 28185–28190.

Russek-Blum, N., Gutnick, A., Nabel-Rosen, H., Blechman, J., Staudt, N., Dorsky, R. I., Houart, C., and Levkowitz, G. (2008). Dopaminergic neuronal cluster size is determined during early forebrain patterning. *Development* **135**, 3401–3413.

Salaneck, E., Larsson, T. A., Larson, E. T., and Larhammar, D. (2008). Birth and death of neuropeptide Y receptor genes in relation to the teleost fish tetraploidization. *Gene* **409**, 61–71.

Sanek, N. A., and Grinblat, Y. (2008). A novel role for zebrafish zic2a during forebrain development. *Dev. Biol.* **317**, 325–335.

Sang, X., Curran, M. S., and Wood, A. W. (2008). Paracrine insulin-like growth factor signaling influences primordial germ cell migration: In vivo evidence from the zebrafish model. *Endocrinology* **149**, 5035–5042.

Schaaf, M. J., Champagne, D., van Laanen, I. H., van Wijk, D. C., Meijer, A. H., Meijer, O. C., Spaink, H. P., and Richardson, M. K. (2008). Discovery of a functional glucocorticoid receptor beta-isoform in zebrafish. *Endocrinology* **149**, 1591–1599.

Schaaf, M. J., Chatzopoulou, A., and Spaink, H. P. (2009). The zebrafish as a model system for glucocorticoid receptor research. *Comp. Biochem. Physiol. A Mol. Integr. Physiol.* **153**, 75–82.

Schlueter, P. J., Royer, T., Farah, M. H., Laser, B., Chan, S. J., Steiner, D. F., and Duan, C. (2006). Gene duplication and functional divergence of the zebrafish insulin-like growth factor 1 receptors. *FASEB J.* **20**, 1230–1232.

Schlueter, P. J., Peng, G., Westerfield, M., and Duan, C. (2007a). Insulin-like growth factor signaling regulates zebrafish embryonic growth and development by promoting cell survival and cell cycle progression. *Cell Death Differ.* **14**, 1095–1105.

Schlueter, P. J., Sang, X., Duan, C., and Wood, A. W. (2007b). Insulin-like growth factor receptor 1b is required for zebrafish primordial germ cell migration and survival. *Dev. Biol.* **305**, 377–387.

Scott, I. C., Masri, B., D'Amico, L. A., Jin, S. W., Jungblut, B., Wehman, A. M., Baier, H., Audigier, Y., and Stainier, D. Y. (2007). The g protein-coupled receptor agtrl1b regulates early development of myocardial progenitors. *Dev. Cell* **12**, 403–413.

Selz, Y., Braasch, I., Hoffmann, C., Schmidt, C., Schultheis, C., Schartl, M., and Volff, J. N. (2007). Evolution of melanocortin receptors in teleost fish: The melanocortin type 1 receptor. *Gene* **401**, 114–122.

Sherwood, N. M., Gray, S. L., and Cummings, K. J. (2003). Consequences of PACAP gene knockout. In: *Pituitary Adenylate Cyclase-Activating Polypeptide* (H. Vaudry and A. Arimura, eds), pp. 347–360. Kluwer Academic Publishers, Norwell, MA.

Sherwood, N. M., and Wu, S. (2005). Developmental role of GnRH and PACAP in a zebrafish model. *Gen. Comp. Endocrinol.* **142**, 74–80.

Shi, Y. (2004). Beyond skin color: Emerging roles of melanin-concentrating hormone in energy homeostasis and other physiological functions. *Peptides* **25**, 1605–1611.

Shimakura, S., Miura, T., Maruyama, K., Nakamachi, T., Uchiyama, M., Kageyama, H., Shioda, S., Takahashi, A., and Matsuda, K. (2008). Alpha-melanocyte-stimulating hormone mediates melanin-concentrating hormone-induced anorexigenic action in goldfish. *Horm. Behav.* **53**, 323–328.

Siegfried, K. R., and Nusslein-Volhard, C. (2008). Germ line control of female sex determination in zebrafish. *Dev. Biol.* **324**, 277–287.

Silverstein, J. T., Breininger, J., Baskin, D. G., and Plisetskaya, E. M. (1998). Neuropeptide Y-like gene expression in the salmon brain increases with fasting. *Gen. Comp. Endocrinol.* **110**, 157–165.

Simon, R., Tietge, J. E., Michalke, B., Degitz, S., and Schramm, K. W. (2002). Iodine species and the endocrine system: Thyroid hormone levels in adult Danio rerio and developing Xenopus laevis. *Anal. Bioanal. Chem.* **372**, 481–485.

So, W. K., Kwok, H. F., and Ge, W. (2005). Zebrafish gonadotropins and their receptors: II. Cloning and characterization of zebrafish follicle-stimulating hormone and luteinizing hormone subunits–their spatial-temporal expression patterns and receptor specificity. *Biol. Reprod.* **72**, 1382–1396.

Soderberg, C., Wraith, A., Ringvall, M., Yan, Y. L., Postlethwait, J. H., Brodin, L., and Larhammar, D. (2000). Zebrafish genes for neuropeptide Y and peptide YY reveal origin by chromosome duplication from an ancestral gene linked to the homeobox cluster. *J. Neurochem.* **75**, 908–918.

Song, H. D., Sun, X. J., Deng, M., Zhang, G. W., Zhou, Y., Wu, X. Y., Sheng, Y., Chen, Y., Ruan, Z., Jiang, C. L., et al. (2004). Hematopoietic gene expression profile in zebrafish kidney marrow. *Proc. Natl Acad. Sci. USA* **101**, 16240–16245.

Song, Y., Golling, G., Thacker, T. L., and Cone, R. D. (2003). Agouti-related protein (AGRP) is conserved and regulated by metabolic state in the zebrafish, Danio rerio. *Endocrine* **22**, 257–265.

Song, Y., and Cone, R. D. (2007). Creation of a genetic model of obesity in a teleost. *FASEB J.* **21**, 2042–2049.

Sreenivasan, R., Cai, M., Bartfai, R., Wang, X., Christoffels, A., and Orban, L. (2008). Transcriptomic analyses reveal novel genes with sexually dimorphic expression in the zebrafish gonad and brain. *PLoS ONE* **3**, e1791.

Starback, P., Lundell, I., Fredriksson, R., Berglund, M. M., Yan, Y. L., Wraith, A., Soderberg, C., Postlethwait, J. H., and Larhammar, D. (1999). Neuropeptide Y receptor subtype with unique properties cloned in the zebrafish: The zYa receptor. *Brain Res. Mol. Brain Res.* **70**, 242–252.

Steven, C., Lehnen, N., Kight, K., Ijiri, S., Klenke, U., Harris, W. A., and Zohar, Y. (2003). Molecular characterization of the GnRH system in zebrafish (Danio rerio): Cloning of chicken GnRH-II, adult brain expression patterns and pituitary content of salmon GnRH and chicken GnRH-II. *Gen. Comp. Endocrinol.* **133**, 27–37.

Sundstrom, G., Larsson, T. A., Brenner, S., Venkatesh, B., and Larhammar, D. (2008). Evolution of the neuropeptide Y family: New genes by chromosome duplications in early vertebrates and in teleost fishes. *Gen. Comp. Endocrinol.* **155**, 705–716.

Takahashi, A., Tsuchiya, K., Yamanome, T., Amano, M., Yasuda, A., Yamamori, K., and Kawauchi, H. (2004). Possible involvement of melanin-concentrating hormone in food intake in a teleost fish, barfin flounder. *Peptides* **25**, 1613–1622.

Takahashi, A., and Kawauchi, H. (2006). Evolution of melanocortin systems in fish. *Gen. Comp. Endocrinol.* **148**, 85–94.

Takayama, S., Hostick, U., Haendel, M., Eisen, J., and Darimont, B. (2008). An F-domain introduced by alternative splicing regulates activity of the zebrafish thyroid hormone receptor alpha. *Gen. Comp. Endocrinol.* **155**, 176–189.

Tan, Q., Balofsky, A., Weisz, K., and Peng, C. (2009). Role of activin, transforming growth factor-beta and bone morphogenetic protein 15 in regulating zebrafish oocyte maturation. *Comp. Biochem. Physiol. A Mol. Integr. Physiol.* **153**, 18–23.

Taylor, J. S., Van de Peer, Y., Braasch, I., and Meyer, A. (2001). Comparative genomics provides evidence for an ancient genome duplication event in fish. *Philos. Trans. R. Soc. Lond. B Biol. Sci.* **356**, 1661–1679.

Tello, J. A., Wu, S., Rivier, J. E., and Sherwood, N. M. (2008). Four functional GnRH receptors in zebrafish: Analysis of structure, signaling, synteny and phylogeny. *Integr. Comp. Biol.* **248**, 570–587.

Tessmar-Raible, K., Raible, F., Christodoulou, F., Guy, K., Rembold, M., Hausen, H., and Arendt, D. (2007). Conserved sensory-neurosecretory cell types in annelid and fish forebrain: Insights into hypothalamus evolution. *Cell* **129**, 1389–1400.

To, T. T., Hahner, S., Nica, G., Rohr, K. B., Hammerschmidt, M., Winkler, C., and Allolio, B. (2007). Pituitary-interrenal interaction in zebrafish interrenal organ development. *Mol. Endocrinol.* **21**, 472–485.

Torgersen, J., Nourizadeh-Lillabadi, R., Husebye, H., and Alestrom, P. (2002). In silico and in situ characterization of the zebrafish (Danio rerio) gnrh3 (sGnRH) gene. *BMC Genomics* **3**, 25.

Toro, S., Wegner, J., Muller, M., Westerfield, M., and Varga, Z. M. (2009). Identification of differentially expressed genes in the zebrafish hypothalamic-pituitary axis. *Gene Expr. Patterns* **9**, 200–208.

Tostivint, H., Joly, L., Lihrmann, I., Ekker, M., and Vaudry, H. (2004). Chromosomal localization of three somatostatin genes in zebrafish. Evidence that the [Pro2]-somatostatin-14 isoform and cortistatin are encoded by orthologous genes. *J. Mol. Endocrinol.* **33**, R1–R8.

Tostivint, H., Joly, L., Lihrmann, I., Conlon, J. M., Ekker, M., and Vaudry, H. (2005). Linkage mapping of the [Pro2]somatostatin-14 gene in zebrafish: Evolutionary perspectives. *Ann. N. Y. Acad. Sci.* **1040**, 486–489.

Tostivint, H., Joly, L., Lihrmann, I., Parmentier, C., Lebon, A., Morisson, M., Calas, A., Ekker, M., and Vaudry, H. (2006). Comparative genomics provides evidence for close evolutionary relationships between the urotensin II and somatostatin gene families. *Proc. Natl Acad. Sci. USA* **103**, 2237–2242.

Tostivint, H., Lihrmann, I., and Vaudry, H. (2008). New insight into the molecular evolution of the somatostatin family. *Mol. Cell. Endocrinol.* **286**, 5–17.

Toyoshima, Y., Monson, C., Duan, C., Wu, Y., Gao, C., Yakar, S., Sadler, K. C., and LeRoith, D. (2008). The role of insulin receptor signaling in zebrafish embryogenesis. *Endocrinology* **149**, 5996–6005.

Tseng, D. Y., Chou, M. Y., Tseng, Y. C., Hsiao, C. D., Huang, C. J., Kaneko, T., and Hwang, P. P. (2009). Effects of stanniocalcin 1 on calcium uptake in zebrafish (Danio rerio) embryo. *Am. J. Physiol. Regul. Integr. Comp. Physiol.* **296**, R549–R557.

Tucker, B., Hepperle, C., Kortschak, D., Rainbird, B., Wells, S., Oates, A. C., and Lardelli, M. (2007). Zebrafish angiotensin II receptor-like 1a (agtrl1a) is expressed in migrating hypoblast, vasculature, and in multiple embryonic epithelia. *Gene Expr. Patterns* **7**, 258–265.

Unger, J. L., and Glasgow, E. (2003). Expression of isotocin-neurophysin mRNA in developing zebrafish. *Gene Expr. Patterns* **3**, 105–108.

van Aerle, R., Kille, P., Lange, A., and Tyler, C. R. (2008). Evidence for the existence of a functional Kiss1/Kiss1 receptor pathway in fish. *Peptides* **29**, 57–64.

Volkoff, H. (2006). The role of neuropeptide Y, orexins, cocaine and amphetamine-related transcript, cholecystokinin, amylin and leptin in the regulation of feeding in fish. *Comp. Biochem. Physiol. A Mol. Integr. Physiol.* **144**, 325–331.

von Hofsten, J., Larsson, A., and Olsson, P. E. (2005). Novel steroidogenic factor-1 homolog (ff1d) is coexpressed with anti-Mullerian hormone (AMH) in zebrafish. *Dev. Dyn.* **233**, 595–604.

Walpita, C. N., Van der Geyten, S., Rurangwa, E., and Darras, V. M. (2007). The effect of 3,5,3′-triiodothyronine supplementation on zebrafish (Danio rerio) embryonic development and expression of iodothyronine deiodinases and thyroid hormone receptors. *Gen. Comp. Endocrinol.* **152**, 206–214.

Wang, D. S., Jiao, B., Hu, C., Huang, X., Liu, Z., and Cheng, C. H. (2008). Discovery of a gonad-specific IGF subtype in teleost. *Biochem. Biophys. Res. Commun.* **367**, 336–341.

Wang, X. G., and Orban, L. (2007). Anti-Mullerian hormone and 11 beta-hydroxylase show reciprocal expression to that of aromatase in the transforming gonad of zebrafish males. *Dev. Dyn.* **236**, 1329–1338.

Wang, Y., and Ge, W. (2003a). Involvement of cyclic adenosine 3′,5′-monophosphate in the differential regulation of activin betaA and betaB expression by gonadotropin in the zebrafish ovarian follicle cells. *Endocrinology* **144**, 491–499.

Wang, Y., and Ge, W. (2003b). Spatial expression patterns of activin and its signaling system in the zebrafish ovarian follicle: Evidence for paracrine action of activin on the oocytes. *Biol. Reprod.* **69**, 1998–2006.

Wang, Y., Wong, A. O., and Ge, W. (2003). Cloning, regulation of messenger ribonucleic acid expression, and function of a new isoform of pituitary adenylate cyclase-activating polypeptide in the zebrafish ovary. *Endocrinology* **144**, 4799–4810.

Wang, Y., and Ge, W. (2004). Developmental profiles of activin betaA, betaB, and follistatin expression in the zebrafish ovary: Evidence for their differential roles during sexual maturation and ovulatory cycle. *Biol. Reprod.* **71**, 2056–2064.

Wang, Y., Li, J., Wang, C. Y., Kwok, A. H., and Leung, F. C. (2007). Identification of the endogenous ligands for chicken growth hormone-releasing hormone (GHRH) receptor: Evidence for a separate gene encoding GHRH in submammalian vertebrates. *Endocrinology* **148**, 2405–2416.

Watanabe, T., Inoue, K., and Takei, Y. (2009). Identification of angiotensinogen genes with unique and variable angiotensin sequences in chondrichthyans. *Gen. Comp. Endocrinol.* **161**, 115–122.

Wendl, T., Lun, K., Mione, M., Favor, J., Brand, M., Wilson, S. W., and Rohr, K. B. (2002). Pax2.1 is required for the development of thyroid follicles in zebrafish. *Development* **129**, 3751–3760.

White, Y. A., Kyle, J. T., and Wood, A. W. (2009). Targeted gene knockdown in zebrafish reveals distinct intraembryonic functions for insulin-like growth factor II signaling. *Endocrinology* **150**, 4366–4375.

Wingert, R. A., Selleck, R., Yu, J., Song, H. D., Chen, Z., Song, A., Zhou, Y., Thisse, B., Thisse, C., McMahon, A. P., et al. (2007). The cdx genes and retinoic acid control the positioning and segmentation of the zebrafish pronephros. *PLoS Genet.* **3**, 1922–1938.

Wittbrodt, J., and Rosa, F. M. (1994). Disruption of mesoderm and axis formation in fish by ectopic expression of activin variants: The role of maternal activin. *Genes Dev.* **8**, 1448–1462.

Wu, S., Adams, B. A., Fradinger, E. A., and Sherwood, N. M. (2006a). Role of two genes encoding PACAP in early brain development in zebrafish. *Ann. N. Y. Acad. Sci.* **1070**, 602–621.

Wu, S., Page, L., and Sherwood, N. M. (2006b). A role for GnRH in early brain regionalization and eye development in zebrafish. *Mol. Cell Endocrinol.* **257–258**, 47–64.

Wu, S., Roch, G. J., Cervini, L. A., Rivier, J. E., and Sherwood, N. M. (2008). Newly-identified receptors for peptide histidine-isoleucine and GHRH-like peptide in zebrafish help to elucidate the mammalian secretin superfamily. *J. Mol. Endocrinol.* **41**, 343–366.

Wu, T., Patel, H., Mukai, S., Melino, C., Garg, R., Ni, X., Chang, J., and Peng, C. (2000). Activin, inhibin, and follistatin in zebrafish ovary: Expression and role in oocyte maturation. *Biol. Reprod.* **62**, 1585–1592.

Yelick, P. C., Abduljabbar, T. S., and Stashenko, P. (1998). zALK-8, a novel type I serine/threonine kinase receptor, is expressed throughout early zebrafish development. *Dev. Dyn.* **211**, 352–361.

Yokogawa, T., Marin, W., Faraco, J., Pezeron, G., Appelbaum, L., Zhang, J., Rosa, F., Mourrain, P., and Mignot, E. (2007). Characterization of sleep in zebrafish and insomnia in hypocretin receptor mutants. *PLoS Biol.* **5**, e277.

Yunker, W. K., Smith, S., Graves, C., Davis, P. J., Unniappan, S., Rivier, J. E., Peter, R. E., and Chang, J. P. (2003). Endogenous hypothalamic somatostatins differentially regulate growth hormone secretion from goldfish pituitary somatotropes in vitro. *Endocrinology* **144**, 4031–4041.

Zecchin, E., Filippi, A., Biemar, F., Tiso, N., Pauls, S., Ellertsdottir, E., Gnugge, L., Bortolussi, M., Driever, W., and Argenton, F. (2007). Distinct delta and jagged genes control sequential segregation of pancreatic cell types from precursor pools in zebrafish. *Dev. Biol.* **301**, 192–204.

Zeng, X. X., Wilm, T. P., Sepich, D. S., and Solnica-Krezel, L. (2007). Apelin and its receptor control heart field formation during zebrafish gastrulation. *Dev. Cell* **12**, 391–402.

Zhou, L., and Irwin, D. M. (2004). Fish proglucagon genes have differing coding potential. *Comp. Biochem. Physiol. B Biochem. Mol. Biol.* **137**, 255–264.

Zhu, Y., Bond, J., and Thomas, P. (2003a). Identification, classification, and partial characterization of genes in humans and other vertebrates homologous to a fish membrane progestin receptor. *Proc. Natl Acad. Sci. USA* **100**, 2237–2242.

Zhu, Y., Rice, C. D., Pang, Y., Pace, M., and Thomas, P. (2003b). Cloning, expression, and characterization of a membrane progestin receptor and evidence it is an intermediary in meiotic maturation of fish oocytes. *Proc. Natl Acad. Sci. USA* **100**, 2231–2236.

Zhu, Y., Song, D., Tran, N. T., and Nguyen, N. (2007). The effects of the members of growth hormone family knockdown in zebrafish development. *Gen. Comp. Endocrinol.* **150**, 395–404.

# 6

# DEVELOPMENTAL PHYSIOLOGY OF THE ZEBRAFISH CARDIOVASCULAR SYSTEM

*JOHN D. MABLY*
*SARAH J. CHILDS*





*Zebrafish: Volume 29*
FISH PHYSIOLOGY


DOI: 10.1016/S1546-5098(10)02906-7

The zebrafish has been widely used for developmental studies. As a result, numerous reviews describe the early morphological events involved in the development of the zebrafish embryo and organ systems. In this review, we attempt to extend this analysis to the potential of the zebrafish as a model for the study of cardiovascular physiology. The zebrafish has many advantages for the study of the cardiovascular system as its optical transparency allows the real-time in vivo examination of the development and function of the heart and vessels. In addition, because zebrafish can survive in the absence of a functioning cardiovascular system for the first few days of life, numerous genetic mutants have been identified, many of which can survive through embryogenesis. Many of these mutants are not available for study in mice as they are lethal. The aquatic environment of zebrafish also allows application of water-soluble drugs to animals in early stages which are taken up into the embryo, for the identification of new chemical compounds affecting cardiovascular formation or function, or for the identification of compounds which reverse pathological conditions. These unique tools have been used to probe the zebrafish cardiovascular system in depth.

## 1. CARDIOVASCULAR DISEASE: MODELS WANTED

The importance of studying cardiovascular physiology in animal models is borne out by the prevalence of cardiovascular disease (CVD) in the North American population. CVD includes a broad variety of conditions including coronary artery disease, high blood pressure, stroke, congestive heart failure and birth defects, and affects 61.8 million U.S. citizens in some form (sources, American Heart Association and Center for Disease Control). In fact, CVD is responsible for more deaths each year than the next most common diseases combined, including cancer, chronic lower respiratory tract diseases, accidents, diabetes mellitus, influenza and pneumonia. With 1 in 5 people having some form of CVD, the importance of better models for these diseases is essential for more accurate and earlier diagnosis. These models can be employed not only to identify the important molecules involved, but also to understand how the symptoms of these conditions can be better treated and possibly eradicated.

The etiology of CVD is complex but there is convincing evidence of a sizeable heritable component in many forms of the disease (Epstein et al. 2002). While the last decade has seen tremendous advances in our understanding of the genetic basis of forms of CVD, such as cardiomyopathies, long-QT, coronary artery disease, aneurysms and stroke, the interactions between genetic factors, epigenetic modifications and environmental exposures remain largely undetermined. Our interpretation of the roles for new genes and non-heritable contributions in CVD will be crucially dependent on the development of relevant animal models that recapitulate features of human cardiac physiology. Ideally, these models will be affordable and facilitate high-throughput analyses for drug development as well.

## 2. CURRENT MODELS FOR CARDIOVASCULAR SYSTEM STUDIES

Animal models for the study of cardiovascular physiology have traditionally included large animals because they more closely mimic the hemodynamic stresses and physiological conditions seen in humans. However, the difficulties in housing large animals and the lack of true high-throughput capabilities created a need for small models to fill this niche, initially occupied by studies utilizing rats. With the advent of targeted gene ablation technology, the mouse became an important player. Difficulties associated with the measurement of cardiovascular parameters in such small animals were partially overcome, but still require substantial expertise. Also, like other small animals, some attributes of mouse cardiac physiology are dramatically different from those of humans and the utility of discoveries in this system may not translate well to larger mammals. For example, the high heart rate of the mouse compared to humans (500–600 vs. 60–90 beats per minute, respectively) necessitates a different profile of sarcomeric and channel protein gene expression. While loss-of-function studies of these genes in mouse will lead to essential insights into their function, the implications for human cardiovascular function have to be assessed carefully and will still require validation in large animal models.

## 3. FILLING A NICHE WITH ZEBRAFISH

Although these concerns might be expected to be further accentuated in zebrafish, these animals possess properties that make them an attractive model system for physiological studies. To draw on the previous example, in zebrafish embryos, the normal heart rate is 120–180 beats per minute, much closer to humans than mouse. In addition, heart development during early stages of zebrafish embryonic development is well conserved with other vertebrates, including the mouse and human. Therefore, the results of studies in zebrafish are relevant to human development and disease. Development proceeds rapidly; the zebrafish heart tube has already formed by 18 hpf (hours post fertilization) and rhythmic, peristaltic contraction has started by 24 hpf. Shortly after contraction is initiated, there is a switch to sequential contractions of the two chambers, characteristic of the vertebrate heart (Baker et al. 1997). This progressive development of form and function in zebrafish resembles that in mammals, although the latter subsequently develop septation between the chambers and the outflow tract, to permit a separate pulmonary circulation.

## 4. UNIQUE ADVANTAGES OF THE ZEBRAFISH FOR CARDIOVASCULAR STUDIES

The zebrafish is a powerful genetic model organism that offers several advantages over rodent models, including the growth of external, transparent embryos and tremendous fecundity (Haffter et al. 1996; Amsterdam et al. 1999). These attributes have been exploited for determination of cell lineages because of the ease of using a transparent, externally developing embryo (Kozlowski et al. 1997; Serluca and Fishman 1999; Vogeli et al. 2006; Jin et al. 2007), and in vivo studies of the effect of labeled transplanted cells (Ho and Kane 1990; Stainier et al. 1995; Serbedzija 1998; Siekmann and Lawson 2007), invaluable when assaying cell autonomy. However, the most important and truly unique property of the zebrafish that makes it an attractive model for studies of cardiovascular physiology is the embryo's lack of dependence upon cardiac function and blood flow for early survival (Chen et al. 1996; Haffter et al. 1996; Stainier et al. 1996; Amsterdam et al. 1999). Zebrafish embryos are able to survive in the absence of cardiac function by oxygen diffusion from the water (Pelster and Burggren 1996; Barrionuevo and Burggren 1999), unlike the mouse which suffers hypoxic deterioration under similar conditions. This key property of zebrafish embryos has made the cardiovascular system one of the very best understood attributes of the zebrafish (Chen and Fishman 2000) and now provides an excellent complement to the mouse for study of the genetic factors regulating both cardiac development and function (Chien 1996, 2001; Christensen et al. 1997).

## 5. VERTEBRATE CARDIAC DEVELOPMENT

Over the past 15 years, much has been learned about the molecular determinants of cardiomyocyte cell fate and vessel growth in zebrafish, but relatively little about the role of the onset of physiological function on morphogenesis in the embryonic heart and the cardiovascular system. As in other vertebrates, cardiomyocyte differentiation begins prior to heart tube formation, where cells in the lateral plate mesoderm express key transcription factors (e.g., GATA4, dHAND, Nkx2.5) as well as genes for ion channels, calcium handling, and myofibrillar structure (Chen and Fishman 2000). The sarcoplasmic reticulum similarly begins assembly prior to contraction (Moorman et al. 1995) but myofibrillar arrays are not evident until the heart-tube stage, with proper sarcomeres developing only after contraction has begun (Ehler et al. 1999). These contractions are at first peristaltic, but soon change to sequential contractions of atrium and ventricle. This occurs prior to the onset of innervation, organized conducting tissue, or definitive sinoatrial or atrioventricular nodes.

## 6. ORIGIN AND LINEAGE OF CELLS IN THE CARDIOVASCULAR SYSTEM

Cardiac function is essential to vascular flow and, in turn, flow through the heart determines aspects of cardiac morphology (Auman et al. 2007). The cardiomyocytes of the zebrafish heart originate in the ventral–marginal zone, involute and migrate through the lateral plate mesoderm and generate a heart tube precursor at the midline, a cone with the pre-ventricular cells pointing apically (Stainier and Fishman 1992; Stainier et al. 1993; Lee et al. 1994). Subsequent work by Stainier and Yelon showed that there is a medial–lateral sequestration of ventricular and atrial precursors respectively, and that the precursors depend upon endodermal signals for proper midline alignment, as in other species (Yelon et al. 1999). The vessels and endocardium (the inner, endothelial lining of the heart) are derived from endothelial precursors in the lateral mesoderm of the embryo, which arise during gastrulation. The ventral mesodermal cells that will form endothelium require bone morphogenetic protein (Bmp) signaling for their specification (Gupta et al. 2006). While the anterior lateral mesoderm of the zebrafish gives rise to the head vasculature and heart, the posterior mesoderm gives rise to trunk vasculature and blood (Fouquet et al. 1997). Anterior lateral plate mesoderm cells acquire either a myocardial or an endothelial fate. The choice between these two fates is controlled by blood and/or vessel precursors (angioblasts) repressing the cardiac specification pathway. Thus in the *cloche* mutant in the absence of blood and angioblasts, the size of the heart is increased (Schoenebeck et al. 2007). Conversely, when the number of vessel precursors is increased by overexpression of a combination of hemangioblast specifying transcription factors Scl/Tal and *Lmo2*, the number of cardiomyocyte precursors is reduced (Gering et al. 2003).

The posterior lateral mesoderm forms vessel and blood precursors. It is adjacent to intermediate mesoderm that forms the kidney. The number of cells acquiring vessel, blood or kidney precursor fates is influenced by several genes. Loss of the transcription factor odd-skipped 1 leads to an expansion of the lateral angioblast population at the expense of kidney precursors (Mudumana et al. 2008). Bmp is also important for this process: expression of a dominant negative Bmp receptor under the hemangioblast *lmo2* promoter leads to an increase in hematopoietic cells and reduced pronephric cells (Gupta et al. 2006). On the other hand, overexpression of the transcription factor Scl leads to lateral and intermediate mesoderm fating towards hematopoietic cells (Gering et al. 1998).

## 7. ZEBRAFISH GENETICS AND DEVELOPMENT OF THE CARDIOVASCULAR SYSTEM

The current era of zebrafish research evolved from the genetic screens performed by Boston and Tubingen which demonstrated the true power of this model as a genetic system to study vertebrate development (Chen et al. 1996; Driever et al. 1996; Stainier et al. 1996; Alexander et al. 1998; Amsterdam et al. 1999). The early groundwork laid by Stainier and Fishman helped launch the zebrafish as a valuable model for studies of the cardiovascular system (Stainier et al. 1993; Lee et al. 1994) with substantial contributions from other investigators along the way. The addition of chemical screens to this repertoire of tools has strengthened the zebrafish in its position as a first-line model organism for developmental studies (Peterson et al. 2000; Kudoh et al. 2001). Whereas the fundamental mechanisms of cardiac and vascular growth and function are highly conserved in zebrafish, key developmental steps in vertebrate cardiovascular development have been determined through genetic and cell biology studies in this system (Fishman and Chien 1997; Stainier 2001; Chen et al. 2005). The strength of zebrafish as a genetic model system enables the systematic dissection of interactions between genes in an intact vertebrate organism and bolsters its potential for screens of genetic or chemical modifiers of cardiovascular physiology. In addition, multiple complementing mutations with similar phenotypes may represent discrete genes in the same molecular pathways (1 phenotype, multiple genes, 1 pathway; Kleaveland et al. 2009; Mably et al. 2003, 2006; or Buchner et al. 2007; Liu et al. 2007). These findings suggest that phenotype-driven screens for genes that disrupt normal cardiovascular growth or function in the zebrafish offer an opportunity to understand the biology of heart and vascular physiology and pathways that may be disrupted in disease.

## 8. GENETIC INFRASTRUCTURE AND TOOLS FOR ZEBRAFISH STUDIES

The genomic infrastructure for cloning in zebrafish includes genetic (Knapik et al. 1996, 1998; Fornzler et al. 1998; Gates et al. 1999) and radiation hybrid maps (Geisler et al. 1999; Hukriede et al. 1999, 2001), large-insert genomic libraries, and an EST compilation (Kelly et al. 2000; Ton et al. 2000). However, the most valuable tool has become the genome sequence and assembly itself (Vogel 2000). Although not yet complete, this resource has greatly changed the speed and ease with which a genetic lesion in a mutant can be identified. The discussion of this resource, as well as tools for gene knockdown with morpholinos (Ekker 2000; Nasevicius and Ekker 2000; Draper et al. 2001) and targeted gene disruption by tilling for ENU-induced mutations (Wienholds et al. 2002, 2003) or by using zinc finger nucleases (Doyon et al. 2008; Meng et al. 2008) are the topics of other chapters in this book.

## 9. ZEBRAFISH CARDIOVASCULAR MUTANTS

Large-scale genetic screens have already been performed in zebrafish to identify mutations affecting the form and function of cardiac (Chen et al. 1996, 2001; Stainier et al. 1996) and skeletal muscle (Granato et al. 1996; Birely et al. 2005), as well as vascular development (Zhong et al. 2001; Roman et al. 2002; Lawson et al. 2003; Parker et al. 2004; Torres-Vazquez et al. 2004; Liu et al. 2007; Covassin et al. 2009) (see Fig. 6.1 and Table 6.1). Much of the experimental work to date has been focused on mutations that perturb the morphology of the heart and vessels. Among these, for example,

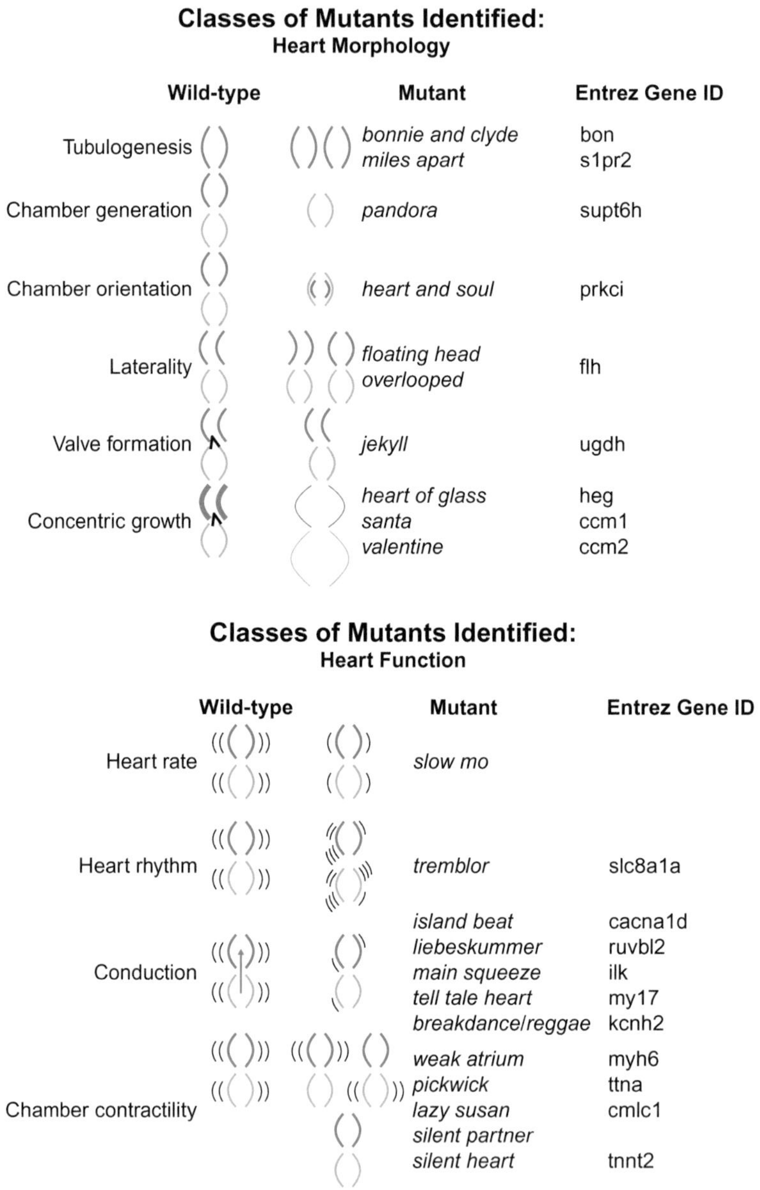

**Classes of Mutants Identified:**
**Heart Morphology**

| | Wild-type | Mutant | Entrez Gene ID |
|---|---|---|---|
| Tubulogenesis | | *bonnie and clyde* | bon |
| | | *miles apart* | s1pr2 |
| Chamber generation | | *pandora* | supt6h |
| Chamber orientation | | *heart and soul* | prkci |
| Laterality | | *floating head* | flh |
| | | *overlooped* | |
| Valve formation | | *jekyll* | ugdh |
| Concentric growth | | *heart of glass* | heg |
| | | *santa* | ccm1 |
| | | *valentine* | ccm2 |

**Classes of Mutants Identified:**
**Heart Function**

| | Wild-type | Mutant | Entrez Gene ID |
|---|---|---|---|
| Heart rate | | *slow mo* | |
| Heart rhythm | | *tremblor* | slc8a1a |
| Conduction | | *island beat* | cacna1d |
| | | *liebeskummer* | ruvbl2 |
| | | *main squeeze* | ilk |
| | | *tell tale heart* | my17 |
| | | *breakdance/reggae* | kcnh2 |
| Chamber contractility | | *weak atrium* | myh6 |
| | | *pickwick* | ttna |
| | | *lazy susan* | cmlc1 |
| | | *silent partner* | |
| | | *silent heart* | tnnt2 |

**Fig. 6.1.** Categories of mutants affecting heart development in zebrafish. Initial phenotypic characterization of these mutants was used to sort them into two broad categories: heart morphology and heart function. Because of the interrelationship between form and function, there is overlap between the two groups. A diagram of the process disrupted in the mutant is indicated, along with its name and the affected gene (where known). See color plate section

**Table 6.1**
Zebrafish genetic mutants with vascular defects. See color plate section

| Phenotype | Wild type | Mutant | Mutant name | Gene name |
|---|---|---|---|---|
| Endothelial specification | 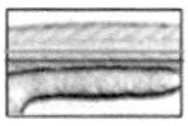 | 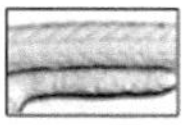 | *cloche* | ? |
| Endothelial specification & ISV sprouting |  | 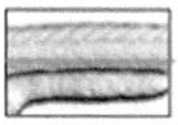 | *Y11* | Ets related protein (etsrp) |
| Artery specification | 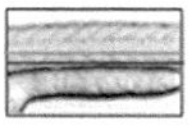 | 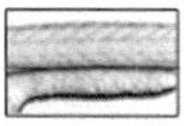 | *gridlock* | Hairy/Enhancer of split with YRPW motif 2 (Hey2) |
| Arteriovenous shunting |  |  | *kurzschluss* | Unc45a |
| ISV sprouting |  | 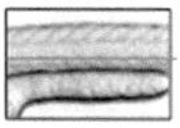 | *t20257* | Flk (VEGFR1) |
| ISV sprouting | 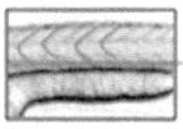 | 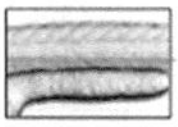 | *Y10* | Phospholipase Cγ1 |
| Venous and lymphatic sprouting | 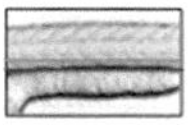 | 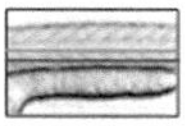 | *full of fluid* | Collagen and calcium-binding EGF domain-1 (ccbe1) |
| Endothelial cell number | 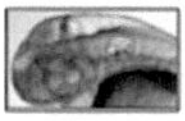 | 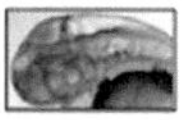 | *violet beauregarde* | Activin like kinase 1 (acvrl1) |
| Vessel patterning | 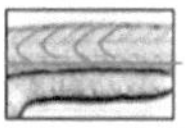 | 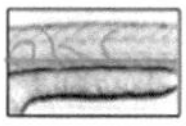 | *out of bounds* | Piexin D1 |
| Vessel dilation & patterning | 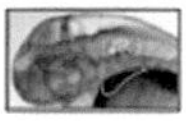 | 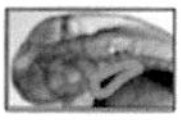 | *adrasteia* | Seryl-tRNA synthetase (sars) |
| Vascular stabilization | 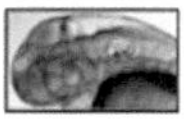 |  | *bubblehead* | Pak-interacting exchange factor (bPix) |
| Vascular stabilization | | 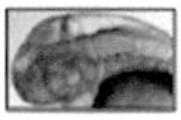 | *redhead* | P21 activated kinase 2a (Pak2a) |
| Vascular stabilization |  |  | *tomato* | Birc2 (ciap 1) |
| Vascular regeneration |  |  | *reg6* | ? |

are mutations that prevent formation of the ventricle, cause cardia bifida (Kikuchi et al. 2000; Kupperman et al. 2000), disorient the position of the chambers so that the ventricle is inside of the atrium (Horne-Badovinac et al. 2001; Peterson et al. 2001), or prevent formation of a normal aorta (Zhong et al. 2000). Hypomorphic mutations, such as described for *bubblehead, redhead* and *gridlock*, might not have been discovered readily by targeted gene mutation, since a null phenotype may sometimes have more pleiotropic and severe defects, with the cardiovascular element being a small component (Zhong et al. 2000; Buchner et al. 2007; Liu et al. 2007).

## 10. MUTANTS AFFECTING RATE, RHYTHM AND CONTRACTILE PROPERTIES OF THE HEART

The utility of zebrafish as a model to study cardiovascular physiology has only recently begun to be exploited, even though some of the mutations identified in the early ENU mutagenesis screens were defects affecting cardiac function (Chen et al. 1996). Distinction between mutants affecting cardiomyocyte contraction and conduction of the electrical wave in the heart were initially made based on early phenotypic characterizations. Therefore, overlap exists between these groups, along with those affecting rate and rhythm. One of these early mutations was *slow mo* (*smo*), a mutant with a low heart rate that was identified as a spontaneous mutant in the Hong Kong/Singapore background. The mutation is autosomal recessive but mutants are viable, unlike most of the cardiovascular mutants, and the heart is functional. Resting heart rate is able to increase as the embryo develops although it remains slower than wild-type from the onset of contraction at 24 hpf. This bradycardia (slower heart rate) is an inherent property of the cardiomyocyte, since dissociation of cells and analysis in culture reveal a similarly decreased contractile rate (Baker et al. 1997). Although the nature of the defect suggests an abnormality in cardiac pacemaking, the genetic lesion remains elusive. However, analysis of the cellular defect revealed that aspects of one pacemaker current is defective, the hyperpolarization-activated cation current ($I_h$). In *smo* hearts, the fast kinetic component of this inward rectifying current is abnormal (no alteration in the slow component was observed), resulting in the altered kinetics. An interesting final observation of the *slow mo* phenotype is that, unlike their wild-type siblings, they show no increase in heart rate in response to temperature elevation, suggesting that the mutation may not be a result of a simple defect in a channel protein (Baker et al. 1997).

While the *smo* mutant possesses a distinctive defect which affects heart rate alone, other mutants, such as *tremblor* (*tre*) and *reggae* (*reg*) affect a related process: heart rhythm and the coordination of the contraction. The *tre* mutant, commonly cited as a model of cardiac fibrillation in zebrafish, results from disruption of the sodium–calcium exchanger 1 (*slc8a1a; ncx1; ncx1h*) (Ebert et al. 2005; Langenbacher et al. 2005). Calcium imaging reveals a clear disruption in calcium transients, and expression of Ncx1 in *tre* embryos is able to rescue the phenotype, although dosage is critical because of the importance of establishing the appropriate calcium homeostasis (Langenbacher et al. 2005). The paralog of this gene, *slc8a1b*, does not appear to be highly expressed in the heart and has been demonstrated to be required for $Ca^{2+}$ extrusion in fish gill cells (Liao et al. 2007).

The mutant *reggae* results from disruption of the ether-à-go-go-related gene channel (*kcnh2*; *erg*) (Hassel et al. 2008). The original description of *reg* was a mutant possessing a cardiac defect that first manifests at 24 hpf with a virtual lack of heart tube contraction and only infrequent waves of contraction detected (Stainier et al. 1996). Patches of beating cardiomyocytes can be detected near the sinus venosus as development proceeds but this activity is rarely propagated to the adjacent atrial cells. On occasion, this pulse is transmitted to the atrium and, because of the uncoordinated nature of the contraction, resembles an atrial fibrillation. Rarely, the ventricle and atrium contract in a normal atrial-ventricular rhythm and the atrioventricular block subsides. This phenotypic description has properties commonly associated with sinus node exit block and suggests that the *reg* mutant could serve as a model for human short-QT syndrome.

When the genetic lesion for *reg* was identified as a missense mutation in the potassium channel *kcnh2*, questions arose regarding the nature of the lesion since a previously described mutation in zebrafish *kcnh2* displayed an alternate phenotype (Langheinrich et al. 2003). The mutant *breakdance* (*bre*), which is characterized by a 2:1 atrioventricular block (two atrial contractions for a single ventricular contraction), similarly results from mutation in the *kcnh2* gene. However, unlike the *reg* lesion, which has been determined to be an activating mutation, the *bre* lesion results in a hypomorphic allele, reducing but not eliminating activity (Langheinrich et al. 2003). Similar to the observed *bre* phenotype, dose-dependent knockdown of *kcnh2* by morpholino injection induces bradycardia and also 2:1 block at low and high morpholino doses, respectively (Langheinrich et al. 2003). The complete loss-of-function phenotype, achieved by complete translational block with morpholino knockdown or loss of function induced by genetic mutation, results in a complete loss of ventricular contraction (Arnaout et al. 2007). The attributes of this second class of defects resulting in compromised or complete loss of Kcnh2 function are being exploited as models for human long-QT syndrome. Recent studies that demonstrate similar biophysical properties for the zebrafish and human Kcnh2 channels reinforce the utility of these mutants and morphants as potential disease models (Scholz et al. 2009).

One of the earliest heart function mutants to be positionally cloned was *island beat* (*isl*), identified as a mutation in the L-type $Ca^{2+}$ channel (cacna1d; cav1.3a) (Rottbauer et al. 2001). The phenotype is similar to those previously described, resembling atrial fibrillation in humans. While the ventricle is non-contractile, individual cells within the atrium beat in an asynchronous fashion. One property of the *isl* hearts that is noteworthy is that while the atrium appears relatively normal morphologically, the ventricle does not possess the normal number of cardiomyocytes (Rottbauer et al. 2001). This defect implicates the *cacna1d* gene in a previously uncharacterized role in the regulation of cellular proliferation, in addition to its canonical role coordinating cardiomyocyte contraction. This is a prime example of how a single gene defect can have crucial roles in the regulation of both form and function of the cardiovascular system.

The mutant *tell tale heart* (*tel*) has a weakly contracting heart resulting from disruption of *regulatory cardiac myosin light chain gene 2* (*myl7; cmlc2*) function (Rottbauer et al. 2006). Similarly, the *lazy susan* (*laz*) mutant displays weak contraction as a result of a mutation in the *atrial essential myosin light chain gene* (*cmlc1*; similar to human *MYL4*) (Meder et al. 2009). Analysis of both *myl4* and *myl7* morphants revealed a similar functional deficit but also a disruption of sarcomere structure induced by the loss of Myl4 and Myl7 protein function. However, sarcomere length was increased as a result of *myl4* loss while it was decreased in *myl7* morphants (Chen et al. 2008).

## 11. Relationship Between Cardiovascular Function and Form

The effect of a morphological defect on organ function appears obvious: without the correct spatial and temporal regulation of cell growth, tissues form abnormally and are unable to carry out their assigned roles. For example, the previously described *isl* mutant has a disruption in a gene typically associated with cardiac function yet reveals its important role in cardiomyocyte growth (Rottbauer et al. 2001). Less obvious is the profound role that abnormal function itself can have on morphological development in the embryo. Contraction of the cardiac chambers and circulation through the vasculature are two distinct sources of force generation that can have dramatic influences on development of the cardiovascular system. The potential role of epigenetic forces on heart morphogenesis is especially intriguing since the onset of cardiac function and its requirement occurs prior to the formation of the mature organ.

Novel genes that promote angiogenesis and vessel patterning in the zebrafish have similarly been identified through the analysis of vascular mutants. The phenotypes of these mutations include a complete absence of endothelial and blood cells in the *cloche* mutant (Stainier et al. 1995), mutants with completely mispatterned vessels as in *out of bounds* (Childs et al. 2002; Torres-Vazquez et al. 2004), a lack of sprouting angiogenesis as seen in Vascular endothelial growth factor (Vegf) pathway mutants and morphants (Covassin et al. 2006, 2009), or missing subsets of the secondary parachordal vessels when the Netrin/Unc5B pathway is perturbed (Wilson et al. 2006). A complete summary of these mutants and their phenotypes can be seen in Table 6.1.

## 12. ENDOTHELIAL CELL SPECIFICATION AND DIFFERENTIATION

The specification of endothelial cells in zebrafish occurs in a very similar manner to other vertebrate species. First, angioblast and primitive erythrocyte lineages are specified from the hemangioblast in the lateral mesoderm, prior to the decision of the angioblast to acquire an artery- or vein-specific fate. Table 6.2 shows the transcription factors and their roles in endothelial specification and differentiation. At the top of the specification hierarchy is the cloche gene which may be caused by mutations in the lysocardiolipin acyltransferase (Lycat) gene. Lycat acts upstream of members of the ets family of transcription factors (*ets1, etsrp, fli1* and *fli1b*) and the stem-cell leukemia gene *scl/tal* (Gering et al. 2003; Sumanas and Lin 2006; Pham et al. 2007) to specify endothelial and blood lineages. The lim-domain-only gene *lmo2* is also at the top of this hierarchy. Etsrp is a dominant ets family member involved in vascular specification, and is sufficient to drive the expression of many endothelial-specific genes such as *scl*, the Vegf-a receptor *flk1*, the Vegf-c receptor *flt4, fli1*, the semaphorin receptor *plexinD1* and vascular endothelial cadherin *cdh5* (Sumanas and Lin 2006; Pham et al. 2007). Despite being sufficient to induce an endothelial program, *etsrp* is redundant with the other ets family members for the specification of endothelium, and quadruple knockdowns of all four genes reveal the strongest phenotype (Sumanas and Lin 2006; Pham et al. 2007).

Ets family proteins co-operate with the forkhead transcription family members Foxc1a and Foxc1b in zebrafish and other vertebrates, synergistically activating endothelial-specific genes (De Val et al. 2008). A "fox-ets" binding motif is found with a 10-fold higher frequency in endothelial-specific genes as opposed to other genes. Loss of *foxc1a* and *foxc1b* together leads to defects in angiogenesis, but not specification of the first cells (De Val et al. 2008). Similarly, *lmo2* and *scl* genetically interact. Loss of *lmo2* in conjunction with *scl* leads to angiogenic defects (Patterson et al. 2007).

## 13. SPECIFICATION OF ARTERY AND VEIN

Angioblasts next acquire an artery-specific or vein-specific fate which occurs prior to the formation of vessels. This involves receptor signaling pathways and downstream transcription factors. Interaction of the sonic hedgehog, notch, and VEGF signaling pathways specifies the artery. The artery expresses numerous "artery-specific genes" including *notch3*, *ephrinB2a, delta like 4* (*dll4*), *neuropilin1a, neuropilin1b, tbx20* and *hey2* (*gridlock*). Fewer vein-specific

**Table 6.2**
Transcriptional control of endothelial development in zebrafish

| Gene | Cardiovascular Expression | Loss of function phenotype | Gain of function phenotype | Reference |
|---|---|---|---|---|
| Ets1 | Endothelium | Moderate ISV angiogenic defects | | (Pham et al. 2007) |
| Etsrp | Endothelium | Severe ISV angiogenic defects | Induction of hemangioblast gene expression | (Pham et al. 2007; Sumanas et al. 2008; Sumanas and Lin, 2006) |
| Fli 1 | Endothelium | Mild ISV angiogenic defects | Induction of hemangioblast gene expression | (Liu et al. 2008; Pham et al. 2007) |
| Fli 1b | Endothelium | Mild ISV angiogenic defects | | (Pham et al. 2007) |
| Foxc1a | Endothelium | Decreased ISV angiogenesis | | (De Val et al. 2008) |
| Foxc1b | Endothelium | Decreased ISV angiogenesis | | (De Val et al. 2008) |
| Foxh1 | Ventral mesoderm | Increased vessel markers | Suppression of vascular development | (Choi et al. 2007) |
| Hey2 (grl) | Endothelium | Loss of artery | | (Zhong et al. 2000, 2001) |
| Hhex | Endothelium | | Induction of vascular genes | (Liao et al. 2000) |
| Lmo2 | Endothelium, blood | Angiogenic defects; reduction of hemangioblast markers | | (Patterson et al. 2007) |
| Scl | Endothelium, blood | | Induction of flk-1, GATA-1, tie-1, fli-1, scl | (Gering et al. 2003; Liao et al. 1998) |
| Sox 7 & Sox 18 | Endothelium | Vein-specific marker decrease; A-V shunts | | (Cermenati et al. 2008; Herpers et al. 2008; Pendeville et al. 2008) |

markers are known in zebrafish, but include the *vegf receptor 3* (*flt4*), *ephb4, neuropilin2a, neuropilin2b, santa* and *valentine* (Ahn et al. 2000; Mably et al. 2003; Mukhopadhyay and Peterson 2008).

To specify the artery, a cascade leads through several signaling pathways. Sonic hedgehog is expressed by the notochord and induces VEGF expression in the adjacent somite tissue. The calcitonin receptor-like receptor, also known as adrenomedullin receptor, is an intermediate signaling step mediating VEGF expression downstream of sonic hedgehog (Nicoli et al. 2008). VEGF in turn acts on angioblasts of the aorta to induce artery specification (Lawson et al. 2002; Cermenati et al. 2008; Herpers et al. 2008; Pendeville et al. 2008). Vein cells are at a greater distance from the notochord and receive less VEGF signal; therefore they do not take on an arterial program. VEGF leads to loss of artery specification through signaling through PLC$\gamma$ (Lawson et al. 2003) and promoting expression of the notch receptor notch 3 (Lawson et al. 2001). Activation of ERK or inhibition of PI3 kinase downstream of VEGF signaling also leads to arterial specification (Hong et al. 2006).

Notch signaling is required for arterial specification. Loss of notch signaling through overexpression of a dominant-negative suppressor of hairless construct or in the mindbomb mutant leads to a reduction in artery markers and ectopic expression of vein markers in the artery (Lawson et al. 2001). Notch signaling activates the transcription factor Hey2 (Gridlock) downstream of the Notch signaling pathway (Zhong et al. 2000, 2001).

Very little is known about vein specification in zebrafish, but two SRY (sex determining region Y)-box genes sox7 and sox18 are required. Loss of both sox7 and sox18 also perturbs some arterial markers (gridlock), but not others (notch3, tbx20, dll4). How the sox genes interact with the artery specification pathway is not well understood (Cermenati et al. 2008; Herpers et al. 2008; Pendeville et al. 2008).

## 14. CELL MIGRATION OF ANGIOBLASTS AND VASCULAR PATTERNING

Blood and angioblast precursors express unique markers in the lateral mesoderm prior to migration to the midline around 12–17 Somite stage (S) (14–16 hpf) (Jin et al. 2005). Both cell populations migrate medially following a path ventral to the somites and dorsal to the endoderm. There are two waves of angioblast migration, one starting at 14 hpf and one at 16 hpf. The endoderm secretes Bmp4 which has an important patterning role in the ventral aorta (Wilkinson et al. 2009) but is not necessary for migration or artery–vein specification of angioblasts, both of which appear to be intrinsic properties of the angioblasts (Jin et al. 2005).

Angioblasts and blood precursors in the anterior embryo start to migrate before cells located in the posterior mesoderm, following the anterior to posterior gradient of maturation of the somites. Thus, staging of the formation of the aorta requires knowledge of the developmental stage as well as the position of the cells along the anterior–posterior axis. At the midline a simple cord of cells coalesces into what will become the dorsal aorta just ventral to the hypochord, and lumenizes (Eriksson and Lofberg 2000).

## 15. ANGIOGENESIS OF THE INTERSEGMENTAL VESSELS

The intersegmental vessels (ISVs) of the zebrafish are one of the most highly studied vessel beds in the zebrafish due to their accessibility and simplicity of patterning (Childs et al. 2002). They first sprout at 20S and migrate dorsally. There are three types of cells in the ISV, a ventral cell which connects the aorta to the ISV, a connector cell and a dorsal cell which connects the ISV to the dorsal longitudinal anastomotic vessel (DLAV) (Childs et al. 2002). Multiple cells overlap and the final ISV is composed of 3–6 cells (Siekmann and Lawson 2007; Blum et al. 2008).

Fine analysis of cell number shows that two cells initially sprout from the aorta, but that one undergoes a cell division at the mid-somite stage (Siekmann and Lawson 2007). The lead cell becomes the "tip cell". The number of tip cells that form is repressed by the notch pathway and excessive tip cells are observed after loss of either the notch ligand delta (Dll4) or the downstream transcription factor Rbpsuh. On the other hand, overexpression of the constitutively active notch intracellular domain leads to a complete absence of tip cells (Leslie et al. 2007; Siekmann and Lawson 2007).

## 16. LUMENIZATION OF VESSELS

Lumenization of the dorsal aorta occurs between 18 and 30 hpf (Jin et al. 2005). The 4–6 cells around the circumference of the aorta hollow out and undergo a morphological change to become more elongated. Cell–cell junctions as detected by ZO-1 immunoreactivity are first seen at 17 hpf. Adherens junctions are also observed at this time. Claudin 5, an essential endothelial claudin, is first apparent at 18 hpf, and is more highly expressed by arterial endothelial cells. Genetic mutations in the Egf-like 7 (Egfl7) protein prevent lumenization from occurring. Egfl7 morphant embryos fail to maintain proper junctions and to assume the correct shape, affecting both cell adhesion and migration (Parker et al. 2004).

Angioblasts from both waves of migration to the midline contribute to the posterior cardinal vein which forms by 24 hpf, and it is lumenized by 30 hpf in a manner similar to the artery (Jin et al. 2005). In contrast, the lumens of smaller vessels such as the intersegmental vessels form by a different mechanism. Lumen formation of the intersegmental vessels has been traced in vivo using an EGFP construct fused to cdc42 (Kamei et al. 2006). In these vessels, small pinocytotic vacuoles form within cells and fuse to form intracellular spaces. The vacuoles in adjacent cells then fuse to form larger intercellular lumens.

## 17. VASCULAR STABILIZATION, MURAL CELLS, AND EXTRACELLULAR MATRIX

Vascular mural cells (mesenchymal cells, pericytes and smooth muscle cells) differentiate in concert with vessel formation. These cells play a critical role in vascular stabilization in zebrafish. Defects in mural cell recruitment or differentiation result in vascular hemorrhage, as observed in genetic mutants of either the βpix/Pak2a pathway, or the cerebral cavernous malformations pathway (Buchner et al. 2007; Liu et al. 2007; Gore et al. 2008).

Markers for mural cells have been developed in zebrafish (Wallace et al. 2005; Georgijevic et al. 2007; Santoro et al. 2009), and include α-smooth muscle actin (αSma), Sm22α, smooth muscle myosin heavy chain, smoothelin, embryonic myosin heavy chain and Cpi-17. Of these αSma and Sm22α are early and robust markers that are also used to recognize early mural cells in addition to the more prevalent visceral smooth muscle (Georgijevic et al. 2007; Santoro et al. 2009). Vascular mural cells are present from 72 hpf around the ventral and anterior dorsal aorta, and become more strongly observed at 96 hpf in the vicinity of the "Y" region where the dorsal aorta bifurcates. These cells are not likely differentiated smooth muscle cells as they only express early markers and do not have the typical elongated morphology of mature smooth muscle cells, or a multilamellar architecture at this time point. They therefore represent an early mural cell phenotype. Pericytes are clearly visualized by transmission electron microscopy in the adult zebrafish eye (Alvarez et al. 2007), but have not been characterized in the embryo at this time point.

Several components of the vascular extracellular matrix have been identified and characterized in zebrafish. Collagen IV surrounds adult zebrafish retinal vessels (Alvarez et al. 2007), and tropoelastin is found around large vessels in the adult (Miao et al. 2007). Laminin is a critical molecule for the migration of endothelial cells in the embryo. In laminin $\alpha_4$ or $\alpha_1$ mutants both the intersegmental vessels of the trunk or hyloid vessel of the eye have highly abnormal morphology (Pollard et al. 2006; Semina et al. 2006).

## 18. ANGIOGENESIS OF ORGANS

After the initial patterning of the vascular system, new vessels start to invade organs to provide a blood supply. In general this occurs after the initial vessels become functional and carry blood flow on the first day of development.

### 18.1. Brain Vessels

The patterning and growth of cranial vessels occurs early and is highly dynamic during the window from 1.2 to 2 dpf when a large number of vessels penetrate the brain (Isogai et al. 2001; Lawson and Weinstein 2002). Once established, the basic pattern of large vessels of the head remains unchanged after 2.5 dpf.

### 18.2. Retinal Vessels

The retina is highly dependent on an early blood supply. The first endothelial cells of the eye are present close to the lens at 48 hpf and form an optic artery and hyloid vessel system by 2.5 dpf (Isogai et al. 2001; Alvarez et al. 2007). These expand and form a "basket" by 15–19 dpf and continue elaborating until 60 dpf when the pattern matures to become equivalent to the adult's. These retinal vessels are covered by a collagen-containing basal lamina and pericytes. Mutations affecting angiogenesis of the early embryo also affect the development of the retinal vasculature including *out of bounds*, which has a mutation in the semaphorin receptor *plexinD1* (Childs et al. 2002; Alvarez et al. 2007). Adult *obd* embryos have one-third more retinal vessels, but fewer branch points.

### 18.3. Kidney Glomerular Vessels

The development of the kidney is highly integrated with the development of blood vessels. Podocytes of the kidney glomerulus recruit blood vessels from the dorsal aorta. Blood flow in the dorsal aorta is required to initiate the invasion into the glomerulus, through the upregulation of the matrix metalloprotease Mmp-2 (Majumdar and Drummond 1999; Serluca et al. 2002).

### 18.4. Endodermal Organ Vessels

In the intestine, the supraintestinal artery and subintestinal vein begin formation at 2 dpf. They undergo growth and remodeling to elaborate into a basket surrounding the gut over the next few days, through 7 dpf. Nearby, vessels forming the hepatic sinusoids of the liver have also begun to form, starting at 3 dpf (Isogai et al. 2001).

### 18.5. Definitive Hematopoietic System

Hematopoietic stem cells (HSC) arise from the ventral dorsal aorta in zebrafish and differentiate into the definitive hematopoietic lineages. This region of the embryo is functionally equivalent to the aorta-gonad-mesonephros (AGM) region in higher vertebrates where hematopoietic stem cells also derive. The dorsal aorta is patterned between opposing gradients of sonic hedgehog (dorsally) and Bmp4 (ventrally) (Gering and Patient 2005; Wilkinson et al. 2009). Bmp4 is directly responsible for the induction of HSCs in the ventral aortic wall around 23 hpf (Wilkinson et al. 2009). From the aorta, the HSCs emerge and migrate to a temporary niche in the posterior blood island before migrating to the mature hematopoietic organ in fish, the kidney (Jin et al. 2007). Interestingly, circulation of blood is required for the induction of HSCs from the aortic wall by inducing nitric oxide (NO) production (North et al. 2009). This may provide one clue as to why circulation begins many days before it is actually required to oxygenate tissues in the fish.

### 18.6. Lymphatic Vessel Development

The lymphatic system in zebrafish is an offshoot of the venous system, originating from the parachordal vessel, which in turn originates from the posterior cardinal vein (Kuchler et al. 2006; Yaniv et al. 2006). In the mouse, the lymphatic system originates by the specification of lymphatic cells from venous cells by the transcription factor Prox1, budding and sprouting off the vein followed by complete separation. In contrast, the zebrafish lymphatics develop by cellular migration followed by tube formation. Flt4, Vegf-c, Prox1 and Ccbe1 are all critical for lymphatic formation in zebrafish and function in parallel (Kuchler et al. 2006; Yaniv et al. 2006; Hogan et al. 2009).

### 18.7. Hypoxia in Zebrafish Angiogenesis

The early vascular pattern of zebrafish embryos raised under hypoxia is not affected before 12 dpf suggesting that hypoxia is not a driving force in early angiogenesis (Schwerte et al. 2003; Grillitsch et al. 2005). However, hypoxia causes a functional change in vessels, and can drive shunting of blood flow to alternative sites. For instance blood flow to the muscle is increased and flow to the gut is decreased when hypoxia is applied to 12–15 dpf embryos (Schwerte et al. 2003). Furthermore, hypoxia can drive angiogenesis in larvae and adults. Capillarization of axial muscle is accelerated by 30% after swim training, particularly in larvae trained between 21 and 32 dpf (Pelster et al. 2003). Hypoxia also plays a significant role in inducing angiogenesis in the adult zebrafish. Adult retinal vessels show changes in neovascularization including tip cell protrusions after 3 days exposed to hypoxia. This translates into an increased density of vessels after 6–12 days of hypoxia (Cao et al. 2008). The changes are dose-dependent on oxygen and responsive to treatment with angiogenesis inhibitors suggesting that they are also VEGF-dependent.

### 18.8. Tumor Angiogenesis

The zebrafish has been recently adopted as a powerful new model for understanding tumor angiogenesis and metastasis (Haldi et al. 2006; Nicoli et al. 2007). Human tumor cells can be xenotransplanted into zebrafish embryos and induce an angiogenic response. The ability to image invading blood vessels at high resolution in real time is a large advantage of the model and has been used to show that migratory tumor cells invade vessels at sites of vascular remodeling, not in intact vessels (Stoletov et al. 2007).

## 19. CHEMICAL BIOLOGY APPROACHES TO STUDY CARDIOVASCULAR BIOLOGY

The zebrafish embryo is accessible to chemical genetic screens. Small molecules are easily added to the medium in which embryos grow, and this technique has been exploited for identification of compounds perturbing zebrafish embryogenesis, and as a potential commercial application of zebrafish biology to identify novel compounds. An array of small molecules are dissolved in embryo medium and plated in 96-well plates. A small number of embryos are added and allowed to develop. The embryos are either analyzed manually or using an automated system. Concentramide was one of the first small molecules identified disrupting epithelial polarity and heart patterning in the embryo. Interestingly the phenotype mimics the genetic mutant *heart and soul*, suggesting that it might act in the same pathway (Peterson et al. 2001). Other molecules which disrupt specific processes in endothelial or heart development have also been described (Huang et al. 2008; Kalen et al. 2009; Molina et al. 2009). Interestingly, screens for chemical suppression of genetic mutant phenotypes have also been successful, such as GS4012 which reverses aortic coarctation in the genetic mutant *gridlock* (Peterson et al. 2004; Hong et al. 2006).

The utility of the zebrafish as a physiological model for small molecule screening is illustrated by the analysis performed by Milan (Milan et al. 2003) for compounds that influence the heart rate in zebrafish embryos. An analysis of 100 molecules was performed using 48 hpf zebrafish embryos distributed in 96-well plates. Twenty-three of the compounds tested were drugs that induced a prolongation of the QT interval in humans and 22 of them demonstrated a consistent bradycardia and atrioventricular block in the zebrafish embryos. In addition, microinjection was able to overcome issues with poor absorption in the zebrafish embryo and produce positive results with four of five initial negatives (Milan et al. 2003).

## 20. FORCES THAT INFLUENCE CARDIOVASCULAR DEVELOPMENT

Contraction of the heart and the onset of blood flow precede the completion of cardiovascular development. The forces on endocardial and myocardial cells in the heart generated by its contractile function (wall stress) have been proposed as a stimulus for its growth and development. Additionally, forces impinge on the endocardial layer of the heart and endothelium lining of the vessels through shear stress. This force, which is exerted in the direction of flow, is generated by movement of blood through the endocardial cavity and vasculature (force exerted at a fluid–solid interface which is typically perpendicular to wall stress).

### 20.1. Shear-stress-induced Signals

Shear stress is absolutely essential for normal heart and vascular development and increases with time as the heart develops. Endothelial cells at the sinus venosus are exposed to low shear stress (1 dyne $cm^{-2}$), while endocardial cells lining the heart and outflow tract (bulbus arteriosus) are exposed to much higher levels: 2.5–10 dynes $cm^{-2}$ in the heart itself and more than 75 dynes $cm^{-2}$ at the bulbo-ventricular valve (Hove et al. 2003). Shear stress is required for looping and bulbus arteriosus formation (Hove et al.

2003). Furthermore, vessels in *silent heart* animals that never have blood circulation (due to absent cardiac function, as previously described) pattern their primary and secondary intersegmental vessel sprouts correctly, but fail to lumenize them, and their morphology is abnormal (Isogai et al. 2003). Lumen diameter of the dorsal aorta and posterior cardinal vein, as well as sub-intestinal vein growth, are also altered as a result of impaired flow (Fig. 6.2; Sehnert et al. 2002; Parker et al. 2004). Reduction in shear stress through removal of approximately 50% of the erythrocytes from a 2 dpf zebrafish embryo results in a small decrease in aorta diameter, and a serious decrease in the development of secondary vessels including the parachordal and vertebral arteries demonstrating that shear stress is an important component driving later vascular patterning and angiogenesis (Kopp et al. 2007).

### 20.2. Transduction of Flow-induced Signals

Signaling between cells in adjacent tissue layers is a common paradigm seen during tissue and organ morphogenesis. For example, endocardial precursor cells direct myocardial cell migration during fusion of the heart tube (Holtzman et al. 2007), and endocardial neuregulin and notch1b are both required for induction of conduction tissue from myocardial cells (Milan et al. 2006). Also, the mutant *heart of glass* (*heg*) is characterized by a thin myocardial wall as a result of mutation in a transmembrane protein which is expressed in the endocardium (Mably et al. 2003). In the heart and vasculature, the endothelial layer is in direct contact with the circulating blood and directly exposed to the flow-induced forces described above. As a result, endothelial cells have been proposed as a source of secreted factors to transduce these flow forces to the cells surrounding them. A role for flow-induced MMP-2 activity in early zebrafish kidney development has been described (Serluca et al. 2002) but little is known about similarly induced genes in the cardiovascular system.

### 20.3. Potential Role for Primary Cilia in Cardiovascular Flow Sensing and Development

The role of cilia in development of the early kidney pronephros and Kupffer's vesicle formation has been well documented (Kramer-Zucker et al. 2005) but monocilia are also present on the luminal surface of the endocardium in other species (Slough et al. 2008; Van der Heiden et al. 2008). The loss of the kinesin Kif3a protein in mouse heart impairs intraflagellar transport (IFT) and cilia formation and has a dramatic effect on myocardial growth. In addition to thinning of the myocardial wall, cells within the endocardial cushions (the valvular precursors) fail to undergo the

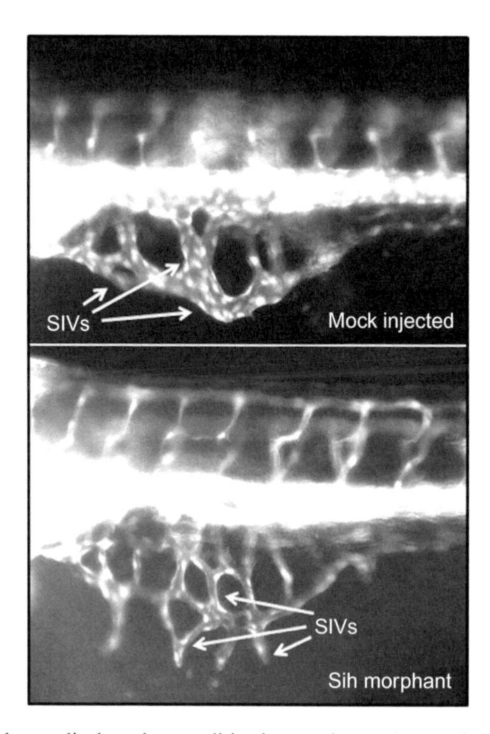


**Fig. 6.2.** *sih* morphants display abnormalities in vessel morphogenesis. Embryos were either mock injected (top panel) or injected with a morpholino to tnnt2. This morpholino results in an embryo with a non-contractile heart and no circulation. In spite of the myocardial-restricted expression of *tnnt2*, the inability to sustain a normal blood flow results in vessel abnormalities, including those illustrated for SIV development (compare SIV formation, indicated by arrows). SIV, sub-intestinal vein.

epithelial-to-mesenchymal transition (EMT) required for their normal development (Slough et al. 2008). The relationship between flow and signaling through primary cilia has not yet been thoroughly investigated during zebrafish cardiovascular morphogenesis.

## 20.4. Wall Stress and Shear Stress in the Heart

The mutants *silent heart* (*sih*) and *weak atrium* (*wea*) each result from mutations in sarcomeric protein genes, cardiac troponin T2 (*tnnt2*) (Sehnert et al. 2002) and atrial myosin heavy chain (*myh6*) respectively (Berdougo et al. 2003). The *sih* mutant has been used extensively as a means to ablate flow in conjunction with loss of function analysis for a number of zebrafish genes. The thin ventricular wall of the *sih* mutant is supportive of a role for contractile force generation in the regulation of myocardial proliferation. However, since both chambers are non-contractile, it is difficult to dissect the distinct roles of contractile and flow-induced forces on the growth of the heart. The mutant *wea*, which lacks only atrial contractility, has been more fully exploited to determine the potential influence of epigenetic factors on cardiac morphogenesis. Loss of atrial function has a severe detrimental effect on growth of the ventricular chamber. While the circulation is maintained in the mutant as a result of ventricular contraction, the myocardial wall of this chamber becomes thickened, resulting in a reduction in lumen diameter (Berdougo et al. 2003). Gene expression changes are also detected in the ventricle (Berdougo et al. 2003).

Extending from these studies, Auman et al. (2007) explored how alterations in organ shape proceed from alterations in cellular morphology. Specifically, elongation of cardiomyocytes facilitates the curvature of the ventricular chamber (Auman et al. 2007). Using a combination of crosses in and between *wea* and *half-hearted* (*haf*) mutants (mutation in ventricular myosin heavy chain, *vmhc*) to inhibit atrial and/or ventricular contractility, both flow and contractility revealed effects on the shape of ventricular cells. In the ventricular wall of hearts from *haf* mutants, loss of contraction in this chamber correlated with abnormal sarcomere formation and loss of restriction of cellular elongation. Loss of flow in conjunction with impaired cardiac contractility was correlated with reduced cardiomyocyte enlargement and elongation. However, by transplanting *haf* ventricular cells into a wild-type host, it was possible to clarify that contractility was the key determinant of this cell size and shape transformation, and that it was independent of the influence of flow (Auman et al. 2007).

## 20.5. Contractility of Vessels

Contractility has been demonstrated in developing zebrafish vessels by manipulation of nitric oxide (NO) (Pelster et al. 2005). Administration of sodium nitroprusside (SNP, an NO donor) to 5 dpf zebrafish leads to arterial diameter increase while administration of L-NAME (a NO inhibitor) leads to vasoconstriction (Fritsche et al. 2000). NO also influences angiogenesis. Chronic administration of SNP leads to early development of the caudal vascular tree and parachordal vessels (Pelster et al. 2005).

Furthermore, embryos clearly have the ability to use this contractility to shunt blood flow depending on need. Digital motion analysis of a fed 8 dpf zebrafish larvae demonstrates that it shunts a large portion of its blood flow away from the intersegmental vessels in muscle to the gut vasculature by blocking flow at the base of these vessels (Schwerte and Pelster 2000).

## 21. ENDOTHELIAL CELL SURVIVAL

Endothelial cells require constant signaling in order to survive. In mouse, autocrine VEGF signaling is a signal for endothelial survival (Lee et al. 2007). A few additional new pathways for endothelial survival have been discovered in zebrafish. One pathway identified in zebrafish involves the microRNA miR-126 (Fish et al. 2008; Wang et al. 2008). miR-126 normally represses two inhibitors of the VEGF pathway, Sprouty-related protein (SPRED1) and phosphoinositol-3 kinase regulatory subunit 2. Loss of miR-126 leads to hemorrhage through inhibition of VEGF signaling. A second genetic pathway was identified from analysis of the zebrafish mutant *tomato*. This mutant has hemorrhages caused by a mutation in the baculoviral IAP repeat-containing protein-2, Birc2. Although some genetic mutants with hemorrhage have defects in endothelial–mural cell interactions, *tomato* mutants have defects in endothelial integrity. In the absence of Birc2 apoptosis and vessel regression occur (Santoro et al. 2007). The function of Birc2 or the pathway through which apoptosis is activated in this mutant is not certain.

## 22. ZEBRAFISH MODELS FOR VASCULAR ANOMALIES

The *santa* (*san*) and *valentine* (*vtn*) mutants display the prominent feature of a dilated heart with an associated failure of the myocardium to thicken beyond a single-cell layer (Mably et al. 2006). The genetic lesions in these two mutants are in the zebrafish homologs of two genes associated with a genetic cerebral vascular disease in humans (cerebral cavernous malformations, CCM) (Revencu and Vikkula 2006). In humans, the lesions are characterized by abnormally dilated vascular channels and are believed to arise as a result of localized defects in blood vessel development. The presence of dominant mutations in the human homologs of *san* and *vtn* associated with a single disease phenotype suggests they may be involved in a common pathway. The phenotypic similarities of the two zebrafish autosomal recessive mutations further support this hypothesis. The *heg* gene has also been implicated in this pathway in both zebrafish and mouse, although mutations in this gene have not been detected in association with CCM (Mably et al. 2003, 2006; Kleaveland et al. 2009).

Similarly to the cardiac defect in the three zebrafish mutants, these vascular anomalies appear to reflect defects in communication between the endothelial cells lining the vessel and the cells surrounding them. Although both *san and vtn* mutant embryos show a reduced arterial and increased venous lumen size, as well as abnormalities in sub-intestinal vein formation (SIVs) (Hogan et al. 2008), similar defects have been observed in mutants lacking flow, as previously described for *sih* (Fig. 6.2; Sehnert et al. 2002; Parker et al. 2004). It is significant that these defects also occur in the *sih* mutants since the gene lesion is in cardiac troponin T, a gene expressed in a heart-restricted pattern throughout development and never in the endothelium of the vasculature (Sehnert et al. 2002). Additionally, these observations, particularly in subintestinal vessel formation, are made well past the first initial detection of the phenotype which makes it unlikely that these defects are a primary aspect of the phenotype (Mably et al. 2006).

A related vascular anomaly, hereditary hemorrhagic telangiectasia type 2 (HHT), is associated with mutations in the activin receptor-like kinase 1 (Acvrl1; Alk1), a TGFbeta type I receptor that is expressed in the vascular endothelium (Roman 2002; Marchuk et al. 2003). In the zebrafish mutant *violet beauregarde* (*vbg*), disruption of Acvrl1 function results in an abnormal circulation pattern detectable by 2 dpf, with flow through the cranial vessels but lack of perfusion in the trunk and tail (Roman et al. 2002). The phenotype has been shown to be a result of increased endothelial cell number within specific cranial vessels, which correlates with expression of the mutated gene (Roman et al. 2002). Since HHT, like the zebrafish mutant, is similarly characterized by cranial vessel malformations, *vbg* is a useful model for this human autosomal dominant disorder.

## 23. TOOLS FOR THE STUDY OF CARDIOVASCULAR PHYSIOLOGY

### 23.1. Blood Pressure

Tools for measurement of cardiovascular parameters in zebrafish embryos still remain impeded by their small size. However, innovative approaches, such as use of a servonull micropressure system for measurement of blood pressure in 2.5–4 dpf zebrafish embryos, have proven successful. Readings from the atrium and ventricle demonstrate that pressure ranges start from 0.1 mmHg in the atrium, 0.4 mmHg in the ventricle and 0.3 mmHg in the ventral aorta, the main outflow vessel of the heart (Pelster and Burggren 1996; Hu et al. 2000; Kopp et al. 2005). Blood pressure increases with developmental stage and in the adult (3 months post fertilization); pressure in the atrium is 0.68 mmHg, while the ventricle has pressures ranging from 0.42 to 2.51 mmHg. There is also a peak systolic pressure gradient from the ventral to the dorsal aorta of 2.16 to 1.51 mmHg (Hu et al. 2001).

### 23.2. Blood Cell Motion Analysis

Laser scanning velocimetry is a non-invasive imaging technique that uses line scans parallel to blood flow to determine the velocity of blood and to infer cardiovascular parameters such as cardiac output, vascular resistance and ventricular contractility (Malone et al. 2007). It can be used to determine morphological abnormalities in embryos, such as the aortic arch constriction in ccm2/*vtn* morphants (morpholino injected embryos). Digital motion analysis is a second image analysis method that examines the differences between subsequent frames of a movie to determine motion of blood cells. It has been used to show the pattern and speed of blood flow in zebrafish embryos (Schwerte and Pelster 2000). This technique can be applied to detecting changes in flow and cardiovascular parameters under physiological conditions.

### 23.3. Generation of Cardiac- and Vessel-specific Reporter Lines

Transgenic zebrafish expressing fluorescent proteins under tissue-specific promoters have become an invaluable tool. In addition to their uses for morphological analyses, such as for the assessment of cell number (Mably et al. 2003, 2006), innovative assay development for screening cardiovascular function has also been performed (Ebert et al. 2005). Using a high-throughput, 96-well format, an automated assay has been developed to determine effects of small molecules on heart rate in a myocardial-specific GFP line (Burns et al. 2005). Other approaches include the generation of a cardiac-specific fluorescent calcium indicator zebrafish transgenic line to allow optical mapping of the heart in live animals (Chi et al. 2008). This tool has already been used to map development of the cardiac conduction system in the zebrafish heart. The use of endothelial-specific GFP lines has also been employed for assessment of endothelial cell growth and morphology. The influence of chemical inhibition of the VEGF receptor in the adult zebrafish tail fin (Bayliss et al. 2006) and the effect of anthrax toxin on vessel morphology under similar treatments (Bolcome et al. 2008) have both been assessed using endothelial GFP transgenic lines.

## 24. VESSEL REGENERATION

Zebrafish have a remarkable ability to regenerate tissues including their highly vascularized caudal fin. Removal of most of the tail fin of an adult fish results in complete regeneration of blood vessels, bone and fin tissue on a timescale of weeks. The caudal fin therefore presents a useful and interesting model of adult angiogenesis. The caudal fin vasculature consists of one artery and two veins associated with each fin ray, along with commissures between the artery and veins, and interray vessels connecting vessels between rays (Huang et al. 2003). Immediately after amputation the vessels are open, but they seal up by 1 day post amputation (dpa). Anastomotic bridges between vessels form by 2 dpa, and a plexus of these connections forms in the next few days. The plexus is resolved into the original pattern, and the length of vessels is regenerated by 35 dpa (Huang et al. 2003). Several genetic pathways modulate fin regeneration. A mutation in the *reg6* gene leads to a complete lack of fin regeneration. *reg6* fish have a mutation in the early growth receptor 1. SKF91488, an inhibitor of histidine methyltransferase, can reverse the *reg6* phenotype suggesting a relationship between angiogenesis and fin regeneration and histidine synthesis (Huang et al. 2003). Fin regeneration has a strong dependence on angiogenesis. Chemical inhibition of VEGF signaling blocks the growth of fin tissue at less than 1 mm. The inhibition is temporary however, and fin growth resumes after the inhibitor is removed (Bayliss et al. 2006).

## 25. CARDIAC REGENERATION

Adult human hearts are unable to regenerate injured myocardial tissue. Instead, the damaged region of the heart is replaced by fibrotic scar, resulting in compromised cardiac function; hypertrophy of the remaining cardiomyocytes is the predominant physiological response employed by the cell. Although commonly used mammalian animal model systems are also refractory to spontaneous myocardial repair in the adult, recent work has challenged the paradigm that the cells within the heart no longer retain proliferative capacity (reviewed in Beltrami et al. 2001; Engel et al. 2005; Anversa et al. 2006). Since evidence from several studies has demonstrated loss of cardiomyocytes by apoptosis (Kang and Izumo 2000), it is sensible to suggest that there is some cardiomyocyte proliferation in the adult heart, if for no other reason than turnover of aged cells during normal growth of the organism. This mechanism is clearly highly regulated and not meant to compensate for the potentially devastating cell damage that occurs as a result of a cardiac pathological process. Therefore, much interest exists in determining if non-functioning cells could be replaced by transplantation or replacement of cells from a non-cardiomyocyte source.

The zebrafish heart, however, retains remarkable regenerative capacity and demonstrates complete myocardial repair within 2 months of 20% ventricular resection (Poss et al. 2002). In this system, the epicardium has been shown to be a source of ventricular cell addition in an FGF-dependent manner (Wills et al. 2008). More recently, the potential of the zebrafish embryo as a model system to identify small molecules that promote cardiomyocyte growth has been exploited. Using a chemical screening approach, an inhibitor of Dusp6 (a dual-specificity phosphatase component of the FGF signaling pathway) was identified and shown to expand cardiac cell lineages in the zebrafish embryo (Molina et al. 2009). Since repair of damaged myocardium rapidly following the insult would have the greatest efficacy for therapy, small molecules would be particularly effective mechanisms for treatment. An approach to recruit and induce an endogenous pool of cells to replace myocardium would be amenable to small molecule therapy and could be rapidly employed upon diagnosis.

## 26. SUMMARY

In summary the zebrafish is a highly adaptable model for exploring both the development and physiology of the cardiovascular system. Future work will include identifying new genes involved in its development, fully characterizing their effects on cardiovascular physiology, and identifying compounds that modulate or reverse clinical phenotypes using the zebrafish as a model.

## REFERENCES


Ahn, D. G., Ruvinsky, I., Oates, A. C., Silver, L. M., and Ho, R. K. (2000). tbx20, a new vertebrate T-box gene expressed in the cranial motor neurons and developing cardiovascular structures in zebrafish. *Mech. Dev.* **95**, 253–258.

Alexander, J., Stainier, D. Y., and Yelon, D. (1998). Screening mosaic F1 females for mutations affecting zebrafish heart induction and patterning. *Dev. Genet.* **22**, 288–299.

Alvarez, Y., Cederlund, M. L., Cottell, D. C., Bill, B. R., Ekker, S. C., Torres-Vazquez, J., Weinstein, B. M., Hyde, D. R., Vihtelic, T. S., and Kennedy, B. N. (2007). Genetic determinants of hyaloid and retinal vasculature in zebrafish. *BMC Dev. Biol.* **7**, 114.

Amsterdam, A., Burgess, S., Golling, G., Chen, W., Sun, Z., Townsend, K., Farrington, S., Haldi, M., and Hopkins, N. (1999). A large-scale insertional mutagenesis screen in zebrafish. *Genes Dev.* **13**, 2713–2724.

Anversa, P., Leri, A., and Kajstura, J. (2006). Cardiac regeneration. *J. Am. Coll. Cardiol.* **47**, 1769–1776.

Arnaout, R., Ferrer, T., Huisken, J., Spitzer, K., Stainier, D. Y., Tristani-Firouzi, M., and Chi, N. C. (2007). Zebrafish model for human long QT syndrome. *Proc. Natl Acad. Sci. USA* **104**, 11316–11321.

Auman, H. J., Coleman, H., Riley, H. E., Olale, F., Tsai, H. J., and Yelon, D. (2007). Functional modulation of cardiac form through regionally confined cell shape changes. *PLoS Biol.* **5**, e53.

Baker, K., Warren, K. S., Yellen, G., and Fishman, M. C. (1997). Defective "pacemaker" current (Ih) in a zebrafish mutant with a slow heart rate. *Proc. Natl Acad. Sci. USA* **94**, 4554–4559.

Barrionuevo, W. R., and Burggren, W. W. (1999). $O_2$ consumption and heart rate in developing zebrafish (Danio rerio): Influence of temperature and ambient $O_2$. *Am. J. Physiol.* **276**, R505–R513.

Bayliss, P. E., Bellavance, K. L., Whitehead, G. G., Abrams, J. M., Aegerter, S., Robbins, H. S., Cowan, D. B., Keating, M. T., O'Reilly, T., Wood, J. M., et al. (2006). Chemical modulation of receptor signaling inhibits regenerative angiogenesis in adult zebrafish. *Nat. Chem. Biol.* **2**, 265–273.

Beltrami, A. P., Urbanek, K., Kajstura, J., Yan, S. M., Finato, N., Bussani, R., Nadal-Ginard, B., Silvestri, F., Leri, A., Beltrami, C. A., et al. (2001). Evidence that human cardiac myocytes divide after myocardial infarction. *N. Engl. J. Med.* **344**, 1750–1757.

Berdougo, E., Coleman, H., Lee, D. H., Stainier, D. Y., and Yelon, D. (2003). Mutation of weak atrium/atrial myosin heavy chain disrupts atrial function and influences ventricular morphogenesis in zebrafish. *Development* **130**, 6121–6129.

Birely, J., Schneider, V. A., Santana, E., Dosch, R., Wagner, D. S., Mullins, M. C., and Granato, M. (2005). Genetic screens for genes controlling motor nerve–muscle development and interactions. *Dev. Biol.* **280**, 162–176.

Blum, Y., Belting, H. G., Ellertsdottir, E., Herwig, L., Luders, F., and Affolter, M. (2008). Complex cell rearrangements during intersegmental vessel sprouting and vessel fusion in the zebrafish embryo. *Dev. Biol.* **316**, 312–322.

Bolcome, R. E., III, Sullivan, S. E., Zeller, R., Barker, A. P., Collier, R. J., and Chan, J. (2008). Anthrax lethal toxin induces cell death-independent permeability in zebrafish vasculature. *Proc. Natl Acad. Sci. USA* **105**, 2439–2444.

Buchner, D. A., Su, F., Yamaoka, J. S., Kamei, M., Shavit, J. A., Barthel, L. K., McGee, B., Amigo, J. D., Kim, S., Hanosh, A. W., et al. (2007). pak2a mutations cause cerebral hemorrhage in redhead zebrafish. *Proc. Natl Acad. Sci. USA* **104**(35), 13996–14000.

Burns, C. E., Traver, D., Mayhall, E., Shepard, J. L., and Zon, L. I. (2005). Hematopoietic stem cell fate is established by the Notch-Runx pathway. *Genes Dev.* **19**, 2331–2342.

Cao, R., Jensen, L. D., Soll, I., Hauptmann, G., and Cao, Y. (2008). Hypoxia-induced retinal angiogenesis in zebrafish as a model to study retinopathy. *PLoS ONE* **3**, e2748.

Cermenati, S., Moleri, S., Cimbro, S., Corti, P., Del Giacco, L., Amodeo, R., Dejana, E., Koopman, P., Cotelli, F., and Beltrame, M. (2008). Sox18 and Sox7 play redundant roles in vascular development. *Blood* **111**, 2657–2666.

Chen, J.-N., Cowan, D. B., and Mably, J. D. (2005). Cardiogenesis and the Regulation of Cardiac-Specific Gene Expression. *Heart Fail. Clin.* **1**, 157–170.

Chen, J.-N., van Bebber, F., Goldstein, A. M., Serluca, F. C., Jackson, D., Childs, S., Serbedzija, G., Warren, K. S., Mably, J. D., Lindahl, P., et al. (2001). Genetic steps to organ laterality in zebrafish. *Comp. Funct. Genomics* **2**, 60–68.

Chen, J. N., and Fishman, M. C. (2000). Genetics of heart development. *Trends Genet.* **16**, 383–388.

Chen, J. N., Haffter, P., Odenthal, J., Vogelsang, E., Brand, M., van Eeden, F. J., Furutani-Seiki, M., Granato, M., Hammerschmidt, M., Heisenberg, C. P., et al. (1996). Mutations affecting the cardiovascular system and other internal organs in zebrafish. *Development* **123**, 293–302.

Chen, Z., Huang, W., Dahme, T., Rottbauer, W., Ackerman, M. J., and Xu, X. (2008). Depletion of zebrafish essential and regulatory myosin light chains reduces cardiac function through distinct mechanisms. *Cardiovasc. Res.* **79**, 97–108.

Chi, N. C., Shaw, R. M., Jungblut, B., Huisken, J., Ferrer, T., Arnaout, R., Scott, I., Beis, D., Xiao, T., Baier, H., et al. (2008). Genetic and physiologic dissection of the vertebrate cardiac conduction system. *PLoS Biol.* **6**, e109.

Chien, K. R. (1996). Genes and physiology: Molecular physiology in genetically engineered animals. *J. Clin. Invest.* **97**, 901–909.

Chien, K. R. (2001). To Cre or not to Cre: The next generation of mouse models of human cardiac diseases. *Circ. Res.* **88**, 546–549.

Childs, S., Chen, J.-N., Garrity, D., and Fishman, M. (2002). Patterning of angiogenesis in the zebrafish embryo. *Development* **129**, 973–982.

Choi, J., Dong, L., Ahn, J., Dao, D., Hammerschmidt, M., and Chen, J. N. (2007). FoxH1 negatively modulates flk1 gene expression and vascular formation in zebrafish. *Dev. Biol.* **304**, 735–744.

Christensen, G., Wang, Y., and Chien, K. R. (1997). Physiological assessment of complex cardiac phenotypes in genetically engineered mice. *Am. J. Physiol.* **272**, H2513–H2524.

Covassin, L., Amigo, J. D., Suzuki, K., Teplyuk, V., Straubhaar, J., and Lawson, N. D. (2006). Global analysis of hematopoietic and vascular endothelial gene expression by tissue specific microarray profiling in zebrafish. *Dev. Biol.* **299**, 551–562.

Covassin, L. D., Siekmann, A. F., Kacergis, M. C., Laver, E., Moore, J. C., Villefranc, J. A., Weinstein, B. M., and Lawson, N. D. (2009). A genetic screen for vascular mutants in zebrafish reveals dynamic roles for Vegf/Plcg1 signaling during artery development. *Dev. Biol.* **329**, 212–226.

De Val, S., Chi, N. C., Meadows, S. M., Minovitsky, S., Anderson, J. P., Harris, I. S., Ehlers, M. L., Agarwal, P., Visel, A., Xu, S. M., et al. (2008). Combinatorial regulation of endothelial gene expression by ets and forkhead transcription factors. *Cell* **135**, 1053–1064.

Doyon, Y., McCammon, J. M., Miller, J. C., Faraji, F., Ngo, C., Katibah, G. E., Amora, R., Hocking, T. D., Zhang, L., Rebar, E. J., et al. (2008). Heritable targeted gene disruption in zebrafish using designed zinc-finger nucleases. *Nat. Biotechnol.* **26**, 702–708.

Draper, B. W., Morcos, P. A., and Kimmel, C. B. (2001). Inhibition of zebrafish fgf8 pre-mRNA splicing with morpholino oligos: A quantifiable method for gene knockdown. *Genesis* **30**, 154–156.

Driever, W., Solnica-Krezel, L., Schier, A. F., Neuhauss, S. C., Malicki, J., Stemple, D. L., Stainier, D. Y., Zwartkruis, F., Abdelilah, S., Rangini, Z., et al. (1996). A genetic screen for mutations affecting embryogenesis in zebrafish. *Development* **123**, 37–46.

Ebert, A. M., Hume, G. L., Warren, K. S., Cook, N. P., Burns, C. G., Mohideen, M. A., Siegal, G., Yelon, D., Fishman, M. C., and Garrity, D. M. (2005). Calcium extrusion is critical for cardiac morphogenesis and rhythm in embryonic zebrafish hearts. *Proc. Natl Acad. Sci. USA* **102**, 17705–17710.

Ehler, E., Rothen, B. M., Hammerle, S. P., Komiyama, M., and Perriard, J. C. (1999). Myofibrillogenesis in the developing chicken heart: Assembly of Z-disk, M-line and the thick filaments. *J. Cell Sci.* **112**(Pt 10), 1529–1539.

Ekker, S. C. (2000). Morphants: A new systematic vertebrate functional genomics approach. *Yeast* **17**, 302–306.

Engel, F. B., Schebesta, M., Duong, M. T., Lu, G., Ren, S., Madwed, J. B., Jiang, H., Wang, Y., and Keating, M. T. (2005). p38 MAP kinase inhibition enables proliferation of adult mammalian cardiomyocytes. *Genes Dev.* **19**, 1175–1187.

Epstein, J. A., Rader, D. J., and Parmacek, M. S. (2002). Perspective: Cardiovascular disease in the postgenomic era–lessons learned and challenges ahead. *Endocrinology* **143**, 2045–2050.

Eriksson, J., and Lofberg, J. (2000). Development of the hypochord and dorsal aorta in the zebrafish embryo (Danio rerio). *J. Morphol.* **244**, 167–176.

Fish, J. E., Santoro, M. M., Morton, S. U., Yu, S., Yeh, R. F., Wythe, J. D., Ivey, K. N., Bruneau, B. G., Stainier, D. Y., and Srivastava, D. (2008). miR-126 regulates angiogenic signaling and vascular integrity. *Dev. Cell* **15**, 272–284.

Fishman, M. C., and Chien, K. R. (1997). Fashioning the vertebrate heart: Earliest embryonic decisions. *Development* **124**, 2099–2117.

Fornzler, D., Her, H., Knapik, E. W., Clark, M., Lehrach, H., Postlethwait, J. H., Zon, L. I., and Beier, D. R. (1998). Gene mapping in zebrafish using single-strand conformation polymorphism analysis. *Genomics* **51**, 216–222.

Fouquet, B., Weinstein, BM, Serluca, FC, and Fishman, MC. (1997). Vessel patterning in the embryo of the zebrafish: Guidance by notochord. *Dev. Biol.* **183**, 37–48.

Fritsche, R., Schwerte, T., and Pelster, B. (2000). Nitric oxide and vascular reactivity in developing zebrafish, Danio rerio. *Am. J. Physiol. Regul. Integr. Comp. Physiol.* **279**, R2200–R2207.

Gates, M. A., Kim, L., Egan, E. S., Cardozo, T., Sirotkin, H. I., Dougan, S. T., Lashkari, D., Abagyan, R., Schier, A. F., and Talbot, W. S. (1999). A genetic linkage map for zebrafish: Comparative analysis and localization of genes and expressed sequences. *Genome Res.* **9**, 334–347.

Geisler, R., Rauch, G. J., Baier, H., van Bebber, F., Brobeta, L., Dekens, M. P., Finger, K., Fricke, C., Gates, M. A., Geiger, H., et al. (1999). A radiation hybrid map of the zebrafish genome. *Nat. Genet.* **23**, 86–89.

Georgijevic, S., Subramanian, Y., Rollins, E. L., Starovic-Subota, O., Tang, A. C., and Childs, S. J. (2007). Spatiotemporal expression of smooth muscle markers in developing zebrafish gut. *Dev. Dyn.* **236**, 1623–1632.

Gering, M., and Patient, R. (2005). Hedgehog signaling is required for adult blood stem cell formation in zebrafish embryos. *Dev. Cell* **8**, 389–400.

Gering, M., Rodaway, AR, Gottgens, B, Patient, RK, and Green, AR. (1998). The SCL gene specifies haemangioblast development from early mesoderm. *EMBO J.* **17**(14), 4029–4045.

Gering, M., Yamada, Y., Rabbitts, T. H., and Patient, R. K. (2003). Lmo2 and Scl/Tal1 convert non-axial mesoderm into haemangioblasts which differentiate into endothelial cells in the absence of Gata1. *Development* **130**, 6187–6199.

Gore, A. V., Lampugnani, M. G., Dye, L., Dejana, E., and Weinstein, B. M. (2008). Combinatorial interaction between CCM pathway genes precipitates hemorrhagic stroke. *Dis. Model Mech.* **1**, 275–281.

Granato, M., van Eeden, F. J., Schach, U., Trowe, T., Brand, M., Furutani-Seiki, M., Haffter, P., Hammerschmidt, M., Heisenberg, C. P., Jiang, Y. J., et al. (1996). Genes controlling and mediating locomotion behavior of the zebrafish embryo and larva. *Development* **123**, 399–413.

Grillitsch, S., Medgyesy, N., Schwerte, T., and Pelster, B. (2005). The influence of environmental PO2 on hemoglobin oxygen saturation in developing zebrafish Danio rerio. *J. Exp. Biol.* **208**, 309–316.

Gupta, S., Zhu, H., Zon, L. I., and Evans, T. (2006). BMP signaling restricts hemato-vascular development from lateral mesoderm during somitogenesis. *Development* **133**, 2177–2187.

Haffter, P., Granato, M., Brand, M., Mullins, M. C., Hammerschmidt, M., Kane, D. A., Odenthal, J., van Eeden, F. J., Jiang, Y. J., Heisenberg, C. P., et al. (1996). The identification of genes with unique and essential functions in the development of the zebrafish, Danio rerio. *Development* **123**, 1–36.

Haldi, M., Ton, C., Seng, W. L., and McGrath, P. (2006). Human melanoma cells transplanted into zebrafish proliferate, migrate, produce melanin, form masses and stimulate angiogenesis in zebrafish. *Angiogenesis* **9**, 139–151.

Hassel, D., Scholz, E. P., Trano, N., Friedrich, O., Just, S., Meder, B., Weiss, D. L., Zitron, E., Marquart, S., Vogel, B., et al. (2008). Deficient zebrafish ether-a-go-go-related gene channel gating causes short-QT syndrome in zebrafish reggae mutants. *Circulation* **117**, 866–875.

Herpers, R., van de Kamp, E., Duckers, H. J., and Schulte-Merker, S. (2008). Redundant roles for sox7 and sox18 in arteriovenous specification in zebrafish. *Circ. Res.* **102**, 12–15.

Ho, R. K., and Kane, D. A. (1990). Cell-autonomous action of zebrafish spt-1 mutation in specific mesodermal precursors. *Nature* **348**, 728–730.

Hogan, B. M., Bos, F. L., Bussmann, J., Witte, M., Chi, N. C., Duckers, H. J., and Schulte-Merker, S. (2009). Ccbe1 is required for embryonic lymphangiogenesis and venous sprouting. *Nat. Genet.* **41**, 396–398.

Hogan, B. M., Bussmann, J., Wolburg, H., and Schulte-Merker, S. (2008). ccm1 cell autonomously regulates endothelial cellular morphogenesis and vascular tubulogenesis in zebrafish. *Hum. Mol. Genet.* **17**, 2424–2432.

Holtzman, N. G., Schoenebeck, J. J., Tsai, H. J., and Yelon, D. (2007). Endocardium is necessary for cardiomyocyte movement during heart tube assembly. *Development* **134**, 2379–2386.

Hong, C. C., Peterson, Q. P., Hong, J. Y., and Peterson, R. T. (2006). Artery/vein specification is governed by opposing phosphatidylinositol-3 kinase and MAP kinase/ERK signaling. *Curr. Biol.* **16**, 1366–1372.

Horne-Badovinac, S., Lin, D., Waldron, S., Schwarz, M., Mbamalu, G., Pawson, T., Jan, Y., Stainier, D. Y., and Abdelilah-Seyfried, S. (2001). Positional cloning of heart and soul reveals multiple roles for PKC lambda in zebrafish organogenesis. *Curr. Biol.* **11**, 1492–1502.

Hove, J. R., Koster, R. W., Forouhar, A. S., Acevedo-Bolton, G., Fraser, S. E., and Gharib, M. (2003). Intracardiac fluid forces are an essential epigenetic factor for embryonic cardiogenesis. *Nature* **421**, 172–177.

Hu, N., Sedmera, D., Yost, H. J., and Clark, E. B. (2000). Structure and function of the developing zebrafish heart. *Anat. Rec.* **260**, 148–157.

Hu, N., Yost, H. J., and Clark, E. B. (2001). Cardiac morphology and blood pressure in the adult zebrafish. *Anat. Rec.* **264**, 1–12.

Huang, C. C., Lawson, N. D., Weinstein, B. M., and Johnson, S. L. (2003). reg6 is required for branching morphogenesis during blood vessel regeneration in zebrafish caudal fins. *Dev. Biol.* **264**, 263–274.

Huang, C. C., Huang, C. W., Cheng, Y. S., and Yu, J. (2008). Histamine metabolism influences blood vessel branching in zebrafish *reg6* mutants. *BMC Dev. Biol.* **8**, 31.

Hukriede, N., Fisher, D., Epstein, J., Joly, L., Tellis, P., Zhou, Y., Barbazuk, B., Cox, K., Fenton-Noriega, L., Hersey, C., et al. (2001). The LN54 radiation hybrid map of zebrafish expressed sequences. *Genome Res.* **11**, 2127–2132.

Hukriede, N. A., Joly, L., Tsang, M., Miles, J., Tellis, P., Epstein, J. A., Barbazuk, W. B., Li, F. N., Paw, B., Postlethwait, J. H., et al. (1999). Radiation hybrid mapping of the zebrafish genome. *Proc. Natl Acad. Sci. USA* **96**, 9745–9750.

Isogai, S., Horiguchi, M., and Weinstein, B. M. (2001). The vascular anatomy of the developing zebrafish: An atlas of embryonic and early larval development. *Dev. Biol.* **230**, 278–301.

Isogai, S., Lawson, N. D., Torrealday, S., Horiguchi, M., and Weinstein, B. M. (2003). Angiogenic network formation in the developing vertebrate trunk. *Development* **130**, 5281–5290.

Jin, H., Xu, J., and Wen, Z. (2007). Migratory path of definitive hematopoietic stem/progenitor cells during zebrafish development. *Blood* **109**, 5208–5214.

Jin, S. W., Beis, D., Mitchell, T., Chen, J. N., and Stainier, D. Y. (2005). Cellular and molecular analyses of vascular tube and lumen formation in zebrafish. *Development* **132**, 5199–5209.

Kalen, M., Wallgard, E., Asker, N., Nasevicius, A., Athley, E., Billgren, E., Larson, J. D., Wadman, S. A., Norseng, E., Clark, K. J., et al. (2009). Combination of reverse and chemical genetic screens reveals angiogenesis inhibitors and targets. *Chem. Biol.* **16**, 432–441.

Kamei, M., Saunders, W. B., Bayless, K. J., Dye, L., Davis, G. E., and Weinstein, B. M. (2006). Endothelial tubes assemble from intracellular vacuoles in vivo. *Nature* **442**, 453–456.

Kang, P. M., and Izumo, S. (2000). Apoptosis and heart failure: A critical review of the literature. *Circ. Res.* **86**, 1107–1113.

Kelly, P. D., Chu, F., Woods, I. G., Ngo-Hazelett, P., Cardozo, T., Huang, H., Kimm, F., Liao, L., Yan, Y. L., Zhou, Y., et al. (2000). Genetic linkage mapping of zebrafish genes and ESTs. *Genome Res.* **10**, 558–567.

Kikuchi, Y., Trinh, L. A., Reiter, J. F., Alexander, J., Yelon, D., and Stainier, D. Y. (2000). The zebrafish bonnie and clyde gene encodes a Mix family homeodomain protein that regulates the generation of endodermal precursors. *Genes Dev.* **14**, 1279–1289.

Kleaveland, B., Zheng, X., Liu, J. J., Blum, Y., Tung, J. J., Zou, Z., Chen, M., Guo, L., Lu, M. M., Zhou, D., et al. (2009). Regulation of cardiovascular development and integrity by the heart of glass-cerebral cavernous malformation protein pathway. *Nat. Med.* **15**, 169–176.

Knapik, E. W., Goodman, A., Atkinson, O. S., Roberts, C. T., Shiozawa, M., Sim, C. U., Weksler-Zangen, S., Trolliet, M. R., Futrell, C., Innes, B. A., et al. (1996). A reference cross DNA panel for zebrafish (Danio rerio) anchored with simple sequence length polymorphisms. *Development* **123**, 451–460.

Knapik, E. W., Goodman, A., Ekker, M., Chevrette, M., Delgado, J., Neuhauss, S., Shimoda, N., Driever, W., Fishman, M. C., and Jacob, H. J. (1998). A microsatellite genetic linkage map for zebrafish (Danio rerio). *Nat. Genet.* **18**, 338–343.

Kopp, R., Pelster, B., and Schwerte, T. (2007). How does blood cell concentration modulate cardiovascular parameters in developing zebrafish (Danio rerio)? *Comp. Biochem. Physiol. A Mol. Integr. Physiol.* **146**, 400–407.

Kopp, R., Schwerte, T., and Pelster, B. (2005). Cardiac performance in the zebrafish breakdance mutant. *J. Exp. Biol.* **208**, 2123–2134.

Kozlowski, D. J., Murakami, T., Ho, R. K., and Weinberg, E. S. (1997). Regional cell movement and tissue patterning in the zebrafish embryo revealed by fate mapping with caged fluorescein. *Biochem. Cell Biol.* **75**, 551–562.

Kramer-Zucker, A. G., Olale, F., Haycraft, C. J., Yoder, B. K., Schier, A. F., and Drummond, I. A. (2005). Cilia-driven fluid flow in the zebrafish pronephros, brain and Kupffer's vesicle is required for normal organogenesis. *Development* **132**, 1907–1921.

Kuchler, A. M., Gjini, E., Peterson-Maduro, J., Cancilla, B., Wolburg, H., and Schulte-Merker, S. (2006). Development of the zebrafish lymphatic system requires VEGFC signaling. *Curr. Biol.* **16**, 1244–1248.

Kudoh, T., Tsang, M., Hukriede, N. A., Chen, X., Dedekian, M., Clarke, C. J., Kiang, A., Schultz, S., Epstein, J. A., Toyama, R., et al. (2001). A gene expression screen in zebrafish embryogenesis. *Genome Res.* **11**, 1979–1987.

Kupperman, E., An, S., Osborne, N., Waldron, S., and Stainier, D. Y. (2000). A sphingosine-1-phosphate receptor regulates cell migration during vertebrate heart development. *Nature* **406**, 192–195.

Langenbacher, A. D., Dong, Y., Shu, X., Choi, J., Nicoll, D. A., Goldhaber, J. I., Philipson, K. D., and Chen, J. N. (2005). Mutation in sodium-calcium exchanger 1 (NCX1) causes cardiac fibrillation in zebrafish. *Proc. Natl Acad. Sci. USA* **102**, 17699–17704.

Langheinrich, U., Vacun, G., and Wagner, T. (2003). Zebrafish embryos express an orthologue of HERG and are sensitive toward a range of QT-prolonging drugs inducing severe arrhythmia. *Toxicol. Appl. Pharmacol.* **193**, 370–382.

Lawson, N. D., Mugford, J. W., Diamond, B. A., and Weinstein, B. M. (2003). phospholipase C gamma-1 is required downstream of vascular endothelial growth factor during arterial development. *Genes Dev.* **17**, 1346–1351.

Lawson, N. D., Scheer, N., Pham, V. N., Kim, C. H., Chitnis, A. B., Campos-Ortega, J. A., and Weinstein, B. M. (2001). Notch signaling is required for arterial-venous differentiation during embryonic vascular development. *Development* **128**, 3675–3683.

Lawson, N. D., Vogel, A. M., and Weinstein, B. M. (2002). sonic hedgehog and vascular endothelial growth factor act upstream of the Notch pathway during arterial endothelial differentiation. *Dev. Cell* **3**, 127–136.

Lawson, N. D., and Weinstein, B. M. (2002). In vivo imaging of embryonic vascular development using transgenic zebrafish. *Dev. Biol.* **248**, 307–318.

Lee, R. K., Stainier, D. Y., Weinstein, B. M., and Fishman, M. C. (1994). Cardiovascular development in the zebrafish. II. Endocardial progenitors are sequestered within the heart field. *Development* **120**, 3361–3366.

Lee, S., Chen, T. T., Barber, C. L., Jordan, M. C., Murdock, J., Desai, S., Ferrara, N., Nagy, A., Roos, K. P., and Iruela-Arispe, M. L. (2007). Autocrine VEGF signaling is required for vascular homeostasis. *Cell* **130**, 691–703.

Leslie, J. D., Ariza-McNaughton, L., Bermange, A. L., McAdow, R., Johnson, S. L., and Lewis, J. (2007). Endothelial signaling by the Notch ligand Delta-like 4 restricts angiogenesis. *Development* **134**, 839–844.

Liao, B. K., Deng, A. N., Chen, S. C., Chou, M. Y., and Hwang, P. P. (2007). Expression and water calcium dependence of calcium transporter isoforms in zebrafish gill mitochondrion-rich cells. *BMC Genomics.* **8**, 354.

Liao, E., Paw, BH, Oates, AC, Pratt, SJ, Postlethwait, JH, and Zon, LI. (1998). SCL/Tal-1 transcription factor acts downstream of *cloche* to specify hematopoietic and vascular progenitors in zebrafish. *Genes Dev.* **12**, 621–626.

Liao, W., Ho, C., Yan, Y. L., Postlethwait, J., and Stainier, D. Y. (2000). Hhex and Scl function in parallel to regulate early endothelial and blood differentiation in zebrafish. *Development* **127**, 4303–4313.

Liu, F., Walmsley, M., Rodaway, A., and Patient, R. (2008). Fli1 acts at the top of the transcriptional network driving blood and endothelial development. *Curr. Biol.* **18**, 1234–1240.

Liu, J., Fraser, S. D., Faloon, P. W., Rollins, E. L., Vom Berg, J., Starovic-Subota, O., Laliberte, A. L., Chen, J. N., Serluca, F. C., and Childs, S. J. (2007). A ®Pix-Pak2a signaling pathway regulates cerebral vascular stability in zebrafish. *Proc. Natl Acad. Sci. USA* **104**, 13990–13995.

Mably, J. D., Chuang, L. P., Serluca, F. C., Mohideen, M. A., Chen, J. N., and Fishman, M. C. (2006). santa and valentine pattern concentric growth of cardiac myocardium in the zebrafish. *Development* **133**, 3139–3146.

Mably, J. D., Mohideen, M. A., Burns, C. G., Chen, J. N., and Fishman, M. C. (2003). heart of glass regulates the concentric growth of the heart in zebrafish. *Curr. Biol.* **13**, 2138–2147.

Majumdar, A., and Drummond, I. A. (1999). Podocyte differentiation in the absence of endothelial cells as revealed in the zebrafish avascular mutant, cloche. *Dev. Genet.* **24**, 220–229.

Malone, M. H., Sciaky, N., Stalheim, L., Hahn, K. M., Linney, E., and Johnson, G. L. (2007). Laser-scanning velocimetry: A confocal microscopy method for quantitative measurement of cardiovascular performance in zebrafish embryos and larvae. *BMC Biotechnol.* **7**, 40.

Marchuk, D. A., Srinivasan, S., Squire, T. L., and Zawistowski, J. S. (2003). Vascular morphogenesis: Tales of two syndromes. *Hum. Mol. Genet.* **12**(Spec No 1), R97–R112.

Meder, B., Laufer, C., Hassel, D., Just, S., Marquart, S., Vogel, B., Hess, A., Fishman, M. C., Katus, H. A., and Rottbauer, W. (2009). A single serine in the carboxyl terminus of cardiac essential myosin light chain-1 controls cardiomyocyte contractility in vivo. *Circ. Res.* **104**, 650–659.

Meng, X., Noyes, M. B., Zhu, L. J., Lawson, N. D., and Wolfe, S. A. (2008). Targeted gene inactivation in zebrafish using engineered zinc-finger nucleases. *Nat. Biotechnol.* **26**, 695–701.

Miao, M., Bruce, A. E., Bhanji, T., Davis, E. C., and Keeley, F. W. (2007). Differential expression of two tropoelastin genes in zebrafish. *Matrix Biol.* **26**, 115–124.

Milan, D. J., Giokas, A. C., Serluca, F. C., Peterson, R. T., and MacRae, C. A. (2006). Notch1b and neuregulin are required for specification of central cardiac conduction tissue. *Development* **133**, 1125–1132.

Milan, D. J., Peterson, T. A., Ruskin, J. N., Peterson, R. T., and MacRae, C. A. (2003). Drugs that induce repolarization abnormalities cause bradycardia in zebrafish. *Circulation* **107**, 1355–1358.

Molina, G., Vogt, A., Bakan, A., Dai, W., de Oliveira, P. Q., Znosko, W., Smithgall, T. E., Bahar, I., Lazo, J. S., Day, B. W., et al. (2009). Zebrafish chemical screening reveals an inhibitor of Dusp6 that expands cardiac cell lineages. *Nat. Chem. Biol.* **5**(9), 680–687.

Moorman, A. F., Vermeulen, J. L., Koban, M. U., Schwartz, K., Lamers, W. H., and Boheler, K. R. (1995). Patterns of expression of sarcoplasmic reticulum Ca(2+)-ATPase and phospholamban mRNAs during rat heart development. *Circ. Res.* **76**, 616–625.

Mudumana, S. P., Hentschel, D., Liu, Y., Vasilyev, A., and Drummond, I. A. (2008). odd skipped related1 reveals a novel role for endoderm in regulating kidney versus vascular cell fate. *Development* **135**, 3355–3367.

Mukhopadhyay, A., and Peterson, R. T. (2008). Deciphering arterial identity through gene expression, genetics, and chemical biology. *Curr. Opin. Hematol.* **15**, 221–227.

Nasevicius, A., and Ekker, S. C. (2000). Effective targeted gene 'knockdown' in zebrafish. *Nat. Genet.* **26**, 216–220.

Nicoli, S., Ribatti, D., Cotelli, F., and Presta, M. (2007). Mammalian tumor xenografts induce neovascularization in zebrafish embryos. *Cancer Res.* **67**, 2927–2931.

Nicoli, S., Tobia, C., Gualandi, L., De Sena, G., and Presta, M. (2008). Calcitonin receptor-like receptor guides arterial differentiation in zebrafish. *Blood* **111**, 4965–4972.

North, T. E., Goessling, W., Peeters, M., Li, P., Ceol, C., Lord, A. M., Weber, G. J., Harris, J., Cutting, C. C., Huang, P., et al. (2009). Hematopoietic stem cell development is dependent on blood flow. *Cell* **137**, 736–748.

Parker, L. H., Schmidt, M., Jin, S. W., Gray, A. M., Beis, D., Pham, T., Frantz, G., Palmieri, S., Hillan, K., Stainier, D. Y., et al. (2004). The endothelial-cell-derived secreted factor Egfl7 regulates vascular tube formation. *Nature* **428**, 754–758.

Patterson, L. J., Gering, M., Eckfeldt, C. E., Green, A. R., Verfaillie, C. M., Ekker, S. C., and Patient, R. (2007). The transcription factors Scl and Lmo2 act together during development of the hemangioblast in zebrafish. *Blood* **109**, 2389–2398.

Pelster, B., and Burggren, W. W. (1996). Disruption of hemoglobin oxygen transport does not impact oxygen-dependent physiological processes in developing embryos of zebra fish (Danio rerio). *Circ. Res.* **79**, 358–362.

Pelster, B., Grillitsch, S., and Schwerte, T. (2005). NO as a mediator during the early development of the cardiovascular system in the zebrafish. *Comp. Biochem. Physiol. A Mol. Integr. Physiol.* **142**, 215–220.

Pelster, B., Sanger, A. M., Siegele, M., and Schwerte, T. (2003). Influence of swim training on cardiac activity, tissue capillarization, and mitochondrial density in muscle tissue of zebrafish larvae. *Am. J. Physiol. Regul. Integr. Comp. Physiol.* **285**, R339–R347.

Pendeville, H., Winandy, M., Manfroid, I., Nivelles, O., Motte, P., Pasque, V., Peers, B., Struman, I., Martial, J. A., and Voz, M. L. (2008). Zebrafish Sox7 and Sox18 function together to control arterial-venous identity. *Dev. Biol.* **317**, 405–416.

Peterson, R. T., Link, B. A., Dowling, J. E., and Schreiber, S. L. (2000). Small molecule developmental screens reveal the logic and timing of vertebrate development. *Proc. Natl Acad. Sci. USA* **97**, 12965–12969.

Peterson, R. T., Mably, J. D., Chen, J. N., and Fishman, M. C. (2001). Convergence of distinct pathways to heart patterning revealed by the small molecule concentramide and the mutation heart-and-soul. *Curr. Biol.* **11**, 1481–1491.

Peterson, R. T., Shaw, S. Y., Peterson, T. A., Milan, D. J., Zhong, T. P., Schreiber, S. L., MacRae, C. A., and Fishman, M. C. (2004). Chemical suppression of a genetic mutation in a zebrafish model of aortic coarctation. *Nat. Biotechnol.* **22**, 595–599.

Pham, V. N., Lawson, N. D., Mugford, J. W., Dye, L., Castranova, D., Lo, B., and Weinstein, B. M. (2007). Combinatorial function of ETS transcription factors in the developing vasculature. *Dev. Biol.* **303**, 772–783.

Pollard, S. M., Parsons, M. J., Kamei, M., Kettleborough, R. N., Thomas, K. A., Pham, V. N., Bae, M. K., Scott, A., Weinstein, B. M., and Stemple, D. L. (2006). Essential and overlapping roles for laminin alpha chains in notochord and blood vessel formation. *Dev. Biol.* **289**, 64–76.

Poss, K. D., Wilson, L. G., and Keating, M. T. (2002). Heart regeneration in zebrafish. *Science* **298**, 2188–2190.

Revencu, N., and Vikkula, M. (2006). Cerebral cavernous malformation: New molecular and clinical insights. *J. Med. Genet.* **43**, 716–721.

Roman, B. L., Pham, V. N., Lawson, N. D., Kulik, M., Childs, S., Lekven, A. C., Garrity, D. M., Moon, R. T., Fishman, M. C., Lechleider, R. J., et al. (2002). Disruption of acvrl1 increases endothelial cell number in zebrafish cranial vessels. *Development* **129**, 3009–3019.

Rottbauer, W., Baker, K., Wo, Z. G., Mohideen, M. A., Cantiello, H. F., and Fishman, M. C. (2001). Growth and function of the embryonic heart depend upon the cardiac-specific L-type calcium channel alpha1 subunit. *Dev. Cell* **1**, 265–275.

Rottbauer, W., Wessels, G., Dahme, T., Just, S., Trano, N., Hassel, D., Burns, C. G., Katus, H. A., and Fishman, M. C. (2006). Cardiac myosin light chain-2: A novel essential component of thick-myofilament assembly and contractility of the heart. *Circ. Res.* **99**, 323–331.

Santoro, M. M., Pesce, G., and Stainier, D. Y. (2009). Characterization of vascular mural cells during zebrafish development. *Mech. Dev.* **126**(8–9), 638–649.

Santoro, M. M., Samuel, T., Mitchell, T., Reed, J. C., and Stainier, D. Y. (2007). Birc2 (cIap1) regulates endothelial cell integrity and blood vessel homeostasis. *Nat. Genet.* **39**, 1397–1402.

Schoenebeck, J. J., Keegan, B. R., and Yelon, D. (2007). Vessel and blood specification override cardiac potential in anterior mesoderm. *Dev. Cell* **13**, 254–267.

Scholz, E. P., Niemer, N., Hassel, D., Zitron, E., Burgers, H. F., Bloehs, R., Seyler, C., Scherer, D., Thomas, D., Kathofer, S., et al. (2009). Biophysical properties of zebrafish ether-a-go-go related gene potassium channels. *Biochem. Biophys. Res. Commun.* **381**, 159–164.

Schwerte, T., and Pelster, B. (2000). Digital motion analysis as a tool for analysing the shape and performance of the circulatory system in transparent animals. *J. Exp. Biol.* **203**(Pt 11), 1659–1669.

Schwerte, T., Uberbacher, D., and Pelster, B. (2003). Non-invasive imaging of blood cell concentration and blood distribution in zebrafish Danio rerio incubated in hypoxic conditions in vivo. *J. Exp. Biol.* **206**, 1299–1307.

Sehnert, A. J., Huq, A., Weinstein, B. M., Walker, C., Fishman, M., and Stainier, D. Y. (2002). Cardiac troponin T is essential in sarcomere assembly and cardiac contractility. *Nat. Genet.* **31**, 106–110.

Semina, E. V., Bosenko, D. V., Zinkevich, N. C., Soules, K. A., Hyde, D. R., Vihtelic, T. S., Willer, G. B., Gregg, R. G., and Link, B. A. (2006). Mutations in laminin alpha 1 result in complex, lens-independent ocular phenotypes in zebrafish. *Dev. Biol.* **299**, 63–77.

Serbedzija, G., Chen, J-N, and Fishman, MC. (1998). Regulation in the heart field of zebrafish. *Development* **125**, 1095–1101.

Serluca, F. C., Drummond, I. A., and Fishman, M. C. (2002). Endothelial signaling in kidney morphogenesis: A role for hemodynamic forces. *Curr. Biol.* **12**, 492–497.

Serluca, F. C., and Fishman, M. C. (1999). Cell lineage tracing in heart development. *Methods Cell Biol.* **59**, 359–365.

Siekmann, A. F., and Lawson, N. D. (2007). Notch signaling limits angiogenic cell behaviour in developing zebrafish arteries. *Nature* **445**, 781–784.

Slough, J., Cooney, L., and Brueckner, M. (2008). Monocilia in the embryonic mouse heart suggest a direct role for cilia in cardiac morphogenesis. *Dev. Dyn.* **237**, 2304–2314.

Stainier, D. Y. (2001). Zebrafish genetics and vertebrate heart formation. *Nat. Rev. Genet.* **2**, 39–48.

Stainier, D. Y., and Fishman, M. C. (1992). Patterning the zebrafish heart tube: Acquisition of anteroposterior polarity. *Dev. Biol.* **153**, 91–101.

Stainier, D. Y., Fouquet, B., Chen, J. N., Warren, K. S., Weinstein, B. M., Meiler, S. E., Mohideen, M. A., Neuhauss, S. C., Solnica-Krezel, L., Schier, A. F., et al. (1996). Mutations affecting the formation and function of the cardiovascular system in the zebrafish embryo. *Development* **123**, 285–292.

Stainier, D. Y., Lee, R. K., and Fishman, M. C. (1993). Cardiovascular development in the zebrafish. I. Myocardial fate map and heart tube formation. *Development* **119**, 31–40.

Stainier, D. Y., Weinstein, B. M., Detrich, H. W., III, Zon, L. I., and Fishman, M. C. (1995). Cloche, an early acting zebrafish gene, is required by both the endothelial and hematopoietic lineages. *Development* **121**, 3141–3150.

Stoletov, K., Montel, V., Lester, R. D., Gonias, S. L., and Klemke, R. (2007). High-resolution imaging of the dynamic tumor cell vascular interface in transparent zebrafish. *Proc. Natl Acad. Sci. USA* **104**, 17406–17411.

Sumanas, S., Gomez, G., Zhao, Y., Park, C., Choi, K., and Lin, S. (2008). Interplay among Etsrp/ER71, Scl, and Alk8 signaling controls endothelial and myeloid cell formation. *Blood* **111**, 4500–4510.

Sumanas, S., and Lin, S. (2006). Ets1-related protein is a key regulator of vasculogenesis in zebrafish. *PLoS Biol.* **4**, e10.

Ton, C., Hwang, D. M., Dempsey, A. A., Tang, H. C., Yoon, J., Lim, M., Mably, J. D., Fishman, M. C., and Liew, C. C. (2000). Identification, characterization, and mapping of expressed sequence tags from an embryonic zebrafish heart cDNA library. *Genome Res.* **10**, 1915–1927.

Torres-Vazquez, J., Gitler, A. D., Fraser, S. D., Berk, J. D., Van, N. P., Fishman, M. C., Childs, S., Epstein, J. A., and Weinstein, B. M. (2004). Semaphorin-plexin signaling guides patterning of the developing vasculature. *Dev. Cell* **7**, 117–123.

Van der Heiden, K., Hierck, B. P., Krams, R., de Crom, R., Cheng, C., Baiker, M., Pourquie, M. J., Alkemade, F. E., DeRuiter, M. C., Gittenberger-de Groot, A. C., et al. (2008).

Endothelial primary cilia in areas of disturbed flow are at the base of atherosclerosis. *Atherosclerosis* **196**, 542–550.

Vogel, G. (2000). Genomics. Sanger will sequence zebrafish genome. *Science* **290**, 1671.

Vogeli, K. M., Jin, S. W., Martin, G. R., and Stainier, D. Y. (2006). A common progenitor for haematopoietic and endothelial lineages in the zebrafish gastrula. *Nature* **443**, 337–339.

Wallace, K. N., Dolan, A. C., Seiler, C., Smith, E. M., Yusuff, S., Chaille-Arnold, L., Judson, B., Sierk, R., Yengo, C., Sweeney, H. L., et al. (2005). Mutation of smooth muscle myosin causes epithelial invasion and cystic expansion of the zebrafish intestine. *Dev. Cell* **8**, 717–726.

Wang, S., Aurora, A. B., Johnson, B. A., Qi, X., McAnally, J., Hill, J. A., Richardson, J. A., Bassel-Duby, R., and Olson, E. N. (2008). The endothelial-specific microRNA miR-126 governs vascular integrity and angiogenesis. *Dev. Cell* **15**, 261–271.

Wienholds, E., Schulte-Merker, S., Walderich, B., and Plasterk, R. H. (2002). Target-selected inactivation of the zebrafish rag1 gene. *Science* **297**, 99–102.

Wienholds, E., van Eeden, F., Kosters, M., Mudde, J., Plasterk, R. H., and Cuppen, E. (2003). Efficient target-selected mutagenesis in zebrafish. *Genome Res.* **13**, 2700–2707.

Wilkinson, R. N., Pouget, C., Gering, M., Russell, A. J., Davies, S. G., Kimelman, D., and Patient, R. (2009). Hedgehog and Bmp polarize hematopoietic stem cell emergence in the zebrafish dorsal aorta. *Dev. Cell* **16**, 909–916.

Wills, A. A., Holdway, J. E., Major, R. J., and Poss, K. D. (2008). Regulated addition of new myocardial and epicardial cells fosters homeostatic cardiac growth and maintenance in adult zebrafish. *Development* **135**, 183–192.

Wilson, B. D., Ii, M., Park, K. W., Suli, A., Sorensen, L. K., Larrieu-Lahargue, F., Urness, L. D., Suh, W., Asai, J., Kock, G. A., et al. (2006). Netrins promote developmental and therapeutic angiogenesis. *Science* **313**, 640–644.

Yaniv, K., Isogai, S., Castranova, D., Dye, L., Hitomi, J., and Weinstein, B. M. (2006). Live imaging of lymphatic development in the zebrafish. *Nat. Med.* **12**, 711–716.

Yelon, D., Horne, S. A., and Stainier, D. Y. (1999). Restricted expression of cardiac myosin genes reveals regulated aspects of heart tube assembly in zebrafish. *Dev. Biol.* **214**, 23–37.

Zhong, T. P., Childs, S., Leu, J. P., and Fishman, M. C. (2001). Gridlock signaling pathway fashions the first embryonic artery. *Nature* **414**, 216–220.

Zhong, T. P., Rosenberg, M., Mohideen, M. A., Weinstein, B., and Fishman, M. C. (2000). gridlock, an HLH gene required for assembly of the aorta in zebrafish. *Science* **287**, 1820–1824.

# 7

# RESPIRATION

*BERND PELSTER*
*BRIAN BAGATTO*



The general principles of gas exchange established for adult fish apply also to embryonic and larval gas exchange. Nevertheless, larval gas exchange is not just adult gas exchange conducted at a smaller scale. Embryonic and early larval gas exchange is cutaneous gas exchange, which can significantly be impaired by unstirred boundary layers. Blood flow achieved by the early onset of cardiac activity is not required for convective oxygen transport; oxygen transport is achieved by bulk diffusion. In the zebrafish, gills develop about two weeks after fertilization and the initial primary function may be ion regulation, and not gas exchange. Accordingly, at this stage of development metabolic activity and cardiac activity are not yet coupled. This connection, which includes an allometric relationship between oxygen consumption, gas exchange area and circulation, is established only in later stages. In spite of the fact that gas exchange is achieved by bulk diffusion, ventilatory as well as cardiac activity is modified in the face of changing oxygen availability even in very early stages. In addition, temperature changes typically provoke severe changes in metabolic rate, probably because larvae are more stenothermic than adult fishes. We



*Zebrafish: Volume 29*
FISH PHYSIOLOGY


DOI: 10.1016/S1546-5098(10)02907-9

are only beginning to understand the plasticity of these systems, the cooperative action of genetic and epigenetic factors on the development of form and function of the ventilatory and cardiovascular system, and the regulatory events behind the adjustments to environmental perturbations. A significant fraction of our current knowledge on these processes has been achieved using the zebrafish, and it promises to be an asset for future research.

## 1. INTRODUCTION

Fish respiratory gas transport has been a research topic for decades. While most studies concentrate on oxygen exchange in water-breathing animals, comparatively little is known about $CO_2$ and $NH_3$ exchange. This may in part be for technical reasons ($NH_3$), but it is likely also related to the fact that both molecules react with water-forming ions which require ion transport mechanisms for the crossing of membranes. As a consequence of this reaction with water, the capacity of water for $CO_2$ and $NH_3$ is very high, facilitating gas exchange tremendously. Oxygen in turn is only transported in physical solution and therefore the oxygen-carrying capacity of water is very low. Accordingly, for water-breathing animals oxygen exchange is much more difficult than $CO_2$ or $NH_3$ exchange.

In most adult fishes, the main organ for the exchange of all three gases is the gills. Additionally, gills are responsible for ion and acid–base balance (Evans et al. 1999, 2005), nitrogenous waste excretion (Randall et al. 1999; Terjesen 2008), hormone production (Zaccone et al. 1996, 2006), and activation or inactivation of circulating metabolites (Olson 1998). Although the circulatory system is typically the first functioning organ during early embryonic development (Rombough and Moroz 1997; Pelster 1999, 2002; Burggren and Bagatto 2008; Schwerte 2009), gills develop much later (Rombough 2002, 2004, 2007). Gas exchange, however, is required from the onset of fertilization. This implies that gas exchange initially must occur through the cellular integument of the developing embryo, having traversed the boundary layer, the chorion and perivitelline fluids (PVF) of the embryo, or the surface epithelium of the hatched larva. The gills can only take over this function once they have developed.

Several studies, ranging from the development of respiratory surfaces to the development of respiratory control and to the effects of hypoxia, for example, have addressed gas exchange in developing zebrafish. This chapter provides a synopsis of these studies, generating an overview of our current knowledge of gas exchange in developing zebrafish. Zebrafish represents a typical member of the cyprinid family and information from other cyprinids or from salmonids will be added occasionally to generate a coherent picture. Finally, possible future directions for research in this area will be highlighted.

## 2. THE PRINCIPLES OF GAS EXCHANGE

Aerobic energy metabolism is dependent on an adequate supply of oxygen, which is required for mitochondrial cytochrome aa3 activity, and on the removal of $CO_2$, which is mainly produced in the citric acid cycle. Consequently, respiratory gas exchange comprises the transport of oxygen and $CO_2$ between the respiratory medium, i.e. the water, and the mitochondria. It involves convectional transport of oxygen and $CO_2$, but also the diffusion of gas molecules. The first step in gas exchange is convective transport via the ventilatory water current. Uptake of oxygen from the water into the organism at the gas exchange areas is achieved by diffusion, which is ultimately driven by the partial pressure gradients between the environmental water and the blood. Inside the blood, convection is responsible for the transport between the gas exchange area and the tissues. Once again, diffusion is required for the transport between capillaries and tissue mitochondria. This transport chain can be visualized as a cascade (Dejours 1981). For water-breathing animals, oxygen transport is much more critical than $CO_2$ transport due to the formation of bicarbonate in water and the resulting high transport capacity for aqueous $CO_2$, which is indeed almost similar to the $CO_2$ transport capacity of air (Dejours 1981).

Considering the decrease in $Po_2$ along the transport cascade, it is apparent that diffusion is the critical process. The diffusive flux of gas across a gas exchanger (either the gas exchange area and the blood or the exchange between capillaries and mitochondria) is described by the Fick equation: $J = K \cdot \Delta p \cdot A/L$, where $J$ is the rate of gas transfer, K represents Krogh's diffusion constant, A is the area available for the gas transfer, L is the length of the pathway that has to be surpassed by diffusion, and $\Delta p$ is the partial pressure gradient along the diffusional pathway.

Krogh's diffusion constant comprises the diffusion coefficient (D) for the gas analyzed, and also the gas solubility ($\alpha$), or, if chemical binding is involved as for $CO_2$ and $O_2$, the transport capacity of the gas ($\beta$) in water or in body fluids: $K = D \cdot \beta$.

Krogh's diffusion constant is dependent on temperature, because an increase in temperature increases Brownian motion and thus the diffusion coefficient. Increasing temperature, however, decreases the solubility of gases in aqueous solutions. As a consequence, Krogh's diffusion constant increases only by a factor of 1.1 for a temperature increase of 10°C (Dejours

1981). Enzymatic reactions, however, typically are characterized by a $Q_{10}$ of about two (i.e., an increase in temperature of 10°C causes a doubling in enzymatic activity). During early development the $Q_{10}$ for some enzymatic reactions may significantly exceed a value of 2 (see below).

While the value for Krogh's diffusion constant is similar in fish gills and in the skin, $K_{O_2}$ for water is about 2.5 times higher than $K_{O_2}$ for tissues (Rombough 2004). Kranenbarg et al. (2000) assume that the diffusion coefficient in tissues is about 33% of the value measured in water. The morphology of the gas exchanger (i.e., its thickness and the area available for the exchange of gases) is crucial for the effectiveness of the gas exchanger. A large surface area combined with a short diffusion distance is the hallmark characteristic of an effective gas exchanger.

Convective transport of gases is achieved by ventilation and by blood flow. In quantitative terms, the amount of gas taken up from the ventilatory water current must be transported by the bloodstream. Ideally, ventilation–perfusion ratios are balanced, so that the amount of gas brought in a certain volume of water to one side of the exchanger is carried away in a similar volume of blood on the other side. Due to the difference in the binding capacity for $CO_2$ and $O_2$ in water, this ideal situation cannot be achieved for both gases at the same time.

Considering embryonic and larval gas exchange, convective gas transport in the blood can only occur after the establishment of blood flow in the circulatory system. As already mentioned, the circulatory system in most larvae (including zebrafish larvae) becomes important for gas transport only in later developmental stages so that bulk diffusion is mainly responsible for gas transport during early development (Kranenbarg et al. 2000; Territo and Altimiras 2001). Depending solely on bulk diffusion may therefore impose limitations on the gas exchange capacity of early larval stages (see below).

Measurements of $Po_2$ in the vicinity of aquatic animals revealed that unstirred boundary layers do exist near the surface of the body (Feder and Pinder 1988; Pinder and Feder 1990; Feder and Booth 1992). In these unstirred layers, a depletion of oxygen is observed due to the removal of oxygen through the body surface. This oxygen is replenished by diffusion from outer water layers, which further slows the process of oxygen transport at the integument. Unstirred layers also reduce the flux of metabolic waste products and ions that are taken up or removed from the body.

Unstirred boundary layers may be particularly disadvantageous for larval fish. Using microelectrodes, intravascular $Po_2$ was measured at various sites in anesthetized rainbow trout larvae in still and in moving water in order to identify possible regions that were especially important for cutaneous gas exchange. In unstirred water, the boundary layer accounted for 95% of the total resistance to oxygen uptake across the skin (Rombough 1992). In flowing water, intravascular $Po_2$ values were significantly elevated demonstrating that convection significantly reduced the magnitude of the unstirred boundary layer. It therefore can be expected that for larvae in free-flowing water, unstirred boundary layers are not such a problem for gas exchange. Experimental immobilization of larvae may, however, rapidly result in oxygen deficiency because the magnitude of the unstirred layers will significantly increase.

## 3. THE SITE OF GAS EXCHANGE

Gills are the main gas exchange organ in adult fish, but in embryonic and some larval stages, the gills have not yet developed. As a consequence, gas exchange of the developing embryo and even of early larval stages initially is through the egg envelope and/or cutaneous surface of the larvae. The egg envelope or chorion represents a physical barrier and provides protection for the developing fertilized egg. However, it is likely a diffusion barrier for respiratory gases. This hypothesis has been addressed and supported many times (Rombough 1988; Ciuhandu et al. 2007; Pelster 2008). $Po_2$ gradients, growth rates and the influence of hypoxia on cardiac activity analyzed in embryos of various species demonstrate that the chorion indeed represents a diffusion barrier, which does not affect (or minimally affects) oxygen supply under normoxic conditions, but may impair oxygen uptake under hypoxic conditions. Diffusion through the perivitelline fluid may be an additional impediment for gas exchange. However, frequent flipping movements of the embryo within the egg assure a mixing of the fluid and thus facilitate gas exchange between the embryo and the chorion (Pelster 2008).

Analysis of gill development in several species revealed that these organs typically form relatively late during development. In chinook salmon (*Oncorhynchus tshawytscha*) for example, at about hatching time (tissue mass ~ 0.04 g) the surface of the yolk-sac accounts for almost 60% of the total surface area, while the gills contribute only 4% (Rombough and Moroz 1990). In zebrafish larvae, the rudiments of gill filaments develop late in the hatching period as buds along the posterior walls of the four branchial arches facing the gill slits (Kimmel et al. 1995; Rombough 2007). Secondary lamellae, the main site of gas exchange in adult fish, are only detected between 12–14 dpf (Rombough 2002).

This clearly means that in zebrafish, like in other fish, gas exchange initially occurs by bulk diffusion via the body surface. A significant portion of the early developing embryo is the yolk, and the yolk-sac has been discussed as a possible site of gas exchange. This assumption is supported by a typically rich vascularization of the yolk surface. In larvae reared under hypoxic conditions, an increase in the vascularization of the yolk-sac has been observed (Garside 1959; Balon 1975; Brooke and Colby 1980), which also suggests that the yolk-sac may contribute to gas exchange. In rainbow trout, however, blood vessel density on the yolk sac had no significant influence on gas exchange (Rombough and Moroz 1997). There actually appears to be little correlation between the efficiency of cutaneous gas exchange and the number of capillaries in the skin or yolk-sac, because diffusion through the unstirred boundary layer represents the main limitation for oxygen exchange (Malvin 1988; Rombough and Moroz 1997). Further evidence shows that cutaneous gas exchange per unit surface area of adult fish is in the same range as the oxygen uptake of larval fish (Kirsch and Nonnotte 1977; Rombough and Ure 1991; Wells and Pinder 1996; Rombough 2004, 2007). The observation that hypoxic larvae increase their capillary supply to the yolk sac or the skin suggests, however, that under conditions of reduced oxygen availability an increased capillarization of cutaneous regions may augment oxygen uptake.

A comparison of surface area and the amount of oxygen taken up by the larvae of six different species of fish including carp revealed that in early developmental stages the exchange capacity of the body surfaces exceeds the metabolic requirements of the tissues (Rombough and Moroz 1997; Rombough 2004, 2007). This may also explain why in early developmental stages no correlation was observed between capillarization of the skin or yolk sac and the efficiency of cutaneous gas exchange. Because the surface area typically expands proportional to body mass$^{2/3}$ (i.e., surface area expands at a slower rate as body mass), at some point in development diffusion is no longer sufficient to meet the metabolic requirements of the growing fish. Based on metabolic rate and surface area, Rombough (2004) predicted that for zebrafish larvae cutaneous respiration should become limiting when the larvae reach a body mass of about 0.3 mg, which is about the body mass at which the secondary lamellae begin to develop.

A detailed analysis of the changes in surface area of gills of walleye larvae revealed that the key to the changes in the contribution of the various structures to total oxygen uptake appears to be the rapid expansion of the surface area of the gill lamellae (Rombough and Moroz 1997). Due to the allometric relation between surface area and body mass, the contribution of cutaneous respiration to total respiration decreases with development, even though cutaneous respiration in many fish remains important throughout their lifetime. The gill arches may also contribute to total gas exchange, but due to the relatively small surface area their contribution to total gas exchange remains limited. Once the secondary lamellae start to form, they rapidly expand and take over the main proportion of total gas exchange. This implies that at this time of development the gill surface area expands more rapidly than metabolic rate. At some point during larval or juvenile stages the excess exchange capacity of the surface areas disappears, so that the total area available for gas exchange matches the metabolic requirements of the species. This has been shown for several species (Rombough 2004, 2007), and applies to zebrafish larvae.

Measurements of oxygen consumption in yolk-sac larvae (alevins) and fry of chinook salmon (*Oncorhynchus tshawytscha*) using a two-chambered respirometer revealed that not only the increase in surface area, but also the concurrent increase in gas exchange efficiency contributes to the rapidly increasing importance of gills as the main site of gas exchange (Rombough and Ure 1991). A comparison of cutaneous and gill oxygen exchange revealed that the area-specific oxygen uptake of the skin increased only slightly during early development, while the area-specific oxygen uptake of the gills increased almost six-fold over the same developmental time period. In conclusion, the increase in branchial oxygen uptake observed with continued development reflects an increase in the branchial surface area and an increase in the efficiency of the gills as a respiratory organ. This increase in efficiency appears to be achieved by a reduction in diffusion distances.

The gill arches appear well before the gills become the main site for gas exchange. This observation has prompted the idea that these organs, well known for their multifunctionality in the adult stage, may serve other functions during early development as well. Gills are crucial for ion exchange and therefore ion homeostasis, and the early development of edema in larvae with various mutations underlines the importance of ion regulation during early development. So-called mitochondria-rich cells or ionocytes (chloride cells) are responsible for ion transport phenomena. Accordingly, ionocytes are detected on the yolk sac and the general body surface during the earliest stages of development. In Mozambique tilapia (*Oreochromis mossambicus*) (Li et al. 1995; Van der Heijden et al. 1999; Hwang et al. 1999), Japanese flounder (*Paralichthys olivaceus*) (Hiroi et al. 1999), brown trout (*Salmo trutta*) (Rojo et al. 1997; Pisam et al. 2000), and rainbow trout (Rombough 1999), large numbers of mitochondria-rich cells appear on the gills well before there is any indication of secondary lamellae. It is therefore possible that initially, gills are more important for ionoregulation than for gas exchange (Rombough 2007). This may perhaps explain the very high mass specific oxygen consumption rate of larval gill tissue (Oikawa and Itazawa 1993; Rombough 2007). Taken together, these data clearly show that initially, gas exchange is achieved by bulk diffusion through the body surface. In zebrafish, under normoxic conditions gills appear to become important for gas exchange at about 2 weeks after fertilization of the eggs, and gills must be ventilated in order to become effective gas exchangers.

## 4. CONVECTIVE TRANSPORT

### 4.1. Gill Ventilation

Using gills as the main site for respiratory gas exchange requires proper ventilation of the gill cavity. In zebrafish larvae, ventilatory movements were first detected at 3 dpf (Jonz and Nurse 2005). Starting with a few beats per minute, the frequency increases to about 50 beats per minute at 7 and 9 dpf. Turesson et al. (2006) recorded 100 or more beats per minute at this stage of development, but initially ventilation appears to be quite irregular. With further development, ventilation rate decreases reaching only a few ventilatory movements per minute at 10 days post fertilization (Turesson et al. 2006).

Control of ventilation typically requires oxygen-dependent receptors detecting variable oxygen availability in the environmental water. Neuroepithelial cells (NEC) located in the gills have been identified as oxygen receptors in the gills of various fish. In zebrafish, NECs were first detected in gill filament primordia of larvae at 5 dpf (Jonz and Nurse 2005). An oxygen-sensitive background potassium channel, which can be inhibited by application of quinidine, appears to be involved in the oxygen-sensing pathway in NEC. In zebrafish larvae at 7 dpf, the hypoxia-induced ventilation was inhibited by quinidine in a dose-dependent manner, suggesting that the oxygen-sensitive neuroepithelial cells were fully innervated by 7 dpf.

Glutamatergic *N*-methyl-D-aspartate (NMDA) receptors are also involved in ventilatory control (Sundin et al. 2003; Turesson and Sundin 2003). In zebrafish larvae, maturation of the glutamatergic system started at about 8 dpf, as indicated by the partial inhibition of the ventilatory response by inhibition of the NMDA receptors. However, only at 13 dpf and older did the larvae completely rely upon a glutamatergic component for the mediation of the ventilatory response to hypoxia (Turesson et al. 2006).

An interesting experiment connected to the control of respiration addressed whether prolonged exposure to hypoxic or hyperoxic conditions during embryonic or early larval stages might affect breathing patterns and responses to acute changes in gas composition in the adult (Vulesevic and Perry 2006). Zebrafish embryos reared under hypoxic conditions for 7 days did not show any modification in the response to acute hypoxia or hypercapnia when tested at the age of 3 month or later. Embryos reared under hyperoxia as an adult showed a blunted response to hypoxia (Fig. 7.1) and no response to hypercapnia. These results clearly show that there is some plasticity in the development of the respiratory control system. It would be interesting to unravel the modifications in the respiratory control system causing these modulations in the ventilatory response to acute changes in the respiratory gas composition.

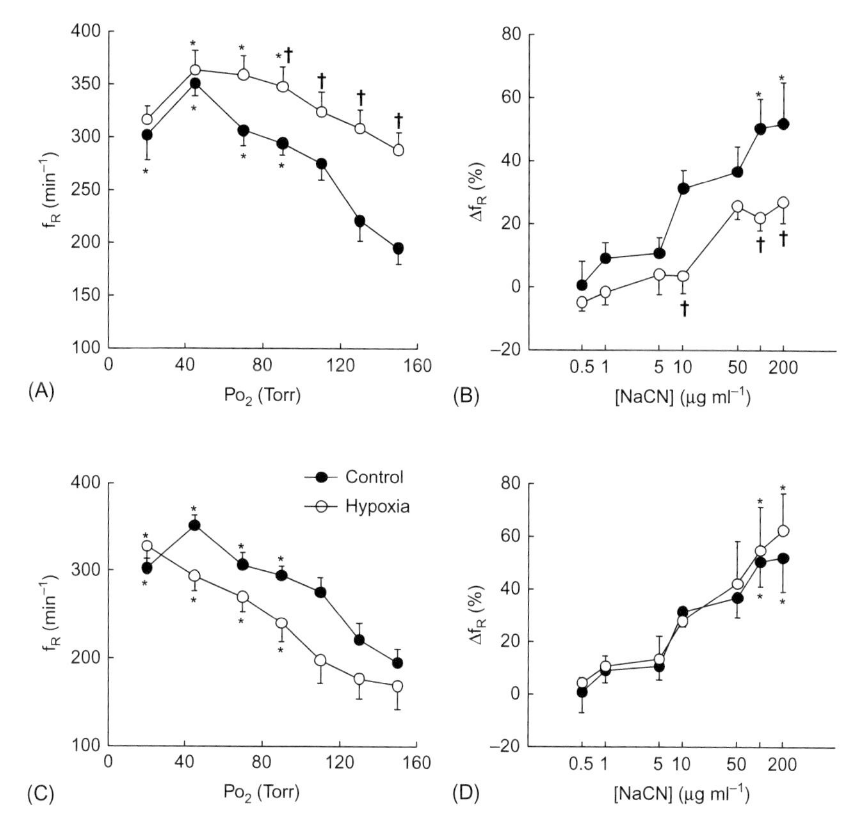


**Fig. 7.1.** (A, B) The respiratory response of adult zebrafish (*Danio rerio*) reared in normoxia (closed symbols) or in hyperoxia ($PO_2 > 46$ kPa, open symbols) for the first 7 days of life to acute hypoxia (A) or sodium cyanide (B). (C, D) The respiratory response of adult zebrafish (*Danio rerio*) raised in normoxia (closed symbols) or in hypoxia ($PO_2$, 4 kPa, open symbols) for the first 7 days of life to acute hypoxia (C) or sodium cyanide (D) (modified after Vulesevic and Perry 2006). Following the hyperoxic episode during early development the ventilatory response to hypoxia was blunted when comparing the rates of change of breathing frequencies with increasing hypoxia, and the effect of external cyanide on breathing frequencies was attenuated. In animals that experienced a hypoxic episode during early development the ventilatory response to acute hypoxia or to cyanide was unaltered.

## 4.2. Circulation

From the site of gas exchange, i.e., from the skin (cutaneous gas exchange) or from the gills, oxygen is typically conveyed to the site of consumption by convective transport in the blood. This is well-established knowledge for adult fishes. Consequently a reduction in the oxygen-carrying capacity of the blood by inactivation of the hemoglobin results in a severe impairment of the oxygen supply to tissues (hypoxemia). Application of CO or phenylhydrazine, for example, blocks or chemically destroys hemoglobin and reduces the oxygen-carrying capacity of the blood, which impairs convective oxygen transport in the bloodstream.

Given the effect of hypoxemia in adult fish, it was quite surprising to see that zebrafish can be raised in an atmosphere containing 1–5% CO, which should be sufficient to essentially inhibit hemoglobin-mediated oxygen transport (Pelster and Burggren 1996; Jacob et al. 2002; Rombough and Drader 2009). These studies demonstrate that oxygen uptake in CO-incubated zebrafish was not different from controls during embryonic and larval development for 42 dpf (Fig. 7.2). Furthermore, lactate metabolism was not activated until 15 dpf (Jacob et al. 2002). Because chronic production of lactate at a higher rate is hardly possible in zebrafish, it can be assumed that mostly aerobic metabolism is sustained until 42 dpf even in presence of CO. This clearly shows that hypoxemia does not provoke a hypoxic response in zebrafish larvae. Therefore, carrier-mediated oxygen transport is not required at this stage of development. Oxygen uptake and thus aerobic metabolism can be sustained by bulk diffusion.

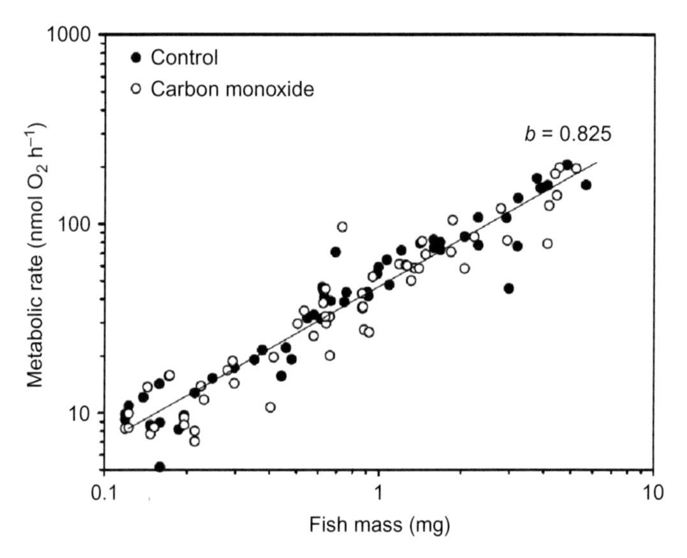


**Fig. 7.2.** Allometric scaling for routine metabolic rate of control and CO-exposed zebrafish larvae (after Rombough and Drader 2009, with permission). Up to a body mass of 6 mg CO exposure has no effect on the rate of oxygen consumption.

## 5. METABOLIC ACTIVITY

Mass-specific oxygen uptake of embryos and larvae is quite variable, but on average it appears to be higher than for adult fish (Rombough 1988). Measurement of oxygen uptake of zebrafish starting with the egg and over the first 100 days of development showed a more than 10-fold increase until 10 days post fertilization (3.6 $\mu L\ g^{-1}\ h^{-1}$ to 39.4 $\mu L\ g^{-1}\ h^{-1}$). Oxygen uptake then decreased to 5.8 $\mu L\ g^{-1}\ h^{-1}$ until 60 days post fertilization, and showed little additional change until 100 days post fertilization (Fig. 7.3) (Barrionuevo and Burggren 1999).

A more detailed analysis of oxygen consumption of zebrafish eggs and larvae between 1 and 9 dpf showed a continuous increase in oxygen uptake from about 3 nmol/h per egg to about 14–16 nmol/h per larva between 5 and 9 dpf (Grillitsch et al. 2005). Eggs at 1 dpf have a body mass of 0.3 mg and larvae at 8 dpf reach a body mass of 2.1 mg at 25°C (Bagatto et al. 2001), while the oxygen consumption in the study of Grillitsch et al. (2005) has been recorded at 28°C. These data fit very well with measurements performed between 9 and 15 dpf in zebrafish larvae either exposed to a water current for many hours a day or raised in stagnant water (Bagatto et al. 2001), and also with the data presented by Rombough and Drader (2009), who measured oxygen consumption in larvae between 5 and 42 dpf (Fig. 7.4). Interestingly,

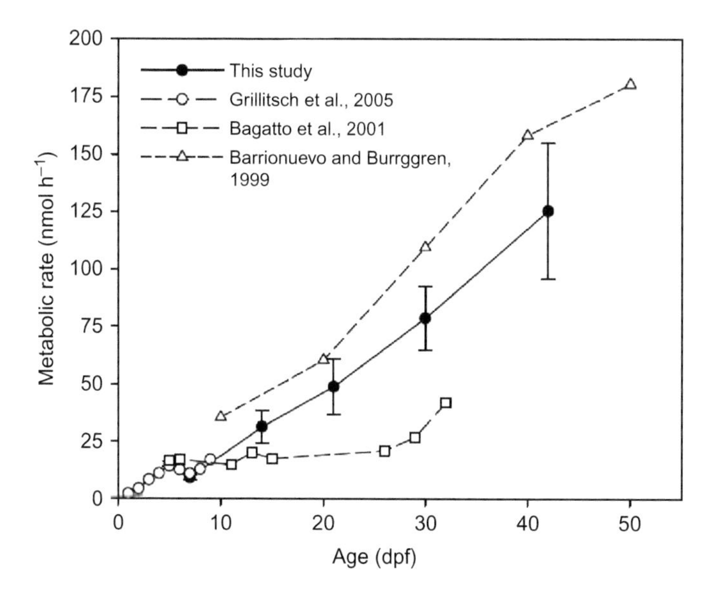


**Fig. 7.3.** Comparison of routine metabolic rates for zebrafish larvae under normoxic control conditions (modified after Rombough and Drader 2009, with permission).

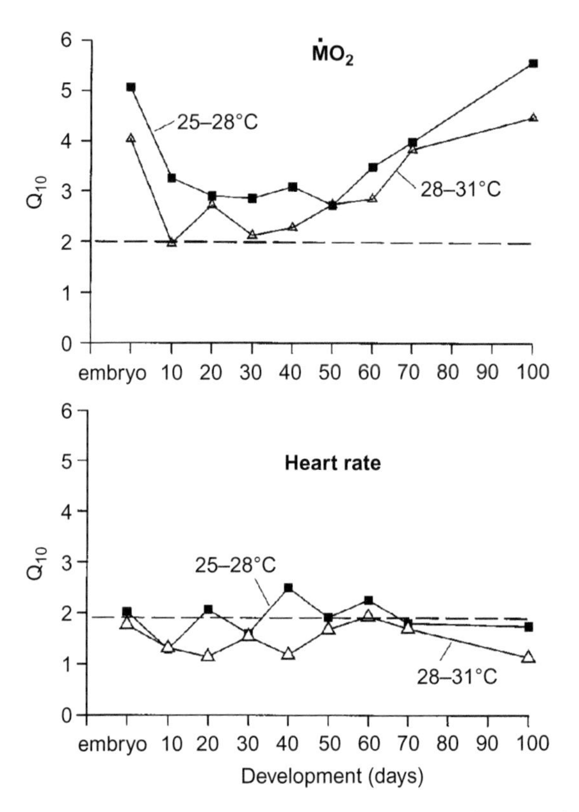


**Fig. 7.4.** $Q_{10}$ values for oxygen consumption and for heart rate as a function of development in zebrafish (Barrionuevo and Burggren 1999, with permission).

between 9 and 15 dpf, chronic swim training did not affect resting oxygen consumption of the larvae, but in animals exposed to a water current of 5 body lengths per second between 21 and 32 dpf resting oxygen consumption was elevated compared to untrained control larvae. The prolonged swimming activity may have resulted in a chronically higher activity level in trained animals, so that the increase in oxygen consumption would be related to a change in resting activity.

Bagatto et al. (2001) even managed to measure oxygen consumption of larvae swimming in a swim tunnel. For any given water velocity between 1 and 5 body lengths per second oxygen consumption of trained larvae was significantly lower than that of untrained larvae. This means that the energy consumption of trained larvae at any given swimming speed is lower than in untrained larvae (i.e., trained larvae swim more efficiently than untrained larvae). This appears to be due to an improvement in the oxygen supply at the level of the muscle tissue (Pelster et al. 2003), since resting cardiac activity was not affected by the chronic swim training. This increased efficiency may be due to an improved efficiency in the ATP-generating reactions, which could be achieved by a reduction in proton leak in the inner mitochondrial membrane. Alternatively, a higher efficiency in the ATP-consuming reactions, a more efficient $Ca^{2+}$ pumping reaction for example, may result in a higher efficiency. Similarly, the efficiency of force generation and myosin ATPase activity in heart and skeletal muscle may increase with development.

## 6. HYPOXIC EFFECTS

Several studies addressed the response of zebrafish larvae to reduced oxygen availability and demonstrate a remarkable resistance of eggs and larvae to hypoxic or in part even anoxic conditions. Padilla and Roth (2001) demonstrated that zebrafish embryos survive 24 hours of anoxia in a status of suspended animation, in which cardiac activity ceases and mitotic activity of the blastomeres is arrested. In later stages, the resistance to anoxia is markedly reduced and the mortality of the embryos in anoxia increases rapidly (Mendelsohn et al. 2008).

The response to reduced environmental oxygen availability is either to reduce oxygen consumption (oxyconformer), or to enhance oxygen uptake in order to sustain the level of aerobic metabolism (oxyregulator). While adult fish are typically oxyregulators, embryos and larvae quite often appear to be oxyconformers (Davenport 1983; Gruber and Wieser 1983). This is especially the case for large eggs, and therefore may be related to the chorion and the perivitelline fluid as a diffusion barrier (Rombough 1988).

In yolk sac larvae at 5 and 6 dpf, oxygen consumption remained constant down to a $P_{O_2}$ of 6.4 kPa. At lower $P_{O_2}$ values oxygen consumption declined, i.e., the critical $P_{O_2}$ (Pcrit) was 6.4 kPa (= 48 mmHg). At a $P_{O_2}$ of 3.3 kPa, oxygen uptake was reduced by approximately 50% compared to control levels (Bagatto et al. 2001). Therefore, zebrafish larvae attempt to sustain the level of oxygen consumption at least initially in the face of declining oxygen availability and thus show oxyregulatory behavior. Although early embryonic stages appear to be quite resistant even against anoxia (Padilla and Roth 2001; Mendelsohn et al. 2008), early larval stages appear to be most sensitive to hypoxia. Larvae at 10 dpf had the highest critical $P_{O_2}$ value (about 7.3 kPa at 25°C), while animals between 60 and 100 dpf had the lowest critical $P_{O_2}$ (Barrionuevo and Burggren 1999). This is at odds with data reported by Rombough and Drader (2009), who determined the lowest critical $P_{O_2}$ values for animals between 5 and 15 dpf, and the highest values for 21 dpf animals. The latter authors propose that the different ontogenetic patterns described for the critical $P_{O_2}$ in the different studies (Barrionuevo and Burggren 1999; Bagatto et al. 2001; Rombough and Drader 2009) may be related to differences in the activity patterns of the larvae. In contrast to the other two studies, Rombough and Drader (2009) used sedated larvae and reported the lowest critical $Po_2$ values. This explanation therefore may be appropriate. In addition, critical $Po_2$ is a very sensitive parameter and technical differences may also have contributed to these different values. If measured in a closed respirometer, the critical $Po_2$ may be affected by the speed with which the $Po_2$ declines in the respirometer chamber. Thus, differences in the volume of the respirometer and in the mass of the respiring tissue may have affected the final result.

The anoxia tolerance of the eggs was quite surprising (Padilla and Roth 2001). A more detailed analysis revealed that in the presence of oxygen, various inhibitors of the respiratory chain elicit the same response as does anoxia, and that the so-called anoxic arrest is accompanied by a rapid activation of the AMP-activated protein kinase pathway (Mendelsohn et al. 2008). This suggests that the energy status and an increased AMP/ATP ratio are the signal inducing arrest, and not the level of oxygen and signaling via hypoxia inducible factor (HIF).

In adults, oxygen deficiency can be achieved not only by environmental hypoxia, but also by reducing the oxygen-carrying capacity of the blood. Recall that in early larval stages hypoxemia does not impair aerobic metabolism because convective oxygen transport is not required (see above), and even the critical $Po_2$ is not affected in animals raised in the presence of 5% CO until 42 dpf (Rombough and Drader 2009). The only parameter affected in larvae raised with a reduced oxygen-carrying capacity is the residual oxygen tension. The residual oxygen tension is defined as the oxygen tension at which the $Po_2$ gradient between the environmental water and the tissues becomes too low to allow further extraction of oxygen from the water. At 7 and 14 dpf, the residual $Po_2$ was significantly higher in CO-treated zebrafish larvae, and it remained elevated until 42 dpf, although the difference was no longer significant (Rombough and Drader 2009). This suggests that although functional hemoglobin is not required for oxygen transport in a normoxic situation, in hypoxia it allows the larvae to extract oxygen from the water down to lower $Po_2$ levels.

Convective oxygen transport under normoxic conditions is not required to sustain aerobic metabolism in zebrafish larvae until several weeks after hatching. Measurements of hemoglobin oxygen saturation in the blood of larvae at various stages revealed, however, that this does not prevent partial unsaturation of the hemoglobin. In zebrafish larvae, intravascular hemoglobin oxygen saturation has been used to determine $Po_2$ in venous blood (Grillitsch et al. 2005). Even in these small larvae, hemoglobin was only partially saturated in certain developmental stages. This demonstrates the presence of partially oxygenated areas in larvae even under normoxic conditions. It could well be that under hypoxic conditions or in a situation with stimulated aerobic metabolism such as exercise, convective oxygen transport significantly contributes to the oxygen supply to tissues even in early larval stages.

Although convective oxygen transport is not required in early developmental stages of fish, sensors to detect environmental oxygen are present in early stages, and zebrafish larvae respond to changes in environmental $Po_2$. Even in very early larvae, a decrease in environmental oxygen availability provokes a stimulation of ventilatory activity and an enhancement of cardiac activity (Jacob et al. 2002; Jonz and Nurse 2005; Turesson et al. 2006). In newly hatched Arctic charr, ventilatory activity and the co-ordination of ventilatory movements increases during hypoxic exposure (McDonald and McMahon 1977). An increase in swimming activity under hypoxic conditions has also been reported for several fish larvae such as speckled trout (*Salvelinus fontinalis*) (Shepard 1955), smallmouth bass (*Micropterus dolomieu*) and largemouth bass (*Micropterus salmoides*) (Spoor 1977, 1984) and northern anchovy (*Engraulis mordax*) (Weihs 1980). An increased swimming activity may indicate attempts by the larvae to escape hypoxic water, but could also be employed to simply reduce the magnitude of possible unstirred layers around the body surface and thus facilitate diffusive gas exchange.

Several studies demonstrate that with the onset of cardiac activity or of ventilation, maturation of the autonomic control systems regulating cardiac activity and ventilation is not yet complete. Hypoxic stimulation of ventilation has been observed in zebrafish as early as 2 dpf and thus cannot be explained by the activity of the glutamatergic system or by activity of the NEC cells. Two non-NMDA receptors are the ionotropic α-amino-3-hydroxy-5-methyl-4-isoxazolepropionate (AMPA) and the γ–aminobutyric acid (GABA)-A receptors, and both may be involved in the control of ventilation in very early stages (Turesson et al. 2006).

Similarly, the autonomic nervous system appears to gain full control over the heart much later than the first hypoxic bradycardia is observed (Schwerte et al. 2006; Schwerte 2009). The heart is responsive to hormones in early developmental stages. It therefore can be assumed that initially during development circulating hormones may be responsible for the regulation of cardiac activity.

## 7. TEMPERATURE EFFECTS

In adult vertebrates, cardiac function is typically viewed as tightly coupled to metabolic activity, but this coupling is not yet established in embryonic or larval fish (Pelster 1999; Pörtner et al. 2006). A common observation on cardiac activity of larval stages is that there is no match between the stimulation of metabolism by an increase in temperature and the stimulation of cardiac activity with increasing temperature (Pelster 2002; Burggren and Bagatto 2008). This clearly is in agreement with the notion that larval respiration is largely independent of convective oxygen transport in the blood.

Many biochemical reactions respond with a doubling in turnover for an increase in temperature of 10°C (i.e., the $Q_{10}$ typically is around 2). A comparison of $Q_{10}$ data for oxygen consumption from about 20 species of fish clearly demonstrated that the $Q_{10}$ of oxygen consumption in fish larvae is quite variable, ranging from 1.49 in *Coregonus lavaretus* to 6.4 in *Pleuronectes platessa* (Rombough 1988). The overall average $Q_{10}$ of about 3 is significantly higher than the much less variable mean $Q_{10}$ of about 2, typically obtained for adult fish (Fry 1971). In addition, temperature sensitivity of larvae is quite variable during development. The earliest stages of zebrafish, for example, are quite sensitive to temperature changes, and metabolism may even have a $Q_{10}$ of 4–5. The $Q_{10}$ then decreases to values between 2 and 3 and increases again in later stages (Barrionuevo and Burggren 1999). Similar results were obtained in the minnow *Phoxinus phoxinus*, although $Q_{10}$ values were not quite as variable as the ones recorded for zebrafish larvae (Schönweger et al. 2000). This may be due to the fact that only early stages (i.e., up to 14 dpf) have been measured, while in zebrafish data have been collected until 100 dpf (Barrionuevo and Burggren 1999). In addition, the minnow is a species of the temperate zone and temperatures between 12.5°C and 25°C have been tested, while zebrafish typically is incubated between 25°C and 32°C. It has been proposed that early life stages in general are more stenothermal than adult stages (Rombough 1988), and stenothermy is accompanied by a higher temperature sensitivity of metabolic reactions, or the higher temperature sensitivity may even be the reason for stenothermic behavior (Pörtner 2006). It has also been discussed that temperature-related changes in activity may contribute to these $Q_{10}$ values (Fuiman and Batty 1997; Wieser and Kaufmann 1998). Enhanced growth or growth efficiency may play a role, and at least in some cases, metabolic activity appears to be somehow correlated with growth rates (Johns and Howell 1980; Rombough 1988).

The high temperature sensitivity of metabolism probably contributes to the strong effect of temperature on the critical $Po_2$. Increasing temperatures coincide with a significant increase in critical $Po_2$, so that at a temperature of 31°C the critical $Po_2$ is already reached at 11 kPa in 10 dpf zebrafish larvae (Barrionuevo and Burggren 1999).

Typically, temperature related stimulation of metabolism far exceeds the activation of cardiac activity, and this is also true for zebrafish. While $Q_{10}$ values for oxygen consumption are about 3 or even higher, the $Q_{10}$ values for cardiac output or heart rate are about 2 or even below (Barrionuevo and Burggren 1999; Jacob et al. 2002).

## 8. FUTURE PERSPECTIVES

Even though the broad topic of respiration in fish has been a topic of research for many decades, we are far from understanding this process in its entirety. There remain many avenues of research which have yet to be explored or are just beginning to be explored. As stated earlier, the zebrafish presents itself as a model organism in not only basic physiological research, but also in applied research. Particularly intriguing is the research into the developmental aspects of the change from diffusion-based respiration to gas exchange requiring circulation for convective transport. Several studies using the transparent zebrafish larvae contributed to our current understanding of the independent early development of metabolic activity and cardiac performance. Due to the existing information on the genome the zebrafish will also be extremely helpful to identify the genes and the expression changes underlying the physiological phenomena and adaptations observed under various environmental conditions.

We are only just beginning to understand the plasticity of this system and how it can accommodate to environmental perturbations. Obvious questions that might be addressed are: Why are very early stages remarkably tolerant towards severe hypoxia or even anoxia? How is ventilation controlled in very early stages? Why is cardiac activity modified in response to hypoxia at a time when hemoglobin oxygen transport is not needed for oxygen supply to tissues? Are cardiac activity and ventilatory activity eventually interlinked, and if so, how is the connection established at the molecular level?

In addition, genetic and non-genetic factors affecting the development of the cardiorespiratory system are just beginning to reveal just how complex this system really is. For example, as stated earlier, larvae typically show an oxyconforming response to hypoxia. However, Moore et al. (2006) have shown that this response is also strongly dependent on the clutch from which the larvae hatched. In fact, some larvae from particular clutches displayed strong oxyregulatory behaviors. While the mechanisms for this remain unknown, the complexity of genetic by environment interactions is a promising area of research.

## REFERENCES

Bagatto, B., Pelster, B., and Burggren, W. W. (2001). Growth and metabolism of larval zebrafish: Effects of swim training. *J. Exp. Biol.* **204**, 4335–4343.

Balon, E. K. (1975). Reproductive guilds of fishes: A proposal and definition. *J. Fish. Res. Board Can.* **32**, 821–864.

Barrionuevo, W. R., and Burggren, W. W. (1999). $O_2$ consumption and heart rate in developing zebrafish (Danio rerio): Influence of temperature and ambient $O_2$. *Am. J. Physiol.* **276**, R505–R513.

Brooke, L. T., and Colby, P. J. (1980). Development and survival of embryos of lake herring at different constant oxygen concentrations and temperatures. *Progressive Fish-Culturist* **42**, 3–9.

Burggren, W. W., and Bagatto, B. (2008). Cardiovascular anatomy and physiology. In: *Fish Larval Physiology* (R.N Finn and B.C. Kapoor, eds), pp. 119–161. Science Publishers, Enfield.

Ciuhandu, C. S., Wright, P. A., Goldberg, J. I., and Stevens, E. D. (2007). Parameters influencing the dissolved oxygen in the boundary layer of rainbow trout (Oncorhynchus mykiss) embryos and larvae. *J. Exp. Biol.* **210**, 1435–1445.

Davenport, J. (1983). Oxygen and the developing eggs and larvae of the lumpfish, Cyclopterus lumpus. *J. Mar. Biol. Assoc. U.K.* **63**, 633–640.

Dejours, P. (1981). *Principles of Comparative Respiratory Physiology* (2nd edn.). Elsevier, Amsterdam.

Evans, D. H., Piermarini, P. M., and Choe, K. P. (2005). The multifunctional fish gill: Dominant site of gas exchange, osmoregulation, acid-base regulation, and excretion of nitrogenous waste? *Physiol. Rev.* **85**, 97–177.

Evans, D. H., Piermarini, P. M., and Potts, W. T. W. (1999). Ionic transport in the fish gill epithelium. *J. Exp. Zool.* **283**, 641–652.

Feder, M. E., and Booth, D. T. (1992). Hypoxic boundary layers surrounding skin-breathing aquatic amphibians: Occurrence, consequences and organismal responses. *J. Exp. Biol.* **166**, 237–251.

Feder, M. E., and Pinder, A. W. (1988). Ventilation and its effect on "infinite pool" exchangers. *Am. Zool.* **28**, 973–983.

Fry, F. E. J. (1971). The effect of environmental factors on the physiology of fish. In: *Fish Physiology VI* (W.S. Hoar and D.J. Randall, eds), pp. 1–98. Academic Press, New York.

Fuiman, L. A., and Batty, R. S. (1997). What a drag it is getting cold: Partitioning the physical and physiological effects of temperature on fish swimming. *J. Exp. Biol.* **200**, 1745–1755.

Garside, E. T. (1959). Some effects of oxygen in relation to temperature on the development of lake trout embryos. *Can. J. Zool.* **37**, 689–698.

Grillitsch, S., Medgyesy, N., Schwerte, T., and Pelster, B. (2005). The influence of environmental $Po_2$ on hemoglobin oxygen saturation in developing zebrafish Danio rerio. *J. Exp. Biol.* **208**, 309–316.

Gruber, K., and Wieser, W. (1983). Energetics of development of the Alpine charr, Salvelinus alpinus, in relation to temperature and oxygen. *J. Comp. Physiol.* **149**, 485–493.

Hiroi, J., Kaneko, T., and Tanaka, M. (1999). In vivo sequential changes in chloride cell morphology in the yolk- sac membrane of Mozambique tilapia (Oreochromis mossambicus) embryos and larvae during seawater adaptation. *J. Exp. Biol.* **202**, 3485–3495.

Hwang, P. P., Lee, T. H., Weng, C. F., Fang, M. J., and Cho, G. Y. (1999). Presence of Na-K-ATPase in mitochondria-rich cells in the yolk-sac epithelium of larvae of the teleost Oreochromis mossambicus. *Physiol. Biochem. Zool.* **72**, 138–144.

Jacob, E., Drexel, M., Schwerte, T., and Pelster, B. (2002). The influence of hypoxia and of hypoxemia on the development of cardiac activity in zebrafish larvae. *Am. J. Physiol.* **283**, R911–R917.

Johns, D. M., and Howell, W. H. (1980). Yolk utilization in summer flounder (Paralichthys dentatus) embryos and larvae reared at two temperatures. *Mar. Ecol. Prog. Ser.* **2**, 1–8.

Jonz, M. G., and Nurse, C. A. (2005). Development of oxygen sensing in the gills of zebrafish. *J. Exp. Biol.* **208**, 1537–1549.
Kimmel, C. B., Ballard, W. W., Kimmel, S. R., Ullman, B., and Schilling, T. F. (1995). Stages of embryonic development of the zebrafish. *Dev. Dyn.* **203**, 253–310.
Kirsch, R., and Nonnotte, G. (1977). Cutaneous respiration in three freshwater teleosts. *Resp. Physiol.* **29**, 339–354.
Kranenbarg, S., Muller, M., Gielen, J. L. W., and Verhagen, J. H. G. (2000). Physical constraints on body size in teleost embryos. *J. Theor. Biol.* **204**, 113–133.
Li, J., Eygensteyn, J., Lock, R., Verbost, P., Heijden, A., Bonga, S., and Flik, G. (1995). Branchial chloride cells in larvae and juveniles of freshwater tilapia Oreochromis mossambicus. *J. Exp. Biol.* **198**, 2177–2184.
Malvin, G. M. (1988). Microvascular regulation of cutaneous gas exchange in amphibians. *Am. Zool.* **28**, 171–177.
McDonald, D. G., and McMahon, B. R. (1977). Respiratory development in Arctic char Salvelinus alpinus under conditions of normoxia and chronic hypoxia. *Can. J. Zool.* **55**, 1461–1467.
Mendelsohn, B. A., Kassebaum, B. L., and Gitlin, J. D. (2008). The zebrafish embryo as a dynamic model of anoxia tolerance. *Dev. Dyn.* **237**, 1780–1788.
Moore, F. B. G., Hosey, M. M., and Bagatto, B. (2006). Cardiovascular system in larval zebrafish responds to developmental hypoxia in a family specific manner. *Front. Zool.* **3**, 4.
Oikawa, S., and Itazawa, Y. (1993). Allometric relationships between tissue respiration and body mass in a marine teleost, porgy Pagrus major. *Comp. Biochem. Physiol.* **105A**, 129–133.
Olson, K. R. (1998). Hormone metabolism by the fish gill. *Comp. Biochem. Physiol.* **119A**, 55–65.
Padilla, P. A., and Roth, M. B. (2001). Oxygen deprivation causes suspended animation in the zebrafish embryo. *Proc. Natl Acad. Sci. USA* **98**, 7331–7335.
Pelster, B. (1999). Environmental influences on the development of the cardiac system in fish and amphibians. *Comp. Biochem. Physiol.* **124A**, 407–412.
Pelster, B. (2002). Developmental plasticity in the cardiovascular system of fish, with special reference to the zebra fish. *Comp. Biochem. Physiol.* **133A**, 547–553.
Pelster, B. (2008). Gas exchange. In: *Fish Larval Physiology* (R.N Finn and B.C. Kapoor, eds), pp. 91–117. Science Publishers, Enfield.
Pelster, B., and Burggren, W. W. (1996). Disruption of hemoglobin oxygen transport does not impact oxygen-dependent physiological processes in developing embryos of zebra fish (Danio rerio). *Circul. Res.* **79**, 358–362.
Pelster, B., Sänger, A. M., Siegele, M., and Schwerte, T. (2003). Influence of swim training on cardiac activity, tissue capillarization, and mitochondrial density in muscle tissue of zebrafish larvae. *Am. J. Physiol.* **285**, R339–R347.
Pinder, A. W., and Feder, M. E. (1990). Effect of boundary layers on cutaneous gas exchange. *J. Exp. Biol.* **154**, 67–80.
Pisam, M., Massa, F., Jemmet, C., and Prunet, P. (2000). Chronology of the appearance of ß, A, and α mitochondria-rich cells in the gill epithelium during ontogenesis of the brown trout (Salmo trutta). *Anat. Rec.* **259**, 301–311.
Pörtner, H.-O. (2006). Climate-dependent evolution of Antarctic ectotherms: An integrative analysis. *Deep Sea Res. Part II: Top. Stud. Oceanogr.* **53**, 1071–1104.
Pörtner, H.-O., Bennett, A. F., Bozinovic, F., Clarke, A., Lardies, M. A., Lucassen, M., Pelster, B., Schiemer, F., and Stillmann, J. H. (2006). Trade-offs in thermal adaptation: The need for a molecular ecological integration. *Physiol. Biochem. Zool.* **79**, 295–313.
Randall, D. J., Wilson, J. M., Peng, K. W., Kok, T. W. K., Kuah, S. S. L., Chew, S. F., Lam, T. J., and Ip, Y. K. (1999). The mudskipper, Periophthalmodon schlosseri, actively transports $NH_4^+$ against a concentration gradient. *Am. J. Physiol.* **277**, R1562–R1567.

Rojo, M. C., Blánquez, M. J., and González, M. E. (1997). Ultrastructural evidence for apoptosis of pavement cells, chloride cells, and hatching gland cells in the developing branchial area of the trout Salmo trutta. *J. Zool. Lond.* **243**, 637–651.

Rombough, P. (2007). The functional ontogeny of the teleost gill: Which comes first, gas or ion exchange? *Comp. Biochem. Physiol.* **148A**, 732–742.

Rombough, P. J. (1988). Respiratory gas exchange, aerobic metabolism and effects of hypoxia during early life. In: *Fish Physiology, Vol. XI, The Physiology of Developing Fish* (W.S Hoar and D.J. Randall, eds), pp. 59–161. Academic Press, San Diego, New York.

Rombough, P. J. (1992). Intravascular oxygen tensions in cutaneously respiring rainbow trout (Oncorhynchus mykiss) larvae. *Comp. Biochem. Physiol.* **101A**, 23–27.

Rombough, P. J. (1999). The gill of fish larvae. Is it primarily a respiratory or an ionoregulatory structure? *J. Fish Biol.* **55**(Suppl. A), 186–204.

Rombough, P. J. (2002). Gills are needed for ionoregulation before they are needed for $O_2$ uptake in developing zebrafish, Danio rerio. *J. Exp. Biol.* **205**, 1787–1794.

Rombough, P. J. (2004). Gas exchange, ionoregulation, and the functional development of the teleost gill. *In* "The Development of Form and Function in Fishes and the Question of Larval Adaptation" (Govoni, J. J., ed.), pp. 47–83. Am. Fish. Soc. Symposium 40, Bethesda, Maryland.

Rombough, P. J., and Moroz, B. M. (1990). The scaling and potential importance of cutaneous and branchial surfaces in respiratory gas exchange in young chinook salmon (Oncorhynchus tshawytscha). *J. Exp. Biol.* **154**, 1–12.

Rombough, P. J., and Moroz, B. M. (1997). The scaling and potential importance of cutaneous and branchial surfaces in respiratory gas exchange in larval and juvenile walleye *Stizostedion vitreum*. *J. Exp. Biol.* **200**, 2459–2468.

Rombough, P. J., and Ure, D. (1991). Partitioning of oxygen uptake between cutaneous and branchial surfaces in larval and young juvenile chinook salmon Oncorhynchus tshawytscha. *Physiol. Zool.* **64**, 717–727.

Rombough, P., and Drader, H. (2009). Hemoglobin enhances oxygen uptake in larval zebrafish (Danio rerio) but only under conditions of extreme hypoxia. *J. Exp. Biol.* **212**, 778–784.

Schönweger, G., Schwerte, T., and Pelster, B. (2000). Temperature-dependent development of cardiac activity in unrestrained larvae of the minnow Phoxinus phoxinus. *Am. J. Physiol.* **279**, R1634–R1640.

Schwerte, T. (2009). Cardio-respiratory control during early development in the model animal zebrafish. *Acta Histochem.* **111**, 230–243.

Schwerte, T., Prem, C., Mairosl, A., and Pelster, B. (2006). Development of the sympatho-vagal balance in the cardiovascular system in zebrafish (Danio rerio) characterized by power spectrum and classical signal analysis. *J. Exp. Biol.* **209**, 1093–1100.

Shepard, M. P. (1955). Resistance and tolerance of young speckled trout (Salvelinus fontinalis) to oxygen lack, with special reference to low oxygen acclimation. *J. Fish Res. Board Can.* **12**, 387–433.

Spoor, W. A. (1977). Oxygen requirements of embryos and larvae of the largemouth bass, Micropterus salmoides (Lacépède). *J. Fish Biol.* **11**, 77–86.

Spoor, W. A. (1984). Oxygen requirements of larvae of the smallmouth bass, Micropterus dolomieui Lacépède. *J. Fish Biol.* **25**, 587–592.

Sundin, L., Turesson, J., and Burleson, M. (2003). Identification of central mechanisms vital for breathing in the channel catfish, Ictalurus punctatus. *Resp. Physiol. Neurobiol.* **138**, 77–86.

Terjesen, B. F. (2008). Nitrogen excretion. In: *Fish Larval Physiology* (R.N Finn and B.C. Kapoor, eds), pp. 263–302. Science Publishers, Enfield.

Territo, P. R., and Altimiras, J. (2001). Morphometry and estimated bulk oxygen diffusion in larvae of Xenopus laevis under chronic carbon monoxide exposure. *J. Comp. Physiol.* **171B**, 145–153.

Turesson, J., Schwerte, T., and Sundin, L. (2006). Late onset of NMDA receptor-mediated ventilatory control during early development in zebrafish (Danio rerio). *Comp. Biochem. Physiol.* **143A**, 332–339.

Turesson, J., and Sundin, L. (2003). N-methyl-D-aspartate receptors mediate chemoreflexes in the shorthorn sculpin Myoxocephalus scorpius. *J. Exp. Biol.* **206**, 1251–1259.

Van der Heijden, A. J. H., van der Meij, J. C. A., Flik, G., and Wendelaar Bonga, S. E. (1999). Ultrastructure and distribution dynamics of chloride cells in tilapia larvae in fresh water and sea water. *Cell Tissue Res.* **297**, 119–130.

Vulesevic, B., and Perry, S. F. (2006). Developmental plasticity of ventilatory control in zebrafish, Danio rerio. *Resp. Physiol. Neurobiol.* **154**, 396–405.

Weihs, D. (1980). Respiration and depth control as possible reasons for swimming of northern anchovy, Engraulis mordax, yolk-sac larvae. *Fish. Bull.* **78**, 109–117.

Wells, P. R., and Pinder, A. W. (1996). The respiratory development of Atlantic salmon. II. Partitioning of oxygen uptake among gills, yolk sac and body surfaces. *J. Exp. Biol.* **199**, 2737–2744.

Wieser, W., and Kaufmann, R. (1998). A note on interactions between temperature, viscosity, body size and swimming energetics in fish larvae. *J. Exp. Biol.* **201**, 1369–1372.

Zaccone, G., Mauceri, A., and Fasulo, S. (2006). Neuropeptides and nitric oxide synthase in the gill and the air-breathing organs of fishes. *J. Exp. Zool.* **305A**, 428–439.

Zaccone, G., Mauceri, A., Fasulo, S., Ainis, L., Lo Cascio, P., and Ricca, M. B. (1996). Localization of immunoreactive endothelin in the neuroendocrine cells of fish gill. *Neuropeptides* **30**, 53–57.

# 8

# IONIC AND ACID–BASE REGULATION

*PUNG-PUNG HWANG*
*STEVE F. PERRY*



Because of their suitability for genetic manipulation, zebrafish have emerged as an attractive model species for research on mechanisms of freshwater fish ionic and acid–base regulation. Recent studies combining translational gene knockdown and histological methods have provided convincing data to support the proposed models, in which three types of ionocytes, $Na^+$-$Cl^-$ co-transporter cells, $Na^+$-$K^+$-ATPase-rich cells, and $H^+$-ATPase-rich cells, express distinct sets of ion transporters and enzymes, and perform the functions of $Cl^-$ uptake, $Ca^{2+}$ uptake, and $Na^+$ uptake/



*Zebrafish: Volume 29*
FISH PHYSIOLOGY


DOI: 10.1016/S1546-5098(10)02908-0

acid secretion and $NH_4^+$ excretion, respectively. Further investigations are now required to elucidate the driving pathways for the apical $Na^+/H^+$ exchanger, $Cl^-/HCO_3^-$ exchanger and NCC in ionocytes. Upon acclimation of fish to harsh environments, functional regulation of ionocytes (and the related ion transporters) is a result of the proliferation and differentiation of newly recruited ionocytes. The underlying mechanisms have been recently dissected by exploring the ionocyte differentiation molecular pathways, in which FOXI3a/b and other transcriptional factors are involved. The proposed models enable zebrafish to serve as a platform to identify the ion transport pathway(s) and the target ion transporter(s) or ionocyte(s) that are controlled by neuroendocrine agents.

## 1. INTRODUCTION

According to FishBase (http://fishbase.sinica.edu.tw/Summary/Species-Summary.php?id=4653), zebrafish inhabit streams, canals, ditches, ponds and beels, occurring in slow-moving to stagnant standing water bodies, particularly rice fields and lower reaches of streams. In a detailed field survey of India and Thailand (McClure et al. 2006), zebrafish were found primarily in warm (24–35°C), moving water of moderate clarity and pH 6.6–8.2. According to a recent review by Spence et al. (2008), zebrafish appear to be a floodplain rather than a true riverine species, and inhabit waters ranging widely in temperatures, from as low as 6°C in winter to 38°C in summer. In laboratory acclimation experiments, zebrafish survive well in extremely soft water ($[Na^+] = 0.03$–$0.1$ mM, $[Cl^-] = 0.03$–$0.04$ mM, $[Ca^{2+}] = 0.02$–$0.05$ mM) (Boisen et al. 2003; Chen et al. 2003; Craig et al. 2007; Yan et al. 2007; Wang et al. 2009), acidic (pH 4) or alkaline (pH 10) water (Horng et al. 2009) and water containing high ammonia (5 mM $NH_4^+$) (Shih et al. 2008). Additionally, adult zebrafish survive and behave normally after gradual acclimation to temperatures as low as 12°C (Chou et al. 2008). Thus, zebrafish must have evolved appropriate mechanisms of ionic and acid–base regulation as well as the control pathways to cope with such diverse environments. While other freshwater (FW) species also exhibit marked habitat diversity, zebrafish offer several advantages over traditional model species (trout, eel, tilapia, or killifish) because of the existence of more-or-less complete genetic databases and their suitability for molecular physiological approaches (Hwang 2009). For these reasons we believe that zebrafish can serve as a powerful model species for physiological research on mechanisms of fish ionic and acid–base regulation.

The topic of ionic acid–base regulation in fish has been summarized in numerous detailed reviews (Marshall 1995; Perry 1997; Perry et al. 2003a,b;

Evans et al. 2005; Claiborne et al. 2002; Marshall, 2002; Marshall and Grosell 2006; Perry and Gilmour 2006; Hwang and Lee 2007; Evans 2008; Hwang 2009; Gilmour and Perry 2009). The reader is encouraged to consult these reviews for a broad overview of ion transport processes in fish. A common theme emerging from many of these review articles is that while mechanisms of ionic regulation are relatively well understood in seawater (SW) fish, there is still considerable uncertainty as to the mechanisms underlying ionic uptake in FW species. The intent of the present chapter is to emphasize some of the recent advances that have emerged from studies on zebrafish and other species which have extended our understanding of FW fish ion regulatory mechanisms.

## 2. IDENTIFICATION OF IONOCYTES IN SKIN/GILLS

### 2.1. Ionocytes in Zebrafish

In the mammalian kidney, different types of ion-transporting cells (ionocytes) in the various segments exhibit specific ion transporter expression patterns and functions. With respect to ion uptake and acid–base regulatory mechanisms, the FW fish gill is in some respects analogous to the mammalian kidney and thus it is not surprising that the gill also possesses various types of ionocytes (also referred to as mitochondrion rich cells; MR cells) to achieve a variety of distinct ion transport functions. In the early literature, subtypes of MR cells were identified mainly based on their morphologies (mainly by electron microscopic observations) and/or whether or not they were enriched with $Na^+$-$K^+$-ATPase (NKA), and their functions were accordingly largely derived from correlated data of MR cell morphologies, NKA expression/activity and/or related ion fluxes. Direct evidence to support the notion that various types of ionocytes possess distinct transport functions emerged only recently with the adoption of molecular physiological approaches. In comparison to other fish species, the zebrafish offers several advantages as a model to explore models of ionic and acid–base regulation. That is because there are: (1) zebrafish-specific genetic databases which facilitate the identification of isoforms of target ion transporters or enzymes; (2) double/triple immunocytochemistry and/or in situ hybridization protocols well established to localize specific transporters or enzymes (as markers of ionocytes) to specific ionocytes; and (3) numerous techniques available to ascribe specific functions to transporters/enzymes and the associated genes as well as the ionocyte subtypes; powerful loss or gain of function techniques (see Chapter 1) are not readily applicable to most other fish species. Pan et al. (2005) first demonstrated that a portion of NKA-rich cells (NaR cells) co-expresses epithelial $Ca^{2+}$ channel (ECaC) mRNA, and subsequently demonstrated the $Ca^{2+}$ uptake function of these cells (Pan et al. 2005; Liao et al. 2007). Ultimately, three types of ionocytes were identified; $H^+$-ATPase-rich (HR) cells, $Na^+$-$K^+$-ATPase-rich (NaR) cells and $Na^+$-$Cl^-$ co-transporter (NCC) cells. HR, NCC, and NaR cells were demonstrated to be involved in $Na^+$ uptake/acid secretion/$NH_4^+$ excretion, $Ca^{2+}$ uptake, and $Cl^-$ uptake, respectively, through the specific expression and functioning of distinct sets of ion transporters and enzymes (Lin et al. 2006, 2008; Esaki et al. 2007; Horng et al. 2007; Liao et al. 2007, 2009; Nakada et al. 2007a; Yan et al. 2007; Shih et al. 2008; Wang et al. 2009). Additionally, glycogen-rich (GR) cells and an unidentified type of ionocyte were also described. Unidentified ionocytes express *ATP1a1a.4* (a NKA α1 subunit subtype) with unknown ion transport function (Liao et al. 2009), while GR cells have abundant glycogen deposits and may provide energy to neighboring ionocytes to carry out ion transport (Tseng et al. 2007, 2009b; Tseng and Hwang 2008). Recent studies have identified several SLC26 anion transporters (Bayaa et al. 2009; Perry et al. 2009; refer to Section 4.1) and ammonia transporters (Braun et al. 2009a,b; refer to Section 6), which are expressed in certain groups of ionocytes in zebrafish skin/gills. Additional studies incorporating double or triple labeling techniques are needed to determine if these ionocytes are different from HR, NCC, and NaR cells. The model for zebrafish skin/gill ionocytes provides an excellent platform to study the detailed ion transport pathways (Hwang 2009; see below).

### 2.2. Comparisons with Other Species

Although the focus of this review is ionic transport mechanisms in zebrafish, to enable comparison with other species, it is important to briefly comment on ionocyte diversity and nomenclature in some other piscine models. Using immunocytochemistry, MR cells in tilapia (*Oreochromis mossambicus*) skin/gill were classified into four types: type I showing only basolateral NKA staining, type II, showing basolateral NKA and apical NCC staining, type III, showing basolateral NKA/ $Na^+$-$K^+$-$2Cl^-$ co-transporter (NKCC1a) and apical $Na^+$/$H^+$ exchanger isoform 3 (NHE3) staining, and type IV, showing basolateral NKA/NKCC1a, and apical cystic fibrosis transmembrane conductance regulator (CFTR)/NHE3 staining (Hiroi et al. 2008; Inokuchi et al. 2009). Based on the distinct expressions of NHE3 and NCC, zebrafish HR and NCC cells are likely analogous to tilapia type II and III MR cells, respectively (Hwang 2009). It was reasonable not to find a counterpart of the tilapia type IV MR cells, a seawater-specific type (Hiroi et al. 2008), in zebrafish, which is a stenohaline FW species. Tilapia type I MR cells express only NKA and do not respond to transfer to different salinities (Hiroi et al. 2008). Similarly in zebrafish, the unidentified type of ionocyte expresses only ATP1a1a.4, and shows no change in cell numbers during acclimation to artificial FWs with different ionic compositions (Liao et al. 2009).

In rainbow trout (*Oncorhynchus mykiss*), two subtypes of MR cells, peanut lectin agglutinin negative ($PNA^-$) and $PNA^+$ MR cells, were identified through cell isolation techniques and were characterized by higher expressions of $H^+$-ATPase and NKA, respectively (Galvez et al. 2002). Serial in vitro physiological and pharmacological studies indicated the distinct ion-transport functions of the two MR cell types: $PNA^-$ cells exhibited a bafilomycin-sensitive acid-activated $Na^+$-uptake (Reid et al. 2003; Parks et al. 2007), while $PNA^+$ cells exhibited $Ca^{2+}$ uptake (Galvez et al. 2006). Subsequent studies employing immunocytochemistry and/or in situ hybridization revealed that $PNA^-$ cells expressed both ECaC mRNA and protein, while $PNA^+$ cells expressed NHE2 mRNA, NHE3 protein and ECaC mRNA (without a protein signal) (Shahsavarani et al. 2006; Ivanis et al. 2008). The transporter expression patterns in the two MR cell types appear to conflict with their respective ion-transport functions that had been previously proposed (Galvez et al. 2002; Reid et al. 2003; Perry et al. 2003a,b). Moreover, Shahsavarani et al. (2006) reported that at least four subtypes of gill cells show differential expression patterns of ECaC and NKA and other flow cytometric and double-labeling experiments further demonstrated the broad expression of ECaC in both $PNA^+$ and $PNA^-$ MR cells. Taken together, the data suggest that there may be more than two types of MR cells (or ionocytes) in gills of rainbow trout; these subtypes of MR cells cannot be identified simply by isolation with PNA.

The variety of subtypes of ionocytes and their transporter expression patterns may not simply reflect species-dependent differences but presumably also reflect the different methodologies (for identifying ionocytes or transporters) used in the different species. The methodologies established in zebrafish may prove useful if applied to other species and may help to identify both conserved and divergent pathways of ionic and acid–base regulation.

## 3. SODIUM UPTAKE AND ACID SECRETION BY SKIN/GILLS

### 3.1. Apical $Na^+$ and $H^+$ Transport

At least two pathways have been proposed for the apical inward transport of $Na^+$ and outward secretion of $H^+$ in fish gill cells: (1) an apical V-type $H^+$-ATPase electrically linked to $Na^+$ absorption via the epithelial

$Na^+$ channel (ENaC) or structural analog and (2) an electroneutral exchange of $Na^+$ and $H^+$ via an apical NHE. The debate as to the feasibility of each pathway has repeatedly been summarized in recent reviews (Evans et al. 2005; Perry and Gilmour 2006; Hwang and Lee 2007; Evans 2008; Hwang 2009). A landmark publication by Avella and Bornancin (1989) questioned the feasibility of the electroneutral $Na^+/H^+$ apical transport pathway in fish gills, and since then, most published models have favored the pathway linking $H^+$-ATPase/ENaC in acid secretion with $Na^+$ uptake mechanisms. However, the role of $Na^+/H^+$ exchange was reevaluated and should not be overlooked because of recently accumulating molecular physiological evidence of NHE (Evans et al. 2005; Hwang and Lee 2007; Evans 2008; Hwang 2009). In zebrafish, bafilomycin inhibited $^{22}Na^+$ influx in soft FW ($[Na^+] = 0.035$ mM, $[Ca^{2+}] = 0.004$ mM, and pH 6.0)-acclimated adult fish (Boisen et al. 2003) and sodium green fluorescence accumulation in embryonic HR cells (Esaki et al. 2007), and translational knockdown of the $H^+$-ATPase A subunit (zATP6V1a) decreased the surface $H^+$ concentration (monitored by a non-invasive scanning ion-selective electrode technique, SIET) in the skin and decreased the whole-body $Na^+$ content in fish acclimated to low-$Na^+$ FW ($[Na^+] = 0.01$ mM, $[Ca^{2+}] =$ 0.19 mM, and pH 6.8) (Horng et al. 2007), indicating an $H^+$-ATPase-dependent $Na^+$ uptake pathway. No orthologs of mammalian ENaC, however, have yet to be found in zebrafish (or other fishes) genomic databases. Moreover, amiloride (10 μM) did not affect $^{22}Na^+$ influx in adults in soft FW (Boisen et al. 2003), nor the sodium green fluorescence accumulation in embryonic HR cells (Esaki et al. 2007). On the other hand, recent molecular physiological studies on zebrafish appear to favor the involvement of $H^+$-ATPase/NHE in the $Na^+$ uptake/acid–base regulatory mechanisms (Hwang 2009), similarly to the current working model for mammalian proximal tubular cells (Wagner et al. 2004). The results of SIET studies (Lin et al. 2006) and sodium green fluorescence accumulation (Esaki et al. 2007) provide in vivo evidence of the functions of $H^+$ secretion/$Na^+$ uptake, which are mainly conducted by HR cells in skin of larval zebrafish. Both amiloride at 100 μM (no effect at 10 μM, see above) and EIPA at 10 μM were found to inhibit the sodium green fluorescence accumulation in HR cells (Esaki et al. 2007), implying the involvement of NHE, instead of ENaC, in $Na^+$ uptake pathways of HR cells. In zebrafish, eight isoforms of the SLC9 (NHE) family, ten isoforms of the carbonic anhyrdase (CA) family, and four paralogs of the NKA α1 subunit (ATP1a1) were identified, and only NHE3b, CA15a, CA2-like a, and ATP1a1a.5 were found to be co-expressed in HR cells (but never in other ionocytes in embryonic skin and/or adult gills) (Yan et al. 2007; Lin et al. 2008; Liao et al. 2009). Translational knockdown of CA15a, CA2-like a, and $H^+$-ATPase A subunits revealed defects of acid secretion and $Na^+$ uptake in zebrafish morphants (Horng et al. 2007; Lin et al. 2008). Acclimation of zebrafish to low-$Na^+$ FW, which stimulates $Na^+$ uptake, was found to specifically enhance the mRNA expression of CA15a, NHE3b, and ATP1a1a.5 (Yan et al. 2007; Lin et al. 2008; Liao et al. 2009), while acidic (pH 4) FW stimulated acid secretion and mRNA expression of CA15a and $H^+$-ATPase with reduced expression of NHE3b (Yan et al. 2007; Lin et al. 2008). These results provide convincing and direct molecular physiological evidence to support a role of these specific isoforms in $Na^+$ uptake/acid secretion pathways in zebrafish HR cells.

### 3.2. Basolateral Transport

The $Na^+$ uptake/acid–base regulation mechanisms in zebrafish HR cells are thought to be similar to those in mammal kidney proximal tubular cells (Yan et al. 2007; Lin et al. 2008). Aided by membrane-bound CA15a4, ambient $HCO_3^-$ and $H^+$ secreted by apical NHE3 (or $H^+$-ATPase) react to form $CO_2$ and $H_2O$ outside the apical membranes of HR cells, and this enables the passive diffusion of $CO_2$ into HR cells (the mechanisms producing a gradient for $CO_2$ back diffusion are currently unkbnown). Then the $CO_2$ is hydrated by cytosolic CA2-like to form $HCO_3^-$ and $H^+$. Protons are recycled back into the ambient environment by NHE3b and $H^+$-ATPase. $HCO_3^-$ may be removed out of the cell across the basolateral membrane via $HCO_3^-$-related transporters. Basolateral NKA (ATP1a1a.5/ATP1b1b) may pump cytosolic $Na^+$ across the basolateral membrane and also provides an intracellular negative potential to drive the electrogenic $Na^+$ transport of basolateral $Na^+$-$HCO_3^-$-co-transporter (NBC1), which is a major basolateral transporter in mammalian proximal tubular cells. NBC has repeatedly been proposed as playing major roles in fish gill $Na^+$ uptake/acid–base regulatory mechanisms, mostly based on gene expression data after acid or hypercapnia treatment (Perry and Gilmour 2006; Parks et al. 2007). Surprisingly, few data are available on the localization of NBC in fish gill ionocytes. In gills of Osorezan dace (*Tribolodon hakonensis*), NBC1 was localized with a homologous antibody in a portion of NKA-labeled MR cells (Hirata et al. 2003). In rainbow trout, an anti-rat kidney NBC was used to detect signals in the basolateral region of a group of gill cells (Parks et al. 2007). However, neither study provided any evidence for the colocalization of NBC with either NHE or $H^+$-ATPase in the same gill ionocytes. On the other hand, in our preliminary experiments, 13 members of SLC4 family were identified from zebrafish, and only two of them, NBC1b (SLC4a4b) and anion exchanger 1b (AE1b, SLC4a1b), were respectively localized by in situ hybridization to a group of ionocytes (Hwang 2009). Further double/triple labeling experiments indicated that NBC1 mRNA was colocalized with neither $H^+$-ATPase nor NKA, but instead AE1 mRNA was colocalized in HR cells (Fig. 8.1). Supporting these localization data, AE1, but not NBC1, mRNA expression in zebrafish gills was stimulated by low-$Na^+$ or acidic (pH 4) FW, which was previously demonstrated to enhance the functions of $Na^+$ uptake/acid secretion by HR cells (Yan et al. 2007; Lin et al. 2008; Horng et al. 2009). Using an anti-tilapia AE1 antibody, AE1 was also localized in the basolateral membrane of NKA-labeled MR cells in spotted green pufferfish (*Tetraodon nigroviridis*) (Tang and Lee 2007). Taken together, AE1b, instead of NBC1b, appears to be involved in basolateral transport pathways in HR cells, although the detailed mechanisms behind them remain to be studied. More species need to be investigated to see if this phenomenon is general to other fish species. As proposed above, zebrafish HR cells are analogous to mammalian proximal tubular cells from the aspect of $Na^+$ uptake/acid secretion mechanisms. However, the two types of ionocytes differ in their basolateral transport pathways (AE1 versus NBC1); the evolutionary and physiological significance of this difference is currently obvious.

### 3.3. Driving Pathways for Apical $Na^+/H^+$ Exchanger

Another issue worthy of further discussion is the driving mechanism of apical NHE3b in zebrafish HR cells. As Avella and Bornancin (1989)

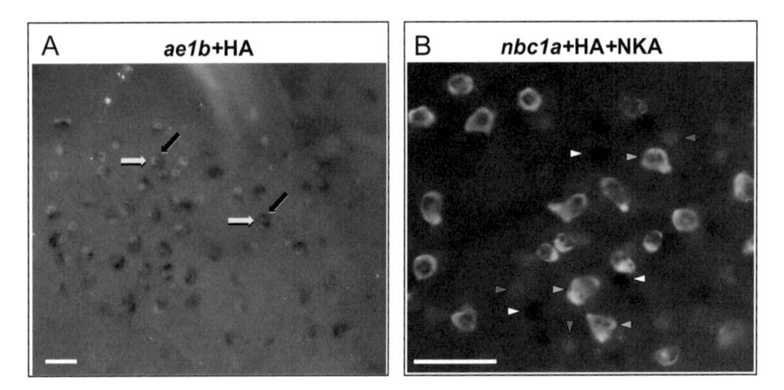


**Fig. 8.1.** Double/triple immunocytochemistry and in situ hybridization of skin ionocytes in zebrafish larvae at 5 d post-fertilization. (A) *slc4a1b* mRNA (*ae1b*, black signal, black arrow) was colocalized with $H^+$-ATPase (HA, green signal, yellow arrow). (B) *slc4a4b* mRNA (*nbc1b*, black signal, white triangle) was not colocalized with HA (red signal, red triangle) nor with $Na^+$-$K^+$-ATPase (NKA, green signal, green triangle). RNA probes for in situ hybridization: *slc4a1b*, nt1363-3217 (FJ211592); *slc4a4b*, nt1363-3217 (EF634453). Antibodies for immunocytochemistry were an anti-avian NKA $\alpha$-subunit monoclonal antibody and an anti-killifish HA A subunit polyclonal antibody. Scale bar: 50 μm. See color plate section

previously questioned, Parks et al. (2008) also emphasized the thermodynamic constraints that might prevent electroneutral apical NHE from functioning in most FW environments. Thermodynamic arguments derived from theoretically calculating the ratios of intracellular/environmental concentrations of $Na^+$ and $H^+$ (Parks et al. 2008) may be too simplified to conclusively exclude the possibility of the operation of apical NHE in FW fish gills. Some recent studies have opened a new avenue to explore this issue. Rhcgl, a putative ammonia transporter, was identified in the apical membrane of zebrafish HR cells (Nakada et al. 2007a). Subsequent molecular physiological experiments using SIET and loss-of-function demonstrated that the proton gradient created by the apical $H^+$-ATPase and/or NHE3b drive facilitative $NH_3$ diffusion by an "acid-trapping" mechanism through Rhcgl in zebrafish HR cells (Shih et al. 2008). A recent study using cultured rainbow trout gill cells proposed a model of an apical "$Na^+/NH_4^+$ exchange complex", in which $H^+$-ATPase, NHE2, ENaC, and Rhcg2 are involved, and this model also affirms the importance of non-ionic diffusion of $NH_3$ in ammonia excretion across FW fish gills (Tsui et al. 2009; see Section 6). On the other hand, in recent structure/function studies (Khademi et al. 2004; Javelle et al. 2008), *Escherichia coli* AmtB, a member of the Rh family, was found to be composed of a vestibule that recruits $NH_4^+/NH_3$, a binding site for $NH_4^+$, and a hydrophobic pore that lowers the $NH_4^+$ pKa to below 6 and conducts $NH_3$, and essential $NH_4^+$ deprotonation takes place at the vestibule before $NH_3$ conduction through the pore. Considering the microenvironment of the HR cell apical region, Rhcgl deprotonates $NH_4^+$ and subsequently protonates $NH_3$ inside and outside the membrane, respectively, and this may create a favorable $H^+$ gradient to drive the operation of NHE3b. Some additional evidence seems to support this notion. Low-$Na^+$ FW stimulated the function (i.e., $Na^+$ uptake) and expression of NHE3b, and also enhanced the function (i.e., $NH_4^+$ secretion) and expression of Rhcgl in zebrafish HR cells (L-Y Lin, T-H Shih, J-L Horng, S-T Liu, P-P Hwang, unpublished data); acute (5–60 min) exposure to acid stimulated both $NH_4^+$ secretion (Shih et al. 2008) and sodium green accumulation (Flynt et al. 2009). A previous study (Esaki et al. 2007) demonstrated that sodium green accumulation is mainly achieved by NHE3b in HR cells (Esaki et al. 2007). According to the thermodynamic constraints proposed by Parks et al. (2008), an acidic environment should prevent electroneutral apical NHE operation, but data of Shih et al. (2008) and Flynt et al. (2009) apparently do not support this notion. Taking all of these facts into account, $H^+$-ATPase and NHE3b respectively contribute >70 and 20% of the proton secretion and the accompanying Rhcgl $NH_4^+$ transport (Shih et al. 2008); however, Rhcgl may play an important role in providing a favorable $H^+$ gradient to drive NHE3b in zebrafish HR cells.

The deprotonation and acid trapping associated with Rhcg1 may shed some light on the long-debated issue of the driving mechanism of electroneutral NHE in FW fish gill cells, but more evidence is urgently needed.

## 4. CHLORIDE UPTAKE AND BASE EXCRETION

Zebrafish possess the highest affinity $Cl^-$ uptake mechanism of any species thus far examined with an affinity constant ($K_m$) of only 8 μM in fish acclimated to ion-poor water (NaCl = 40 μM) (Boisen et al. 2003). It is the combination of high $Cl^-$ uptake affinity and transport capacity ($J_{MAX}$) that allows zebrafish to inhabit extremely dilute environments. Surprisingly little, however, is known about the nature of $Cl^-$ uptake in zebrafish or in any other fish species for that matter. It would appear that at least two different pathways contribute to $Cl^-$ uptake, one involving some type of $Cl^-/HCO_3^-$ exchange process (Krogh 1937, 1938) and another incorporating $Na^+$-$Cl^-$ co-transport.

### 4.1. $Cl^-$ Uptake via $Cl^-/HCO_3^-$ Exchange

The prior evidence implicating $Cl^-/HCO_3^-$ exchange as a mechanism for $Cl^-$ uptake in FW fish is extensive and probably undisputed (Krogh 1937, 1938; Maetz and Garcia Romeu 1964; De Renzis and Maetz 1973; De Renzis 1975; Perry et al. 1981) (reviewed by Marshall 1995, 2002; Claiborne 1998; Evans et al. 1999, 2005; Perry et al. 2003a,b; Perry and Gilmour 2006; Marshall and Grosell 2006; Tresguerres et al. 2006; Hwang and Lee 2007; Evans 2008). Members of two gene families have been implicated: the SLC4 bicarbonate transporters (also referred to as the AE gene family; reviewed by Alper et al. 2002; Romero et al. 2004; Pushkin and Kurtz 2006; Alper 2006) and the SLC26 anion transporters (see reviews by Mount and Romero 2004; Romero et al. 2006; Soleimani and Xu 2006; Sindic et al. 2007; Ohana et al. 2008; Bayaa et al. 2009). Members of the SLC4 family specifically move $HCO_3^-$ (or $CO_3^{2-}$) (Pushkin and Kurtz 2006) while the SLC26 proteins promiscuously transfer a variety of anions including $Cl^-$, $HCO_3^-$, $OH^-$, $SO_4^{2-}$, formate, oxalate and iodide.

While there is some indirect evidence suggesting an involvement of SLC4 genes in apical $Cl^-/HCO_3^-$ exchange (Wilson et al. 2000), current models of ionic regulation in fish tend to favor a role for SLC26 genes as first proposed by Piermarini et al. (2002). The results of experiments performed on larval (Bayaa et al. 2009) and adult (Perry et al. 2009) zebrafish suggest a possible role for three SLC26 family members, A3, A4 and A6. Specifically, targeted gene knockdown using antisense oligonucleotide morpholinos demonstrated a significant role for SLC26A3 (accession FJ170816.1) in $Cl^-$ uptake in larval fish raised in control water and roles for A3, A4 (accession NM_000441) and A6c (accession FJ170818) in fish raised in water containing low [$Cl^-$]. Prolonged (7 days) or acute (24 h) exposure of fish to elevated (2 or 5 mM) ambient [$HCO_3^-$] caused marked increases in $Cl^-$ uptake capacity that were accompanied by elevated levels of SLC26 mRNA. The increases in $J_{in}Cl^-$ associated with high ambient [$HCO_3^-$] were not observed in the SLC26 morphants (Bayaa et al. 2009). Net base excretion was markedly inhibited in the SLC26A3 and A6c morphants thereby implicating these genes in $Cl^-/HCO_3^-$ exchange. Thus, it was proposed by Bayaa et al. (2009) that under normal conditions, $Cl^-$ uptake in zebrafish larvae is mediated by SLC26A3 $Cl^-/HCO_3^-$ exchangers but under conditions necessitating higher rates of high affinity $Cl^-$ uptake, SIC26A4 and SLC26A6c may assume a greater role.

In adult zebrafish it was demonstrated that SLC26 anion transporter mRNA and protein (SLC26a3 only) were present in the gill where they appeared to be regulated by ambient $Cl^-$ levels (Perry et al. 2009). Owing to their localization to cells on the filament and at the bases of lamellae as well as the correspondence between transporter mRNA levels, $Cl^-$ uptake rates and net acid excretion, it was suggested that A3, A4 and A6 are expressed in ionocytes where they function as $Cl^-/HCO_3^-$ exchangers (Perry et al. 2009).

The ability of an electroneutral $Cl^-/HCO_3^-$ exchanger to function on the apical membrane of MRCs in FW fish is constrained by an unfavorable $Cl^-$ gradient that is unlikely to be overcome (at least at the macroscopic level) by the slight prevailing $HCO_3^-$ gradient (Perry 1997; Tresguerres et al. 2006). In this regard, the involvement of V-type $H^+$-ATPase to energize $Cl^-$ uptake can be envisaged in two different ways. First, as discussed by Tresguerres et al. (2006), cytosolic carbonic anhydrase (Lin et al. 2008) working in concert with a V-type $H^+$-ATPase (to prevent acid accumulation) may serve to create microenvironments enriched with $HCO_3^-$ as $CO_2$ is hydrated in close proximity to apical membrane $Cl^-/HCO_3^-$ exchangers. For this scheme to work with a basolateral localization of the V-type $H^+$-ATPase, it is essential that the basolateral and apical membranes are near each other which indeed is a likely scenario in the NaR MRCs because the basolateral membranes are highly infolded and extend throughout the cytoplasm (Kessel and Beams 1962). The resultant high levels of $HCO_3^-$ formed from $CO_2$ hydration could serve to drive electroneutral $Cl^-/HCO_3^-$ exchange while the $H^+$ could be removed from the cell by basolateral $H^+$ secretion via the V-type $H^+$-ATPase. A second potential mechanism linking V-type $H^+$-ATPase and $Cl^-/HCO_3^-$ exchange at the gill is the pumping of $H^+$ across the apical membrane into the water which could serve to lower $HCO_3^-$ levels in an external microenvironment to potentially enlarge the outwardly directed $HCO_3^-$ gradient (Marshall 2002). Similarly, Boisen et al. (2003) suggested that an apically positioned V-type $H^+$-ATPase could serve to increase intracellular $HCO_3^-$ levels which would also serve to enhance the $HCO_3^-$ gradient. If similar to SLC26A6 in toadfish (*Opsanus beta*) gut which is electrogenic (Grosell et al. 2009), $Cl^-/HCO_3^-$ exchange across the gill apical membrane could be facilitated by increased negative potential inside the cell (Perry et al. 2009).

### 4.2. $Cl^-$ Uptake via $Na^+$-$Cl^-$ Co-transport

Although first identified many years ago in flounder urinary bladder (Renfro 1975), only recently was a member of the SLC12 gene family, the $Na^+$-$Cl^-$ co-transporter (NCC), demonstrated to reside on the apical membrane of gill MR cells in tilapia, *Oreochromis mossambicus* (Hiroi et al. 2008; Inokuchi et al. 2008, 2009). In zebrafish, a specific variant, SLC12a10.2 (NCC-like 2, see Section 2.1) was found to be localized on the gill and yolk sac where it was confined to a specific group of ionocytes (NCC cells, Section 2.1) (Wang et al. 2009). Translational knockdown of SLC12a10.2 using an antisense morpholino oligonucleotide resulted in a reduction in $Cl^-$ uptake and a fall in whole body $Cl^-$ levels in larvae thereby implicating NCC-like 2 as a mechanism of $Cl^-$ uptake (Wang et al. 2009). NCC is an electroneutral transporter, and needs a $Na^+$ gradient to drive the cotransport of $Na^+/Cl^-$ (Renfro 1975). The $Na^+$ concentration in FW would appear to be too low to permit the operation of NCC-like mechanism in zebrafish NCC cells for the transepithelial uptake of $Cl^-$ (Wang et al. 2009; Hwang 2009); this is an issue that will need to be addressed in the future.

## 5. CALCIUM UPTAKE

The main source of calcium in fish, unlike in terrestrial vertebrates, is via its absorption from the environment rather than by dietary gains. In adult fish, the predominant route of $Ca^{2+}$ entry from the environment is across the gill epithelium while in larvae, cutaneous $Ca^{2+}$ absorption is a significant route of uptake prior to full development of the gill. The long standing model for branchial transepithelial $Ca^{2+}$ uptake (Perry and Flik 1988) contends that entry across the apical membrane is by diffusion through epithelial calcium channels (ECaC) while movement into the blood is mediated by plasma membrane $Ca^{2+}$-ATPase (PMCA) and sodium–calcium exchanger (NCX) (see also reviews by Flik and Verbost 1993; Flik et al. 1995; Marshall 2002; Marshall and Grosell 2006).

Studies on zebrafish, and in particular, those utilizing reverse genetics techniques (see Chapter 1), have been instrumental in confirming the branchial model of $Ca^{2+}$ uptake in fish and for elucidating mechanisms of hormonal control of $Ca^{2+}$ movement across the gill (Pan et al. 2005; Tseng et al. 2009a). Not only have the results of these studies confirmed the essential involvement of apical membrane ECaC, but they have also demonstrated that the effects of the potent inhibitory hormone, stanniocalcin (Wendelaar Bonga and Pang 1991) on $Ca^{2+}$ uptake are likely mediated at the level of ECaC (Tseng et al. 2009a). While NCX and PMCA certainly contribute to overall transepithelial $Ca^{2+}$ uptake, it would appear that the rate-limiting step in the overall process is the diffusive movement of $Ca^{2+}$ through ECaC. For example, the stimulatory effect of low $Ca^{2+}$ exposure on $Ca^{2+}$ uptake in zebrafish larvae is associated with increased expression of zECaC mRNA while expression of PMCA and NCX are unaltered (Liao et al. 2007).

There is extensive indirect and correlative evidence showing that the MR cell is the predominant site of $Ca^{2+}$ uptake at the gill (Perry and Wood 1985; Ishihara and Mugiya 1987; McCormick et al. 1992; Marshall et al. 1992; Perry et al. 1992; Shahsavarani et al. 2006; Galvez et al. 2006). In zebrafish, a subset of MR cells express ECaC, NCX and PMCA; these cells have been proposed to be the sites of $Ca^{2+}$ uptake in zebrafish (Liao et al. 2007). In rainbow trout, pavement cells and MR cells have been proposed as the site of $Ca^{2+}$ uptake because both cell types appear to express ECaC (Shahsavarani et al. 2006).

## 6. AMMONIA EXCRETION

The mechanisms underlying the excretion of ammonia (referred to here as the sum of gaseous $NH_3$ and ionic $NH_4^+$) across the fish gill have been the subject of numerous excellent reviews (Evans and Cameron 1986; Randall and Wright 1989; Walsh and Henry 1991; Mommsen and Walsh 1992; Wood 1993; Wright 1995; Wilkie 2002; Weihrauch et al. 2009; Wright and Wood 2009). Our intent here is not to duplicate the exhaustive content of these previous reviews but instead to focus on those studies that have utilized zebrafish to provide evidence supporting a revision in our understanding of branchial ammonia excretion. Specifically, revised models (Nakada et al. 2007b; Wright and Wood 2009) is de-emphasizing the movement of $NH_3$ by passive diffusion through a lipid bilayer as well as the movement of $NH_4^+$ by electroneutral $Na^+/NH_4^+$ exchange while arguing for an important role of ammonia channels in $NH_3/NH_4^+$ translocation. These ammonia channels are produced by a gene family which encode for a variety of Rh glycoproteins which in fish include Rhag, Rhbg and Rhcg. Following the pioneering study of Nakada et al. (2007b) demonstrating Rh glycoproteins in the gill of pufferfish (*Takifugu rubripes*), they have subsequently been identified in zebrafish (Nakada et al. 2007a; Shih et al. 2008; Braun et al. 2009a,b), mangrove killifish (*Kryptolebias marmoratus*) (Nawata et al. 2007), rainbow trout (Nawata et al. 2007; Hung et al. 2008; Tsui et al. 2009), toadfish (*Opsanus beta*) (Weihrauch et al. 2009) and longhorn sculpin (*Myoxocephalus octodecemspinosus*) (Claiborne et al. 2008). Based on the two species that have been examined (zebrafish and pufferfish), it would appear that the Rh proteins in the gill are spatially segregated (Braun et al. 2009b; Nakada et al. 2007b) to form a distinct transport pathway. For example, Rhag appears to be restricted to the basolateral and apical membranes of pillar cells, Rhbg to the basolateral membranes of epithelial cells and Rhcg to the apical membrane of pavement cells (Rhcg2) or MR cells (Rhcg1). Thus, it has been proposed (Nakada et al. 2007b) that the net transfer of ammonia across the gill epithelium involves its sequential movement from the blood channel across the pillar cell (facilitated by Rhag), across the basolateral membranes of epithelial cells (Rhbg) and finally across the apical membrane of pavement cells (Rhcg2) or MR cells (Rhcg1). Currently, it is not known which of these processes constitutes the rate-limiting step in the overall transepithelial flux of ammonia.

Studies on zebrafish have been instrumental in providing the first direct evidence that the Rh glycoproteins are required for normal rates of ammonia excretion (Shih et al. 2008; Braun et al. 2009a) and for confirming the acid-trapping hypothesis of ammonia excretion first proposed by Wright et al. (1989). In separate studies, morpholino knockdown techniques were used to provide evidence for a role of Rhcg1 (Shih et al. 2008) or all three Rh glycoproteins in ammonia excretion in zebrafish larvae (Braun et al. 2009a). As originally proposed (Wright et al. 1989), the acid-trapping hypothesis of ammonia excretion suggested that the acidification of an external boundary layer adjacent to the gill epithelium (owing to $CO_2$ excretion) assists the diffusive movement of $NH_3$ because the acidic environment facilitates the conversion of $NH_3$ to $NH_4^+$. Thus, the chemical conversion of $NH_3$ to $NH_4^+$ in the boundary layer serves to maintain a favorable blood-to-water $PNH_3$ gradient. While the earlier experiments of Wright et al. (1989) on rainbow trout provided evidence for a link between $CO_2$ excretion and $NH_3$ diffusion, a more recent study on zebrafish larvae (Shih et al. 2008) demonstrated that direct $H^+$ secretion (via V-type $H^+$-ATPase) provides an alternate source of external acidification to promote the acid-trapping mechanism (see Section 3.3). Specifically, it was shown that inhibition of $H^+$ secretion using the proton pump inhibitor bafilomycin or by translational knockdown of $H^+$-ATPase caused inhibition of ammonia excretion by the

HR cells which are known to be enriched with apical membrane $H^+$-ATPase and Rhcg1. Ultimately, at least three sources of external acidification likely contribute to the acid-trapping mechanism of ammonia excretion: $CO_2$ diffusion and its hydration to $H^+$ and $HCO_3^-$, $H^+$ secretion via $H^+$-ATPase and $H^+$ addition to the water via NHE. With respect to external acidification via the hydration of $CO_2$, it has been debated (at least in rainbow trout) whether or not an externally oriented CA isoform is involved (Wright et al. 1986; Perry et al. 1999). A CA4-like isoform (termed CA15a by the authors) was recently demonstrated to be present on the apical membrane of HR cells of zebrafish larvae (Lin et al. 2008). Further studies are required to determine a role (if any) for this CA isoform in promoting ammonia excretion via acid-trapping.

## 7. DIFFERENTIATION AND FUNCTIONAL REGULATION OF IONOCYTES

### 7.1. Short-term Functional Regulation of Ion Transporters

Ionocytes in gills and skin must regulate their transport functions and capacities for maintaining homeostasis in aquatic environments which generally fluctuate in temperature, pH, ion levels and salinity. Upon acute environmental challenge, ionocytes can adjust their transporter activity and function through complicated signaling pathways in the short term, of minutes to hours (Marshall 2003; Marshall et al. 2005; Shaw et al. 2008). In killifish (*Fundulus heteroclitus*) operculum, for example, active $Cl^-$ secretion via the apical CFTR in MR cells was reduced within 30 min after a hypotonic shock (Marshall 2003), while hypertonic shock increased CFTR $Cl^-$ secretion within 1 h (Shaw et al. 2008). Exposure of tilapia embryos to low-$Cl^-$ artificial fresh water caused an increase in active MR cells with apical openings (from inactive ones without openings) within 4 h, an increase in the apical opening size of MR cells within 10 h, and stimulation of $Cl^-$ uptake function within 24 h before MR cells began to increase in number (beyond 48 h) (Lin and Hwang 2001, 2004). Similarly, ionocytes in zebrafish larvae respond rapidly to acute changes in environmental pH. HR cells in zebrafish embryonic skin modulate the apical surface structure (labeled with ConA) and acid secretion function ($H^+$ activity measured by SIET) within 10–120 min after acute transfer from pH 7 to pH 4 or the reverse direction (J. L. Horng, P-. P. Hwang, Z. H. Wen, C. S. Lin, H. W. Chen, L. Y. Lin, unpublished data). This rapid alteration of the apical structure and acid secretion function appears to be critical for the zebrafish to cope with acute pH disturbances in the environment. These rapid changes may involve the trafficking of $H^+$-ATPase molecules between intracellular compartments and apical cell membranes as reported for intercalated cells of the mammalian kidney (Brown et al. 2009). A recent study on zebrafish micro(mi)RNA 8 (Flynt et al. 2009) has shed some light on this issue. The highly conserved *miR-8* family is specifically expressed in HR cells and enables control of $Na^+$ uptake function by modulating the expression of Nherf1 (Flynt et al. 2009), which is a regulator of apical trafficking of transmembrane ion transporters including NHEs (Yun et al. 1997). Loss- and gain-of-function experiments demonstrated that disruption of *miR-8* family function leads to an inability to absorb $Na^+$ (monitored by sodium green accumulation) and blocks the ability to properly traffic and/or cluster transmembrane glycoproteins (monitored by ConA staining) at the apical surface of HR cells (Flynt et al. 2009).

### 7.2. Long-term Functional Regulation of Ionocytes

With long-term (days to weeks) acclimation to harsh environments, increased numbers and/or function of MR cells are well documented in many species (Evans et al. 2005; Hwang and Lee 2007). Whether the increase in the number of cells is a result of the proliferation and differentiation of newly recruited ionocytes is an important issue in related fields. Early studies (Conte and Lin 1967; Tondeur and Sargent 1979; Chretien and Pisam 1986) using $^3$H-thymidine as a tracer reported that acclimation to seawater stimulated the turnover of gill cells, and consequently a higher production of MR cells in several euryhaline species. A flow cytometry study on isolated gill cells also reported higher mitotic activity in SW-adapted eel (*Anguilla japonica*) than in gills of FW-adapted fish (Wong and Chan 1999). A transformation from a pre-existing FW type of MR cell to a SW type was also noted (Wong and Chan 1999), and the transitions between MR types with different transporter expressions or morphologies were also reported in tilapia embryos studied by in vivo sequential tracing with confocal microscopy (Hiroi et al. 1999; Lin and Hwang, 2004). However, in tilapia embryos, most (75–90%) skin MR cells survived 24–96 h after transfer from FW to SW or from FW water to different $Cl^-$ levels, suggesting that upon acclimation to a new environment, the turnover of MR cells does not significantly change and only $<$10–25% of MR cells originate from presumed undifferentiated cells (Hiroi et al. 1999; Lin and Hwang 2004). According to those previous studies, MR cells were proposed to arise from stem cells or an undifferentiated cell population. To uncover the mechanisms behind these proposed cellular events, it is necessary to explore the origin of stem cells, how stem cells differentiate to ionocytes (or MR cells), and how these differentiation pathways are regulated during acclimation to a changing environment. The pathways underlying the differentiation of ionocytes were not known until recent studies were performed on zebrafish.

### 7.3. Differentiation Pathways of Ionocytes

Using mutants, loss-of-function, gain-of-function, transgenic zebrafish and other molecular/cellular approaches, Hsiao et al. (2007) first proposed a pathway for the differentiation and specification of zebrafish ionocytes, which was mostly supported and further explored in subsequent studies (Janicke et al. 2007; Esaki et al. 2009) (Fig. 8.2). The process of epidermal development relies on balanced signaling gradients secreted from non-neural (BMPs) and the neural ectoderm (including the chordin and noggin), and is evolutionarily conserved among vertebrates (De Robertis and Kuroda 2004; Moreau and Leclerc 2004). Under stimulation by BMPs, downstream targets of the transcription factor, p63, are activated in epidermal stem cells, which thereafter undergo the subsequent process of terminal differentiation (Bakkers et al. 2002; Hsiao et al. 2007; Janicke et al. 2007) (Fig. 8.2). At tail-bud stage (10 h post-fertilization (hpf)), Foxi3a is initially expressed in a subgroup of p63-positive epidermal stem cells, and this subgroup gives rise to specifications for ionocyte progenitor cells (Hsiao et al. 2007) (Fig. 8.2). Concurrently, DeltaC is co-expressed with Foxi3a in ionocyte progenitors, initiating the Delta-Notch competitive lateral inhibitory pathway (Hsiao et al. 2007). Through this Delta-Notch-mediated cell–cell interaction, epidermal ionocyte progenitors compete to inhibit one another's adoption of an ionocyte fate, ultimately creating a balanced population of epidermal ionocytes and keratinocytes in zebrafish skin (Hsiao et al. 2007) (Fig. 8.2).

On the other hand, in ionocyte progenitors, Foxi3a activates the expression of Foxi3b at a later (5-somite) stage (at 11.7 hpf), and Foxi3b also increasees Foxi3a expression through a positive feedback loop (Hsiao et al. 2007) (Fig. 8.2). Foxi3a and Foxi3b activate ATP1b1b ($Na^+$-$K^+$-ATPase β subunit) at the 11-somite stage (14.3 hpf) and shortly thereafter (in the 15-somite stage at 16.5 hpf), ATP6v1a ($H^+$-ATPase subunit A) triggers the differentiation of ionocytes (Hsiao et al. 2007) (Fig. 8.2). Thereafter, the overlapping expressions of Foxi3a and Foxi3b by ionocyte progenitors are differentially regulated by unknown factor(s) (see below), giving rise to different ionocyte lineages; NaR cells appear to be set to the primary differentiation fate by a higher level of Foxi3b expression, while HR cells need high Foxi3a/low Foxi3b function to activate their secondary differentiation fate (Hsiao et al. 2007; Esaki et al. 2009) (Fig. 8.2).

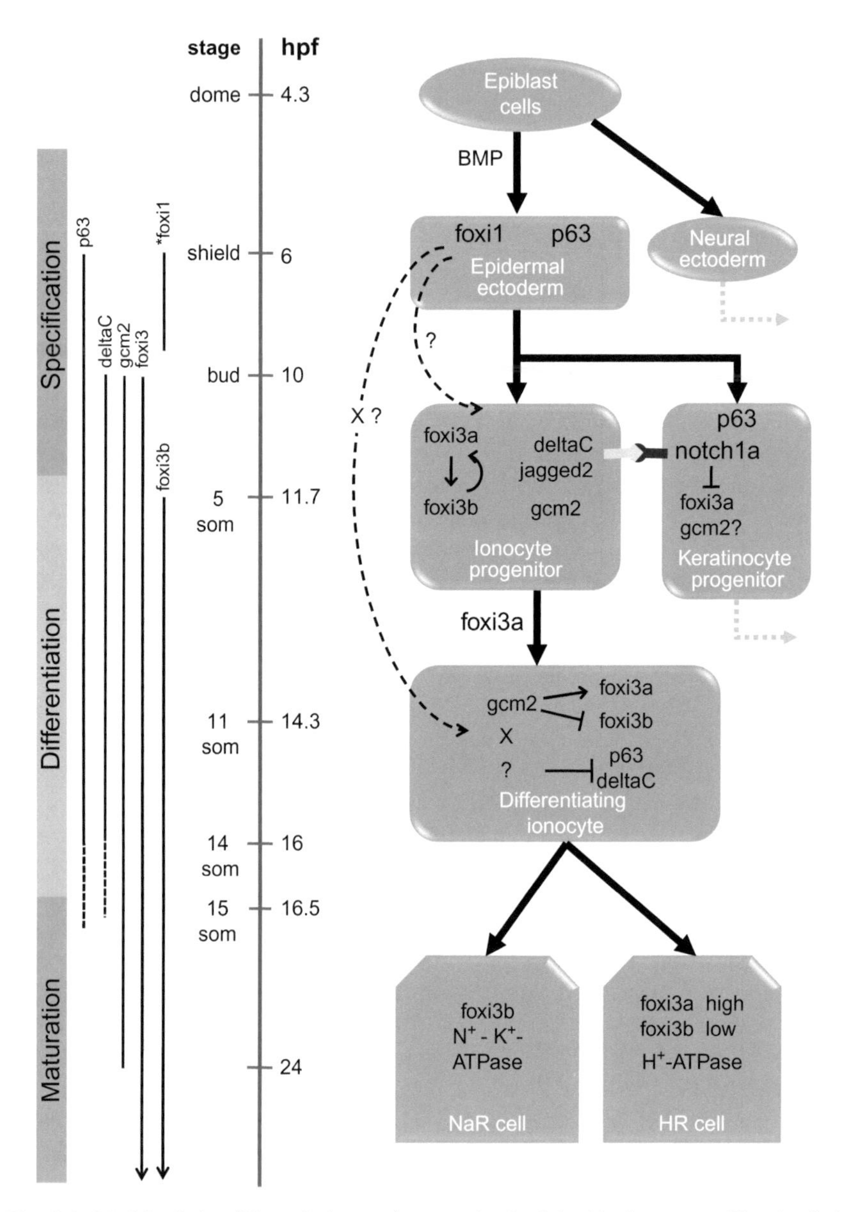


**Fig. 8.2.** Model of the differentiation pathways of zebrafish skin ionocytes. The detailed explanation of the pathways can be found in the text. The solid line indicates the expression duration of the related gene in ionocyte (or its progenitor), while dotted line shows downregulation of the gene. Foxi1 is expressed in the epidermal ectoderm only during 6–9 h post fertilization (hpf). "X" and "?" indicate unknown factor and pathway, respectively. BMP, bone morphogenetic protein; HR, $H^+$-ATPase-rich; NaR, $Na^+$-$K^+$-ATPase-rich.

A cell fate-related transcription factor, GCM2, was found to be expressed at almost the same time as Foxi3a in the zebrafish ectoderm and its mRNA is confined to HR cells. Gain- and loss-of-function experiments demonstrated a specific role for GCM2 in HR cell differentiation only (Chang et al. 2009). A subsequent study by Esaki et al. (2009) further indicated that knockdown of GCM2 did not affect Foxi3a expression at 12 hpf, but thereafter at 24 hpf, it suppressed Foxi3a expression in zebrafish morphants. This implies that GCM2 is initially expressed almost simultaneously with Foxi3a, but it appears that its effect on the differential regulation of Foxi3a/Foxi3b begins at a later stage (the 11-somite stage) when ionocyte progenitors are proceeding to different ionocyte lineages (Fig. 8.2). Interestingly in Esaki's study (2009), a *quadro* mutant deficient in the Foxi1 gene failed to differentiate HR cells (with normal NaR cells), suggesting a role of Foxi1 in HR cell differentiation (Fig. 8.2).

### 7.4. Proliferation, Differentiation and Functional Regulation of Ionocytes

Ionocyte differentiation is not only a subject of development biology, but also an important issue related to integrative and regulatory physiology of transepithelial transport. A recent study (Horng et al. 2009) indicated a compensatory enhancement in acid secretion by zebrafish embryos after acclimation to an acidic (pH 4) FW. This functional enhancement by HR cells was achieved not only by increasing the cell number but also by increasing the acid-secreting function ($H^+$ activity measured by the SIET) of single cells. The stimulation of function in each HR cell is mediated by morphological adjustment of the apical openings (Horng et al. 2009). Immunostaining results using p63 (a marker of epidermal stem cells) or PCNA (a marker of cell proliferation) further demonstrated that additional HR cells in pH 4 water may be differentiated not only from ionocyte precursor cells but also from newly proliferating epithelial stem cells (Horng et al. 2009), a conclusion that was recently supported by Chang et al. (2009). Concomitant increases in the expression of $H^+$-ATPase subunit A and GCM2 in zebrafish gills were found 4 d after acclimation to pH 4 FW (Chang et al. 2009). On the other hand, enhancement of ionocyte function via increased ionocyte differentiation was also reported in zebrafish gills during acclimation to a low temperature (Chou et al. 2008). Acute exposure to 12°C obviously impaired branchial $Ca^{2+}$ influx (compared to the control at 28°C), while acclimation for 30 d caused compensatory recovery of $Ca^{2+}$ influx; this recovery of $Ca^{2+}$ influx resulted from enhancement of ECaC mRNA expression and an increase in the number of ECaC-expressing cells, which was mediated by stimulation by Foxi3a (Chou et al. 2008). Thus, to cope with harsh environments, zebrafish appear to develop different compensatory strategies, including (1) acute modulation existing ionocytes and (2) for long-term acclimation, increasing cell proliferation and/or differentiation of ionocytes resulting in an enhancement of the overall ion regulation capacity.

## 8. NEUROENDOCRINE CONTROL OF ION UPTAKE AND ACID–BASE REGULATION

To cope with fluctuating salt levels in the environment, fish regulate their ionoregulatory functions by rapid modulation of existing mechanisms (transport and ionocytes) followed by subsequent differentiation of ionocytes and synthesis of new transport proteins, as discussed above. The neuroendocrine system plays a critical role in controlling these molecular/ cellular events during acclimation and recently, Evans et al. (2005) published a comprehensive and informative review on this issue. Although limited in its salinity tolerance compared to other traditional euryhaline model species, zebrafish can be used to examine aspects of neuroendocrine control particularly in acclimation to ion poor water. However, recent studies on zebrafish have largely focused on the development and ontogeny of the neuroendocrine system including the hypothalamus (Blechman et al. 2007), adenohyophysis (Pogoda and Hammerschmidt 2007), neurohypophysis (Eaton and Glasgow 2007; Eaton et al. 2008), the hypothalamus-pituitary-interrenal axis (Alderman and Bernier 2009; Alsop and Vijayan, 2009), the thyroid (Wendl et al. 2007) and parathyroid (Okabe and Graham 2004). Few studies have addressed the neuroendocrine control of ion regulation mechanisms in zebrafish. Liu et al. (2006) generated germline transgenic zebrafish co-expressing red fluorescent protein directed by prolactin (PRL) regulatory elements (PRL-RFP) and used gain- or loss-of-function approaches to study the osmotic response of PRL cells. Zebrafish embryos showed a greater number of larger PRL cells in 0.025% salt medium than in 0.1% medium, and this osmotic response by PRL cells appeared to be dopamine-independent based on a pharmacological experiment (Liu et al. 2006). Translational knockdown of the PRL receptor (PRLR) impaired the osmotic response by PRL cells, and over-expression of PRL resulted in PRL-REF suppression (i.e., a decrease in the PRL cell mass), suggesting that osmotic feedback on PRL cells is mediated by PRL signaling via the PRLR in the pituitary (Liu et al. 2006). Localization of PRLR mRNA in the pituitary supports the above notion (Liu et al. 2006), but PRLR mRNA localization in skin/gill ionocytes appears to have been overlooked. Hoshijima and Hirose (2007) demonstrated that mRNAs for atrial natriuretic peptide (ANP), renin, PRL, growth hormone (GH) and parathyroid hormone 1 (PTH 1) in zebrafish embryos showed differential responses to environmental ionic composition. Compared to FW ($[Na^+] = 686\ \mu M$, $[Cl^-] = 804\ \mu M$, $[Ca^{2+}] = 16.2\ \mu M$), mRNA expression for all genes was increased in $1/20 \times$ FW and decreased (all genes except GH) in $100 \times$ FW (Hoshijima and Hirose 2007). Compared to the fish in the $1/20 \times$ FW group which lacked a swimbladder, supplementation with $CaCl_2$ (final $[Ca^{2+}] = 16.2\ \mu M$; $[Cl^-] = 71\ \mu M$) rescued the swimbladder defect and decreased the effects on renin, PRL, and GH, but reversed the effect on PTH 1. Supplementation with NaCl ($[Na^+] = 686\ \mu M$; $[Cl^-] = 696\ \mu M$) caused a more severe defect of edema, with no effect on the genes (Hoshijima and Hirose 2007). Previous studies on FW-acclimated fish indicated the complicated effects of environmental $[Ca^{2+}]$ on gill $Na^+/Cl^-$ uptake (Avella et al. 1987; Chang et al. 2001), and ambient $[Na^+/Cl^-]$ on $Ca^{2+}$ balance (Chang et al. 2001). Future studies incorporating an experimental design which carefully considers the water ionic composition may allow one to examine the effect of a specific ion without influence from other ions. Adding supplementary NaCl to $1/20 \times$ FW (Hoshijima and Hirose 2007) appeared to be a harsher ionic situation for zebrafish embryos, and thus induced a synergistic impact on the hydromineral balance. Studies by Liu et al. (2006) and Hoshijima and Hirose (2007) supported the roles of those target hormones in iono- and/or osmoregulatory responses in zebrafish as did previous studies in other species (Evans et al. 2005).

Stanniocalcin (STC), a hormone secreted by the corpuscles of Stannius, is known to exert potent hypocalcemic actions in fish by its inhibitory effects on branchial $Ca^{2+}$ uptake (Lafeber and Perry 1988; Lafeber et al. 1988; Hanssen et al. 1989; Perry et al. 1989). However, the target of action by STC on $Ca^{2+}$ uptake mechanisms was unknown (Evans et al. 2005; Hoenderop et al. 2005) until a recent loss-of-function study on zebrafish (z) STC was performed (Tseng et al. 2009a). zSTC 1 mRNA was first detected in cells near the presumptive location of the corpuscles of Stannius at 1 dpf, indicating the early development of this organ in zebrafish (Tseng et al. 2009a). Incubating zebrafish embryos in low-$Ca^{2+}$ (0.02 mM) FW stimulated $Ca^{2+}$ influx and ECaC mRNA expression, while decreasing zSTC1mRNA expression. Translational knockdown of zSTC resulted in increases in the $Ca^{2+}$ content, $Ca^{2+}$ influx, and zECaC mRNA, but with no effect on the mRNA expressions of zPMCA2 or zNCX1b (Tseng et al. 2009a), which are co-expressed in ECaC-expressing ionocytes (Liao et al. 2007). These data not only demonstrate that zSTC-1 controls $Ca^{2+}$ homeostasis by regulating the expression and function of zECaC, but also support the consensus view that zECaC is the gatekeeper of transepithelial $Ca^{2+}$ transport in fish gills (Hwang and Lee 2007; Liao et al. 2007) and in mammalian kidney (Hoenderop et al. 2005).

Evans et al. (2005) comprehensively reviewed the neural control of fish gills and reported that many signaling agents including those derived from cholinergic, adrenergic, serotonergic and nitrergic neurons, vasoactive intestinal polypeptides and endothelin, elicit both hemodynamic and ionic transport effects. Recent studies by Jonz and Nurse (2003, 2006) examined details of the innervation of ionocytes in zebrafish. Based on immunocytochemical experiments, gill ionocytes (labeled with NKA as the marker) were intimately associated with nerve fibers (labeled with the zebrafish neuronal marker, zn-12) originating from outside the filaments, and embryonic skin ionocytes are situated among a dense network of varicose nerve fibers (Jonz and Nurse 2006). Previous studies have demonstrated that a variety of neuropeptides are present in the nerve fibers of fish gills (Evans et al. 2005). It will be challenging but potentially informative to determine if the neurons innervating zebrafish ionocytes contain different neuropeptides which are involved in regulating the functions of the various types of ionocytes.

## 9. RENAL IONIC AND ACID–BASE REGULATION

Similarly to the mammalian nephron, the FW fish kidney also plays an important role in ionic and acid–base regulation. Approximately 95% of the filtered NaCl is reabsorbed by the kidneys in FW fish (reviewed by Perry et al. 2003b). The kidney is essential for acid–base regulation in FW fish because of its ability to regulate the extent of $HCO_3^-$ reabsorption from urine and thus assists in setting plasma $HCO_3^-$ levels (Perry et al. 2003b; Perry and Gilmour 2006). Typically, however, the kidney contributes $<10\%$ of whole-body net excretion of acid–base equivalents compared to the gill, which accounts for $>90\%$ (Claiborne et al. 2002). Ion uptake and acid–base regulatory mechanisms in the fish kidney have received much less attention than those in the gill (Perry et al. 2003b; Perry and Gilmour 2006). In the zebrafish, several studies have focused on development of the kidney (Drummond 2005; Wingert and Davidson 2008; Lyons et al. 2009; Vasilyev et al. 2009). In different segments of the zebrafish nephron, expression patterns of ion transporters, including SLC4a2 (AE2; an anion exchanger), SLC4a4 (a $Na^+$-$HCO_3^-$ cotransporter), SLC9a3 (an NHE), SLC12a1 (a NKCC), SLC12a3 (an NCC), SLC13a3 (a $Na^+$/dicarboxylate transporter), SLC20a1 (a $Na^+$-$PO_4$ cotransporter), SGLT (a sodium glucose cotransporter), CLCK (a $Cl^-$ channel), and ROMK (a renal outer medullary $K^+$ channel), were identified (Shmukler et al. 2005; Nichane et al. 2006; Van Campenhout et al. 2006; Wingert et al. 2007). However, no studies have yet been performed to explore the roles of these transporters in ion uptake and acid–base regulatory mechanisms in the zebrafish kidney. In zebrafish embryos, SLC9a3 has 2 paralogs, NHE3a and NHE3b (Yan et al. 2007), and they are differentially expressed in proximal tubules and the pronephric duct, respectively (Fig. 8.3). Two isoforms of carbonic anhydrase (CA), CA2-like a, CA6 (Lin et al. 2008), and CA4c are also expressed in zebrafish embryonic kidneys. Preliminary experiments (Fig. 8.3) showed that acclimation to pH 4 FW affected the mRNA expressions of NHE3a,

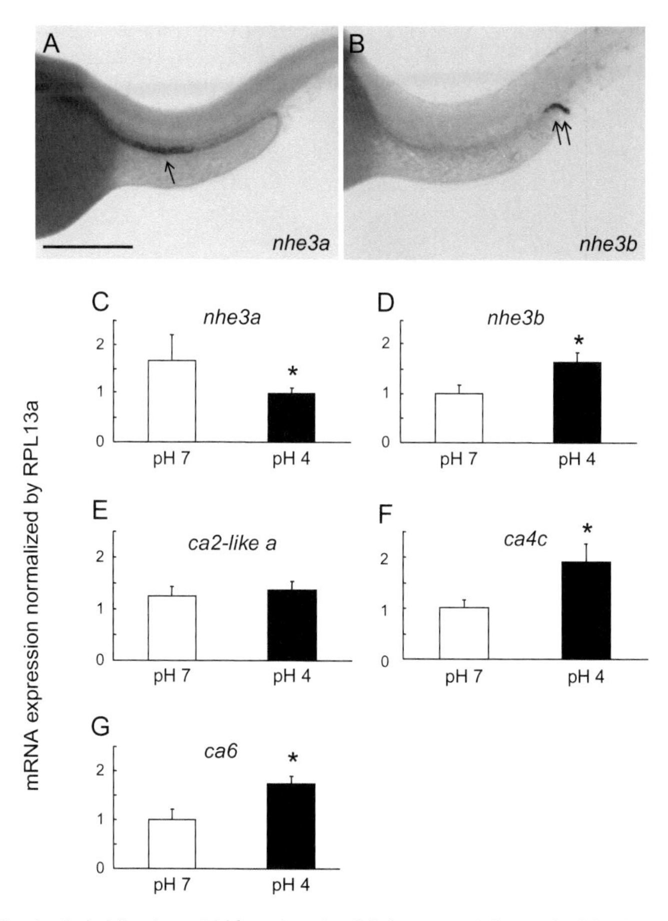


**Fig. 8.3.** In situ hybridization of kidney in zebrafish larvae at 5 d post-fertilization (A, B) and quantitative real-time PCR analysis in adult kidney (C–G). *nhe3a* (A) and *nhe3b* (B) mRNAs are differentially expressed in the proximal tubules (arrow) and the pronephric duct (double arrow), respectively. RNA probes for in situ hybridization: *nhe3a*, nt2686-3417 (NM_001113473.1); *nhe3b*, nt1807-2646 (NM_001113479.1). Scale bar: 50 μm. Asterisk indicates significant difference from pH 7 control ($p<0.05$, Student $t$-test). See color plate section

NHE3b, CA4c, and CA6 in the kidney, implying a role for these transporters and enzymes in zebrafish acid–base regulatory mechanisms. These preliminary data support the proposed model of Perry and Gilmour (2006) in which fish renal cells achieve the functions of $H^+$ secretion/$Na^+$ and $HCO_3^-$ uptake via apical NHE, $H^+$-ATPase, and CA4, cytosolic CA2-like, and basolateral NBC (Perry et al. 2003a; Georgalis et al. 2006; Ivanis et al. 2008). Like its mammalian counterpart, the pronephric nephron in zebrafish is segmented and expresses specific sets of markers (mainly ion transporters, as described above) in distinct regions (Wingert and Davidson 2008). Zebrafish may be a better model (compared to other fish species) for exploring the functions of ion uptake and acid–base regulation in respective segments of the kidney, assuming that in vivo methodologies to assay the zebrafish renal functions can be established in the future. Recently, an in vivo method based on clearance of a 70-kDa fluorescent dextran was established to assay the glomerular filtration in zebrafish embryos (Hentschel et al. 2007).

## 10. CONCLUSIONS AND PERSPECTIVES

Recent studies on ion regulation in zebrafish have provided convincing data to support the proposed models (Figs 8.2 and 8.4). In the model for ionic uptake, 4 types of ionocytes express distinct sets of ion transporters and enzymes, and 3 of them, NCC cells, NaR cells and HR cells, perform the functions of $Cl^-$ uptake, $Ca^{2+}$ uptake and $Na^+$ uptake/acid secretion and $NH_4^+$ excretion, respectively (Fig. 8.4). This model should serve as a platform for further studies on the neuroendocrine control pathways and the functional regulation and differentiation of ionocytes during acclimation to fluctuating environments. Moreover, knowledge gained from studies of zebrafish ion regulation may also enhance our understanding of mammalian renal transport physiology by providing an in vivo working model because the transporter expression patterns and functions of zebrafish ionocytes are analogous to those of proximal tubular cells, intercalated cells, $Ca^{2+}$ reabsorption cells and distal convoluted cells. Several issues remain to be explored in the future: (1) elucidating the mechanisms that drive the apical NHE in HR cells and NCC in NCC cells, (2) identification and functional analysis of the basolateral transporters in HR cells and NCC cells, (3) establishing the functions of the unidentified type of ionocytes.

Upon acclimation of fish to harsh environments, increased numbers and altered function of ionocytes are well documented in many species. Whether the increase in the number of cells is a result of the proliferation and

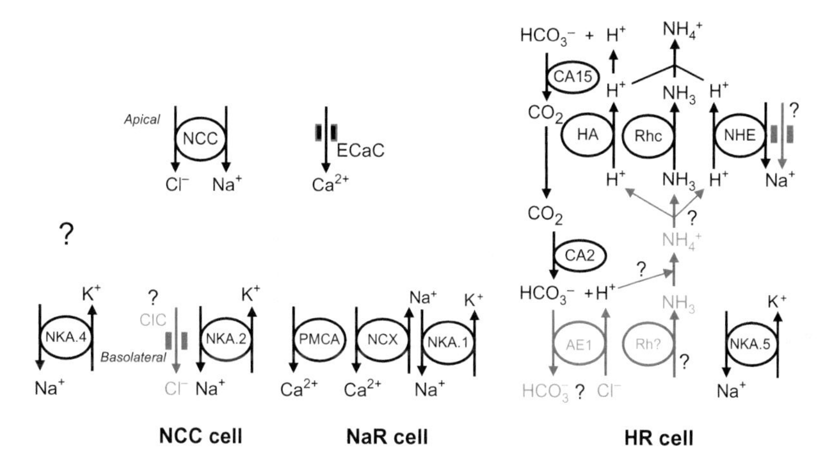


**Fig. 8.4.** Model of the ion regulation mechanism in zebrafish skin/gill ionocytes. Four types of ionocytes, $H^+$-ATPase-rich (HR) cell, $Na^+$-$K^+$-ATPase-rich (NaR) cell, $Na^{+-}Cl^-$ cotransporter (NCC) cell, and an unidentified type of ionocyte, express different sets of ion transporters and/or enzymes, and are respectively responsible for acid secretion/$Na^+$ uptake/$NH_4^+$ excretion, $Ca^{2+}$ uptake, $Cl^-$ uptake, and unknown functions. A question mark indicates an unidentified transporter or pathway. AE1, anion exchanger 1b; CA2, carbonic anhydrase 2-like a; CA15, carbonic anhydrase 15a; ClC, $Cl^-$ channel; ECaC, epithelial $Ca^{2+}$ channel; HA, $H^+$-ATPase; NCC, $Na^+$-$Cl^-$ cotransporter (SLC12a10.2); NCX, $Na^+/Ca^{2+}$ exchanger 1b; NHE, $Na^+/H^+$ exchanger 3b; NKA.1-5, $Na^+$-$K^+$-ATPase $\alpha$1 subunit subtypes (ATP1a1a.1–5, respectively); PMCA, plasma membrane $Ca^{2+}$-ATPase; Rh, Rhesus protein; Rhc, Rh type C glycoprotein 1.

differentiation of newly recruited ionocytes is an important issue in related fields and now may be addressed using the knowledge of the molecular pathways through Foxi3a/*b* and other transcriptional factors involved in ionocyte specification and differentiation (Fig. 8.2). Differentiation pathways of ionocytes in zebrafish appear to be extremely complicated and thus warrant additional studies.

The involvement of the neuroendocrine system in controlling fish ionic and acid–base regulation during environmental acclimation has been an important area of research with a long history in the related fields. Thus far, zebrafish studies have focused mostly on the development and ontogeny of the neuroendocrine system, and only very few studies have addressed the neuroendocrine control of ion regulation mechanisms. Similarly, although several studies have focused on the development of the kidney in zebrafish, none has so far been designed to explore the roles of the renal transporters in ion uptake and acid–base regulatory mechanisms. Future studies on neuroendocrine control of ion regulation mechanisms in zebrafish should not merely confirm the well-known pathways in other species. Instead, the proposed model for ion regulation (Fig. 8.4) and the differentiation pathways of ionocytes (Fig. 8.2) enable zebrafish to serve as a powerful model to identify the ion transport pathway(s) and the target ion transporter(s) or ionocyte(s) that are controlled by neuroendocrine agent(s), as well as to explore the neuroendocrine control of ionocyte differentiation. On the other hand, to explore the mechanism of ion uptake and acid–base regulation in the zebrafish kidney, there is an urgent need to establish in vivo methodologies for analyzing zebrafish renal function.

## ACKNOWLEDGMENTS

Our original research was financially supported by NSERC grants to SFP and grants to PPH from the National Science Council and Academia Sinica of Taiwan, ROC. We thank Y. C. Tung for her technical and secretarial assistance.

## REFERENCES

Alderman, S. L., and Bernier, N. J. (2009). Ontogeny of the corticotropin-releasing factor system in zebrafish. *Gen. Comp. Endocrinol.* **164**, 61–69.

Alper, S. L. (2006). Molecular physiology of SLC4 anion exchangers. *Exp. Physiol.* **91**, 153–161.

Alper, S. L., Darman, R. B., Chernova, M. N., and Dahl, N. K. (2002). The AE gene family of $Cl^-/HCO_3^-$ exchangers. *J. Nephrol.* **15**(Suppl. 5), S41–S53.

Alsop, D., and Vijayan, M. M. (2009). Molecular programming of the corticosteroid stress axis during zebrafish development. *Comp. Biochem. Physiol.* **153**, 49–54.

Avella, M., and Bornancin, M. (1989). A new analysis of ammonia and sodium transport through the gills of the freshwater rainbow trout (*Salmo gairdneri*). *J. Exp. Biol.* **142**, 155–175.

Avella, M., Masoni, A., Bornancin, M., and Mayer-Gostan, N. (1987). Gill morphology and sodium influx in the rainbow trout (*Salmo gairdneri*) acclimated to artificial freshwater environments. *J. Exp. Zool.* **241**, 159–169.

Bakkers, J., Hild, M., Kramer, C., Furutani-Seiki, M., and Hammerschmidt, M. (2002). Zebrafish Delta Np63 is a direct target of bmp signaling and encodes a transcriptional repressor blocking neural specification in the ventral ectoderm. *Dev. Cell* **2**, 617–627.

Bayaa, M., Vulesevic, B., Esbaugh, A., Braun, M., Ekker, M., Grosell, M., and Perry, S. F. (2009). The involvement of SLC26 anion exchangers in $Cl^-/HCO_3^-$ exchange in zebrafish (*Danio rerio*) larvae. *J. Exp. Biol.* **212**, 1940–1948.

Blechman, J., Borodovsky, N., Eisenberg, M., Nabel-Rosen, H., Grimm, J., and Levkowitz, G. (2007). Specification of hypothalamic neurons by dual regulation of the homeodomain protein Orthopedia. *Development* **134**, 4417–4426.

Boisen, A. M., Amstrup, J., Novak, I., and Grosell, M. (2003). Sodium and chloride transport in soft water and hard water acclimated zebrafish (*Danio rerio*). *Biochim. Biophys. Acta* **1618**, 207–218.

Braun, M., Steele, S. L., Ekker, M., and Perry, S. F. (2009a). Nitrogen excretion in developing zebrafish (*Danio rerio*): A role for Rh proteins and urea transporters. *Am. J. Physiol. Renal Physiol.* **296**, F994–F1005.

Braun, M. H., Steele, S. L., and Perry, S. F. (2009b). The responses of adult and juvenile zebrafish to high external ammonia and urea transporter inhibition: Nitrogen excretion and expression of Rh and UT proteins. *J. Exp. Biol.* **212**, 3846–3856.

Brown, D., Paunescu, T. G., Breton, S., and Marshansky, V. (2009). Regulation of the V-ATPase in kidney epithelial cells: dual role in acid/base homeostasis and vesicle trafficking. *J. Exp. Biol.* **212**, 1762–1772.

Chang, I. C., Lee, T. H., Yang, C. H., Wei, Y. Y., Lu, F. I., and Hwang, P. P. (2001). Modulation in morphology and function of gill mitochondria-rich cells in fish acclimated to different environments. *Physiol. Biochem. Zool.* **74**, 111–119.

Chang, W. J., Horng, J. L., Yan, J. J., Hsiao, C. D., and Hwang, P. P. (2009). The transcription factor, glial cell missing 2, is involved in differentiation and functional regulation of $H^+$-ATPase-rich cells in zebrafish (*Danio rerio*). *Am. J. Physiol.* **296**, R1192–R1201.

Chen, Y. Y., Lu, F. I., and Hwang, P. P. (2003). Comparisons of calcium regulation in fish larvae. *J. Exp. Zool.* **295A**, 127–135.

Chou, M. Y., Hsiao, C. D., Chen, S. C., Chen, I. W., Liu, S. T., and Hwang, P. P. (2008). Hypothermic effects on gene expressions in zebrafish gills: Up-regulations in differentiation and function of ionocytes as compensatory responses. *J. Exp. Biol.* **211**, 3077–3084.

Chretien, M., and Pisam, M. (1986). Cell Renewal and differentiation in the gill epithelium of freshwater-adapted or saltwater-adapted euryhaline fish as revealed by $[H^3]$ thymidine autoradiography. *Biol. Cell* **56**, 137–150.

Claiborne, J. B. (1998). Acid-base regulation. In: *The Physiology of Fishes*. Vol. 2. (D.H. Evans, ed.), pp. 171–198. CRC Press, Boca Raton.

Claiborne, J. B., Edwards, S. L., and Morrison-Shetlar, A. I. (2002). Acid–base regulation in fishes: Cellular and molecular mechanisms. *J. Exp. Zool.* **293**, 302–319.

Claiborne, J., Kratochvilova, H., Diamanduros, A. W., Hall, C., Phillips, M. E., Hirose, S., and Edwards, S. (2008). Expression of branchial Rh glycoprotein ammonia transporters in the marine longhorn sculpin (*Myoxocephalus octodecemspinosus*). *Bull. Mt. Desert Isl. Biol. Lab. Salisb. Cove Maine* **47**, 67–68.

Conte, F. P., and Lin, D. H. Y. (1967). Kinetics of cellular morphogenesis in gill epithelium during sea water adaptation of *Oncorhynchus* (Walbaum). *Comp. Biochem. Physiol.* **23**, 945–957.

Craig, P. M., Wood, C. M., and McClelland, G. B. (2007). Gill membrane remodeling with soft-water acclimation in zebrafish (*Danio rerio*). *Physiol. Genomics* **30**, 53–60.

De Renzis, G. (1975). The branchial chloride pump in the goldfish *Carassius auratus*: Relationship between $Cl^-/HCO_3^-$ and $Cl^-/Cl^-$ exchanges and the effect of thiocyanate. *J. Exp. Biol.* **63**, 587–602.

De Renzis, G., and Maetz, J. (1973). Studies on the mechanism of chloride absorption by the goldfish gill: Relation with acid–base regulation. *J. Exp. Biol.* **59**, 339–358.

De Robertis, E. M., and Kuroda, H. (2004). Dorsal-ventral patterning and neural induction in *Xenopus* embryos. *Annu. Rev. Cell Dev. Biol.* **20**, 285–308.

Drummond, I. A. (2005). Kidney development and disease in the zebrafish. *J. Am. Soc. Nephrol.* **16**, 299–304.

Eaton, J. L., and Glasgow, E. (2007). Zebrafish orthopedia (otp) is required for isotocin cell development. *Dev. Genes Evol.* **217**, 149–158.

Eaton, J. L., Holmqvist, B., and Glasgow, E. (2008). Ontogeny of vasotocin-expressing cells in zebrafish. Selective requirement for the transcriptional regulators orthopedia and single-minded 1 in the preoptic area. *Dev. Dyn.* **237**, 995–1005.

Esaki, M., Hoshijima, K., Nakamura, N., Munakata, K., Tanaka, M., Ookata, K., Asakawa, K., Kawakami, K., Wang, W., Weinberg, E. S., and Hirose, S. (2009). Mechanism of development of ionocytes rich in vacuolar-type $H^+$-ATPase in the skin of zebrafish larvae. *Dev. Biol.* **329**, 116–129.

Esaki, M., Hoshijima, K., Kobayashi, S., Fukuda, H., Kawakami, K., and Hirose, S. (2007). Visualization in zebrafish larvae of $Na^+$ uptake in mitochondria-rich cells whose differentiation is dependent on *foxi3a*. *Am. J. Physiol.* **292**, R470–R480.

Evans, D. H. (2008). Teleost fish osmoregulation: What have we learned since August Krogh, Homer Smith, and Ancel Keys. *Am. J. Physiol.* **295**, R704–R713.

Evans, D. H., and Cameron, J. N. (1986). Gill ammonia transport. *J. Exp. Zool.* **239**, 17–23.

Evans, D. H., Piermarini, P. M., and Choe, K. P. (2005). The multifunctional fish gill: Dominant site of gas exchange, osmoregulation, acid-base regulation, and excretion of nitrogenous waste. *Physiol. Rev.* **85**, 97–177.

Evans, D. H., Piermarini, P. M., and Potts, W. T. W (1999). Ionic transport in the fish gill epithelium. *J. Exp. Zool.* **283**, 641–652.

Flik, G., and Verbost, P. M. (1993). Calcium transport in fish gills and intestine. *J. Exp. Biol.* **184**, 17–29.

Flik, G., Verbost, P. M., and Wendelaar Bonga, S. E. (1995). Calcium transport processes in fishes. In: *Cellular and Molecular Approaches to Fish Ionic Regulation* (C.M. Wood and T.J. Shuttleworth, eds), pp. 317–336. Academic Press, New York.

Flynt, A. S., Thatcher, E. J., Burkewitz, K., Li, N., Liu, Y., and Patton, J. G. (2009). miR-8 microRNAs regulate the response to osmotic stress in zebrafish embryos. *J. Cell Biol.* **185**, 115–127.

Galvez, F., Reid, S. D., Hawkings, G., and Goss, G. G. (2002). Isolation and characterization of mitochondria-rich cell types from the gill of freshwater rainbow trout. *Am. J. Physiol.* **282**, R658–R668.

Galvez, F., Wong, D., and Wood, C. M. (2006). Cadmium and calcium uptake in isolated mitochondria-rich cell populations from the gills of the freshwater rainbow trout. *Am. J. Physiol.* **291**, R170–R176.

Georgalis, T., Gilmour, K. M., Yorston, J., and Perry, S. F. (2006). Roles of cytosolic and membrane-bound carbonic anhydrase in renal control of acid-base balance in rainbow trout, *Oncorhynchus mykiss*. *Am. J. Physiol.* **291**, F407–F421.

Gilmour, K. M., and Perry, S. F. (2009). Carbonic anhydrase and acid–base regulation in fish. *J. Exp. Biol.* **212**, 1647–1661.

Grosell, M., Mager, E. M., Williams, C., and Taylor, J. R. (2009). High rates of $HCO_3^-$ secretion and $Cl^-$ absorption against adverse gradients in the marine teleost intestine: The involvement of an electrogenic anion exchanger and $H^+$-pump metabolon? *J. Exp. Biol.* **212**, 1684–1696.

Hanssen, R. G., Lafeber, F. P., Flik, G., and Wendelaar Bonga, S. E. (1989). Ionic and total calcium levels in the blood of the European eel (*Anguilla anguilla*): Effects of stanniectomy and hypocalcin replacement therapy. *J. Exp. Biol.* **141**, 177–186.

Hentschel, D. M., Mengel, M., Boehme, L., Liebsch, F., Albertin, C., Bonventre, J. V., Haller, H., and Schiffer, M. (2007). Rapid screening of glomerular slit diaphragm integrity in larval zebrafish. *Am. J. Physiol.* **293**, F1746–F1750.

Hirata, T., Kaneko, T., Ono, T., Nakazato, T., Furukawa, N., Hasegawa, S., Wakabayashi, S., Shigekawa, M., Chang, M. H., Romero, M. F., and Hirose, S. (2003). Mechanism of acid adaptation of a fish living in a pH 3.5 lake. *Am. J. Physiol.* **284**, R1199–R1212.

Hiroi, J., Kaneko, T., and Tanaka, M. (1999). In vivo sequential changes in chloride cell morphology in the yolk-sac membrane of Mozambique tilapia (*Oreochromis mossambicus*) embryos and larvae during seawater adaptation. *J. Exp. Biol.* **202**, 3485–3495.

Hiroi, J., Yasumasu, S., McCormick, S. D., Hwang, P. P., and Kaneko, T. (2008). Evidence for an apical Na-Cl cotransporter involved in ion uptake in a teleost fish. *J. Exp. Biol.* **211**, 2584–2599.

Hoenderop, J. G., Nilius, B., and Bindels, R. J. (2005). Calcium absorption across epithelia. *Physiol. Rev.* **85**, 373–422.

Horng, J. L., Lin, L. Y., and Hwang, P. P. (2009). Functional regulation of $H^+$-ATPase-rich cells in zebrafish embryos acclimated to an acidic environment. *Am. J. Physiol.* **296**, C682–C692.

Horng, J. L., Lin, L. Y., Huang, C. J., Katoh, F., Kaneko, T., and Hwang, P. P. (2007). Knockdown of V-ATPase subunit A (*atp6v1a*) impairs acid secretion and ion balance in zebrafish (*Danio rerio*). *Am. J. Physiol.* **292**, R2068–R2076.

Hoshijima, K., and Hirose, S. (2007). Expression of endocrine genes in zebrafish larvae in response to environmental salinity. *J. Endocrinol.* **193**, 481–491.

Hsiao, C. D., You, M. S., Guh, Y. J., Ma, M., Jiang, Y. J., and Hwang, P. P. (2007). A positive regulatory loop between *foxi*3a and *foxi*3b is essential for specification and differentiation of zebrafish epidermal ionocytes. *PLoS ONE* **2**, e302.

Hung, C. C., Nawata, C. M., Wood, C. M., and Wright, P. A. (2008). Rhesus glycoprotein and urea transporter genes are expressed in early stages of development of rainbow trout (*Oncorhynchus mykiss*). *J. Exp. Zool.* **309**, 262–268.

Hwang, P. P. (2009). Ion uptake and acid secretion in zebrafish (*Danio rerio*). *J. Exp. Biol.* **212**, 1745–1752.

Hwang, P. P., and Lee, T. H. (2007). New insights into fish ion regulation and mitochondrion-rich cells. *Comp. Biochem. Physiol. A* **148**, 479–497.

Inokuchi, M., Hiroi, J., Watanabe, S., Lee, K. M., and Kaneko, T. (2008). Gene expression and morphological localization of NHE3, NCC and NKCC1a in branchial mitochondria-rich cells of Mozambique tilapia (*Oreochromis mossambicus*) acclimated to a wide range of salinities. *Comp. Biochem. Physiol. A* **151**, 151–158.

Inokuchi, M., Hiroi, J., Watanabe, S., Hwang, P. P., and Kaneko, T. (2009). Morphological and functional classification of ion-absorbing mitochondria-rich cells in the gills of Mozambique tilapia. *J. Exp. Biol.* **212**, 1003–1010.

Ishihara, A., and Mugiya, Y. (1987). Ultrastructural evidence of calcium uptake by chloride cells in the gills of the goldfish, *Carassius auratus*. *J. Exp. Biol.* **242**, 121–129.

Ivanis, G., Braun, M., and Perry, S. F. (2008). Renal expression and localization of SLC9A3 sodium/hydrogen exchanger and its possible role in acid-base regulation in freshwater rainbow trout (*Oncorhynchus mykiss*). *Am. J. Physiol.* **295**, R971–R978.

Janicke, M., Carney, T. J., and Hammerschmidt, M. (2007). *foxi3* transcription factors and Notch signaling control the formation of skin ionocytes from epidermal precursors of the zebrafish embryo. *Dev. Biol.* **307**, 258–271.

Javelle, A., Lupo, D, Ripoche, P., Fulford, T., Merrick, M., and Winkler, F. (2008). Substrate binding, deprotonation, and selectivity at the periplasmic entrance of the *Escherichia coli* ammonia channel AmtB. *Proc. Natl Acad. Sci. USA* **105**, 5040–5045.

Jonz, M. G, and Nurse, C. A. (2003). Neuroepithelial cells and associated innervation of the zebrafish gill: A confocal immunofluorescence study. *J. Comp. Neurol.* **461**, 1–17.

Jonz, M. G., and Nurse, C. A. (2006). Epithelial mitochondria-rich cells and associated innervation in adult and developing zebrafish. *J. Comp. Neurol.* **497**, 817–832.

Kessel, R. G., and Beams, H. W. (1962). Electron microscope studies on the gill filaments of *Fundulus heteroclitus* from sea water and fresh water with special reference to the ultrastructural organization of the "chloride cell". *J. Ultrastruct. Res.* **6**, 77–87.

Khademi, S., O'Connell, J., III, Remis, J., Robles-Colmenares, Y., Miercke, L. J. W., and Stroud, R. M. (2004). Mechanism of ammonia transport by Amt/MEP/Rh: Structure of AmtB at 1.35 Å. *Science* **305**, 1587–1594.

Krogh, A. (1937). Osmotic regulation in freshwater fishes by active absorption of chloride ions. *Z. Vergl. Physiol.* **24**, 656–666.

Krogh, A. (1938). The active absorption of ions in some freshwater animals. *Z. Vergl. Physiol.* **25**, 335–350.
Lafeber, F. P., Flik, G., Wendelaar Bonga, S. E., and Perry, S. F. (1988). Hypocalcin from Stannius corpuscles inhibits gill calcium uptake in trout. *Am. J. Physiol.* **254**, R891–R896.
Lafeber, F. P., and Perry, S. F. (1988). Experimental hypercalcemia induces hypocalcin release and inhibits branchial $Ca^{2+}$ influx in freshwater trout. *Gen. Comp. Endocrinol.* **72**, 136–143.
Liao, B. K., Chen, R. D., and Hwang, P. P. (2009). Expression regulation of Na-K-ATPase alpha 1 subunit subtypes in zebrafish gill ionocytes. *Am. J. Physiol.* **296**, R1897–R1906.
Liao, B. K., Deng, A. N., Chen, S. C., Chou, M. Y., and Hwang, P. P. (2007). Expression and water calcium dependence of calcium transporter isoforms in zebrafish gill mitochondrion-rich cells. *BMC Genomics.* **8**, 354.
Lin, L. Y., and Hwang, P. P. (2001). Modification of morphology and function of integument mitochondria-rich cells in tilapia larvae (*Oreochromis mossambicus*) acclimated to ambient chloride levels. *Physiol. Biochem. Zool.* **74**, 469–476.
Lin, L. Y., and Hwang, P. P. (2004). Mitochondria-rich cell activity in the yolk-sac membrane of tilapia (*Oreochromis mossambicus*) larvae acclimatized to different ambient chloride levels. *J. Exp. Biol.* **207**, 1335–1344.
Lin, L. Y., Horng, J. L., Kunkel, J. G., and Hwang, P. P. (2006). Proton pump-rich cell secretes acid in skin of zebrafish larvae. *Am. J. Physiol.* **290**, C371–C378.
Lin, T. Y., Liao, B. K., Horng, J. L., Yan, J. J., Hsiao, C. D., and Hwang, P. P. (2008). Carbonic anhydrase 2-like a and 15a are involved in acid-base regulation and $Na^+$ uptake in zebrafish $H^+$-ATPase-rich cells. *Am. J. Physiol.* **294**, C1250–C1260.
Liu, N. A., Liu, Q., Wawrowsky, K., Yang, Z., Lin, S., and Melmed, S. (2006). Prolactin receptor signaling mediates the osmotic response of embryonic zebrafish lactotrophs. *Mol. Endocrinol.* **20**, 871–880.
Lyons, J. P., Miller, R. K., Zhou, X., Weidinger, G., Deroo, T., Denayer, T., Park, J. I., and Ji, H. (2009). Requirement of Wnt/beta-catenin signaling in pronephric kidney development. *Mech. Dev.* **126**, 142–159.
Maetz, J., and Garcia Romeu, F. (1964). The mechanism of sodium and chloride uptake by the gills of a fresh-water fish, *Carassius auratus*: II. Evidence for $NH_4^+/Na^+$ and $HCO_3^-$/ $Cl^-$exchanges. *J. Gen. Physiol.* **47**, 1209–1227.
Marshall, W. S. (1995). Transport processes in isolated teleost epithelia: Opercular epithelium and urinary bladder. In: *Cellular and Molecular Approaches to Fish Ionic Regulation* (C.M. Wood and T.J. Shuttleworth, eds), pp. 1–23. Academic Press, New York.
Marshall, W. S. (2002). $Na^+$, $Cl^-$, $Ca^{2+}$ and $Zn^{2+}$ transport by fish gills: Retrospective review and prospective synthesis. *J. Exp. Zool.* **293**, 264–283.
Marshall, W. S. (2003). Rapid regulation of NaCl secretion by estuarine teleost fish: Coping strategies for short-duration freshwater exposures. *Biochim. Biophys. Acta* **1618**, 95–105.
Marshall, W. S., Bryson, S. E., and Wood, C. M. (1992). Calcium transport by isolated skin of rainbow trout. *J. Exp. Biol.* **166**, 297–316.
Marshall, W. S., and Grosell, M. (2006). Ion transport, osmoregulation and acid-base balance. In: *The Physiology of Fishes* (D.H. Evans and J.B. Claiborne, eds), pp. 177–230. CRC Press, Boca Raton.
Marshall, W. S., Ossum, C. G., and Hoffmann, E. K. (2005). Hypotonic shock mediation by p38 MAPK, JNK, PKC, FAK, OSR1 and SPAK in osmosensing chloride secreting cells of killifish opercular epithelium. *J. Exp. Biol.* **208**, 1063–1077.
McClure, M. M., McIntyre, P. B., and McCune, A. R. (2006). Notes on the natural diet and habitat of eight danionin fishes, including the zebrafish *Danio rerio*. *J. Fish. Biol.* **69**, 553–570.

McCormick, S. D., Hasegawa, S., and Hirano, T. (1992). Calcium uptake in the skin of a freshwater teleost. *Proc. Natl Acad Sci USA* **89**, 3635–3638.

Mommsen, T. P., and Walsh, P. J. (1992). Biochemical and environmental perspectives on nitrogen metabolism in fishes. *Experientia* **48**, 583–593.

Moreau, M., and Leclerc, C. (2004). The choice between epidermal and neural fate: A matter of calcium. *Int. J. Dev. Biol.* **48**, 75–84.

Mount, D. B., and Romero, M. F. (2004). The SLC26 gene family of multifunctional anion exchangers. *Pflugers Arch.* **447**, 710–721.

Nakada, T., Hoshijima, K., Esaki, M., Nagayoshi, S., Kawakami, K., and Hirose, S. (2007a). Localization of ammonia transporter Rhcg1 in mitochondrion-rich cells of yolk sac, gill, and kidney of zebrafish and its ionic strength-dependent expression. *Am. J. Physiol.* **293**, R1743–R1753.

Nakada, T., Westhoff, C. M., Kato, A., and Hirose, S. (2007b). Ammonia secretion from fish gill depends on a set of Rh glycoproteins. *FASEB J* **21**, 1067–1074.

Nawata, C. M., Hung, C. C., Tsui, T. K., Wilson, J. M., Wright, P. A., and Wood, C. M. (2007). Ammonia excretion in rainbow trout (*Oncorhynchus mykiss*): Evidence for Rh glycoprotein and $H^+$-ATPase involvement. *Physiol. Genomics* **31**, 463–474.

Nichane, M., Van Campenhout, C., Pendeville, H., Voz, M. L., and Bellefroid, E. J. (2006). The $Na^+/PO_4$ cotransporter slc20a1 gene labels distinct restricted subdomains of the developing pronephros in *Xenopus* and zebrafish embryos. *Gene Expr. Patterns* **6**, 667–672.

Ohana, E., Yang, D., Shcheynikov, N., and Muallem, S. (2008). Diverse transport modes by the Solute Carrier 26 family of anion transporters. *J. Physiol.* **587**, 2179–2185.

Okabe, M., and Graham, A. (2004). The origin of the parathyroid gland. *Proc. Natl Acad. Sci. USA* **101**, 17716–17719.

Pan, T. C., Liao, B. K., Huang, C. J., Lin, L. Y., and Hwang, P. P. (2005). Epithelial $Ca^{2+}$ channel expression and $Ca^{2+}$ uptake in developing zebrafish. *Am. J. Physiol.* **289**, R1202–R1211.

Parks, S. K., Tresguerres, M., and Goss, G. G. (2007). Interactions between $Na^+$ channels and $Na^+$-$HCO_3^-$ cotransporters in the freshwater fish gill MR cell: A model for transepithelial $Na^+$ uptake. *Am. J. Physiol.* **292**, C935–C944.

Parks, S. K., Tresguerres, M., and Goss, G. G. (2008). Theoretical considerations underlying $Na^+$ uptake mechanisms in freshwater fishes. *Comp. Biochem. Physiol. C* **148**, 411–418.

Perry, S. F. (1997). The chloride cell: Structure and function in the gill of freshwater fishes. *Annu. Rev. Physiol.* **59**, 325–347.

Perry, S. F., and Flik, G. (1988). Characterization of branchial transepithelial calcium fluxes in freshwater trout, *Salmo gairdneri*. *Am. J. Physiol.* **254**, 491–498.

Perry, S. F., Furimsky, M., Bayaa, M., Georgalis, T., Nickerson, J. G., and Moon, T. W. (2003a). Integrated involvement of $Na^+/HCO_3^-$ cotransporters and V-type $H^+$-ATPases in branchial and renal acid-base regulation in freshwater fishes. *Biochem. Biophys. Acta* **1618**, 175–184.

Perry, S. F., and Gilmour, K. M. (2006). Acid-base balance and $CO_2$ excretion in fish: Unanswered questions and emerging models. *Respir. Physiol. Neurobiol.* **154**, 199–215.

Perry, S. F., Gilmour, K. M., Bernier, N. J., and Wood, C. M. (1999). Does gill boundary layer carbonic anhydrase contribute to carbon dioxide excretion: A comparison between dogfish (*Squalus acanthias*) and rainbow trout (*Oncorhynchus mykiss*). *J. Exp. Biol.* **202**, 749–756.

Perry, S. F., Goss, G. G., and Fenwick, J. C. (1992). Interrelationships between gill chloride cell morphology and calcium uptake in freshwater teleosts. *Fish Physiol. Biochem.* **10**, 327–337.

Perry, S. F., Haswell, M. S., Randall, D. J., and Farrell, A. P. (1981). Branchial ionic uptake and acid-base regulation in the rainbow trout, *Salmo gairdneri*. *J. Exp. Biol.* **92**, 289–303.

Perry, S. F., Seguin, D., Lafeber, F. P., Wendelaar Bonga, S. E. W., and Fenwick, J. C. (1989). Depression of whole-body calcium uptake during acute hypercalcemia in American eel, *Anguilla rostrata*, is mediated exclusively by corpuscles of Stannius. *J. Exp. Biol.* **147**, 249–261.

Perry, S. F., Shahsavarani, A., Georgalis, T., Bayaa, M., Furimsky, M., and Thomas, S. (2003b). Channels, pumps, and exchangers in the gill and kidney of freshwater fishes: Their role in ionic and acid–base regulation. *J. Exp. Zool. A* **300**, 53–62.

Perry, S. F., Vulesevic, B., and Bayaa, M. (2009). Evidence that SLC26 anion transporters mediate branchial chloride uptake in adult zebrafish (*Danio rerio*). *Am. J. Physiol. Regul. Integr. Comp. Physiol.* **297**, R988–R997.

Perry, S. F., and Wood, C. M. (1985). Kinetics of branchial calcium uptake in the rainbow trout: Effects of acclimation to various external calcium levels. *J. Exp. Biol.* **116**, 411–433.

Piermarini, P. M, Verlander, J. W., Royaux, I. E., and Evans, D. H. (2002). Pendrin immunoreactivity in the gill epithelium of a euryhaline elasmobranch. *Am. J. Physiol.* **283**, R983–R992.

Pogoda, H. M, and Hammerschmidt, M. (2007). Molecular genetics of pituitary development in zebrafish. *Semin. Cell Dev. Biol.* **18**, 543–558.

Pushkin, A., and Kurtz, I. (2006). SLC4 base ($HCO_3^-$, $CO_3^{2-}$) transporters: Classification, function, structure, genetic diseases, and knockout models. *Am. J. Physiol.* **290**, F580–F599.

Randall, D. J., and Wright, P. A. (1989). The interaction between carbon dioxide and ammonia excretion and water pH in fish. *Can. J. Zool.* **67**, 2936–2942.

Reid, S. D., Hawkings, G. S., Galvez, F., and Goss, G. G. (2003). Localization and characterization of phenamil-sensitive $Na^+$ influx in isolated rainbow trout gill epithelial cells. *J. Exp. Biol.* **206**, 551–559.

Renfro, J. L. (1975). Water and ion transport by the urinary bladder of the teleost *Pseudopleuronectes americanus*. *Am. J. Physiol.* **228**, 52–61.

Romero, M. F., Chang, M. H., Plata, C., Zandi-Nejad, K., Mercado, A., Broumand, V., Sussman, C. R., and Mount, D. B. (2006). Physiology of electrogenic SLC26 paralogues. *Novartis Found. Symp.* **273**, 126–128.

Romero, M. F., Fulton, C. M., and Boron, W. F. (2004). The SLC4 family of $HCO_3^-$ transporters. *Pflügers Arch.* **447**, 495–509.

Shahsavarani, A., McNeill, B., Galvez, F., Wood, C. M., Goss, G. G., Hwang, P. P., and Perry, S. F. (2006). Characterization of a branchial epithelial calcium channel (ECaC) in freshwater rainbow trout (*Oncorhynchus mykiss*). *J. Exp. Biol.* **209**, 1928–1943.

Shaw, J. R., Sato, J. D., VanderHeide, J., LaCasse, T., Stanton, C. R., Lankowski, A., Stanton, S. E., Chapline, C., Coutermarsh, B., Barnaby, R., Karlson, K., and Stanton, B. A. (2008). The role of SGK and CFTR in acute adaptation to seawater in *Fundulus heteroclitus*. *Cell. Physiol. Biochem.* **22**, 69–78.

Shih, T. H., Horng, J. L., Hwang, P. P., and Lin, L. Y. (2008). Ammonia excretion by the skin of zebrafish (*Danio rerio*) larvae. *Am. J. Physiol.* **295**, C1625–C1632.

Shmukler, B. E., Kurschat, C. E., Ackermann, G. E., Jiang, L., Zhou, Y., Barut, B., Stuart-Tilley, A. K., Zhao, J., Zon, L. I., Drummond, I. A., Vandorpe, D. H., Paw, B. H., and Alper, S. L. (2005). Zebrafish slc4a2/ae2 anion exchanger: cDNA cloning, mapping, functional characterization, and localization. *Am. J. Physiol.* **289**, F835–F849.

Sindic, A., Chang, M. H., Mount, D. B., and Romero, M. F. (2007). Renal physiology of SLC26 anion exchangers. *Curr. Opin. Nephrol. Hypertens.* **16**, 484–490.

Soleimani, M., and Xu, J. (2006). SLC26 chloride/base exchangers in the kidney in health and disease. *Semin. Nephrol.* **26**, 375–385.

Spence, R., Gerlach, G., Lawrence, C., and Smith, C. (2008). The behaviour and ecology of the zebrafish, *Danio rerio*. *Biol. Rev.* **83**, 13–34.

Tang, C. H., and Lee, T. H. (2007). The effect of environmental salinity on the protein expression of $Na^+/K^+$-ATPase, $Na^+/K^+/2Cl^-$ cotransporter, cystic fibrosis transmembrane conductance regulator, anion exchanger 1, and chloride channel 3 in gills of a euryhaline teleost, *Tetraodon nigroviridis*. *Comp. Biochem. Physiol. A* **147**, 521–528.

Tondeur, F., and Sargent, J. R. (1979). Biosynthesis of macromolecules in chloride cells in the gills of the common eel, *Anguilla anguilla*, adapting to sea water. *Comp. Biochem. Physiol. B* **62B**, 13–16.

Tresguerres, M., Katoh, F., Orr, E., Parks, S. K., and Goss, G. G. (2006). Chloride uptake and base secretion in freshwater fish: A transepithelial ion-transport metabolon? *Physiol. Biochem. Zool.* **79**, 981–996.

Tseng, D. Y., Chou, M. Y., Tseng, Y. J., Hsiao, C. D., Huang, C. J., Kaneko, T., and Hwang, P. P. (2009a). Effects of stanniocalcin 1 on calcium uptake in zebrafish (*Danio rerio*) embryo. *Am. J. Physiol.* **296**, R549–R557.

Tseng, Y. C., Chen, R. D., Lee, J. R., Liu, S. T., Lee, S. J., and Hwang, P. P. (2009b). Specific Expression and regulation of glucose transporters in zebrafish ionocytes. *Am. J. Physiol.* **297**, R275–R290.

Tseng, Y. C., and Hwang, P. P. (2008). Some insights into energy metabolism for fish osmoregulation. *Comp. Biochem. Physiol. C* **148**, 419–429.

Tseng, Y. C., Huang, C. J., Chang, J. C., Teng, W. Y., Baba, O., Fann, M. J., and Hwang, P. P. (2007). Glycogen phosphorylase in glycogen-rich cells is involved in energy supply for ion regulation in fish branchial epithelia. *Am. J. Physiol.* **283**, R482–R491.

Tsui, T. K. N., Hung, C. Y. C., Nawata, C. M., Wilson, J. M., Wright, P. A., and Wood, C. M. (2009). Ammonia transport in cultured gill epithelium of freshwater rainbow trout: The importance of Rhesus glycoproteins and the presence of an apical $Na^+/NH_4^+$ exchange complex. *J. Exp. Biol.* **212**, 878–892.

Van Campenhout, C., Nichane, M., Antoniou, A., Pendeville, H., Bronchain, O. J., Marine, J. C., Mazabraud, A., Voz, M. L., and Bellefroid, E. J. (2006). Evi1 is specifically expressed in the distal tubule and duct of the *Xenopus* pronephros and plays a role in its formation. *Dev. Biol.* **294**, 203–219.

Vasilyev, A., Liu, Y., Mudumana, S., Mangos, S., Lam, P. Y., Majumdar, A., Zhao, J., Poon, K. L., Kondrychyn, I., Korzh, V., and Drummond, I. A. (2009). Collective cell migration drives morphogenesis of the kidney nephron. *PLoS Biol.* **7**(1), e9.

Wagner, C. A., Finberg, K. E., Breton, S., Marshansky, V., Brown, D., and Geibel, J. P. (2004). Renal vacuolar-ATPase. *Physiol. Rev.* **84**, 1263–1314.

Walsh, P. J., and Henry, R. P. (1991). Carbon dioxide and ammonia metabolism and exchange. In: *Biochemistry and Molecular Biology of Fishes* (P.W. Hochachka and T.P. Mommsen, eds), Vol. 1, pp. 181–207. Elsevier Science Publishers, Amsterdam.

Wang, Y. F., Tseng, Y. C., Yan, J. J., Hiroi, J., and Hwang, P. P. (2009). Role of SLC12A10.2, a Na-Cl cotransporter-like protein, in a $Cl^-$ uptake mechanism in zebrafish (*Danio rerio*). *Am. J. Physiol.* **296**, R1650–R1660.

Weihrauch, D., Wilkie, M. P., and Walsh, P. J. (2009). Ammonia and urea transporters in gills of fish and aquatic crustaceans. *J. Exp. Biol.* **212**, 1716–1730.

Wendelaar Bonga, S. E. W., and Pang, P. K. T. (1991). Control of calcium regulating hormones in the vertebrates–parathyroid hormone, calcitonin, prolactin, and stanniocalcin. *Int. Rev. Cytol.* **128**, 139–213.

Wendl, T., Adzic, D., Schoenebeck, J. J., Scholpp, S., Brand, M., Yelon, D., and Rohr, K. B. (2007). Early developmental specification of the thyroid gland depends on han-expressing surrounding tissue and on FGF signals. *Development* **134**, 2871–2879.

Wilkie, M. P. (2002). Ammonia excretion and urea handling by fish gills: Present understanding and future research challenges. *J. Exp. Zool.* **293**, 284–301.

Wilson, J. M., Laurent, P., Tufts, B. L., Benos, D. J., Donowitz, M., Vogl, A. W., and Randall, D. J. (2000). NaCl uptake by the branchial epithelium in freshwater teleost fish: An immunological approach to ion-transport protein localization. *J. Exp. Biol.* **203**, 2279–2296.

Wingert, R. A., and Davidson, A. J. (2008). The zebrafish pronephros: A model to study nephron segmentation. *Kidney Int.* **73**, 1120–1127.

Wingert, R. A., Selleck, R., Yu, J., Song, H. D., Chen, Z., Song, A., Zhou, Y., Thisse, B., Thisse, C., McMahon, A. P., and Davidson, A. J. (2007). The cdx genes and retinoic acid control the positioning and segmentation of the zebrafish pronephros. *PLoS Genet.* **3**, e189.

Wong, C. K. C., and Chan, D. K. O. (1999). Chloride cell subtypes in the gill epithelium of Japanese eel *Anguilla japonica*. *Am. J. Physiol.* **277**, R517–R522.

Wood, C. M. (1993). Ammonia and urea metabolism and excretion. In: *The Physiology of Fishes* (D.H. Evans, ed.), pp. 379–425. CRC Press, Boca Raton.

Wright, P. A. (1995). Nitrogen excretion: Three end products, many physiological roles. *J. Exp. Biol.* **198**, 273–281.

Wright, P. A., Heming, T., and Randall, D. J. (1986). Downstream pH changes in water flowing over the gills of rainbow trout. *J. Exp. Biol.* **126**, 499–512.

Wright, P. A., Randall, D. J., and Perry, S. F. (1989). Fish gill water boundary layer: A site of linkage between carbon dioxide and ammonia excretion. *J. Comp. Physiol. B* **158**, 627–635.

Wright, P. A., and Wood, C. M. (2009). A new paradigm for ammonia excretion in aquatic animals: Role of Rhesus (Rh) glycoproteins. *J. Exp. Biol.* **212**, 2303–2312.

Yan, J. J., Chou, M. Y., Kaneko, T., and Hwang, P. P. (2007). Gene expression of $Na^+/H^+$ exchanger in zebrafish $H^+$-ATPase-rich cells during acclimation to low-$Na^+$ and acidic environments. *Am. J. Physiol.* **293**, C1814–C1823.

Yun, C. H., Oh, S., Zizak, M., Steplock, D., Tsao, S., Tse, C. M., Weinman, E. J., and Donowitz, M. (1997). cAMP-mediated inhibition of the epithelial brush border $Na^+/H^+$ exchanger, NHE3, requires an associated regulatory protein. *Proc. Natl Acad. Sci. USA* **94**, 3010–3015.

# 9

# THE ZEBRAFISH AS A MODEL FOR HUMAN DISEASE

*CONG XU*
*LEONARD I. ZON*



The zebrafish has emerged as a useful model for human disease. Zebrafish embryos develop externally and are available for observation and manipulation at all developmental stages. Transparency of zebrafish embryos and larvae allows real-time imaging of internal organs. The zebrafish is conducive to large-scale forward genetic screens, which have generated models of a variety of human diseases. This chapter reviews the achievements of the zebrafish in modeling human disease including cancer, hematopoietic disorders, and cardiovascular disease. It surveys approaches that are available to zebrafish researchers to generate models for human disease and examines the expanding field of chemical screening in zebrafish.



*Zebrafish: Volume 29*
FISH PHYSIOLOGY


DOI: 10.1016/S1546-5098(10)02909-2

## 1. INTRODUCTION

Model organisms have been extremely useful for the study of human disease and biology. As there is a surprising degree of evolutionary conservation of basic cellular processes among all organisms, both invertebrates and vertebrates have contributed to our understanding of human disease. The power of forward genetics has made invertebrates like *C. elegans* and *Drosophila* popular models. Due to the lack of vertebrate-specific structures and organ systems, such as the skeleton, liver, kidney, multichambered heart, multilineage hematopoietic system, and notochord, many human diseases cannot be studied in invertebrates. Vertebrate models have been used extensively to study these disorders.

Mice share striking similarities with humans at many levels spanning from genomic homology to anatomy and physiology. Targeting a defined gene by homologous recombination in mouse embryonic stem (ES) cells allows researchers to create mouse models that are theoretically capable of mimicking most human diseases caused by genetic mutations. Additionally, transgenic technology enables the generation of mouse models that overexpress disease-causing alleles. Combining targeted transgene activation techniques allows spatial and temporal control of disease-causing alleles. Transgenic mouse models have been used to create models for many human acquired diseases, remarkable examples including cancer models with tissue-specific activation of oncogenes (Frese and Tuveson 2007) and neural degenerative models that express polyglutamine expansions in specific neurons (Price et al. 1998). Despite the advantages of mouse models, the prohibitive cost of mouse colonies can limit many experiments, particularly large-scale genetic screens.

An alternative vertebrate organism, the zebrafish, is conducive to large-scale genetics. Once known mostly as a tropical fish pet, the zebrafish has developed into a powerful model organism for studying development and organogenesis. In comparison with the mouse, there are several advantages to using zebrafish to model human diseases. Firstly, a pair of adult zebrafish is capable of producing hundreds of fertilized eggs per week, and only a small space is required to maintain a large number of zebrafish. The high fecundity and low maintenance cost enable most zebrafish labs to perform large-scale forward genetic screens. Secondly, mouse embryos develop in utero, whereas zebrafish embryos develop externally and are readily available for observation and manipulation right after fertilization. Lastly, for many diseases, it is difficult to examine disease progression in a mouse without surgery and postmortem examination. Transparency of zebrafish embryos allows real-time imaging of internal organs. In addition, generation of a transparent zebrafish line named *casper* has enabled real-time monitoring in adults. *casper* can be used for monitoring of hematopoietic and tumor cell engraftment after transplantation (White et al. 2008).

This chapter discusses the use of zebrafish as a model system for the study of human disease. It surveys approaches that are available to zebrafish researchers to generate models for human diseases. This chapter does not intend to cover all disease models but focuses on cancer, hematopoietic disorders, and cardiovascular disease, fields where the zebrafish has proven particularly useful. Finally, this chapter examines the expanding field of chemical screening in the zebrafish, which aids in the development of therapeutic drugs.

## 2. CANCER

The zebrafish has been a useful cancer model for several reasons. Like humans, zebrafish develop tumors spontaneously (Kent et al. 2002; Smolowitz et al. 2002; Matthews 2004). Comparison of the zebrafish and human genomes revealed that most oncogenes and tumor suppressors are evolutionarily conserved (Postlethwait et al. 2000; Liu et al. 2002). The zebrafish has enabled the application of forward genetics to cancer research, in part because a large number of animals with cancer can be generated. The relative transparency of the zebrafish significantly helps in identifying events involved in carcinogenesis and tumor progression.

Several tumor types have been generated, including melanoma (Patton et al. 2005), rhabdomyosarcoma (Langenau et al. 2007), kidney cancer, liver tumors, malignant peripheral nerve sheath tumors (MPNSTs) (Berghmans et al. 2005), and leukemia (Langenau et al. 2003). Strategies used to study cancer in zebrafish include carcinogenic tumor induction, forward genetic screens, targeted tumor suppressor inactivation, spatial and temporal control of oncogene activation, and xenotransplantion of human tumor cells.

### 2.1. Transgenic Models Overexpressing Oncogenes

Cancer onset and progression usually require activation of oncogenes. Transgenic animals expressing activated oncogenes in a tissue-specific manner recapitulate human tumors. The success of mouse models propelled the development of similar approaches in zebrafish. Most zebrafish cancer models express mammalian oncogenes.

The first transgenic cancer model developed in the zebrafish was a myc-induced T-cell leukemia (Langenau et al. 2003). The *rag2* promoter was used to drive the mouse c-myc (*mMyc*) oncogene in zebrafish lymphoid cells. To visualize zebrafish cells carrying the oncogene in vivo, a chimeric

*EGFP-mMyc* transgene was generated. Wild-type fish embryos were microinjected with the *rag::mMyc* or *rag2::EGFP-mMyc* transgene. Five per cent of $F_0$ fish developed T-cell leukemias. Tumors arose in the thymus, spread locally to gills and the retro-orbital area, and then disseminated into abdominal organs and muscles. To assess the transplantability of zebrafish leukemic cells, *rag2::mMyc* was injected into embryos from a stable transgenic *rag2::GFP* line, so that the lymphoblasts would be GFP positive when leukemia developed. GFP-positive lymphoblasts were transplanted intraperitoneally into irradiated wild-type adult zebrafish. The homing and progression of the transplanted leukemic cells were detected in vivo in recipients by 26 days post transplantation. The ability to examine the progression of T-cell leukemia in vivo and in real-time is a particular strength of this model.

Early onset of leukemia in *mMyc* transgenic fish before sexual maturity prohibited their propagation, and thus prevented the establishment of a stable *mMyc* transgenic line. To overcome this problem, a conditional transgene was created in which the *EGFP-mMyc* oncogene is preceded by a floxed *dsRed* gene, labeling thymocytes with red fluorescence (Langenau et al. 2005a). Injection of *Cre* RNA into one-cell stage embryos from this line led to induction of T-cell acute lymphoblastic leukemia (T-ALL) in 100% of the offspring (see Chapter 1 on "Genetic Tools" for details). A further improvement was made by crossing the conditional line to a transgenic line expressing *Cre* under the control of a heat shock promoter (Feng et al. 2007). After heat shock treatment at 3 days post fertilization (dpf), 81% of double transgenic fish developed T-lymphoblastic lymphoma, which rapidly progressed to T-ALL.

A transgenic line overexpressing bcl2 fused to EGFP under the *rag2* promoter was developed to block apoptosis in lymphoid cells (Langenau et al. 2005b). To test the ability of bcl2 to block irradiation-induced apoptosis in malignant cells, double transgenic fish carrying both *rag::EGFP-mMyc* and *rag2::EGFP-bcl2* were generated. T-cell leukemias induced by *rag2::EGFP-mMyc* alone were ablated by irradiation, whereas T-cell leukemias in fish expressing both *mMyc* and *bcl2* were resistant to irradiation-induced apoptosis.

A human pre-B ALL zebrafish model was generated by ubiquitously overexpressing the *TEL-AML1* fusion oncogene, which is present in 25% of childhood pre-B acute lymphoblastic leukemias. TEL-AML1 overexpression induced B-cell differentiation and 3% of transgenic fish developed oligoclonal B-lineage ALL leukemia (Sabaawy et al. 2006). Another leukemia model was developed by Chen and colleagues based on the observation that about 60% of T-cell ALL patients have activating mutations in the *NOTCH1* gene. Constitutively active human *NOTCH1* under the *rag2* promoter was injected into wild-type embryos, and seven of 16 mosaic fish developed a T-cell lymphoproliferative disease at about 5 months. When the stable transgenic line was crossed to the *rag2::EGFP-bcl2* line, the leukemia onset was dramatically accelerated, suggesting synergy between the Notch pathway and the bcl2-mediated anti-apoptotic pathway (Chen et al. 2007).

Several transgenic models for solid tumors have also been created. A zebrafish model for melanoma was developed in our laboratory (Patton et al. 2005). On the basis that BRAF activation is one critical event in melanoma pathogenesis, a transgenic zebrafish line overexpressing activated $BRAF^{V600E}$ was generated. To restrict the expression in melanocytes, $BRAF^{V600E}$ expression was driven by the melanocyte-specific *mitf* promoter. When injected into wild-type one-cell embryos, mutant $\mathrm{BRAF}^{\mathrm{V600E}}$ led to dramatic patches of ectopic melanocytes, similar to human nevi. Remarkably, when injected into *p53*-deficient fish embryos, melanocyte lesions rapidly developed into invasive melanomas in 7% of injected fish within 4 months. In addition, melanoma cells were serially transplantable when injected into irradiated recipients. This study was the first to provide direct evidence that the BRAF and p53 pathways interact genetically to promote melanoma formation.

Though tumor transplantation is utilized as an established assay for tumor malignancy, the opacity of adult fish has limited visualization of transplanted cells in vivo. Recent generation of a transparent zebrafish line named *casper* can address this problem (White et al. 2008). *casper* fish lack both melanocytes and iridophores, and are thus almost entirely transparent. After transplantation of 200 000 $BRAF^{V600E}$*;p53−/−* melanoma cells into *casper*, a large mass was seen 14 days post transplantation at the injection site, and metastatic melanoma cells were detected 5 to 28 days post transplantation in 37.5% of all recipients. This elegant technique allowed imaging of tumor development in vivo, and could be utilized as a useful platform to answer questions involving tumor growth, invasion, metastasis, and angiogenesis at an anatomic resolution not readily achievable in murine or other systems.

In addition to the melanoma model, a rhabdomyosarcoma model has been developed (Langenau et al. 2007). When the *rag2* promoter was used to drive *mMyc* expression in lymphocytes, surprisingly, there was also expression in undifferentiated skeletal muscle cells. The *rag2* promoter was utilized to express activated human $kRAS^{G12D}$, a common mutated form of *RAS* seen in human tumors, in undifferentiated muscle cells. Fifty per cent of wild-type fish injected with *rag2::*$kRAS^{G12D}$ at the one-cell stage developed rhabdomyosarcoma by 80 days of life. Microarray analysis revealed that zebrafish rhabdomyosarcoma was similar to the human embryonal subtype of the disease. To examine the roles of different subpopulations in the tumor mass, a dual fluorescently labeled rhabdomyosarcoma was created by co-injecting *rag2::*$kRAS^{G12D}$ and *rag2::DsRed* constructs into one-cell stage embryos from an *α-actin::GFP* line that ubiquitously expresses GFP. By fluorescence activated cell sorting (FACS), the tumor mass was separated into four populations, GFP single positive, DsRed single positive, GFP and DsRed double positive, and double negative population. Serial transplantation and limiting dilution analysis on these populations suggested that the DsRed single positive population exhibited the features of cancer stem cells. Further gene expression study demonstrated that this population shared a similar self-renewal program with non-transformed muscle satellite cells. The existence of cancer stem cells in solid tumors is still controversial, but this study has provided a tool to examine the cancer stem cell hypothesis in zebrafish.

## 2.2. Forward Genetic Screens

Forward genetic screens in the zebrafish have been used to study cancer biology. Based on the observation that many cancer patients exhibit cell cycle defects, a forward genetic screen was performed in our laboratory to identify mutants with mitotic defects, indicated by change of phosphorylated histone 3 (pH3) staining. To mutagenize the genome, wild-type male zebrafish were treated with ethylnitrosourea (ENU), a chemical mutagen that causes genomic mutations within the premeiotic germ cells. To produce F1 offspring carrying genomic mutations, ENU-treated males were then bred to wild-type females. Eggs were collected by squeezing an F1 female and fertilized with UV-inactivated sperm to produce haploid F2 embryos (Fig. 9.1). Immunohistochemistry was performed on 36 hours post fertilization (hpf) embryos to assess pH3 expression level. Nineteen mutants were identified, and several genes with loss-of-function mutations have been mapped. Mutations were found in *bmyb*, a transcriptional regulator and member of a putative proto-oncogene family (Shepard et al. 2005), and *separase*, a mitotic regulator (Shepard et al. 2007). Both mutants have defects in mitotic progression and spindle formation, and exhibit genome instability. Both homozygous mutants are embryonic lethal, and a carcinogenesis study revealed increased cancer susceptibility in heterozygous adults. Prior to this screen, neither *bmyb* nor *separase* had been definitively implicated in cancer, underlining the use of forward genetic screens in understanding new mechanisms of cancer biology.

Tumor initiation and progression usually require genetic mutations and/or chromosomal translocations, resulting in the activation of oncogenes and inactivation of tumor suppressors. Genomic instability has been viewed as a risk factor for tumor initiation and progression. To identify genes that safeguard genomic stability, Moore and colleagues (Moore et al. 2004, 2006) performed a forward genetic screen for mutants affecting genomic stability using the *golden* zebrafish. Zebrafish homozygous for this mutation exhibit a

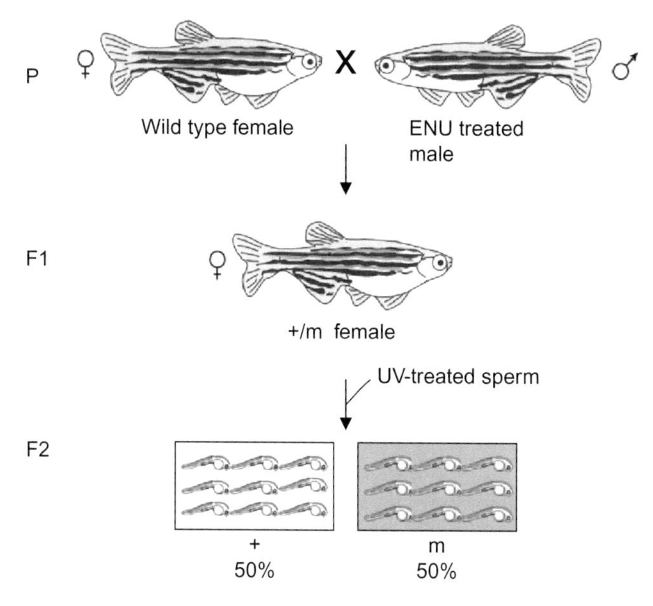


**Fig. 9.1.** Schematic of a haploid screen. In a haploid screen, wild-type male fish are treated with ENU, a chemical mutagen that generates point mutations in the spermatogonia. ENU-treated males are then mated to wild-type females to produce F1 progeny heterozygous for a specific mutation (m). Eggs are collected by squeezing F1 females and fertilized with UV-treated sperm, whose DNA has been destroyed. Resulting haploid clutch from a heterozygous female contains 50% mutant (m) and 50% wild-type (+) progeny. See color plate section

light-colored eye phenotype, whereas heterozygotes have the wild-type dark eye. Some pigmented eye cells of the heterozygous zebrafish will lose the wild-type allele, resulting in a mosaic eye phenotype that acts as a measurement of genomic stability. Zebrafish heterozygous for the *golden* mutation were treated with ENU, and in the subsequent generation 12 mutants with increased genomic instability were obtained. As predicted, all 12 mutant lines showed sensitivity to cancer in heterozygous adults, corroborating the strong connection between genomic instability and cancer. The underlying mutations have not yet been mapped.

A large-scale retroviral insertional mutagenesis screen was conducted in Nancy Hopkins' lab (Amsterdam et al. 2004a) to identify genes essential for embryogenesis. While maintaining the heterozygous founders they observed 12 lines with elevated incidence of malignant peripheral nerve sheath tumors (MPNSTs) (Amsterdam et al. 2004b). One line had a mutation in the zebrafish paralog of the mammalian tumor suppressor gene, neurofibromatosis type 2. Surprisingly, the remaining 11 out of 12 lines carried insertions in ribosomal protein (*rp*) genes. The observation that *p53* homozygous mutants develop the same rare tumor linked *p53* to *rp* deficient MPNSTs (Berghmans et al. 2005). Later study found that although wild-type *p53* is transcribed in MPNST cells, its protein does not get synthesized. These findings demonstrated that many ribosomal protein genes act as haploinsufficient tumor suppressors potentially by modulating the p53 pathway.

## 3. HEMATOPOIETIC DISORDERS

Blood development is highly conserved among vertebrates. Similar to mammals, the zebrafish also has two waves of hematopoiesis, primitive and definitive, regulated by the same group of transcription factors (de Jong and Zon 2005). Unlike mammals, zebrafish embryos develop externally and are optically transparent. This allows direct visualization of circulating blood cells and the beating heart with a simple dissecting microscope.

Pioneering forward genetic studies performed in Boston and Tubingen have generated many interesting zebrafish mutants with hematopoietic phenotypes that mirror human diseases (Ransom et al. 1996). This unbiased approach is particularly useful for identifying novel disease-related genes not previously indicated by human genetics. In this screen, male adults were treated with ENU, a chemical mutagen that results in many germline mutations. These fish were then mated to wild-type females to produce F1 fish that carry heterozygous mutations at a rate of approximately 100–200 mutations per fish (Patton and Zon 2001). F1 fish were mated to wild-type fish to produce F2 siblings carrying the same mutations in their genome. By incrossing F2 siblings, F3 fish harboring homozygous mutations were generated and scored for hematopoietic phenotypes (Fig. 9.2). More than 40 mutants consisting of 26 complementation groups with hematopoietic defects were identified in these screens (de Jong and Zon 2005). The resulting blood mutants can be divided into four groups: no red blood cells (bloodless), progressive anemia, hypochromic anemia, and photosensitivity (Ransom et al. 1996).

### 3.1. Bloodless

The first group of mutants has no red blood cells. Several mutants fail to express gata1, an erythroid zinc finger transcription factor expressed in erythroid progenitors. The *vlad tepes* (*vlt*) mutant harbors a mutation in the *gata-1* gene itself (Lyons et al. 2002). Homozygous *vlt* mutant embryos show complete lack of circulating erythrocytes after 24 hpf, but have normal expression of hematopoietic progenitor markers and normal development of myeloid and lymphoid lineages. Studies in *vlt* demonstrate that *gata-1* plays

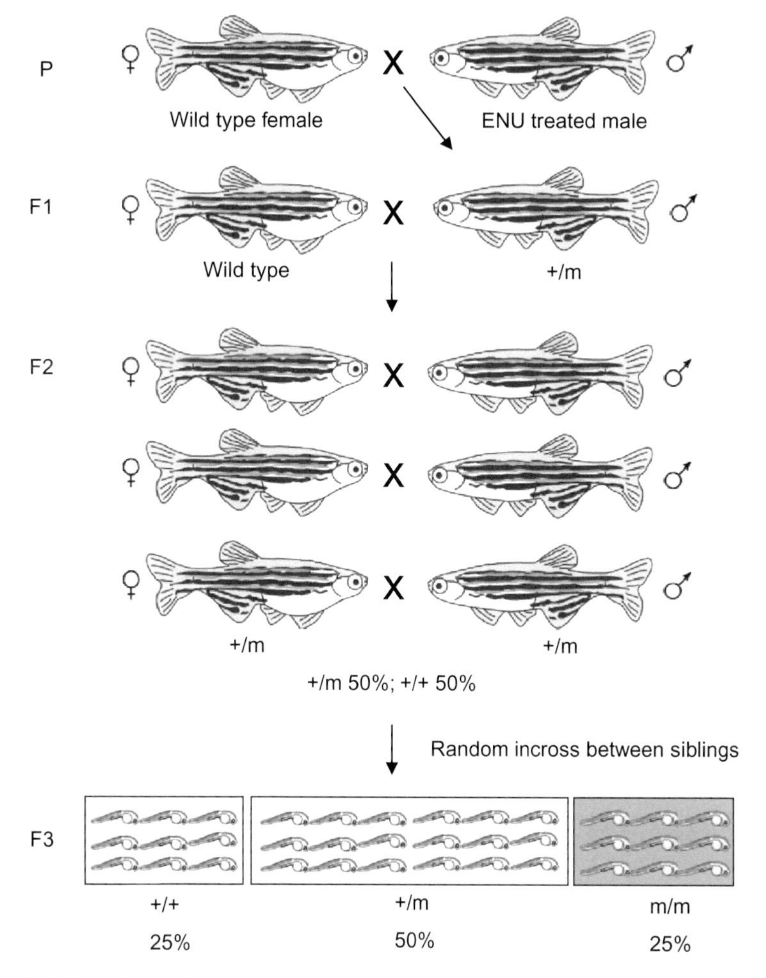


**Fig. 9.2.** Schematic of a large-scale F2 screen. In F2 screen, male adults are treated with ENU. These fish are mated to wild-type females to produce heterozygous F1 progeny. F1 fish are mated to wild-type fish to produce F2 families, in which half of siblings are heterozygous for a specific mutation (m) and the other half are wild-type (+). By incrossing F2 siblings, F3 progeny are generated as 25% wild-type (+/+), 50% heterozygous (+/m) and 25% homozygous (m/m). See color plate section

an essential role in zebrafish hematopoiesis, suggesting significant conservation of its function between mammals and zebrafish.

*moonshine* (*mon*) has a mutation in transcriptional intermediary factor 1γ (TIF1γ) (Ransom et al. 2004). During development, hematopoietic progenitor cells in *mon* mutants fail to express normal levels of hematopoietic transcription factors, including *gata-1*, and undergo apoptosis. *mon* mutant embryos have several dozen blood cells at the start of circulation but never produce more than 100 cells, whereas wild-type embryos have 1000–3000 circulating cells. In addition, the pro-erythroblast stage is severely blocked in *mon* homozygous mutants. Most *mon* mutant animals die with severe anemia by 10–14 dpf. However, rare homozygous *mon* mutant fish (approximately 1 in 500 embryos) can survive to adulthood. Presumably, the severe anemic defect in *mon* adults requires a high-output state of circulating blood to compensate for low blood cell number, and this leads to marked cardiomegaly. This demonstrates that TIF1γ is required for both primitive and definitive erythropoiesis.

### 3.2. Progressive Anemia

The second blood mutant group can be identified by a decreased number of blood cells after normal onset of primitive hematopoiesis. Three of these mutants, *merlot/chablis* (*mot/cha*) (Shafizadeh et al. 2002), *riesling* (*ris*) (Liao et al. 2000), and *retsina* (*ret*) (Paw et al. 2003), have been characterized at the molecular level. Successful positional mapping of *mot/cha* and *ris* revealed mutations in two genes that encode components of the red cell membrane cytoskeleton. To tolerate physical and chemical insults, red blood cells develop a unique cell membrane structure that integrates with blood-specific cytoskeleton molecules. Zebrafish carrying mutations in these genes develop progressive anemia.

The *mot/cha* mutant animals have mutations in a gene encoding erythrocyte protein 4.1 (also known as band4.1 or 4.1R) (Shafizadeh et al. 2002). Erythrocyte protein 4.1 is an erythroid-specific structural membrane protein which anchors the spectrin–actin-based cytoskeleton to the plasma membrane. In humans, red cells that are completely deficient in Erythrocyte protein 4.1 have abnormal elliptical cell morphology and a fragile membrane. The *mot/cha* mutant fish have normal onset of primitive hematopoiesis but later exhibit severe hemolytic anemia around 4 dpf. A small percentage of *mot* mutant fish can survive to adulthood. These fish suffer severe anemia, and their red blood cells are arrested in the basophilic erythroblast stage. The erythrocytes of these adult fish have abnormal membrane projections and increased osmotic fragility (Shafizadeh et al. 2002).

Another progressive anemia mutant *ris* was found to have a mutation in the erythroid β-spectrin gene (Liao et al. 2000). β-spectrin is one of the most abundant components of the erythrocyte cytoskeleton. Although hematopoietic development is unaffected, hemolytic anemia appears by 3–4 dpf in *ris* mutant fish. Mutant red cells exhibit abnormal spherical morphology, undergo apoptosis, and display defective aggregation of microtubule filaments. These cells are reminiscent of those seen in the human erythroid disorder hereditary spherocytosis, caused by a mutation in human β-spectrin. Patients with this disease have increased red cell progenitors in the bone marrow, and adult *ris* zebrafish also have increased hematopoietic progenitors in the kidney marrow (Liao et al. 2000).

*ret* mutants have a mutation in an erythroid-specific cytoskeletal protein, anion exchanger 1 (AE1 also known as band3) (Paw et al. 2003). Similar to *mot/cha* and *ris* mutants, the *ret* mutants develop anemia around 4 dpf after normal hematopoietic development. *ret* mutants exhibit an erythroid-specific defect in cell division with marked dyserythropoiesis, similar to the congenital dyserythropoietic anemia type II in humans. Erythroblasts of *ret* show binuclearity and undergo apoptosis due to a failure to complete chromosome segregation and cytokinesis. Band3 was demonstrated to play a critical role in chromosomal segregation during anaphase, as lack of band3 in *ret* mutants impairs cytokinesis in zebrafish erythroblasts (Paw et al. 2003).

### 3.3. Hypochromia

The hypochromic zebrafish mutants are characterized by abnormally small erythrocytes with pale cytoplasm resulting from impaired hemoglobin synthesis. Hemoglobin is the iron-containing oxygen-transport metalloprotein in the red blood cells of vertebrates. Since the production of hemoglobin is a multi-step process, mutations in many genes can lead to hypochromia.

The *sauternes* (*sau*) mutant fish was found to have a mutation in the erythroid-specific isoform of δ-aminolevulinate synthase (*ALAS2*) gene, encoding an enzyme required for the first step in heme biosynthesis (Brownlie et al. 1998). *sau* mutant fish exhibit hypochromic microcytic anemia, indicating perturbed heme production. As in human patients, mutations in *ALAS2* lead to congenital sideroblastic anemia (CSA), and *sau* represents the first animal model of this disease (Brownlie et al. 1998).

The zebrafish mutant *zinfandel* (*zin*) has hypochromic microcytic anemia during embryonic development, but recovers in adulthood (Brownlie et al. 2003). Utilizing positional mapping, *zin* mutants were found to carry a mutation close to the zebrafish major globin locus on chromosome 3, though not in the coding region of the gene. This mutant could serve as a model for thalassemia caused by a mutation in a *cis* regulatory element of the major globin locus (Brownlie et al. 2003).

Several mutants, including *weissherbst* (*weh*) (Donovan et al. 2000), *chardonnay* (*cdy*) (Donovan et al. 2002), and *chianti* (*cia*) (Wingert et al. 2004), have defects of iron metabolism as well as hypochromic anemia. Iron metabolism is an essential component of heme synthesis. In humans, defects in iron uptake, storage and distribution can lead to hypochromic anemias. The *weh* zebrafish mutation maps to the *ferroportin 1* gene, which encodes a novel iron transporter (Donovan et al. 2000). In zebrafish, *ferroportin 1* transports iron from the yolk sac into circulation. Zebrafish with the *ferroportin 1* mutation have decreased mean corpuscular hemoglobin levels but a nearly normal red blood cell count. Human Ferroportin 1 is expressed at the basal surface of placental syncytiotrophoblasts, suggesting its similar role of iron transportation from mother to fetus (Donovan et al. 2000). Subsequently, ferroportin was found to be defective in patients with hemochromatosis type IV (Njajou et al. 2001; Devalia et al. 2002; Roetto et al. 2002; Wallace et al. 2002). This was the first time that a zebrafish gene had led to the discovery of a new human disease gene.

Another hypochromic mutant, *cdy*, has normal onset of circulation but no hemoglobin expression by 48 hpf (Donovan et al. 2002). *Cdy* was mapped to the gene coding for divalent metal transporter 1 (DMT1), a transmembrane protein that is required for release of transferrin-transported iron in the cytoplasm of erythroid precursors (Donovan et al. 2002).

*cia* is another mutant defective in iron metabolism (Wingert et al. 2004). This mutant starts to exhibit hypochromic anemia by 36 hpf. The erythrocytes of the *cia* mutant are hypochromic and microcytic, and the kidney marrow of these fish is hypercellular, with an increased proportion of erythroid progenitors. *transferrin receptor 1* (*tfr1a*) has been identified as the gene mutated in *cia*. The expression of *tfr1a* is restricted to erythrocytes, suggesting a critical role for iron acquisition in these cells (Wingert et al. 2004).

### 3.4. Photosensitivity

Mutants in this category are characterized by autofluorescence and photoablation of erythrocytes when subjected to light exposure. This phenotype is also seen in patients with congenital erythropoietic porphyrias.

The zebrafish *yquem* (*yqe*) mutant exhibits a photosensitive porphyria syndrome (Wang et al. 1998). A mutation in the gene encoding uroporphyrinogen decarboxylase (UROD) was identified in *yqe*. In humans, homozygous deficiency of this enzyme leads to hepatoerythropoietic porphyria (HEP). *yqe* was the first animal model of HEP (Wang et al. 1998).

Another zebrafish photosensitive mutant *dracula* (*drc*) was found to bear mutations in ferrochelatase (Childs et al. 2000). Similarly to *yqe*, *drc* mutant animals exhibit strong fluorescence in erythrocytes and lack red blood cells when exposed to normal light. In human, ferrochelatase has been shown to mediate incorporation of ferrous iron into protoporphyrin IX. Defects in this enzyme lead to erythropoietic protoporphyria (Childs et al. 2000).

## 4. CARDIOVASCULAR DISEASES

Cardiovascular disease has become the most common cause of death in the US. Because of disease heterogeneity, late onset, variable penetrance and high mortality, it is extremely difficult to identify important disease genes by family linkage analysis. Moreover, multiple elements like genetic mutations, extrinsic injury, and environmental exposures all contribute to cardiovascular disorders and these factors may vary widely in different patients. Animal models that recapitulate individual disease factors are important for understanding pathogenesis.

The zebrafish is an ideal model for studying cardiovascular disease. First, having a closed cardiovascular system, the development of the cardiovascular system is highly conserved between fish and mammals. Secondly, zebrafish embryos can tolerate absence of blood flow because its oxygen is delivered by diffusion rather than by the cardiovascular system. This unique feature enables direct attribution of cardiac defects to particular genes by eliminating the secondary effect of hypoxia. Thirdly, zebrafish embryos develop externally and are transparent, allowing direct observation of the beating heart and circulating blood under a simple dissecting microscope. Lastly, high fecundity and fast development allow forward genetic screens for cardiovascular disorders to be conducted in zebrafish (see Chapter 6 on the "Developmental Physiology of the Zebrafish Cardiovascular System" and Chapter 7 on "Respiration" for more information).

### 4.1. Cardiomyopathies

Cardiomyopathy is an important type of cardiovascular disease caused by defects in the myocardium. Based on anatomical and physiological criteria, cardiomyopathy has been classified as either dilated or hypertrophic cardiomyopathy. Dilated cardiomyopathy is defined as having an impaired systolic contraction usually leading to an enlarged ventricle. Hypertrophic cardiomyopathy is characterized by inappropriate myocardial hypertrophy, myofibrillar disarray and impaired diastolic function. In many cases there is a dynamic systolic obstruction to outflow from the ventricle. Examples of zebrafish mutants in this category include *pickwick* (*pik*) (Xu et al. 2002), *tel tale heart* (*tel*) (Rottbauer et al. 2006), *dead beat* (*ded*) (Lawson et al. 2003), and *main squeeze* (*msq*) (Bendig et al. 2006).

The heart of the zebrafish mutant *pik* develops normally but is poorly contractile (Xu et al. 2002). Using positional mapping, *pik* mutants were found to harbor a mutation in an alternatively spliced exon of the *titin* gene. Titin is the largest known protein and spans the half-sarcomere from the

Z-disc to the M-line in heart and skeletal muscle. *pik* mutant fish can form myofibrils, but lack normal sarcomeres. Absence of Titin results in the blockage of sarcomere assembly and causes a functional disorder. Mutations in the alternatively spliced exon of *titin* have also been found in dilated cardiomyopathy patients (Xu et al. 2002).

Another mutant *tel* has a similar cardiac phenotype to the *pik* mutant (Rottbauer et al. 2006). By positional cloning, a mutation in the cardiac regulatory myosin light-chain gene *mlc2* was identified. *tel* mutants fail to assemble functional cardiac sarcomeres due to a complete loss of organized thick myofilaments. In humans, it is known that mutations in the *mlc2* gene cause hypertrophic cardiomyopathy. The *tel* model represents the first in vivo model for cardiomyopathy caused by the *mlc2* mutation (Rottbauer et al. 2006).

The zebrafish *ded* mutant exhibits a cardiac contractility defect by 48–60 hpf (Lawson et al. 2003). A mutation in the gene encoding phospholipase C gamma-1 (Plcg1) was identified in *dead beat*. Plcg1 transduces the signal from vascular endothelial growth factor (VEGF), and its receptor FLT1. Further dissection of this pathway in rat ventricular cardiomyocytes suggested that blockage of VEGF-PLCG1 signaling decreases calcium transients. This study revealed that cardiomyocytes control the strength of the heart beat through the VEGF-PLCG1 cascade (Lawson et al. 2003).

*msq* is another mutant that has a heart contractility defect. This mutant has a missense mutation in the *integrin-linked kinase* (*ilk*) gene that impairs its kinase activity and disrupts binding of ILK to the Z-disc adaptor protein β-parvin (Affixin) (Bendig et al. 2006). Homozygous *msq* mutant fish are characterized by loss of cardiac contractility from 60–72 hpf. Lack of *atrial natriuretic factor* (*anf*) expression in *msq* suggested that the heart may use the Integrin-ILK-β-parvin network as a mechanical stretch sensor, as *anf* expression can be induced by stretch (Bendig et al. 2006). Subsequently, *ILK* was found to be defective in patients with dilated cardiomyopathy (Knoll et al. 2007).

### 4.2. Arrhythmias

Arrhythmia is another type of cardiovascular disease characterized by abnormal heart rhythm. This disease has been difficult to model in vivo due to our incomplete understanding of the etiology of most clinical rhythm disorders. The zebrafish is a powerful model used to dissect molecular pathways of cardiovascular disease. Because fundamental electrical properties of the zebrafish heart are remarkably similar to those of the human heart, the zebrafish is an appropriate model for studying human arrhythmias. Mutagenesis screens have identified several mutants with arrhythmia, including *island beat* (*isl*) (Rottbauer et al. 2001), *tremblor* (*tre*) (Langenbacher et al. 2005), *breakdance* (Arnaout et al. 2007), and *reggae* (*reg*) (Hassel et al. 2008).

*Isl* has a null mutation in the *α1C* L-type calcium channel subunit (*C-LTCC*) gene (Rottbauer et al. 2001). The *isl* atrium is relatively normal in size, but individual cells contract chaotically in a pattern resembling cardiac fibrillation. The ventricle is completely silent. Further characterization revealed that *isl* fails to acquire the normal number of cardiomyocytes. Study on the *isl* mutant revealed separate roles of calcium signaling via C-LTCC in regulating heart growth as well as heart contraction (Rottbauer et al. 2001).

*tre* is another mutant with a cardiac fibrillation phenotype (Langenbacher et al. 2005). The *tre* mutation was mapped to the gene encoding cardiac-specific sodium–calcium exchanger 1 (*NCX1*). The hearts of the *tre* mutant fish exhibited chaotic movements and failed to develop synchronized contractions. The loss of calcium extrusion and subsequent cellular calcium overloading have been identified as the causes of the cardiac fibrillation phenotype in *tre* mutant fish. This work signifies the essential role of calcium homeostasis in establishing normal rhythmic contraction (Langenbacher et al. 2005).

Two classes of mutants have been found to carry mutations in the gene *kcnh2*, which encodes the channel responsible for the rapidly activating delayed rectifier $K^+$ current (Arnaout et al. 2007; Hassel et al. 2008). *breakdance* mutant fish exhibit a variable atrioventricular block that results from null or hypomorphic kcnh2 alleles. In humans, autosomal dominant kcnh2 mutations account for approximately 45% of mutation-positive long QT syndrome, a repolarization disorder associated with sudden cardiac death. *breakdance* represents the first in vivo model for this disease (Arnaout et al. 2007). By contrast, *reg* mutant fish harbor activating mutations in *kcnh2* that lead to accelerated repolarization. Mutant fish exhibit only intermittent exit of impulses from the sinus venosus, and a failure of atrial or ventricular escape (Hassel et al. 2008).

## 5. CHEMICAL SCREENING IN ZEBRAFISH

Conventional target-based drug discovery involves in vitro biochemical screens, requiring a previously identified and validated therapeutic target. A high-throughput small-molecule screen is designed based on target binding or function, and it identifies molecules that modify the activity of the target protein. Many target-based drug screens have successfully provided lead compounds that were later developed as therapeutic drugs. But most complex diseases caused by multiple defects have more than one therapeutic target, and many in vitro biochemical assays are poor surrogates for these diseases. In contrast to target-based approaches, screens utilizing zebrafish are guided by desired phenotype. In these screens, thousands of embryos are arrayed in multi-well plates, and are treated with chemicals during certain developmental stages. The embryos are allowed to grow and then scored for desired phenotype (Fig. 9.3). Phenotype-based screens enable drug discovery

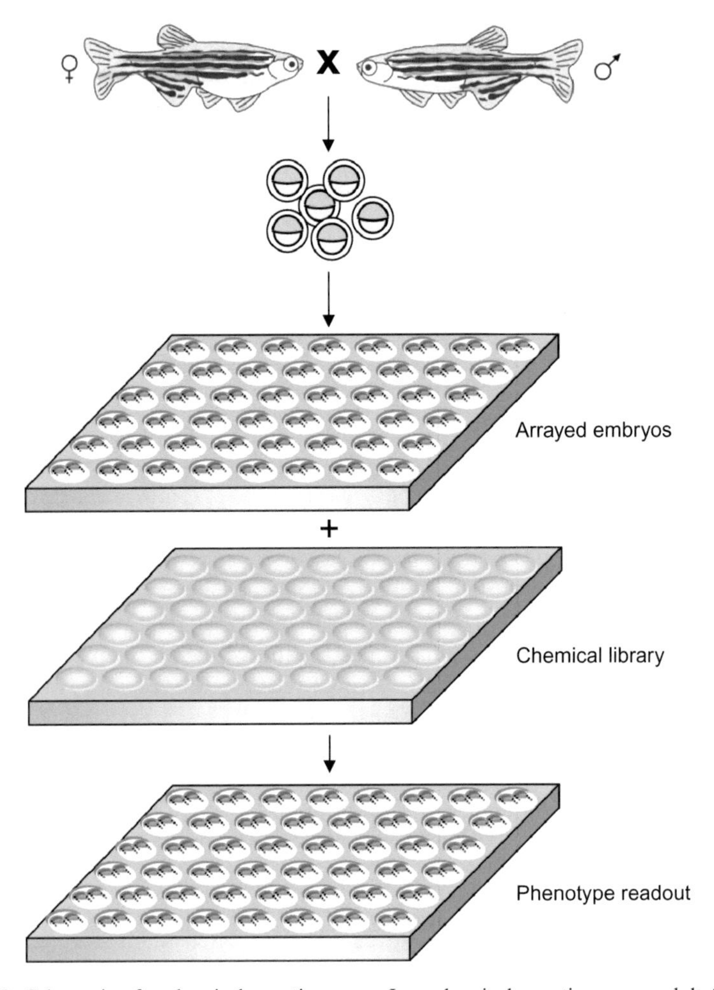


**Fig. 9.3.** Schematic of a chemical genetic screen. In a chemical genetic screen, adult fish are crossed to produce hundreds of embryos. Embryos are arrayed in multi-well plates. Compounds from a chemical library are added into fish water at a certain time point. Embryos are allowed to develop and scored for desired phenotype. See color plate section

without knowing the specific molecular targets and mechanisms. This approach overcomes the problem of complex diseases involving several targets. While assaying for chemicals' bioavailability, in vivo screens also examine their toxicity, allowing elimination of the ones with strong side effects. The feasibility to perform small-molecule screens in zebrafish has been demonstrated by several chemical genetic screens described below.

One example was a screen for suppressors of the *gridlock* mutation. Despite normal perfusion of the head, homozygous *gridlock* mutant fish have no circulation to the trunk and tail due to dysmorphogenesis of the dorsal aorta. The *gridlock* mutation maps to the *hey2* gene, encoding a bHLH transcriptional repressor (Peterson et al. 2004). The *gridlock* embryos were transferred to 96-well plates, and then exposed to small molecules from a structurally diverse chemical library and examined for normal circulation after 48 hpf. After screening 5000 small molecules, two were found to suppress the *gridlock* phenotype, restoring circulation to the tail. Restoration of tail circulation involved the rescue of the aorta morphological defect. The hits, GS4012 and GS3999, represented a novel class of compounds that were not previously known to influence vasculogenesis or angiogenesis. Treatment with these chemicals leads to upregulated expression of vascular endothelial growth factor (VEGF), a growth factor triggering vasculogenesis and angiogenesis. GS4012 was found to promote the formation of human endothelial tubules, suggesting its conserved role in humans (Peterson et al. 2004).

Another example is a chemical screen for the suppressor of *crb* mutation. *crb* has a mutation in the *bmyb* gene. The mutant fish have an increased number of pH3-positive cells and exhibit increased cancer susceptibility (Stern et al. 2005). To identify small molecules that interact with the bmyb pathway, a strategy was designed to screen for chemical suppressors that rescue the pH3 phenotype. Sixteen thousand compounds were tested, and a previously unknown compound named persynthamide was identified that specifically suppressed the elevated pH3 staining of *crb* embryos. As it specifically suppresses cell-cycle defect in mutant cells without affecting wild-type cells, this compound could be potentially useful as an anticancer drug (Stern et al. 2005).

One concern of conducting chemical screens in zebrafish is that evolutionary distance between fish and mammals may render a lead chemical identified in zebrafish not functional in humans. The study on prostaglandin (PG) E2 as an activator of hematopoietic stem cells (HSCs) showed that chemicals identified in zebrafish may also function in mammals (North et al. 2007). To identify new pathways modulating definitive HSC formation, a chemical genetic screen was conducted on zebrafish embryos. Wild-type embryos were incubated with individual chemicals and were examined for alterations of *runx1* and *cmyb* expression, two genes that are required for definitive HSC development. Of the 2357 chemicals screened, ten affected the prostaglandin pathway. The chemicals that enhance PGE2 synthesis increased HSC numbers and those that block PGE2 synthesis decreased stem cell numbers. To examine if PGE2 can also increase HSC numbers in mammals, murine embryonic stem cell differentiation assay and spleen colony-forming assays were conducted with a stable derivative of PGE2. Both assays suggested that PGE2 exposure provides stimulation for the murine HSC expansion (North et al. 2007). Because PGE2 exposure stimulates the reconstitution of hematopoietic lineages from transplanted HSCs, this could be used therapeutically to improve the effectiveness of bone marrow transplantation.

Chemical screens using zebrafish embryos eliminate hits that have strong side effects, low uptaking efficiency, and poor in vivo bioactivity. These screens do not require validated molecular targets, thus enabling drug discovery for diseases whose etiology is poorly understood. This screen system makes small molecule discovery feasible in academic settings. With a chemical library, any zebrafish lab can perform chemical screens.

## 6. SUMMARY

This chapter discusses zebrafish models in the fields of cancer, blood disorders, and cardiovascular diseases. In the other fields including muscle diseases, infection, and inflammation, the zebrafish also plays an important role in modeling these diseases. The zebrafish has already provided a wealth of understanding of embryonic development, and should play a larger role in disease study and drug discovery.

## REFERENCES

Amsterdam, A., Nissen, R. M., Sun, Z., Swindell, E. C., Farrington, S., and Hopkins, N. (2004a). Identification of 315 genes essential for early zebrafish development. *Proc. Natl Acad. Sci. USA* **101**, 12792–12797.

Amsterdam, A., Sadler, K. C., Lai, K., Farrington, S., Bronson, R. T., Lees, J. A., and Hopkins, N. (2004b). Many ribosomal protein genes are cancer genes in zebrafish. *PLoS Biol.* **2**, E139.

Arnaout, R., Ferrer, T., Huisken, J., Spitzer, K., Stainier, D., Tristani-Firouzi, M., and Chi, N. (2007). Zebrafish model for human long QT syndrome. *PNAS* **104**, 11316–11321.

Bendig, G., Grimmler, M., Huttner, I., Wessels, G., Dahme, T., Just, S., Trano, N., Katus, H., Fishman, M., and Rottbauer, W. (2006). Integrin-linked kinase, a novel component of the cardiac mechanical stretch sensor, controls contractility in the zebrafish heart. *Genes Dev.* **20**, 2361–2372.

Berghmans, S., Murphey, R. D., Wienholds, E., Neuberg, D., Kutok, J. L., Fletcher, C. D. M., Morris, J. P., Liu, T. X., Schulte-Merker, S., Kanki, J. P., et al. (2005). tp53 mutant zebrafish develop malignant peripheral nerve sheath tumors. *Proc. Natl Acad. Sci. USA* **102**, 407–412.

Brownlie, A., Donovan, A., Pratt, S., Paw, B., Oates, A., Brugnara, C., Witkowska, H., Sassa, S., and Zon, L. (1998). Positional cloning of the zebrafish sauternes gene: A model for congenital sideroblastic anaemia. *Nat. Genet.* **20**, 244–250.

Brownlie, A., Hersey, C., Oates, A., Paw, B., Falick, A., Witkowska, H., Flint, J., Higgs, D., Jessen, J., Bahary, N., et al. (2003). Characterization of embryonic globin genes of the zebrafish. *Dev. Biol.* **255**, 48–61.

Chen, J., Jette, C., Kanki, J. P., Aster, J. C., Look, A. T., and Griffin, J. D. (2007). NOTCH1-induced T-cell leukemia in transgenic zebrafish. *Leukemia* **21**, 462–471.

Childs, S., Weinstein, B., Mohideen, M., Donohue, S., Bonkovsky, H., and Fishman, M. (2000). Zebrafish dracula encodes ferrochelatase and its mutation provides a model for erythropoietic protoporphyria. *Curr. Biol.* **10**, 1001–1004.

de Jong, J., and Zon, L. (2005). Use of the zebrafish system to study primitive and definitive hematopoiesis. *Annu. Rev. Genet.* **39**, 481–501.

Devalia, V, Carter, K., Walker, A. P., Perkins, S. J., Worwood, M, May, A, and Dooley, J. S. (2002). Autosomal dominant reticuloendothelial iron overload associated with a 3-base pair deletion in the ferroportin 1 gene. *Blood* **100**, 695–697.

Donovan, A., Brownlie, A., Dorschner, M., Zhou, Y., Pratt, S., Paw, B., Phillips, R., Thisse, C., Thisse, B., and Zon, L. (2002). The zebrafish mutant gene chardonnay (cdy) encodes divalent metal transporter 1 (DMT1). *Blood* **100**, 4655–4659.

Donovan, A., Brownlie, A., Zhou, Y., Shepard, J., Pratt, S., Moynihan, J., Paw, B., Drejer, A., Barut, B., Zapata, A., et al. (2000). Positional cloning of zebrafish ferroportin1 identifies a conserved vertebrate iron exporter. *Nature* **403**, 776–781.

Feng, H., Langenau, D. M., Marge, J., Quinkertz, A., Gutierrez, A., Neuberg, D., Kanki, J. P., and Look, A. T. (2007). Heat-shock induction of T-cell lymphoma/leukaemia in conditional Cre/lox-regulated transgenic zebrafish. *Br. J. Haematol.* **138**, 169–175.

Frese, K., and Tuveson, D. (2007). Maximizing mouse cancer models. *Nat. Rev. Cancer* **7**, 645–658.

Hassel, D., Scholz, E., Trano, N., Friedrich, O., Just, S., Meder, B., Weiss, D., Zitron, E., Marquart, S., Vogel, B., et al. (2008). Deficient zebrafish ether-à-go-go-related gene channel gating causes short-QT syndrome in zebrafish reggae mutants. *Circulation* **117**, 866–875.

Kent, M., Bishop-Stewart, J., Matthews, J., and Spitsbergen, J. (2002). Pseudocapillaria tomentosa, a nematode pathogen, and associated neoplasms of zebrafish. *Comp. Med.* **52**, 354–358.

Knoll, R., Postel, R., Wang, J., Kratzner, R., Hennecke, G., Vacaru, A., Vakeel, P., Schubert, C., Murthy, K., Rana, B., et al. (2007). Laminin-alpha4 and integrin-linked kinase mutations cause human cardiomyopathy via simultaneous defects in cardiomyocytes and endothelial cells. *Circulation* **116**, 515–525.

Langenau, D. M., Feng, H., Berghmans, S., Kanki, J. P., Kutok, J. L., and Look, A. T. (2005a). Cre/lox-regulated transgenic zebrafish model with conditional myc induced T cell acute lymphoblastic leukemia. *Proc. Natl Acad. Sci. USA* **102**, 6068–6073.

Langenau, D. M., Jette, C., Berghmans, S., Palomero, T., Kanki, J. P., Kutok, J. L., and Look, A. T. (2005b). Suppression of apoptosis by bcl-2 overexpression in lymphoid cells of transgenic zebrafish. *Blood* **105**, 3278–3285.

Langenau, D. M., Keefe, M. D., Storer, N. Y., Guyon, J. R., Kutok, J. L., Le, X., Goessling, W., Neuberg, D. S., Kunkel, L. M., and Zon, L. I. (2007). Effects of RAS on the genesis of embryonal rhabdomyosarcoma. *Genes Dev.* **21**, 1382–1395.

Langenau, D. M., Traver, D., Ferrando, A. A., Kutok, J. L., Aster, J. C., Kanki, J. P., Lin, S., Prochownik, E., Trede, N. S., Zon, L. I., et al. (2003). Myc-induced T cell leukemia in transgenic zebrafish. *Science* **299**, 887–890.

Langenbacher, A. D., Dong, Y., Shu, X., Choi, J., Nicoll, D. A., Goldhaber, J. I., Philipson, K. D., and Chen, J.-N. (2005). Mutation in sodium–calcium exchanger 1 (NCX1) causes cardiac fibrillation in zebrafish. *PNAS* **102**, 17699–17704.

Lawson, N., Mugford, J., Diamond, B., and Weinstein, B. (2003). phospholipase C gamma-1 is required downstream of vascular endothelial growth factor during arterial development. *Genes Dev.* **17**, 1346–1351.

Liao, E., Paw, B., Peters, L., Zapata, A., Pratt, S., Do, C., Lieschke, G., and Zon, L. (2000). Hereditary spherocytosis in zebrafish riesling illustrates evolution of erythroid beta-spectrin structure, and function in red cell morphogenesis and membrane stability. *Development* **127**, 5123–5132.

Liu, T., Zhou, Y., Kanki, J., Deng, M., Rhodes, J., Yang, H., Sheng, X., Zon, L., and Look, A. (2002). Evolutionary conservation of zebrafish linkage group 14 with frequently deleted regions of human chromosome 5 in myeloid malignancies. *Proc. Natl Acad. Sci. USA* **99**, 6136–6141.

Lyons, S., Lawson, N., Lei, L., Bennett, P., Weinstein, B., and Liu, P. (2002). A nonsense mutation in zebrafish gata1 causes the bloodless phenotype in vlad tepes. *Proc. Natl Acad. Sci. USA* **99**, 5454–5459.

Matthews, J. (2004). Common diseases of laboratory zebrafish. *Meth. Cell Biol.* **77**, 617–643.

Moore, J. L., Gestl, E., and Cheng, K. C. (2004). Mosaic eyes, genomic instability mutants, and cancer susceptibility. *Meth. Cell Biol.* **76**, 555–668.

Moore, J. L., Rush, L. M., Breneman, C., Mohideen, M.-A. P. K., and Cheng, K. C. (2006). Zebrafish genomic instability mutants and cancer susceptibility. *Genetics* **174**, 585–600.

Njajou, O., Vaessen, N., Joosse, M., Berghuis, B., van Dongen, J., Breuning, M., Snijders, P., Rutten, W., Sandkuijl, L., Oostra, B., et al. (2001). A mutation in SLC11A3 is associated with autosomal dominant hemochromatosis. *Nat. Genet.* **28**, 213–214.

North, T., Goessling, W., Walkley, C., Lengerke, C., Kopani, K., Lord, A., Weber, G., Bowman, T., Jang, I., Grosser, T., et al. (2007). Prostaglandin E2 regulates vertebrate haematopoietic stem cell homeostasis. *Nature* **447**, 1007–1011.

Patton, E. E., Widlund, H. R., Kutok, J. L., Kopani, K. R., Amatruda, J. F., Murphey, R. D., Berghmans, S., Mayhall, E. A., Traver, D., Fletcher, C. D. M., et al. (2005). BRAF mutations are sufficient to promote nevi formation and cooperate with p53 in the genesis of melanoma. *Curr. Biol.* **15**, 249–254.

Patton, E. E., and Zon, L. (2001). The art and design of genetic screen: Zebrafish. *Nat. Rev. Genet.* **2**, 956–966.

Paw, B., Davidson, A., Zhou, Y., Li, R., Pratt, S., Lee, C., Trede, N., Brownlie, A., Donovan, A., Liao, E., et al. (2003). Cell-specific mitotic defect and dyserythropoiesis associated with erythroid band 3 deficiency. *Nat. Genet.* **34**, 59–64.

Peterson, R. T., Shaw, S. Y., Peterson, T. A., Milan, D. J., Zhong, T. P., Schreiber, S. L., MacRae, C. A., and Fishman, M. C. (2004). Chemical suppression of a genetic mutation in a zebrafish model of aortic coarctation. *Nat. Biotechnol.* **22**, 595–599.

Postlethwait, J., Woods, I., Ngo-Hazelett, P., Yan, Y., Kelly, P., Chu, F., Huang, H., Hill-Force, A., and Talbot, W. (2000). Zebrafish comparative genomics and the origins of vertebrate chromosomes. *Genome Res.* **10**, 1890–1902.

Price, D., Sisodia, S., and Borchelt, D. (1998). Genetic neurodegenerative diseases: The human illness and transgenic models. *Science* **282**, 1079–1083.

Ransom, D., Bahary, N., Niss, K., Traver, D., Burns, C., Trede, N., Paffett-Lugassy, N., Saganic, W., Lim, C., Hersey, C., et al. (2004). The zebrafish moonshine gene encodes transcriptional intermediary factor 1gamma, an essential regulator of hematopoiesis. *PLoS Biol.* **2**, 237.

Ransom, D. G., Haffter, P., Odenthal, J., Brownlie, A., Vogelsang, E., Kelsh1, R. N., Brand, M., van Eeden, F. J. M., Furutani-Seiki, M., Granato, M., et al. (1996). Characterization of zebrafish mutants with defects in embryonic hematopoiesis. *Development* **123**, 311–319.

Roetto, A., Merryweather-Clarke, A., Daraio, F., Livesey, K., Pointon, J., Barbabietola, G., Piga, A., Mackie, P., Robson, K., and Camaschella, C. (2002). A valine deletion of ferroportin 1: A common mutation in hemochromastosis type 4. *Blood* **100**, 733–744.

Rottbauer, W., Baker, K., Wo, Z., Mohideen, M., Cantiello, H., and Fishman, M. (2001). Growth and function of the embryonic heart depend upon the cardiac-specific L-type calcium channel alpha1 subunit. *Dev. Cell* **1**, 265–275.

Rottbauer, W., Wessels, G., Dahme, T., Just, S., Trano, N., Hassel, D., Burns, C., Katus, H., and Fishman, M. (2006). Cardiac myosin light chain-2: A novel essential component of thick-myofilament assembly and contractility of the heart. *Circ. Res.* **99**, 323–331.

Sabaawy, H., Azuma, M., Embree, L., Tsai, H., Starost, M., and Hickstein, D. (2006). TEL-AML1 transgenic zebrafish model of precursor B cell acute lymphoblastic leukemia. *Proc. Natl Acad. Sci. USA* **103**, 15166–15171.

Shafizadeh, E., Paw, B., Foott, H., Liao, E., Barut, B., Cope, J., Zon, L., and Lin, S. (2002). Characterization of zebrafish merlot/chablis as non-mammalian vertebrate models for severe congenital anemia due to protein 4.1 deficiency. *Development* **129**, 4359–4370.

Shepard, J. L., Amatruda, J. F., Finkelstein, D., Ziai, J., Finley, K. R., Stern, H. M., Chiang, K., Hersey, C., Barut, B., Freeman, J. L., et al. (2007). A mutation in separase causes genome instability and increased susceptibility to epithelial cancer. *Genes Dev.* **21**, 55–59.

Shepard, J. L., Amatruda, J. F., Stern, H. M., Subramanian, A., Finkelstein, D., Ziai, J., Finley, K. R., Pfaff, K. L., Hersey, C., Zhou, Y., et al. (2005). A zebrafish bmyb mutation causes genome instability and increased cancer susceptibility. *Proc. Natl Acad. Sci. USA* **102**, 13194–13199.

Smolowitz, R., Hanley, J., and Richmond, H. (2002). A three-year retrospective study of abdominal tumors in zebrafish maintained in an aquatic laboratory animal facility. *Biol. Bull.* **203**, 265–266.

Stern, H., Murphey, R., Shepard, J., Amatruda, J., Straub, C., Pfaff, K., Weber, G., Tallarico, J., King, R., and Zon, L. (2005). Small molecules that delay S phase suppress a zebrafish bmyb mutant. *Nat. Chem. Biol.* **1**, 366–370.

Wallace, D., Pedersen, P., Dixon, J., Stephenson, P., Searle, J., Powell, L., and Subramaniam, V. (2002). Novel mutation in ferroportin1 is associated with autosomal dominant hemochromatosis. *Blood* **100**, 692–694.

Wang, H., Long, Q., Marty, S., Sassa, S., and Lin, S. (1998). A zebrafish model for hepatoerythropoietic porphyria. *Nat. Genet.* **20**, 239–243.

White, R., Mark, Sessa, A., Burke, C., Bowman, T., LeBlanc, J., Ceol, C., Bourque, C., Dovey, M., Goessling, W., Burns, C., Erter, et al. (2008). Transparent adult zebrafish as a tool for in vivo transplantation analysis. *Cell Stem Cell* **2**, 183–189.

Wingert, R., Brownlie, A., Galloway, J., Dooley, K., Fraenkel, P., Axe, J., Davidson, A., Barut, B., Noriega, L., Sheng, X., et al. (2004). The chianti zebrafish mutant provides a model for erythroid-specific disruption of transferrin receptor 1. *Development* **131**, 6225–6235.

Xu, X., Meiler, S., Zhong, T., Mohideen, M., Crossley, D., Burggren, W., and Fishman, M. (2002). Cardiomyopathy in zebrafish due to mutation in an alternatively spliced exon of titin. *Nat. Genet.* **30**, 205–209.

# 10

# PERSPECTIVES ON ZEBRAFISH AS A MODEL IN ENVIRONMENTAL TOXICOLOGY

*JOHN J. STEGEMAN*
*JARED V. GOLDSTONE*
*MARK E. HAHN*



Zebrafish are being used as a model in all areas of environmental toxicology, for example to discern mechanisms, as a tool in testing, in monitoring the toxic potential in particular environments, in qualitative or quantitative screening of effluents for the presence of effective levels of



*Zebrafish: Volume 29*
FISH PHYSIOLOGY


DOI: 10.1016/S1546-5098(10)02910-9

toxicants, or in screening compounds for particular activities in hazard identification. There is enormous potential for that use to grow. In this chapter we emphasize the mechanistic aspects, including new information on the genes that comprise the "defensome". Toxicogenomic studies are uncovering the involvement of new genes, and pointing to the possibility of common mechanisms and pathways that are shared at least partly in different toxicological processes, such as wnt signaling involved in craniofacial defects in embryos and altered fin regeneration in adults. Other pathways are less well known. Placing defensome genes in gene-regulatory networks is an open area, rich with possibilities. The metabolism and disposition of chemicals is strongly related to their toxicity. While we are learning much about the expression of genes that are implicated in these processes, study of the proteins themselves has lagged substantially. Strong inferences regarding effects of particular chemicals from results in this model organism require knowledge of the function of these proteins, whether enzymes, receptors or other transcription factors. Studies of many of the relevant genes and proteins in zebrafish are still emerging from the descriptive phase relative to the studies in mammals, but new tools and approaches in zebrafish are enabling a growing number of mechanistic advances. Examining similarities and differences between zebrafish and other species will lead to new fundamental understanding of the defense against and susceptibility to chemicals.

## 1. INTRODUCTION

The world is awash in chemicals that have biological activity and that can adversely affect the health of organisms. In the past century, the planet has become chemically different from the condition at any preceding time, with a chemical "envelope" that includes many compounds hazardous to human and animal health and life. These are the purview of environmental toxicology.

Addressing the role of zebrafish as a model in environmental toxicology requires an understanding of the concerns in the field, and a consideration of how any model may be used in addressing such concerns. Environmental toxicology, broadly considered, involves the study of the fate and biological effects of toxic compounds in the natural environment, and includes effects on humans, domestic resource animals and wildlife, individuals and populations. Zebrafish have proven to be a superb model for vertebrate development and disease processes, even helping to unravel mechanisms in mysterious diseases such as Alzheimer's and progeria. Are they an equally good model for environmental toxicology?

Non-mammalian models have been used in research on basic biological mechanisms and processes for more than 100 years. During that time, the field of toxicology arose and matured. The Society of Toxicology was founded in 1961, as an outgrowth of the Toxicology and Safety Evaluation Gordon Research Conference. Toxicology at that time was, to a large degree, the flip side of pharmacology. The importance of chemicals in the environment as potential toxicants affecting wildlife jumped into view with *Silent Spring* (Carson 1962). The founding of the National Institute of Environmental Health Sciences in 1969 and the Environmental Protection Agency in 1970 ushered in a new era of environmental toxicology.

The central paradigm in toxicology encompasses the sources of chemicals, and the processes of adsorption, distribution, metabolism, elimination, and effects of toxicants. Effects depend on the inherent biological activity and toxicity of a compound related to its structure, and the influence of metabolism on that structure and activity, as well as on the internal dose of parent compound and metabolite. Toxicology is an integrative science that falls at the intersection of organismal biology, cell and organ physiology, and molecular biology, as well as chemistry. Environmental toxicology extends to include ecology and ecosystem function, as well as environmental chemistry. Over the past 40 years, the field of environmental toxicology has evolved rapidly. It includes toxicity testing of chemicals and drugs, and the chemical basis of disease. And now, research on molecular mechanisms of chemical action is providing fundamental insights into basic control processes in cell biology and physiology.

Assessing the risk and detecting the effects of chemicals requires not only an understanding of the basic mechanisms involved, but also how these mechanisms contribute to effects in organisms under conditions of exposure in their particular environments. Input sources and amounts, distribution, and biotic and abiotic fate of chemicals all are relevant to effects in ecosystems, as they are to effects in individuals. All constituent species of ecosystems, including humans, are of concern, and this extends to individual as well as population level effects. Assessing risk to component species of an ecosystem often is based on assays with surrogate species, animal models for the species of concern.

So what is the value of zebrafish as a model? One might ask equally, what is the value of any biological model in environmental toxicology, given that in principle, any species may be a species of concern? How can one species be used as a surrogate for the many? This of course depends on the question being asked, on the intended use of a biological model. Perhaps most often, models are used for discerning fundamental mechanisms underlying the adverse effects of particular toxicants, which then may be extrapolated to other species or used to guide studies to determine the nature of those effects directly in a species of concern. A model also may be used as a tool in testing, in monitoring the toxic potential in particular environments, in qualitative or quantitative screening of effluents for the presence of effective levels of toxicants, or screening compounds for particular activities in hazard identification. In fact, zebrafish are being used at some level in all areas of environmental toxicology, and there is enormous potential for that use to grow.

In this chapter we emphasize the mechanistic components and aspects more strongly than chemical screening and environmental assessment with zebrafish. This includes new information on the genes that comprise the "defensome". Certainly, risk assessment and identifying chemical toxicity are important. However, the molecular participants in chemical effect and protection are still being described. The use of zebrafish as tools for drug discovery, in assessing the risk associated with specific novel or legacy chemicals, and assessing environmental quality will all be enhanced by a greater understanding of the defensome, and mechanisms of chemical effect in this species. There have been a number of fine reviews of the use of zebrafish in various aspects of toxicology, including pathology, developmental toxicology, and others (Spitsbergen and Kent 2003; Teraoka et al. 2003; Carney et al. 2006b; Carvan et al. 2007). We do not wish to repeat the material in those reviews and the reader is urged to consult those sources. Rather, our purpose is to provide some substantial detail on the molecular underpinnings for use of zebrafish in toxicology, and to provide some perspectives on recent successes, and to suggest areas of future research. The relevance of this chapter extends beyond zebrafish; the discussion is a guide to issues and questions that are pertinent to the development or use of any animal model in environmental toxicology.

## 2. CURRENT ISSUES IN ENVIRONMENTAL TOXICOLOGY

The over-riding issues in environmental toxicology have been with us for decades: understanding the molecular mechanisms of chemically induced disease or dysfunction, and determining risk associated with real exposures. This pertains to long known and to newly identified chemicals of concern, and includes unforeseen kinds of effects.

Organic pollutants of concern in environmental toxicology include so-called legacy pollutants, as well as many newly recognized and newly emerging contaminants. Legacy pollutants generally are persistent chemicals known for years as problems, but often with mechanisms of toxicity that still are not understood. Emerging problems involve chemicals that are more recently found to be hazardous or pose risk. The nature of effects and the extent of risk are often uncertain. Some major types of chemicals of concern are listed in Table 10.1. More complete lists can be found in other sources (e.g., Reddy 2008). Environmental chemicals of concern include those that are synthetic, made for industrial (plasticizers, nanomaterials), agricultural (insecticides, herbicides, fungicides), and personal use (cosmetics, pharmaceuticals, etc.). Natural products that are of concern include antibiotics, hydrocarbons, and natural toxins (e.g., harmful algal toxins and allelochemicals). Some of these chemicals are illustrated in Figure 10.1, to indicate the diversity of structures that are potentially of concern. This diversity emphasizes the magnitude of the problem facing the environmental toxicology community.

For both humans and animals, the concerns are with fundamental understanding of mechanisms, as well as direct assessment of environmental exposure and its consequences, and predictive toxicity testing of chemicals. Some of the more difficult and long-standing questions involve subtle effects. These include determination of features controlling susceptibility or resistance to effects, pathway discovery linking mechanisms and disease, low dose long-term effects, thresholds and the balance of adaptation and adverse outcome, interactive effects of chemical mixtures, adult effects of exposure

**Table 10.1**
Selected chemicals of concern in environmental toxicology

| Chemical class | Example |
|---|---|
| Industrial | Polyhalogenated aromatic hydrocarbons (PHAHs; e.g., dioxins and biphenyls) |
| | Polycyclic aromatic hydrocarbons (PAHs; e.g., pyrene, benzo[a] pyrene, naphthalenes, anthracenes, phenanthrenes) |
| | Phthalates |
| | Polybrominated diphenyl ethers (PBDEs) |
| | Estrogen mimics (e.g., bisphenol A) |
| | Heavy metals |
| Agricultural | Insecticides |
| | Herbicides |
| | Fungicides |
| Personal care | Pharmaceuticals |
| | Sunscreens |
| | Fragrances (musks) |
| | Sanitizers (triclosan) |
| | Cosmetics |

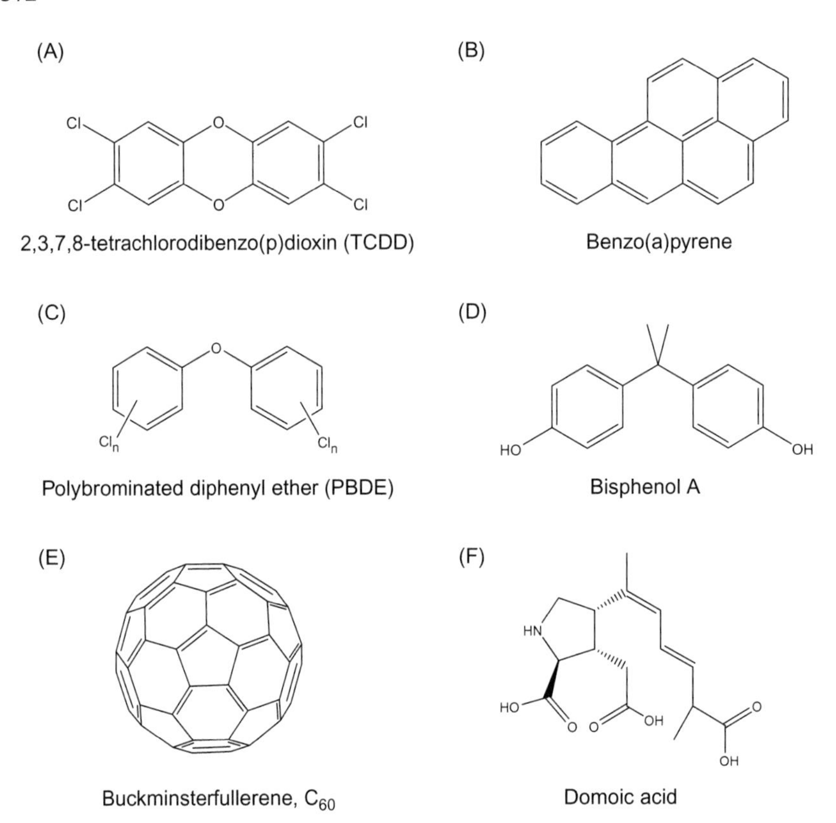


**Fig. 10.1.** Illustrative chemical structures. The diversity of chemicals of concern in environmental toxicology is extremely broad, and illustrated here with a few examples. These include legacy pollutants such as (A) chlorinated dioxins and (B) benzo(a)pyrene as well as emerging pollutants such as (C) polybrominated diphenyl ethers and (D) bisphenol A. More recently, nanomaterials such as (E) buckminsterfullerene ($C_{60}$) and natural products such as (F) domoic acid have gained attention

during development, trans-generational effects (genetic and epigenetic), and species differences in the occurrence and regulation of genes and function of genes and proteins in toxicological processes. The other major focus in environmental toxicology is the assessment of hazard and risk associated with chemicals that are in use. The use of models in assessing risk and predictive toxicity testing may depend on the mechanism or mode of action, and the degree to which the mechanisms can be extrapolated to other species. Issues in predictive and regulatory toxicology include development of high-throughput and reliable assays, and application of these to analysis of environmental matrices contaminated with legacy pollutants, testing of new and emerging chemicals and commercial products, and assessing the safety of drugs and food products.

## 3. THE ZEBRAFISH TOXICOLOGY LANDSCAPE: EMERGENCE OF A NEW MODEL

The literature on zebrafish is growing rapidly; as of fall 2009 there were more than 12 000 papers. The record on zebrafish goes back to 70 years or more; Vascotto reviewed the rise of zebrafish to its current position, focusing on the early papers that set the stage (Vascotto et al. 1997). In searching for papers on zebrafish and toxicology, we used more than 25 combinations of search terms, including redundant searches with different terms. More than 800 unique, relevant papers were retrieved as of summer of 2009. By comparison, there are 12 400 papers on trout (all trout species), going back to 1872 (ca. 450 in toxicology), and about 1800 papers on medaka, going back to 1951 (ca. 120 in toxicology). The searches no doubt missed some papers, and some only deal with zebrafish in a tangential way, for example their being mentioned in a paper focusing on some other species. We regret that constraints to do not permit us to address all papers in this broad and inclusive field.

The potential of zebrafish embryos in developmental toxicology was recognized early and many of the early papers used zebrafish embryos as bioassay material in testing chemicals for effects on cell division and differentiation, or other endpoints (Jones and Huffman 1957). By 1960, most of the early papers that dealt with chemical effects involved developmental toxicity (see Vascotto et al. 1997).

While zebrafish have strong advantages exploited extensively to investigate vertebrate biology, physiology, molecular biology and disease processes, especially in development, they are by no means the only species with such advantages. Since the 1960s, environmental toxicology and carcinogenesis studies have been pursued in a wide range of fish models, including rainbow trout (*Oncorhynchus mykiss*), medaka (*Oryzias latipes*), fathead minnow (*Pimephales promelas*), swordtails (*Xiphophorus* spp.), *Rivulus rivulus*, *Poeciliopsis* spp., killifish (*Fundulus heteroclitus*), scup (*Stenotomus chrysops*), and roach (*Rutilus rutilus*), as well as zebrafish and others. Many species were represented in various meetings emphasizing comparative toxicology. A benchmark example, which is recommended reading, is the "Use of Small Fish in Carcinogenicity Testing" (NCI 1984). The reason zebrafish rose to prominence, superseding most other aquatic vertebrate models, is in part because of the persistence of George Streisinger.

With George as a champion, the potential of zebrafish became recognized ever more broadly, until the zebrafish emerged as one of the most important vertebrate model species in embryology and developmental biology (Fishman 2001).

The value of zebrafish as a model in toxicology – and in biology generally – stems from features that can be classified as phylogenetic, logistical, and technical. As a vertebrate sharing a variety of evolutionarily conserved genes and pathways with other vertebrates, the zebrafish is well positioned *phylogenetically* for extrapolation of results to humans and other vertebrates. Consistent with this, one study showed that 36 of 41 known mammalian teratogens also were teratogenic in zebrafish (Nagel 2002 as cited in Parng 2005). Zebrafish also possess many *logistical* advantages, especially for developmental studies. Attributes well known to all (and so often repeated) include the external development of nearly transparent zebrafish embryos allowing direct observation of all stages of embryonic development. The small size, rapid embryonic development (~60 hr from fertilization to hatching), and short generation time (4 mo) of zebrafish make it an ideal laboratory subject. Large numbers of animals can be generated and housed efficiently, and multigenerational studies can be conducted in a matter of months to years. Finally, because of the initial and subsequent interest in zebrafish as a model (because of its phylogenetic and logistical advantages), much effort has gone into developing techniques to enhance the facility with which zebrafish can be subjected to genetic analysis and manipulation. These *technical* advantages now include a sequenced genome, thousands of EST sequences, various DNA libraries, microarray resources, and a variety of methods for generating transgenic fish, knocking down gene expression in embryos, and targeting loci for inactivation (see Sections 5 and 7). While some of these advantages are shared with other fish models (and other model species beyond fish), together the collection of features that now characterize zebrafish make it the preeminent fish model for fundamental studies in many areas. Consequently, the National Research Council of the US National Academy of Sciences has recommended increased use of zebrafish and other alternative models for research in developmental toxicology (NRC 2000).

Other fish species continue to be used in toxicology and carcinogenesis, including some of the aforementioned, especially rainbow trout (Bailey et al. 1996), medaka (Hinton et al. 2005), fathead minnow (Ankley et al. 2009) and killifish (*F. heteroclitus*; Burnett et al. 2007). Each has a place in environmental toxicology, and many offer advantages over zebrafish in being amenable to studies in the environment that are not possible with zebrafish. Thus, killifish have been extensively used in environmental toxicology studies for decades, and have been the focus of studies on mechanisms of a resistance adaptation to the toxicity of important environmental toxicants, which developed in populations of this species exposed over generations to extremely high levels of contaminants (Burnett et al. 2007). The fathead minnow is widely used as a model for studies of endocrine disruption caused by hormone mimics (Hinton et al. 2005; Watanabe et al. 2007; Ankley et al. 2009; Garcia-Reyero et al. 2009). It is still true today that in some cases more information on toxicological processes in fish has been derived from species other than zebrafish. That is changing rapidly, profoundly so. However, these other species are not to be ignored, especially in the realm of environmental toxicology.

Given the attributes that make zebrafish useful as a developmental model, it is no surprise that studies of developmental toxicology are more prominent than those of the adult animal. As of summer 2009 we retrieved over 200 papers by searching "zebrafish and developmental toxicology"; understandably this chapter will emphasize developmental effects of chemicals. Success in using embryo-larval stages in assessing toxicity has been robust enough that in 2005, the "zebrafish embryo test" was accepted in Germany as a standardized assay for testing toxicity of waste-water, an alternative to traditional tests with adult fish (Nagel 2002; Braunbeck et al. 2005).

A chapter attempting to catalog and summarize progress in a rapidly changing field such as environmental toxicology quickly can become outdated. Thus, we have selected some studies that have brought us here, and point to directions for progress in the future, which may apply for some time.

## 4. MOLECULAR ENVIRONMENTAL TOXICOLOGY: THE DEFENSOME CONCEPT

The strength of inference from studies in any model in environmental toxicology will depend on the degree to which the genes, proteins and pathways that participate in a particular outcome are similar between the model species and the species of concern. Do the different species have homologs of genes that are relevant? Are they regulated in similar ways? Do protein products function similarly? Answering these questions concerning the conservation of genes and proteins involved in toxicology is essential in the development and use of any model and in the interpretation of results obtained.

The "chemical defensome" concept (Goldstone et al. 2006) includes genes, proteins and processes that protect against injury from toxicants such as xenobiotics (Table 10.1), as well as endogenous compounds (bile acids, heme breakdown products), and chemically and endogenously generated radicals including reactive oxygen species. The structures of concern are highly diverse, requiring either non-specific enzymatic responses or a broad array of specific enzymatic reactions. The chemical defensome is comprised of several classes of proteins that function coordinately to protect the cell (Figure 10.2). The suites of defensome genes and proteins include soluble receptors and ligand-activated transcription factors that act as sensors of toxicants or cellular damage, proteins that transport chemicals from the cell, and various enzymes that metabolize xenobiotics. The metabolism of xenobiotic chemicals can determine their half-life and effects in the body. These often are grouped into phase I, phase II and phase III processes. Phase I enzymes catalyze oxidative and reductive transformation of chemicals. Phase II conjugating enzymes add functional groups and form more readily excreted metabolites. Phase III comprises efflux transporters that actively eliminate toxicants, often immediately, and eliminate biotransformation products. The defensome also includes antioxidant enzymes that protect against externally and internally generated reactive oxygen species (ROS) or other radicals.

### 4.1. Molecular Participants and Targets

Toxic effects depend not only on the function of the defensome genes and proteins but also on the non-defensome genes and proteins that may be targets. It is helpful to consider the genes and proteins in different categories based on the type of involvement, whether as participants in the metabolism and disposition of chemicals, or as targets of chemicals. One might consider three broad categories of such genes and proteins. The first group includes defensome genes and proteins that directly participate in modifying the structure, concentration and disposition of a chemical, i.e., the phase I, II, and III processes, as well as the genes encoding the proteins that regulate their expression (Figs 10.2 and 10.3). The second category includes members of the same gene families as comprising the first category, but that may directly increase the biological activity of a chemical or favor distribution to a target site, thereby enhancing the toxicity of the chemical. Many of the same genes or proteins that are part of the defensome also can increase the activity of a chemical or increase distribution to a target site. This is perhaps a slightly different way to think of the "double-edged sword" that describes many enzymes that metabolize xenobiotics. Thirdly, there are genes and proteins that are targets that are involved in the adverse effects of chemicals, but that are not part of the defensome. The third group includes genes that may function in normal processes including those that govern development, reproduction and normal organ physiology. Genes and proteins in all three

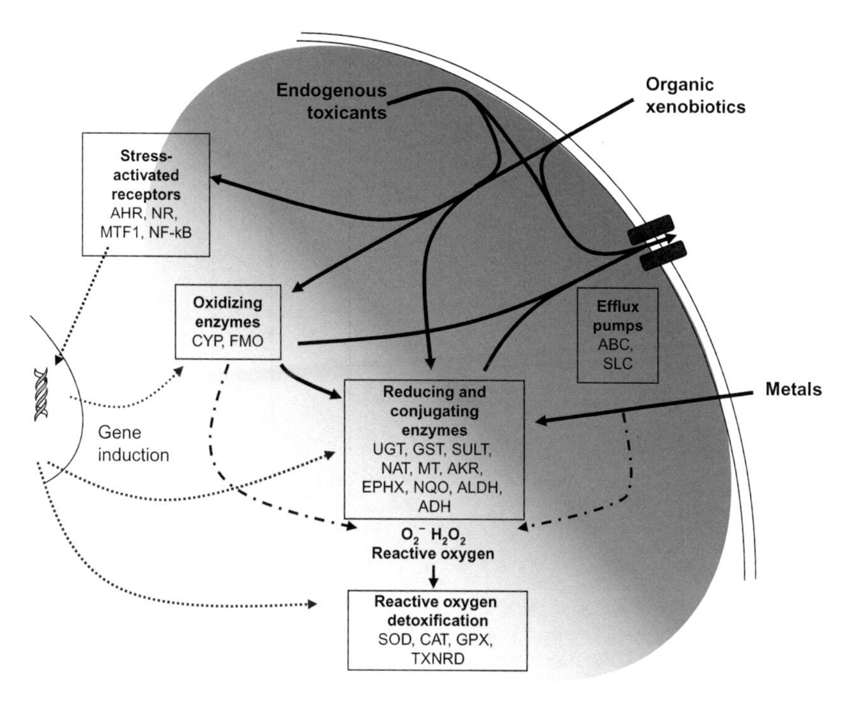


**Fig. 10.2.** The chemical defensome. This generalized scheme indicates the complex integration of pathways involved in oxidation, reduction, conjugation, and excretion of toxic compounds. Both exogenous and endogenous toxicants are metabolized and/or excreted by members of the gene families detailed here and in the text. Abbreviations are noted in the text. Modified from Goldstone et al. (2006); Goldstone (2008).

groups can be targets of chemicals, which may act to increase expression or function, or to inhibit or decrease expression and function.

Functions of defensome genes in zebrafish may sometimes differ from those in species of concern in environmental assessment, whether humans, other fish, or other types of organisms. However, the molecular similarities and differences between zebrafish and other species potentially may yield fundamental insight into mechanisms underlying toxicological processes. We now know that basic metabolic pathways and networks are shared (evolutionarily conserved) among vertebrates and are profoundly similar between fish and humans. Thus, the value of zebrafish as a model for humans and other vertebrates, including other fish, is based on broad similarity of genes involved in the effects of chemicals. But there are limits to that similarity, and our current understanding of it, even though many of these defensome genes, enzymes and processes have been studied for decades in other species of fish.

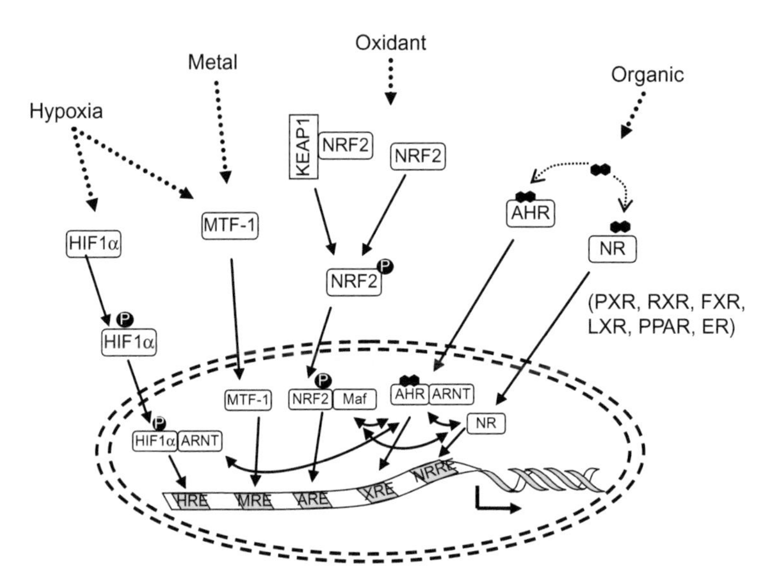


**Fig. 10.3.** Some of the stress response transcription factor pathways illustrating stressors, targets, and known cross-talk. Hypoxia activates both the HIF1α and MTF-1 pathways, metal stress activates MTF-1, oxidative stress activates the NRF2 pathway (as well as others, not shown), and organic xenobiotics activate AHR or various NRs. These transcription factors have specific response elements (REs) in the regulatory regions of responsive genes, including hypoxia RE (HRE; HIF1α/ARNT), metal RE (MRE; MTF-1), antioxidant RE (ARE; NRF2), xenobiotic RE (XRE; AHR/ARNT), and specific NR-REs (e.g., estrogen response elements, PXR response elements). Known receptor cross-talk connections are show with arrows; most recent studies have focused on AHR/ARNT cross-talk. Modified from Goldstone (2008).

As yet, there are very few defensome genes that have been studied in detail (identity, transcript expression, protein expression, substrate or ligand specificity) in zebrafish. Among the defensome cytochrome P450s, only CYP1A has been studied in great detail. Some genes and gene products that are targets but that are not part of the defensome (e.g., CYP19) also have been studied extensively. But for many genes, our knowledge of specific functions of the proteins in fish is known from other species, and inferred for zebrafish. This is related to the fact that these proteins were studied in other species for many years prior to the adoption of zebrafish by the larger biomedical community. This can be expected to change dramatically in the next few years, although those other models will not be abandoned, which is reasonable given that many such models are wildlife species of more direct concern than zebrafish in assessing the state of the environment. However, even when orthologous genes and proteins that are participants or targets are established, studies show that the function of the orthologs can differ substantially in different species. (Orthology refers to an evolutionary relationship in which the two genes (orthologs) in different species are both descended from the same gene that existed in the most recent common ancestor of those species (Fitch 1970). Orthology does not require that the two genes be functionally equivalent). Thus, gene loss, gene gain, and ortholog differences all mean that extrapolation is not always straightforward.

### 4.2. Zebrafish Defensome Gene Families

The current draft assembly of the zebrafish genome produced by the Wellcome Trust Sanger Institute is available at (http://www.ensemble.org/Danio_rerio); this is a foundation for studies to identify molecular participants and targets. Genes that are targets involved in chemical effects are still being identified in zebrafish in experiments, expression profiling, and gene knockdown studies; this is addressed in subsequent sections. However, by scouring the genome database we have obtained a largely complete complement of "defensome" gene families and gene numbers (Table 10.2). Thus, in the following sections there is somewhat greater detail than in other sections of the chapter, given the importance of the defensome genes to the use of a model.

#### 4.2.1. Phase I, II, and III Gene Families

*Cytochrome P450*: The CYP superfamily includes enzymes that catalyze oxidative transformation of many endogenous and exogenous chemicals, with crucial roles in normal physiology and disease processes. Animal CYPs can be separated into two major functional classes. The first group includes those CYPs involved principally in synthesis or activation of endogenous regulatory molecules, including steroids and other lipoidal regulatory molecules. The second group includes those defensome genes involved in the metabolism of xenobiotics; many of these also function in the metabolism of endogenous compounds. Thus, CYP enzymes can determine the persistence and action of endogenous regulatory molecules as well as many drugs and other toxicants and carcinogens.

Searching the zebrafish genome has uncovered 89 CYP genes (see Table 10.2; Goldstone et al., unpublished data). However, at present there is functional or regulatory information for relatively few of these. As of 2009 we could find fewer than 200 papers that report on zebrafish CYP genes or proteins and their function or regulation prominently enough to be retrieved in our search. Among those papers, more than a third concern genes in CYP family 1 (CYP1s), most involving CYP1A as an indicator of AH receptor agonist action. Most of the remaining publications concern CYP19 (aromatase; 44 papers), CYP26s (retinoid hydroxylases; 27 papers), or

**Table 10.2**
Gene family names and numbers in the zebrafish and human defensome

| Gene Family | Number of genes in family | | Reference |
|---|---|---|---|
| | Human | Zebrafish | |
| *Transporters* | | | |
| ABC | 49 | 52 | (Dean and Annilo 2005; Annilo et al. 2006) |
| ABCB | 11 | 15 | |
| ABCC | 13 | 11 | |
| ABCG | 5 | 5 | |
| OAT/SLC21 | 5 | 24 | |
| OATP/SLCO1-6 | 10 | 7 | (Meier-Abt et al. 2005) |
| *Monooxygenases* | | | |
| CYP | 57 | 89 | |
| CYP1 | 3 | 5 | |
| CYP2 | 16 | 42 | |
| CYP3 | 3 | 5 | |
| FMO | 6 | 8 | |
| *Reductases* | | | |
| AKR | 14 | 18 | |
| AKR1 | 10 | 5 | |
| AKR7 | 2 | 2 | |
| ALDH | 19 | 20 | (Lassen et al. 2005; Pittlik et al. 2008) |
| ALDH1 | 7 | 5 | |
| ADH | 8 | 14 | (Reimers et al. 2004b) |
| NQO | 2 | 1 | |
| EPHX | 4 | 3 | |
| 3,17-beta-HSD | 17b:12 | 3b:19, 17b:9 | (Mindnich and Adamski 2009) |
| *Transferases* | | | |
| SULT | 13 | 37 | |
| SULT1,2,3 | 10 | 16 | (Liu et al. 2005; Yasuda et al. 2005a, 2008) |
| GST | 18 | 15 | (Pearson 2005) |
| GST$\alpha,\pi,\mu$ | 11 | 7 | (Suzuki et al. 2005) |
| MAPEG | 3 | 7 | (Bresell et al. 2005) |
| UGT | 13 | 45 | (George and Taylor 2002; Leaver et al. 2007; Huang and Wu 2010) |
| NAT | 2 | 7 | (Begay et al. 1998; Sim et al. 2008) |
| *Metal detoxification* | | | |
| MT | 4 | 3 | (Chen et al. 2004; Yan and Chan 2004; Wu et al. 2008) |
| *Receptors* | | | |
| AHR | 2 | 3 | (Tanguay et al. 1999; Andreason et al. 2002a; Karchner et al. 2005) |
| AHRR | 1 | 2 | (Evans et al. 2005) |
| NFE2-related | 4 | 5 | (Kobayashi et al. 2002; Hahn et al. 2007) |
| KEAP1 | 1 | 2 | (Li et al. 2008) |
| NR | 48 | 70 | (Bertrand et al. 2007) |
| NR1A,B,C,H,I | 14 | 17 | |
| NR2A,B | 7 | 10 | |
| NR3 | 9 | 12 | |
| MTF-1 | 1 | 1 | (Chen et al. 2007; Hogstrand et al. 2008) |

CYP11A1 (P450 side chain cleavage enzyme, 11 papers). An additional few deal with other steroid metabolizing CYPs (e.g., CYP17) or bile acid hydroxylases. Fewer than ten papers concern any of the remaining 75 or so zebrafish CYPs. Clearly, there is an enormous gap in our understanding of zebrafish CYP functions, relative to our understanding of those in human.

Much of what we "know" about the catalytic functions of zebrafish CYPs is by inference only; for now the function of CYP proteins is better known in other fish species, which were studied as model organisms before zebrafish came to the fore. However, this situation is changing, as attention to zebrafish as a toxicological model is increasing, along with the recognition of the need for specific information about the metabolism of xenobiotics.

An example of current understanding, or lack thereof, is in benzo[a]-pyrene (BaP) metabolism. The regiospecific metabolism of polycyclic aromatic hydrocarbons (PAHs) can determine their toxicity, mutagenicity, and carcinogenicity. While metabolism of PAH, principally BaP, has been examined extensively in fish for more than 25 years (e.g. Stegeman et al. 1984), there have been few such studies in zebrafish. Nevertheless, BaP metabolism appears to be catalyzed extensively by zebrafish CYP1A, with a regiospecificity like that seen with other fish. Miranda et al. (2006) found that heterologously expressed zebrafish CYP1A protein formed BaP-phenols, -quinones and -diols, as does CYP1A from other species. With added epoxide hydrolase, the formation of benzo-ring dihydrodiols (7,8- and 9,10-diols) was prominent. The expressed protein also was able to catalyze formation of products binding to DNA and protein, indicating a role in mutagenesis (Miranda et al. 2006). Activation of BaP to a mutagen in zebrafish is also indicated by the studies of Ananuma et al. showing that BaP significantly increased the transversion frequency in the gill and liver of fish containing the rspL mutational target, consistent with the known base changes elicited by BaP in other systems (Amanuma et al. 2000, 2002, 2008).

Complicating this seemingly simple story are observations that implicate toxicokinetic processes and other enzymes in PAH toxicity. Jonsson et al. localized CYP1A induction and irreversible binding of waterborne 7,12-dimethylbenz[a]anthracene (DMBA) in zebrafish gills (Jonsson et al. 2009a). They observed that treatment with inducers ß-naphthoflavone (BNF) or BaP enhanced both CYP1A levels and DMBA binding, suggesting activation of the DMBA. However, while CYP1A expression was present broadly in the gill, DMBA binding was observed at the exposed leading edges. While this suggests that toxicokinetic processes are involved, the role of catalysts other than CYP1A cannot be excluded, especially given the fact that in mammals it is primarily CYP1B1 that activates DMBA (Kleiner et al. 2004). The role of other enzymes in PAH toxicity is also supported by observations that knocking down expression of CYP1A could enhance the toxicity of BNF (Billiard et al. 2006). Which other CYP, or other pathways, lead to toxicity have yet to be determined in zebrafish.

Conservation and diversity of vertebrate CYP genes: Genome sequencing, coupled with traditional searching for genes in other fish species, has disclosed that in some CYP gene families there is a high degree of conservation between zebrafish and other vertebrates. CYP genes in families 7 to 51 in fish correspond, usually directly, to CYP gene families and subfamilies in mammals. It is reasonable to assume that strong conservation of sequence also reflects a conservation of function (although this is by no means certain). Accordingly, the steroidogenic enzymes (e.g., CYP19s) and others with narrow catalytic function (e.g., CYP 26s), show a high degree of conservation of sequence as well as of enzymatic activities and physiological function (e.g., Hernandez et al. 2007).

CYP gene families 1–4, those most involved with xenobiotic metabolism, are more diverse than other CYP families. Zebrafish (and other fish) CYP1 and CYP3 sequences can be recognizably classified to subfamilies that occur also in mammals, although the fish CYP1 and CYP3A genes occupy clades divergent from the mammalian clades for these genes. Qiu et al. (2008) found that most CYP3 genes in vertebrates are located within two different genomic regions. All mammalian CYP3s are in a single subfamily, CYP3A, which occurs in fish as well. Unlike other fish CYP3As, however, zebrafish CYP3A65 is not located in either of the two genomic regions (Qiu et al. 2008). Zebrafish have four CYP3Cs that share synteny with the mammalian CYP3As, while fugu and possibly other fishes have a CYP3B subfamily, about which almost nothing is known (McArthur et al. 2003).

Both mammals and fish have multiple CYP1 subfamilies. Most fish including zebrafish have only one CYP1A gene that is phylogenetically neither CYP1A1 nor CYP1A2 (hence "CYP1A"), although fish CYP1As generally are more CYP1A1-like in structure (Morrison et al. 1998) and regulation (i.e., strongly induced in extrahepatic organs [Stegeman et al. 1991]). Zebrafish and mammals both have a single CYP1B gene, CYP1B1 (Godard et al. 2005). In addition, zebrafish have two novel CYP1 subfamilies, CYP1C and CYP1D. CYP1C genes have not been found in mammals, but the CYP1D subfamily does occur, and is a pseudogene in human (Godard et al. 2005; Goldstone et al. 2009).

Despite the fact that fish CYP1 and CYP3 clades and mammalian CYP1 and CYP3 clades are distinct in molecular phylogeny, there are functional similarities in different taxa consistent with the relatively limited diversity in these CYP families. Thus, CYP3As are the primary catalysts of testosterone 6ß-hydroxylase in fish and mammals (Celander et al. 1996), and mammalian and fish CYP1As are prominent in metabolism and activation of many PAH pro-carcinogens. The roles of homologous CYP in metabolism of particular compounds can differ between taxa, however. For example, the regio-specific oxidation of BaP and the metabolism of planar HAH appear to differ in degree between fish and mammalian CYP1As (Stegeman 1981; Schlezinger et al. 2000), apparently reflecting species differences in CYP1A structures (considered in a later section). The functions are only beginning to be defined with zebrafish enzymes.

In contrast to the CYP1 and CYP3 families, in the CYP2 and CYP4 families there is an enormous gulf between zebrafish and mammals, almost completely obscuring the nature of the homologous relationships that may exist. There are 11 CYP2 subfamilies in humans, and also 11 CYP2 subfamilies in zebrafish. However, in the sequence-based CYP nomenclature, only two CYP2 subfamilies occur in common between zebrafish (or other fish) and human (CYP2R and CYP2U). Members of some CYP2 subfamilies that are distinct in mammals and fish nevertheless show molecular phylogenetic relatedness, and have similar catalytic functions; i.e., killifish CYP2P3 and human CYP2J2 show nearly identical regio- and stereoselectivity for oxidation of arachidonic acid (Oleksiak et al. 2003). More recently, analysis of shared synteny revealed a large cluster of zebrafish genes, all of which appear to be co-orthologs of CYP2J2 (Goldstone et al., unpublished data) (Fig. 10.4). Thus, while identifying strict orthologs may be difficult, it is likely that co-orthologous "sets" of genes may be identified, reflecting independent diversification in the two lineages. (As a note, automated annotation of CYP2s, as at ZFIN, often is incorrect [Wang et al. 2007]). *Sensu strictu* a CYP2J subfamily does not occur in fish; although based on our studies, multiple fish CYP2s occur in a cluster co-orthologous to human CYP2J2.)

While many CYPs participate directly in altering chemical reactivity and toxicity, a large number do not. However, all CYPs, whether directly involved in xenobiotic responses or not, can be targets of exogenous chemicals that

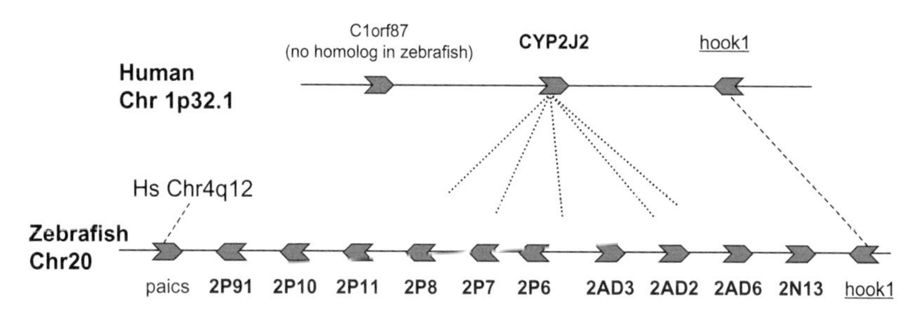


**Fig. 10.4.** Tandemly duplicated zebrafish CYP2 genes that are co-orthologs of human CYP2J2. The multiple tandem duplications observed at the CYP2J locus in vertebrates illustrates the difficulty in inferring functional similarities based on sequence identity (or indeed in this case based also on synteny). (Goldstone et al., unpublished data).

disrupt their function. It is also suspected that many, perhaps most CYP isoforms, also function to generate regulatory molecules or morphogen gradients (Nebert and Dalton 2006), but in vivo confirmation of such roles has been obtained for few CYPs. Moreover, multiple CYPs could act in concert to create temporal and spatial regions of morphogen action, but such possible intersecting roles have seldom been examined in vivo. The potential involvement of CYP in developmental toxicity is considered in a later section.

*Other oxidoreductases*: Other important oxidoreductase enzymes include the flavoprotein monooxygenases (FMO [Ziegler 2002]), aldo-keto reductases (AKR [Jin and Penning 2007]), aldehyde dehydrogenases (ALDHs), NADPH-dependent quinone oxidoreductase (NQO), and epoxide hydrolase (EPHX). In contrast to the CYPs, much less is known about substrate specificity of many of these enzymes, even in humans (Krueger and Williams 2005; Penning and Drury 2007). Zebrafish have various numbers of homologs of all of these genes (Table 10.2), and deciphering the precise evolutionary relationships to mammalian genes may shed light on the functions of these genes both in xenobiotic metabolism and in normal homeostasis.

Aldo-keto reductases (AKRs) are a broadly conserved protein superfamily that consists of generally soluble, monomeric NADPH-linked oxidoreductases (Jin and Penning 2007). AKRs often convert carbonyl-containing compounds to alcohols, aldehydes to primary alcohols, and ketones to secondary alcohols. Several AKR families are involved in xenobiotic metabolism. In particular, members of the AKR1 subfamily catalyze the NADP-dependent oxidation of PAH trans-dihydrodiols to the *o*-quinones, and the AKR7As are aflatoxin dialdehyde reductases. PAH ortho-quinones participate in redox cycling leading to radical formation as well as forming bulky DNA adducts (Bolton et al. 2000).

Zebrafish have twelve AKR genes: three are ancient yet not related to mammalian AKRs, three are AKR6 (potassium voltage-gated channel, shaker-related) homologs, one is an AKR7 (aflatoxin reductase) homolog, and five are AKR1s. The zebrafish AKR1s have been cloned and sequenced in our labs (Karchner et al., unpublished data).

Aldehyde dehydrogenases (ALDHs) are also termed retinaldehyde dehydrogenases, as one of the most important ALDH reactions in vertebrate development is the irreversible oxidation of retinal to retinoic acid (Lee et al. 1991). Retinoids play very important roles in vertebrate patterning, and zebrafish ALDH genes have been investigated primarily in the context of development (Dobbs-McAuliffe et al. 2004; Pittlik et al. 2008). ALDH enzymes may also help maintain the cellular redox balance via ROS scavenging and the production of reducing equivalents as NADPH or NADH. ALDH2 has also been investigated in zebrafish in the context of acetaldehyde metabolism and ethanol toxicity (Lassen et al.

2005). ALDH2 has tandemly duplicated in zebrafish, as it has in other teleosts, and this tandem duplicate has been perhaps misleadingly named ALDH2b based on the assumption that this gene duplication is the result of the teleost whole genome duplication (Song et al. 2006). There appear to be 7 ALDH1 or ALDH2 genes in zebrafish, and a total of 20 ALDH genes; other ALDH family enzymes may have roles in reducing oxidative stress but there is far less evidence for their involvement in other "defensome" mechanisms.

*Conjugating enzymes*: Conjugating enzymes (Phase II enzymes) include glutathione-S-transferase (GST), sulfotransferase (SULT), UDP-glucuronosyltransferase (UGT), and N-acetyl transferase (NAT). Biotransformation by Phase II enzymes generally results in detoxification, but N-acetylation, sulfate, or glutathione conjugation can instead lead to toxic metabolites in a chemical-specific manner.

Glutathione S-transferases are divided into two superfamilies, the soluble GSTs and the membrane-associated GSTs involved in eicosanoid and glutathione metabolism (MAPEG). GSTs catalyze the conjugation of glutathione to various ligands, including both xenobiotics and endogenous substrates such as steroid hormones. Soluble GSTs are divided into six families, alpha, pi, mu, theta, zeta, and omega, while the MAPEG are divided into 2 large families (Pearson 2005). Zebrafish have 7 GSTs in the xenobiotic-metabolizing GSTalpha, pi, and mu classes (Table 10.2), and also 7 MAPEG proteins.

Zebrafish pi class GSTs (GSTp1 and GSTp2) have been shown to be part of the oxidative stress response, and are in part regulated by NFE2L2 (NRF2; see below) via the antioxidant-responsive element (Suzuki et al. 2005). At least one of the upstream glutathione synthesis genes, glutamate cysteine ligase catalytic subunit (GCLC; see below) is also regulated via the conserved ARE (Almeida et al. 2009).

*Sulfotransferase*: Cytosolic SULTs catalyze the transfer of sulfonate from 3′-phosphoadenosine-5′-phosphosulfate (PAPS) to a variety of substrates containing hydroxyl or amino groups, including hormones, neurotransmitters, and drugs and other environmental contaminants (reviewed in Coughtrie 2002). Sulfated products are generally more water soluble and thus are more readily excreted.

Zebrafish SULTs have been investigated primarily in the context of thyroid- and steroid-hormone sulfating transferases (Liu et al. 2005; Yasuda et al. 2005a,b, 2008). Liu and coworkers have characterized 5 SULT1 and 2 SULT3 transferases. In addition to distinct activities on different hormones, Liu and coworkers found different expression patterns for the various SULT1 and SULT3 isoforms during normal zebrafish development. Zebrafish have 37 different SULT homologs, versus 13 in human, and the diversity of sulfotransferase activity has not been fully explored.

UDP-glucuronosyltransferases (UGTs) mainly conjugate UDP-glucuoronic acid with a range of lipophilic substrates including steroids, thyroid hormones, bile acids, bilirubin, and a variety of drugs and environmental contaminants. In mammals, the UGTs are divided into four subfamilies, with somewhat different substrate specificities: UGT1s conjugate phenols, amines, carboxylic acids and bilirubins, while UGT2s are primarily involved in the conjugation of steroids and bile acids (Tukey and Strassburg 2000; Mackenzie et al. 2005). Similar to the CYPs, SULTs, and GSTs, some UGT isoforms exhibit narrow substrate specificity, while others have broader, overlapping substrate specificity (Mackenzie et al. 2005). UGT3s were recently described as UDP-N-acetylglucosaminyltransferases (Mackenzie et al. 2008), with activity on bile acids and estradiol, while UGT8s exhibits UDP galactosetransferase activity during glycosophingolipid and cerebroside synthesis (Bosio et al. 1996).

There are at least 45 UGT genes in the zebrafish genome (Table 10.2), in the UGT1, UGT2, UGT3, and UGT8 families. Zebrafish UGTs have been studied in some detail (Huang and Wu 2010). Fishes in general have a UGT1B with multiple substrate-determining first exons, similar to the UGT1A found in mammals (George and Taylor 2002; Leaver et al. 2007). This genetic organization, with the substrate (aglycone)-binding region encoded on one or more N-terminal exons and the carboxy-terminal half containing the binding site for UDP-glucoronic acid, allows for independent transcriptional control for multiple transcripts originating from differentially spliced products. The zebrafish genome has several UGT loci exhibiting this multiple splicing, and also has at least three tandemly duplicated single exon genes.

Metallothioneins (MTs) are the primary metal-detoxifying genes in vertebrates. MTs are sulfur-rich chelators important in metal ion homeostasis which are inducible by a wide variety of toxic metals, including $Zn^{2+}$, $Cd^{2+}$, $Cu^{2+}$, $As^{3+}$, $As^{5+}$, $Cr^{3+}$, $Cr^{5+}$, $Ni^{2+}$, and $Hg^{2+}$. MT protein expression in adult fish tissues has been used for a long time as a marker of metal pollution in aquatic ecosystems. MT protects cells from toxicity by binding metals. MTs are regulated primarily by binding of MTF1 to metal response elements (MREs) in the 5′ regulatory regions, although the transcription factors SP1 and AP1 may also be involved. Three metallothionein genes are found in zebrafish: mt1, mt2, and similar to metallothionein-B (smt-B) (Yan and Chan 2002, 2004; Chen et al. 2004; Wu et al. 2008). The induction of various mt forms in zebrafish embryos may differ depending on the developmental stage (Chan et al. 2006). Wu et al (2008) identified 10 MREs in allelic variants in the promoter region of zebrafish MT, which affected the expression in zebrafish cell lines.

*ABC transporters*: ABC transporters actively export chemicals from the cell (Phase III). There are 52 ABC transporter genes in zebrafish. Analysis of the molecular evolution of ABC genes shows that the subfamilies in zebrafish correspond to those in humans (Annilo et al. 2006), with the exception of an ABCH subfamily gene found only in zebrafish, described further by Popovic (Popovic et al. 2009). The evidence for gene duplication and deletion suggests an ongoing process of ABC gene evolution (Annilo et al. 2006).

*Antioxidant defenses*: Studies assessing effects of chemicals and other exposures (e.g., UV radiation) frequently identify oxidative stress (reactive oxygen species and other radicals) as possibly contributing to toxicity. Thus, antioxidant defenses are part of the chemical defensome. The glutathione system plays important roles in defense against oxidative stress.

Glutathione is a tripeptide thiol synthesized from glutamate, cysteine, and glycine that is present in concentrations approaching 5 mM in many cells. Reduced glutathione (GSH) itself can react with free radicals to form glutathionil radicals, which can dimerize to form oxidized glutathione, GSSG. GSH oxidation by reactive oxygen species and other free radicals can also be catalyzed by glutathione peroxidases (GPX); the resulting GSSG is recycled back to reduced GSH using NADPH as a reductant. Another major role for glutathione is carried out by the GSTs, as noted above. Zebrafish appear to have only one copy of the basic glutathione synthesis and recycling genes, including GCLC, GCLM, GS and GSR.

The glutathione peroxidase (GPX) enzymes, which use reduced GSH to detoxify reactive species, represent an important protective antioxidant mechanism. Vertebrate GPX evolution is complicated, with several independent gene duplication events in vertebrate evolution as well as gene loss in the teleost lineage and zebrafish in particular. GPX1 and GPX4 are duplicated as a result of the teleost whole genome duplication (i.e., ohnologs, see Section 4.3), while GPX2 and GPX7 are present in other fish but not in zebrafish. GPX5 and 6 appear to be eutherian innovations arising from duplication of GPX3, as only homologs of GPX3 are present in fish genomes.

### 4.2.2. Transcription Factors: Xenobiotic Receptors and Regulation of Defensome Genes

Nuclear and cytosolic protein receptors and other conditional (ligand- or stress-activated) transcription factors are intimately involved in the regulation of defensome genes and in mechanisms of chemical toxicity (Ma 2008). Such transcription factors serve as sensors of the cellular environment, initiating genomic (transcriptional) and non-genomic responses that usually result in compensatory changes in gene expression or protein function. A variety of proteins serve this sensor function in mammals and homologs of almost all are present also in zebrafish, often as duplicates. They include several proteins that are activated by small organic ligands, by metals, and by other environmental stressors.

The AHR is perhaps the most extensively studied transcription factor that mediates responses to environmental xenobiotics in vertebrates (Okey 2007). AHR ligands include some of the most common pollutants of longstanding concern, the halogenated and polycyclic aromatic hydrocarbons, as well as numerous natural products and endogenous chemicals (Denison and Nagy 2003). AHR acts as a dimer with an AHR nuclear translocator (ARNT) protein to regulate gene expression in response to these ligands.

In contrast to the single AHR (AHR1) in most mammals, there are 3 AHR homologs (AHR1a, AHR1b and AHR2) in zebrafish (Tanguay et al. 1999; Andreasen et al. 2002a; Karchner et al. 2005). AHR multiplicity is a characteristic of most fish, the result of retaining one or more copies of the AHR2 gene – which was lost in most mammalian lineages – combined with the effects of the fish-specific whole-genome duplication, which resulted in multiple AHR1 and AHR2 paralogs in many fish species (Hahn et al. 2006). Like mammals, zebrafish possess two ARNT genes (ARNT1 and ARNT2), although the zebrafish genes undergo alternative splicing to yield several transcripts and proteins that differ in functional properties (Tanguay et al. 2000; Prasch et al. 2006).

The three zebrafish AHR paralogs possess distinct, although incompletely understood, functions in vitro and in vivo. AHR2 and AHR1b both function in vitro as high-affinity receptors for TCDD that activate the transcription of an artificial reporter gene construct, with AHR1b being slightly less sensitive than AHR2 to activation by TCDD (Tanguay et al. 1999; Karchner et al. 2005). In contrast, AHR1a neither binds TCDD in vitro nor is capable of activating transcription of a reporter gene in response to a variety of typical AHR ligands (Andreasen et al. 2002b). In embryos, studies employing morpholinos targeting AHR2 have demonstrated that AHR2 is required for the embryotoxicity of AHR ligands such as TCDD, PCB-126, and some PAHs (Prasch et al. 2003; Incardona et al. 2005, 2006; Jonsson et al. 2007) and regulates the induction of CYP1A and other CYP1 forms by some of the same compounds (Prasch et al. 2003; Incardona et al. 2006; Jonsson et al. 2007; Yin et al. 2008). The role of AHR1b is not well understood but by inference may have a more restricted role in chemical toxicity; it has been hypothesized that AHR1b may be involved in physiological rather than toxicological functions (Karchner et al. 2005). AHR1a, although apparently inactive in vitro, has been suggested as mediating some aspects of embryotoxicity for some small PAH compounds (Incardona et al. 2006).

An AHR-related repressor of AHR signaling has been identified in mammals, and two paralogs (AHRRa and AHRRb) exist in zebrafish. The ability to repress AHR is conserved in both AHRR forms (Evans et al. 2005, 2008). However, the two zebrafish AHRR paralogs may have undergone subfunction partitioning, as they appear to regulate either physiological (AHRRa) or toxicological (AHRRb) functions of zebrafish AHRs (Jenny et al. 2009).

*Nuclear receptors (NRs)*: Members of the nuclear receptor superfamily play important roles in embryonic development and hormonal signaling, and several of them are also involved in regulating responses to exogenous and endogenous toxicants, or are targets for xenobiotic chemicals. An exhaustive review of nuclear receptors and their toxicological roles in zebrafish is beyond the scope of this chapter; more information can be found elsewhere (Bertrand et al. 2007). We mention two examples to illustrate the relationship between NRs in zebrafish and humans.

The NR1I subfamily of nuclear receptors in vertebrates includes the pregnane X receptor (PXR), constitutive androgen receptor (CAR), and vitamin D receptor (VDR). PXR (NR1I2), CAR (NR1I3), and peroxisome proliferator activated receptors (PPARs; NR1C) form heterodimers with the retinoid x receptors (RXRs, NR2B) and regulate members of the CYP2, CYP3 and CYP4 gene families, respectively in mammals (Waxman 1999; Wei et al. 2002; Yamamoto et al. 2003). While there is a preference in these linkages, it also is well known that there is cross talk, or mutual participation of multiple receptors in regulating expression of genes in some CYP subfamilies. Thus, both CAR and PXR are involved in the regulation of CY2B and CYP3A genes in mammals (Xie et al. 2000; Wei et al. 2002). In contrast to mammals, however, fish (including zebrafish) lack a CAR ortholog; studies by several groups indicate that CAR diverged as a new NR1 line *after* the divergence of fish and mammals (Moore et al. 2002; Krasowski et al. 2005; Reschly et al. 2007). This indicates a fundamental difference between fish and humans in regulation of CYP expression, known for years but only recently beginning to be understood at a mechanistic level. Thus, understanding the function of PXR as a xenobiotic receptor and CYP transcriptional activator in zebrafish (and other fish) gains importance.

PXRs in general are known for their ability to bind a wide array of endogenous and exogenous compounds. PXR and related NRs farnesoid X receptor (FXR, NR1H4) and liver X receptors (LXR; NR1H2/3), are thought to serve endogenously as receptors for bile salts (Krasowski et al. 2005). PXRs are also known to bind steroids, and xenobiotics including phenobarbital, clotrimazole, and 1,4-bis-[2-(3,5,-dichloropyridyloxy)] benzene TCPOBOP (Maglich et al. 2002; Wei et al. 2002). Using the ligand binding domain, Moore et al. (2002) found that the zebrafish PXR is activated by some, but not all, of the same chemicals as mammalian PXRs (Moore et al. 2002), an observation expanded upon by Krasowski and co-workers (Reschly et al. 2007; Ekins et al. 2008).

Another toxicologically important group of NRs are the estrogen and androgen receptors (ERs and ARs, respectively), the focus of intense interest because of their roles in mediating endocrine disruption caused by environmental xenobiotics. Humans possess two ER genes, ERα and ERβ (NR3A1 and NR3A2). Zebrafish, like most other teleosts, possess one ERα and two ERβ paralogs (ERβa and ERβb, also sometimes called ERβ1 and ERβ2) (Bardet et al. 2002; Menuet et al. 2002); the two ERβ forms are thought to be products of the fish-specific whole-genome duplication (Bertrand et al. 2007). Several studies have demonstrated that the three zebrafish ERs, and in particular the two ERβ paralogs, have distinct expression patterns and functional characteristics, including different ligand and response element specificities (Menuet et al. 2002; Le Page et al. 2006; Bertrand et al. 2007; Froehlicher et al. 2009a). Like ERβ, the androgen receptor (NR3C4) was also duplicated in the fish-specific whole-genome duplication (Bertrand et al. 2007; Douard et al. 2008). However, zebrafish differ from many other teleosts in having lost one of the AR paralogs (Douard et al. 2008).

While discussing steroid receptors like ER and AR, it is important to note a distinct family of steroid receptors that has been discovered only recently, but could be of importance as a target of xenobiotics. Unlike the NR3 subfamily of soluble steroid receptors, these novel steroid receptors are membrane-bound, members of the G-protein coupled receptor (GPR) family. For example, a membrane estrogen receptor, GPR30, has been identified in several species of fish, including zebrafish, and shown to be involved in regulating oocyte maturation (Pang et al. 2008; Pang and Thomas 2009). Studies in zebrafish could be instrumental in elucidation of the role of these receptors in toxicology.

*NRF2:* In mammals, oxidant and pro-oxidant chemicals elicit an anti-oxidant response or oxidative stress response (OSR), which involves the increased expression of genes whose products act to mitigate the oxidant challenge. Oxidants, electrophiles, and sulfhydryl-reactive agents initiate this response by activating NF-E2-related factor 2 (NRF2, also called NFE2-like 2 (NFE2L2)), a cap'n'collar (CNC)-basic-leucine zipper (bZIP) family protein. NRF2 is normally found in the cytoplasm, where an interaction with Kelch-like-ECH-associated protein (Keap1), an adapter for a Cul3 ubiquitin ligase complex, targets it for rapid proteasomal degradation. Oxidative stress disrupts the interaction between NRF2 and Keap1, after which NRF2 enters the nucleus and forms a heterodimer with one of several small Maf proteins; NRF2-Maf dimers bind to anti-oxidant response elements (ARE) and activate transcription of genes encoding antioxidant enzymes such as SOD, GSTs, NQO1, and many others (Kensler and Wakabayashi 2009; Nguyen et al. 2009).

The molecular mechanisms underlying the regulation of the oxidative stress response in zebrafish are just beginning to be uncovered. Carvan et al. (2001) demonstrated that zebrafish oxidant-responsive transcription factors are able to recognize mammalian ARE sequences (Carvan et al. 2001). The most direct evidence for an oxidative stress response in fish and its mechanistic similarity to that in mammals comes from the work of Kobayashi, Yamamoto, and colleagues, who cloned the cDNAs for zebrafish NRF2 and Keap1 and showed that these proteins mediated the induction of GSTP1, NQO1, and GCLC by tBHQ in zebrafish embryos (Kobayashi et al. 2002; Suzuki et al. 2005; Li et al. 2008; Kobayashi et al. 2009). Zebrafish also possess the small Maf proteins that form heterodimers with NRF2 (Takagi et al. 2004). Knock-down of NRF2 expression by microinjection of morpholino oligonucleotides into embryos demonstrated the essential role of this protein in mediating the oxidative stress response in embryos (Kobayashi et al. 2002; Timme-Laragy et al. 2009).

Zebrafish possess two Keap1 paralogs (Keap1a and Keap1b) (Li et al. 2008), which both target NRF2 for degradation but exhibit oxidant specificity (Kobayashi et al. 2009). The differential sensitivity of the zebrafish Keap1 paralogs was exploited by Kobayashi et al. (2009) to help identify specific cysteines that are targeted in a compound-specific way, providing new mechanistic insight into NRF2 activation by oxidants. Together, these studies provide an excellent example of how new fundamental understanding can be obtained through detailed investigation of paralogous genes in zebrafish.

Metal transcription factor 1 (MTF1) is the primary metal-sensing transcription factor in vertebrates. MTF1 is activated metal binding to a cluster of five cysteine residues near the C-terminus (He and Ma 2009), and zinc may serve a central role in activation of MTF1. MTF1 binds to metal response elements (MREs) in the 5′ regulatory regions of metal-responsive genes, including metallothioneins (MTs) (Chen et al. 2007). MTF1 also plays essential roles in hepatic development during embryogenesis, as targeted disruption of the Mtf1 gene in mice results in lethal liver degeneration by gestation day 14 (Wang et al. 2004). MTF1 has been suggested to be a master regulator of gene expression during zebrafish embryogenesis (Hogstrand et al. 2008). In zebrafish, MTF1 has also been shown to control the expression of zinc exporters (ZnT1) and zinc importers (ZnT5, ZIP4, ZIP10), thus managing zinc homeostasis (Hogstrand et al. 2008; Zheng et al. 2008).

### 4.3. Gene Duplication: Orthologs, Paralogs, and Ohnologs in Xenobiotic Sensing and Metabolism

Extrapolation of results from one species to another, for example from zebrafish to humans, requires consideration of differences in the diversity, regulation, and function of toxicologically relevant genes between the two species. Although the basic genetic pathways important in toxicology are evolutionarily conserved among vertebrates generally, it has become clear over the last decade that there are important differences in gene diversity between fish and mammals (Table 10.3). These differences have arisen in several ways: (1) as a result of the whole genome duplication (R3) that occurred at about the time of the teleost radiation, approximately 350 million years ago, after the divergence of lineages leading to modern fish and mammals (Amores et al. 1998; Christoffels et al. 2004; Postlethwait et al. 2004), (2) from earlier tandem or whole-genome duplications (R1, R2) (Dehal and Boore 2005), followed by lineage-specific gene loss in zebrafish or humans, or (3) from gene duplications occurring in either or both lineages since their divergence (lineage-specific diversification).

In considering the homologous relationships between zebrafish and human genes, it is important to define the type of homology, and especially whether genes are orthologs or paralogs (Fitch 1970) and, if the latter, whether they are ohnologs (paralogs resulting from a whole-genome duplication (Wolfe 2000; Postlethwait 2006a). Often, these relationships cannot be ascertained from sequence comparisons alone (for example, by reciprocal best BLAST hit or phylogenetic analysis), because of loss of genes in some lineages and independent duplication in others. Gene structure and shared synteny increasingly are being used to determine the evolutionary history, and thus the homologous relationships, of genes in two species (Postlethwait 2006, 2007).

An issue of practical significance raised by differences in gene diversity is that of nomenclature, which influences how we think about the genes in question. There is resistance by some to accord clearly orthologous genes in different species the same name; the argument is that with evolutionary distance, the orthologs could have different functions. However, as noted earlier, orthology refers to evolutionary relationships and is independent of function, although orthologs often do have identical or highly similar functions. Regardless, it is important to continue to work towards a system of gene nomenclature that illuminates rather than obscures orthologous and paralogous relationships.

The difference in gene number between fish and humans, and in particular the existence of zebrafish paralogs (ohnologs) that are co-orthologs of a single mammalian gene, provides an opportunity for new

**Table 10.3**
Zebrafish orthologs and paralogs of human genes involved in toxicology: Selected examples

| Human genes | Zebrafish genes | Evolutionary relationships | Functional relationships* | References |
|---|---|---|---|---|
| AHR | AHR1a, AHR1b; AHR2 | Co-orthologs (ohnologs); paralog, lost in humans | AHR2 regulates response to HAHs; AHR1a and AHR1b functions unknown | (Tanguay et al. 1999; Andreasen et al. 2002a; Karchner et al. 2005) |
| ARNT1, ARNT2 | ARNT1, ARNT2 | Orthologs | ARNT1, with AHR2, regulates response to HAHs. | (Tanguay et al. 2000; Prasch et al. 2006) |
| AHRR | AHRRa, AHRRb | Co-orthologs (ohnologs) | Distinct functions in regulation of endogenous (AHRRa) and ligand-induced (AHRRb) AHR activity | (Evans et al. 2005; Jenny et al. 2009) |
| PXR | PXR | Ortholog | | (Moore et al. 2002) |
| CAR | (not present) | Paralog of PXR | | |
| ERα; ERβ | ERα; ERβa, ERβb | Orthologs; co-orthologs (ohnologs) | | (Bardet et al. 2002; Menuet et al. 2002) |
| NRF2 (NFE2L2) | NRF2a, NRF2b | Co-orthologs (ohnologs) | NRF2a regulates response to oxidant chemicals; NRF2b function unknown | (Kobayashi et al. 2002; Hahn et al. 2007; Timme-Laragy et al unpublished) |
| Keap1 | Keap1a, Keap1b | Co-orthologs (ohnologs) | Keap1a and Keap1b have different chemical specificity | (Kobayashi et al., 2002, 2009; Li et al., 2008) |
| CYP1A1 | CYP1A | Ortholog | | |
| CYP1A2 | — | | | |
| CYP1B1 | CYP1B1 | Ortholog | | |
| — | CYP1C1, CYP1C2 | Paralogs, lost in humans | | (Godard et al. 2005) |
| CYP1D1p | CYP1D1 | Ortholog | | (Goldstone et al. 2009) |
| CYP19 | CYP19A1, CYP19A2 | Co-orthologs (ohnologs) | Tissue-specific expression patterns in ovary (CYP19A1) and brain (CYP19A2) | (Kishida and Callard 2001) |
| ADH1, ADH4, ADH6, ADH7 | ADH8a, ADH8b | independent duplications in the two lineages | | (Reimers et al. 2004a) |

*Function of zebrafish genes in relation to human homologs.

mechanistic insights in toxicology. In general, the study of zebrafish ohnologs has potential to yield new information about the function of their single human counterpart (Postlethwait et al. 2004). The duplication, degeneration, complementation model of gene evolution (Force et al. 1999; Lynch and Conery 2000) predicts that the multiple functions of a human gene may be partitioned between the fish ohnologs [subfunction partitioning (Postlethwait et al. 2004)]. Most of the evidence in support of this model comes from studies of zebrafish paralogs that have distinct patterns of expression that together sum to the expression pattern of their mammalian ortholog (Postlethwait et al. 2004). This has been referred to as "regulatory partitioning" (Hahn and Hestermann 2007). For other genes, rather than (or in addition to) the partitioning of expression patterns, the fish genes may diverge with regard to specific functions – for example differential loss of certain functional domains or specialization for subsets of ligands ("functional partitioning") (de Souza et al. 2005; Hawkins et al. 2005). In either case, paralog-specific studies involving knock-down or other approaches in zebrafish may reveal novel functional aspects of their mammalian ortholog. An example is a recent study showing distinct functional roles of zebrafish AHRR paralogs in development and the response to TCDD exposure (Jenny et al. 2009).

The existence of ohnologs also may make zebrafish a good model for other fish, many of which can be expected to retain similar sets of ohnologs from the fish-specific whole-genome duplication. However, there also may be differences in the numbers of the duplicated genes retained, and there may be additional duplication in some lineages. Fish, as a species-rich vertebrate lineage with an array of species-specific gene losses and duplications following R3, can be expected to possess a sometimes bewildering diversity of genes (orthologs and paralogs) that may result in varied responses to chemicals. Nevertheless, it now is clear from studies in a variety of fish species that the fundamental elements of the response machinery and a deep similarity of responses will be present in most or all fishes; superimposed on that is variation in some of the details, which may affect the potency and efficacy of toxicants in different species.

## 5. ADVANCES AND INSIGHTS WITH ZEBRAFISH

Studies in zebrafish have provided substantial new information on genes involved in toxic mechanisms, including new genes, new understanding of known genes, and new knowledge of gene regulation. Zebrafish also are yielding new insights into pathways associated with phenotypes of particular toxicants. This section presents notable examples of the progress in these areas.

### 5.1. Gene Discovery and Characterization

As implied in Table 10.3, scouring genome databases has uncovered novel genes in the various defensome gene families, including biotransformation enzymes and transcriptional regulators. As indicated above, evolutionary distance and gene loss, gain, and conversion, mean that some of these genes identified in zebrafish differ from those in humans or in other fish. The CYP1s are an emphatic example. Halogenated and polynuclear aromatic hydrocarbon pollutants are metabolized primarily by members of the CYP1 family. The CYP1 gene family in mammals consists of two CYP1A genes, CYP1A1 and CYP1A2, and CYP1B1. Multiple knockouts in mice show that mammalian CYP1A1, CYP1A2 and CYP1B1 play overlapping yet different roles in PHAH- and PAH-induced toxic responses (e.g., Dragin et al. 2008).

Although phylogenetically neither CYP1A1 nor CYP1A2 (hence "CYP1A"), zebrafish CYP1A is more CYP1A1-like in structure (Morrison et al. 1995), regulation (e.g., strongly induced by AHR agonists in extrahepatic organs) and function, including in metabolism of BaP. Zebrafish CYP1B1 is much less well known than the CYP1A. Transcriptional activation by AHR agonists and basal expression occurs in multiple sites during development (Yin et al. 2008). The paralogous CYP1Cs were identified in scup and zebrafish (Godard et al. 2005), and the newer gene, CYP1D1, which is not induced by halogenated AHR agonists, was described in zebrafish (Goldstone et al. 2009). The absence of the CYP1Cs and of a functional CYP1D1 in humans raises questions about how these novel genes participate in toxic effects of AHR agonists or other substrates. The CYP1C and CYP1D1s expressed in yeast all act on alkoxyresorufin substrates (Goldstone et al. 2009 and unpublished). They also have activity with xenobiotic substrates (unpublished). There are many zebrafish CYP genes about which we know almost nothing, such as those in the group of 10 CYP2AAs; the field is poised for discoveries.

Retinoic acid (RA) is essential for normal development, and high levels can be teratogenic. White et al. (1997) identified 4-oxo-RA and 4-OH-RA as major products of retinoic acid metabolism by a zebrafish CYP enzyme (P450RAI) that was the first CYP26 described (White et al. 1997). CYP26 is an RA-inducible enzyme that is a critical factor in RA signaling, controlling the location and concentration of this vital morphogen. CYP26 was later identified in humans, and as seen in Table 10.3 there are now multiple CYP26s, in zebrafish and humans. Dobbs-McAuliffe et al. (2004) examined the response of zebrafish embryos to RA challenge and showed that enzyme expression levels and patterns vary with the age of the embryo and timing of treatment with RA. RA signaling and CYP26 expression are targets for chemical effects (Dobbs-McAuliffe et al. 2004). Tiboni et al. showed that in mouse, CYP26A1 and CYP26B1 transcript level are affected by fluconazole (Tiboni et al. 2009). While a role of CYP26 in RA signaling is clear, CYP26 as a target for chemical effects has yet to be fully established; zebrafish will play a role in that.

### 5.2. Expression Profiling: Toxicogenomics

An approach to unraveling and inferring mechanisms is expression profiling, by qPCR or by microarray, or by proteomic approaches. There is a rush to do global profiling of gene expression changes in zebrafish embryos and adults, to determine whether there are expression signatures that can be attributed to different chemical exposure, to discern adaptive from adverse changes, to identify genes that may be targets leading to adverse effects, and to discover gene regulatory networks involved in toxicity. There are often confounding issues in distinguishing primary from secondary responses and interpreting low dose versus high dose effects. Yet much is being learned.

Profiling by qPCR: In general, these studies have built a solid foundation of knowledge and understanding of the dose–response, tissue distribution, developmental stage, and structure-activity relationships for how particular genes respond to effectors. Profiling with PCR usually is preferred when there are questions of only a few genes, for example, determining the responses of the novel CYP1 genes (CYP1C1, CYP1C2, CYP1D1) to PHAH or other AHR agonists, and whether and which AHRs are involved in the regulation of these (Jonsson et al. 2007; Goldstone et al. 2009). Likewise, examining the effects of various suspected endocrine active substances on the expression of the CYP19 genes is more approachable by PCR (Cheshenko et al. 2007). With current understanding, however, screening of chemicals for effects on expression profiles of one or even several genes will not always reveal a mechanism; different agonists for the same receptor may affect expression of the same genes in different ways (e.g., Jonsson et al. 2009b).

Some studies have examined the response of small sets of genes for coordinate regulation. Bainy and colleagues (Bresolin et al. 2005) examined the expression of PXR, CYP3A and MDR1 in zebrafish treated with pregnenolone 16$\alpha$-carboninitrile (PCN), a PXR agonist, and determined that there is coordinate regulation of these genes in zebrafish, as there is in mammals. Hoffmann and Oris (2006) examined the effects of waterborne BaP on eight genes involved in steroidogenesis in head, liver and ovary of adult zebrafish (CYP11A1, CYP17, CYP19A1, CYP19A2, 20β-HSD, ERβ, FSHβ, LHβ, and vitellogenin, with CYP1A as a marker and because it metabolizes BaP). The objective was to assess possible mechanisms by which the PAH, not an obvious endocrine disruptor, could be acting to affect reproduction (Hoffmann and Oris 2006). The results showed that the PAH exposure did alter reproductive output and expression of genes involved in reproduction, but variably in the various organs. There are differing results in other studies of BaP effects (Segner 2009), perhaps understandable with a compound like BaP that is not an obvious estrogen mimic.

*Microarray profiling:* Studies to date employing microarrays for expression profiling related to toxic chemical or pollutant effects are listed in Table 10.4. In addition to addressing larger objectives, many microarray studies confirm that a few genes are most responsive to particular toxicants, supporting observations from earlier expression studies. Thus, induction of vitellogenin and CYP19A2 gene expression reflect estrogenic action of chemicals (e.g., Kausch et al. 2008; Ruggeri et al. 2008), and CYP1A induction is among the most prominent responses to AHR agonists (Handley-Goldstone et al. 2005).

Microarray studies to date have examined expression in whole embryos, adult liver, brain and gonad, with a variety of toxicants and to meet a variety of objectives (Table 10.4). Some of the studies are directed at understanding mechanisms of action where such mechanisms are not fully known, including TCDD and other AHR agonists, BPDE, domoic acid and fullerenes ($C_{60}$) (see Table 10.4). Others seek to determine the extent of gene expression change in response to chemicals that have a presumptive mode of action, or to classify a toxicant, as for example whether it is an estrogenic compound or an AHR agonist. Identifying slight effects on expression of some known genes, such as the 5–6-fold induction of CYP1A by 3,4-dichloroaniline (Voelker et al. 2007), may be important in seeking the consequences of that lesser response. A few key studies addressed below exemplify the success in uncovering mechanisms. (Array studies for biomarker discovery are considered in Section 6.)

One of the first studies to profile transcriptional changes due to dioxin in zebrafish embryos used a cDNA microarray with ca. 3000 unique features, derived from a zebrafish heart cDNA library (Handley-Goldstone et al. 2005). That study identified several patterns consistent with effects on pathways involved in cardiac function and effects. The array was an uncharacterized cDNA array, and sequencing was required to identify up-regulated transcripts. A large number of probes detected induced transcripts that when sequenced revealed a novel retrotransposon that was induced by TCDD, possibly as a secondary result of oxidative stress (Goldstone et al. unpublished data).

**Table 10.4**
Zebrafish toxicological microarray studies

| Type of array | Chemicals | Organs/tissues examined | Major findings | Reference |
|---|---|---|---|---|
| cDNA array | Hypoxia | Embryos | Adaptive changes to hypoxia | (Ton et al. 2003) |
| Custom cDNA array (heart library) | TCDD | Embryos | Cardiomyopathy, novel retroelement | (Handley-Goldstone et al. 2005) |
| Affymetrix GeneChip Genome Array | Valproic acid, trichostatin A | Embryos 25 hpf | Valproic acid teratogenesis linked to inhibition of histone deacetylase | (Gurvich et al. 2005) |
| Brain-specific cDNA array | Mianserin, chlorpromazine | Brain | Neuroendocrine gene expression disruption | (van der Ven et al. 2005, 2006b) |
| Brain-specific cDNA array | Mianserin | Brain, gonad | Induction of estrogen biomarkers | (van der Ven et al. 2006a) |
| Affymetrix GeneChip | TCDD | Regenerating fin | TCDD impaired cellular differentiation and extracellular matrix composition | (Andreasen et al. 2006) |
| Affymetrix GeneChip | TCDD | Embryo hearts | Rapid gene induction in heart | (Carney et al. 2006a) |
| Affymetrix GeneChip | 7alpha-ethynylestradiol (EE2) | Adult female liver | Effects on steroid binding and metabolism | (Hoffmann et al. 2006) |
| Custom oligo | $Na_2HAsO_4$ | Liver | Increased gene expression relating to metabolism and damage response | (Lam et al. 2006) |
| Affymetrix GeneChip Genome Array | $C_{60}$, tetrahydrofuran | Larvae | Solvent upregulates oxidative stress genes | (Henry et al. 2007) |
| 17k oligo (Sigma Genosys) | Aroclor 1254 | Embryo | HSC70 and other neuronal genes downregulated | (Kreiling et al. 2007) |
| Qiagen LongOligo | EE2 | Liver, telencephalon | Organ differences; Vtg strongly upregulated | (Martyniuk et al. 2007) |

| | | | | |
|---|---|---|---|---|
| 14k oligo (MWG) | Pharmaceutical mixture | ZFL cells | Repression of primary metabolism genes, increase of DNA repair | (Pomati et al. 2007) |
| 17k oligo (Sigma Genosys) | EE2 | Adult gonad | Cell cycle progression and GPX genes altered | (Santos et al. 2007) |
| 14k oligo (MWG) | 3,4-dichloroaniline | Embryos | AHR-mediated effects | (Voelker et al. 2007) |
| 16K oligo (Sigma Genosys) | 4-chloroaniline, DDT, $CdCl_2$, TCDD valproic acid, $As_2O_3$, $CH_3ClHg$, $PbCl_2$ Aroclor 1254, tBHQ, acrylamide | Embryos | Barcode signatures of diverse chemicals | (Yang et al. 2007) |
| Affymetrix GeneChip Genome Array | Retinoic acid, TCDD | Embryonic heart | NR2F5-dependent cell cycle gene downregulation | (Chen et al. 2008) |
| Affymetrix GeneChip Genome Array | Tetrabromobisphenol A (TBBPA) | Liver | TBBPA interferes with thyroid and vitamin A homeostasis | (De Wit et al. 2008) |
| Affymetrix GeneChip Genome Array | TCDD | Ovary | TCDD attenuates gonadotropin responsiveness | (Heiden et al. 2008) |
| Affymetrix GeneChip | 17a-methylDHT | Adult female liver | Alteration in steroid and retinoic acid regulation | (Hoffmann et al. 2008) |
| Custom oligo | Zn | ZF4 cells | Identification of MTF1-dependent genes | (Hogstrand et al. 2008) |
| 17k oligo (Sigma Genosys) | Produced water (PAH mixture) | Adult liver | Downregulation of cell cycle genes | (Holth et al. 2008) |
| 14K oligo (OciChip) | E2, BPA, genistein | Adult male liver | BPA profile unique; Vtg in common | (Kausch et al. 2008) |
| Affymetrix GeneChip Genome Array | Ethanol, nicotine | Adult brain | Conservation of drug-dependence pathways with mammals | (Kily et al. 2008) |
| Affymetrix GeneChip Genome Array | TCDD | Regenerating fin | R-spondin1 induced; Sox9 down-reg. (indirect) | (Mathew et al. 2008) |
| 17k oligo (Sigma Genosys) | E2, 4-nonylphenol | Male liver | Vtg, proteolysis genes upregulated | (Ruggeri et al. 2008) |

(*Continued*)

**Table 10.4** (*continued*)

| Type of array | Chemicals | Organs/tissues examined | Major findings | Reference |
|---|---|---|---|---|
| 14k oligo (MWG) | C-60 | Embryos | Oxidative stress genes upregulated | (Usenko et al. 2008) |
| Affymetrix GeneChip Genome Array | 6-OH-BPDE 6-MeO-BPDE, BPDE | Embryonic fibroblasts | 6-OH-BPDE uncouples oxidative phosphorylation | (van Boxtel et al. 2008) |
| Agilent 22k oligo array | Fadrozole, EE2, 17ß-trenbolone | Brain, gonad | Greater similarity of Fad to Tren than either to EE2 | (Wang et al. 2008a,b) |
| Affymetrix GeneChip Genome Array | TCDD | Embryo–larval jaw | chondrogenic genes affected; Sox9b down-reg. | (Xiong et al. 2008) |
| Custom multispecies oligo | E2 and nonylphenol | Adult liver | Cross-species array validation | (Baker et al. 2009) |
| 17k oligo (Sigma Genosys) | $CuSO_4$ (water) | Adult liver | MT up-reg.; indirect responses exceeded direct | (Craig et al. 2009) |
| Affymetrix GeneChip Genome Array | Domoic acid | Adult brain | Neuropathic genes expressed at non-neurobehavioral doses | (Lefebvre et al. 2009) |
| Affymetrix GeneChip Genome Array | $NaAsO_2$ | Embryos | Low dose As downregulation of immune response | (Mattingly et al. 2009) |
| 17k oligo (Sigma Genosys) | Natural pollutant mixtures (PCB, PDBE, DDT) | Male liver, testis | Changes in hormone regulators, insulin signaling, p53 | (Nourizadeh-Lillabadi et al. 2009) |
| Agilent 22k oligo array | Fadrozole | Brain, ovary | Neurodegeneration, ovarian disruption | (Villeneuve et al. 2009) |
| Agilent 44k oligo array | Amphetamine | Brain | Brain development transcription factors upregulated | (Webb et al. 2009) |
| Agilent 22k oligo array | TCDD, tBHQ | Embryos 4 dpf | Oxidative stress response gene battery | (Hahn et al. 2007, unpublished) |

Many of the array studies now employ commercial arrays, which have the benefit of including 15–20 000 unique features. The information resulting from expression profiling studies is improving; as the number of genes represented has increased, so has the number of genes identified as affected by a toxicant. However, there are varying numbers of features on arrays developed and used by different groups, including commercial arrays and often genes important to chemical response are missing. We have, for example, found that commercial arrays usually lack a substantial number of CYP genes (unpublished information). Embellishing existing arrays or designing focused arrays is necessary to assess changes in a full set of related genes. For example, Bertrand et al. (2007) isolated all 70 NR genes in zebrafish plus genes for 31 of the coactivators and corepressors involved with the NR. They examined the expression of the full suite of nuclear receptors during development, blending array analysis and high throughput in situ hybridization. This approach suggested links between the expression of NR genes and their coregulators. They determined that NRs are mainly expressed during organogenesis, and that a majority of NR genes are expressed during central nervous system (CNS) and retina development. While this is not directly revealing of any toxicological insights, knowledge of these expression patterns is certain to bear on interpretations that may derive from toxicological studies.

Array analysis can prompt searching promoter regions, to discern direct from indirect effects of toxicants. Using a 16 730 oligonucleotide zebrafish microarray with probes related to metal metabolism and toxicity, Craig et al (2009) found that 573 genes in the zebrafish liver responded significantly to copper exposure, many with distinct up-regulation at moderate Cu exposure, and significant down-regulation under high Cu exposure. Analysis of promoter regions of these significantly regulated genes for glucocorticoid (GRE) and metal (MRE) response elements showed that 30% contained only a GRE sequence, and 2.5% contained only a consensus MRE. This suggested that indirect effects of Cu exposure on gene expression could be more important than direct effects. Similarly, analysis showed no dioxin response elements in the promoter region of the novel retroelement discovered by Goldstone et al., implying that induction of that retroelement is an indirect effect of TCDD (Goldstone et al. unpublished data).

Proteomic analysis is being applied (e.g., De Souza et al. 2009), which complements transcriptomic profiling, although as yet there is only a handful of proteomic studies. Forlin and colleagues (Kling et al. 2008; Kling and Forlin 2009) determined proteomic changes in zebrafish liver cells and adult liver in response to multiple brominated flame retardants, identifying sex differences in response. De Wit et al. (2008) also examined the flame retardant tetrabromobisphenol A and Wang et al. (2009) determined proteomic changes in liver of fish exposed to the cyanobacterial toxin microcystin-LR (De Wit et al. 2008; Wang et al. 2009). Shi et al. (2009) examined embryo responses to perfluorooctane sulfonate. Each of the studies identified suites of responsive proteins, including some involved in detoxification, energy metabolism, lipid and steroid metabolism, cell structure, signal transduction, and apoptosis.

*Expression profiling by deep sequencing:* An alternative or complement to microarray analyses is deep sequencing or pyrosequencing using high-throughput methods such as those developed recently by 454 Life Sciences (now Roche Applied Sciences) (Margulies et al. 2005) and Solexa (now Illumina) (Bentley et al. 2008). An advantage of such methods is that they provide an unbiased view of transcriptional response, identifying transcripts (including novel splice variants) that might not be represented on microarrays. While deep sequencing has yet to be applied to zebrafish in the context of toxicology, its potential is illustrated by recent studies using deep sequencing to characterize the response of zebrafish to bacterial infection (Hegedus et al. 2009) and to characterize microRNAs in zebrafish (Soares et al. 2009).

*Localization of effects:* Whole embryo or whole organ PCR or microarray studies do not permit conclusions about the tissue location of gene expression changes. Such concerns about localization extend back decades. Ultimately, the organs, cell types, and the intracellular location where protein–protein and protein–DNA interactions occur must be known in order to fully approach the mechanism of action. These are only now being addressed in toxicological studies in zebrafish.

Microarray analyses within the same labs comparing different organs in embryos (e.g., heart and jaw of developing fish [Chen et al. 2008; Xiong et al. 2008]) or adults (e.g. liver and telencephalon [Martyniuk et al. 2007]), establishes that expression profiles can differ markedly between different organs. The same can be true of cell types. Thus, isolating different organs and accompanying array analysis with in situ hybridization is necessary to establish mechanisms underlying a particular phenotypic effect.

General features have been established for some genes. AHR2 and ARNT2 are expressed widely, and CYP1A is induced by TCDD in cells and tissues throughout developmental stages and adult zebrafish (Andreasen et al. 2002b). Immunohistochemical analysis has shown that various AHR agonists induce CYP1A in similar organs and cell types in other fish (e.g., Stegeman et al. 1991; Van Veld et al. 1997). This implies that CYP1A expression in one fish may extrapolate to other fish species as well. However, differences in sensitivity of induction can be pronounced, even in the same species, as in killifish that have acquired a resistance to the toxicity of

PHAH, reflected as well in the lack of induction of CYP1A (Elskus et al. 1999; Bello et al. 2001).

### 5.3. Gene Knockdown with Morpholine-substituted Oligonucleotides

Disrupting translation with morpholine-substituted oligonucleotides (Nasevicius and Ekker 2000) has been an important approach to address mechanisms by which particular genes are regulated by toxicants, or the role of particular genes in the toxicity of a given chemical in the developing embryo. As in mammals, gene knockdown studies confirm that in zebrafish the toxicity of dioxin and related AHR agonists is dependent on AHR and ARNT, although in zebrafish the toxicologically relevant AHR form is AHR2, rather than AHR1 as in mammals (Prasch et al. 2003, 2006; Antkiewicz et al. 2006). The questions that remain unanswered have to do with how this toxicity is accomplished.

More than half of the studies to date have examined knockdown of components of the AHR response system (AHR, ARNT, AHRR, CYP1 genes) or other genes implicated in TCDD effects. That can be expected to change dramatically in the next few years, as the technology is widely available, and the environmental toxicology community expands their use of it.

### 5.4. Mechanistic Insights from Zebrafish and Converging Mechanisms

The approaches above are yielding important new insights into mechanisms of toxicity, and identification of chemicals that are ligands for transcriptional regulators, or those that may act indirectly. One of the major success stories, still unfolding, is in the understanding of the effects of TCDD and other planar halogenated aromatic hydrocarbons. Profound developmental effects of these AHR agonists are well known, and similar among vertebrates. Evident in Figure 10.5, these include edema or effusion effects and craniofacial defects. Other more subtle effects include a variety of other vascular structural and circulatory defects (Fig. 10.6). There has been an aggressive effort to understand the cardiovascular defects associates with TCDD or other AHR agonist exposure. These are reviewed elsewhcre (Heideman et al. 2005; Carney et al. 2006b; Goldstone and Stegeman 2006).

While morpholino targeting has established that these effects are mediated largely by AHR2, the downstream effects have presented a greater challenge. The broad expression of CYP1A is likely a compensatory response. Whether it is also path of pathways leading to toxicity is possible, but not clear. The possibility that CYP1s that are strongly induced by PHAH may be involved in their toxicity is still open. The obvious

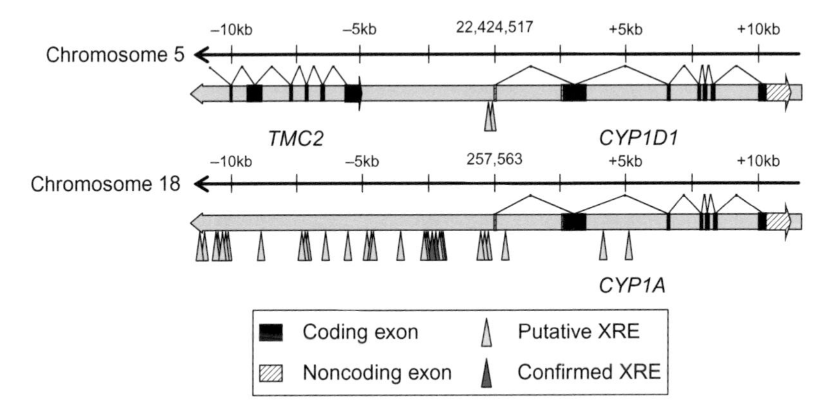


**Fig. 10.5.** Gene structure of zebrafish CYP1D1 and CYP1A. Zebrafish CYP1A and CYP1D1 are structurally similar and closely related phylogenetically, yet CYP1D1 is a pseudogene in humans. Zebrafish CYP1D1 transcript is highly expressed during early development and is expressed in embryonic and adult neural tissues. CYP1D1 provides an example of gene-finding in zebrafish that may shed light on mammalian gene subfunctionalization. Expressed exons are indicated by black boxes, while striped boxes indicate untranslated regions. Shown also are the location of calculated XRE sequences (KNGCGTG); note that not all of the XRE sequences indicated for CYP1A have been shown to be active. Modified from Goldstone et al. (2009).

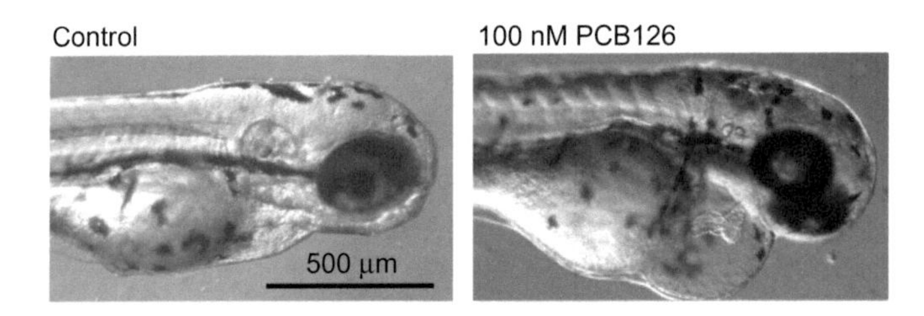


**Fig. 10.6.** Morphological effects caused by 3,3′,4,4′,5-pentachlorobiphenyl, PCB126, in 80-hpf zebrafish embryos. One of the significant strengths of zebrafish as a developmental and toxicological model is the ability to readily define phenotypes. Pericardial edema, hemorrhage and craniofacial defects are among the most obvious phenotypes of AHR agonist exposure, although there are also other causes. Embryos were exposed to 100 ppm acetone (control) or 100 nM PCB126 for 24 h starting at 8 h post fertilization (hpf). (Jonsson and Stegeman, unpublished). See color plate section

phenotypic effects do not appear to depend on CYP1A (Carney et al. 2004). Likewise, knockdown of CYP1B1 did not appear to affect the more overt phenotypes of TCDD exposure (Yin et al. 2008). But more subtle effects, such as on blood flow in the mesencephalic vein, could be affected. Recent work suggests that CYP1Cs may be involved in this PHAH effect in the brain (Kubota et al. unpublished data).

Microarray analysis and gene knockdown studies of effects of TCDD on craniofacial development in embryos, and fin regeneration in adults, have converged on common pathways involving effects on connective tissue, an exciting development. Tanguay and colleagues have published a series of papers on the ability of TCDD to inhibit tissue regeneration and the development of the zebrafish as a model to investigate the underlying mechanisms. Starting with the initial observations that TCDD blocks fin regeneration in adult and larval zebrafish (Zodrow et al. 2004; Mathew et al. 2006), Tanguay's group went on to identity disruptions in extracellular matrix linked to TCDD-dependent down-regulation of Sox9b (Andreasen et al. 2006). They then traced this effect back to the ability of TCDD to stimulate wnt signaling through induction of R-Spondin1, showing through gene knock-down studies that the induction of R-Spondin1 and interaction with its receptor LRP6 were required steps in the inhibition of tissue regeneration by TCDD (Mathew et al. 2008). These studies provide one of the most detailed descriptions of a mechanism underlying a specific toxic effect of dioxin, in *any* system. An elegant study of jaw tissue has revealed that an AHR-dependent downregulation of Sox9b is a proximal cause of the jaw defect (Xiong et al. 2008). This convergence appears to be coming close to an ultimate mechanism. As wnt signaling is broadly important (Verkade and Heath 2008) we can expect that it will be involved in other effects of TCDD in zebrafish and in mammalian systems, for example in the effects of TCDD on prostate development (Vezina et al. 2009).

Another series of studies using zebrafish to investigate the mechanism of TCDD toxicity has focused on the AHR repressor (AHRR), an AHR-regulated, TCDD-inducible repressor of AHR function. The role of AHRR in the mechanism of TCDD toxicity and whether it serves physiological roles outside of toxicology are not known. Zebrafish possess two AHRR paralogs (co-orthologs of the single mammalian AHRR) (Evans et al. 2005). Jenny et al. (2009) hypothesized that the two zebrafish AHRRs had undergone subfunction partitioning since the divergence of fish and mammalian lineages. Using morpholino knock-down experiments they showed that whereas AHRRb modulates the gene expression response to TCDD (knock-down causes an enhanced response to TCDD), knock-down of AHRRa in the absence of exogenous chemical exposure caused embryonic phenotypes closely resembling those characteristic of TCDD effects (edema and craniofacial abnormalities). Thus, AHRRa appears to be involved in regulating a physiological role of AHR during development.

An other study that has yielded important mechanistic insights using expression profiling in zebrafish is the report by Gurvich et al. (2005). In this study, the mechanism of teratogenesis caused by the anti-epileptic drug valproic acid was explored using zebrafish embryos exposed to this compound and to less active analogs. Valproic acid caused changes in gene expression resembling those of the histone deacetylase (HDAC) inhibitor Trichostatin A. Subsequent studies showed that the activity of valproic acid analogs as teratogens was related to their ability to act as inhibitors of HDAC. The results are highly suggestive that HDAC inhibition is an important component of the mechanism of valproate teratogenesis.

Another possible convergence of mechanisms involves oxidative stress. Data from many sources suggest that oxidative stress, whether as reactive oxygen species, or other radicals and alteration of redox balance involving sulfhydryl groups, may be a path to toxicity common to many contaminants. Oxidative stress has been implicated in PHAH toxicity (Cantrell et al. 1996; Hassoun et al. 1997; Latchoumycandane and Mathur 2002), including in effects of TCDD in developing zebrafish (Dong et al. 2002; Goldstone and Stegeman 2006). Fullerenes ($C_{60}$) also elicit oxidative stress (Usenko et al. 2008), and PBDE appears to be an uncoupler of oxidative phosphorylation (van Boxtel et al. 2008). This indicates the potential importance of the Nrf2 pathway described in Section 4. While reactive oxygen species can be derived from a variety of cellular processes, few reports have directly addressed the question of the precise source(s) of ROS resulting from chemical exposure, or how much different sources contribute to the mechanisms of toxicity. One possibility is uncoupling of CYP catalytic cycles (Schlezinger et al. 1999, 2006). No doubt CYP1A principally plays a protective role against rapidly metabolized substrates such as many PAHs, but the issues with PHAH are more complex, and the degree to which uncoupling contributes to toxicity is unresolved.

## 6. TOXICITY AND ENVIRONMENTAL ASSESSMENT WITH ZEBRAFISH

Zebrafish are increasingly being employed in predictive toxicology, in screening of chemicals and drugs, and in assessment of environmental matrices and environmental inputs. Traditional approaches to environmental screening tend to use other species, and have relied on analysis of biomarkers of exposure or effect, often measured in tissue samples from whole individuals. Predictive screening for human and animal toxicity also relies on use of human or other cell lines. For example, the Biomolecular Screening Branch of the National Toxicology Program, in collaboration with the EPA and the NIH Chemical Genomics Center, has undertaken an effort to screen 10 000 compounds of toxicological concern. Zebrafish will be important in providing an in vivo context for this in vitro screening effort. Efficiency, specificity, and higher predictive power from a model all are needed. The in vivo context, with intact xenobiotic metabolism and defensome functions aids predictability although, as implied above, there is still a dearth of explicit knowledge regarding the functions of the majority of defensome genes. The application of zebrafish for assessing chemicals to predict toxicity and the nature and significance of environmental contamination is a very active area of research. The current literature includes studies of pesticides, pharmaceuticals, personal care products, metals, organic chemicals, arsenicals, alcohols, and hormonally active agents [e.g., estrogens (Henry et al. 2009; Jin et al. 2009), pharmaceuticals (Pomati et al. 2007), metals (Hogstrand et al. 2008) and organics (Holth et al. 2008)].

### 6.1. Molecular Biomarkers

Molecular biomarkers based on changes in gene expression have been used for decades to assess exposure in aquatic systems (e.g., Stegeman et al. 1992). The wealth of data from many species indicates that some genes can be used reliably as biomarkers in most vertebrates. Thus, the induction of CYP1A is used as a biomarker of exposure to AHR agonists, and induction of vitellogenins or CYP19a2 is a commonly used marker for exposure to active doses of estrogenic chemicals, and MT induction of exposure to metals. In zebrafish, these are used as markers to confirm activation of the AHR, ER or MRE pathways, in studies to identify mechanisms or in screening of chemicals for likely mode of action. Development of non-destructive methods for obtaining biomarker information is also being pursued, for example in analysis of CYP1A in gill biopsies (Jonsson et al. 2009a), or CYP1A assessed (by EROD activity) in intact embryos (Bohonowych et al. 2008). While induction of proteins or transcripts is used as a marker in environmental assessment with many species, studies of zebrafish exposed directly in the environment are unusual. However, it is increasingly common for zebrafish embryos to be used in assessment of chemicals, wastewaters or environmental extracts.

Zebrafish embryo–larval tests are actively being developed for screening of chemicals and wastewaters for chemicals with endocrine disrupting and other types of actions (e.g. Raldua and Babin 2009; Segner 2009). The Fish Embryo Test has been a mandatory component in routine whole effluent testing in Germany since 2005, and is being further developed for chemical testing (Lammer et al. 2009a). Identifying new biomarkers and approaches with zebrafish embryo–larval tests is an active pursuit. Toxicogenomic studies are proving the vaunted utility of expression profiling in identifying and validating genes and gene networks as new biomarkers of exposure or adverse outcome. Strähle and coworkers (Yang et al. 2007) exposed zebrafish embryos to a suite of environmental toxicants including organic compounds and metals, and profiled expression patterns using an oligonucleotide microarray. They obtained specific expression profiles for each of the chemicals examined, and could predict with high probability the identity of the toxicant from the expression profiles. Changes in gene expression were observed at toxicant concentrations that did not cause morphological effects, and although fewer genes responded at lower toxicant concentrations, the changes in transcript levels were robust. The study highlights the potential for use of zebrafish embryos in systematic, large-scale analysis of chemical effects on the developing vertebrate embryo. Similar gene "classifiers" have been obtained in comparison of smaller numbers of chemicals in adult zebrafish (Wang et al. 2008a).

The value of zebrafish as a model for environmental toxicity will be enhanced by the comparative approach, in which the common responses and locations of gene expression changes are related to phenotypes in different species. Cheng and colleagues (Chen et al. 2009) have approached this using antibody probes proven in zebrafish to transfer the technology to a marine sentinel species. Based on results with zebrafish, Chen et al. (2009) screened marine medaka embryos immunohistochemically with antibodies to 61 known proteins, identifying 17 that gave reliable results with the medaka. With ISH they also identified 11 mRNA probes that were informative in medaka. Ankley and co-workers likewise are using the comparative approach, establishing similarities between zebrafish and fathead minnow (Ankley et al. 2009).

Interestingly, Baker et al. (2009) used a microarray designed with probes based on sequences in several unrelated fish species, and detected effects of nonylphenol in zebrafish. The array included genes that are involved in the actions of adrenal and sex steroids, thyroid hormone, and xenobiotic responses. Such a microarray could provide a tool for screening environments for the presence of chemicals that affect endocrine systems in diverse fish species. The utility of zebrafish arrays to screen other species is equally possible. Extrapolating to potential sentinel species is certain to expand, and approaches to cross species comparisons such as those in 4DXpress will eventually prove the utility of zebrafish to identify changes that may be generic among vertebrates (Haudry et al. 2008).

### 6.2. Transgenic Approaches

The use of zebrafish engineered with reporter genes linked to promoter regions of chemical response genes has been a goal for a decade or more. There has been growing success (Udvadia and Linney 2003), and now the promise for mechanistic studies and for screening seems on the verge of being met. Amanuma et al. (2000) generated a line carrying a shuttle vector plasmid containing the rpsl gene of *E. coli*, and a kanamycin resistance gene for recovering the plasmid, with the objective of using the line for detecting mutagen aquatic environments. The initial studies showed that direct acting mutagens and also model promutagens, including BaP, increased the frequency of mutations in the transgene. Subsequent studies with the rpsL line detected mutagenesis by BaP in the transgene in gill and liver of adult fish (Amanuma et al. 2002). In both organs the base changes were consistent with those known to result from PAH such as BaP in other systems.

Carvan and colleagues (Kusik et al. 2008) generated a line with luciferase and GFP linked to the electrophile response element, for use in detecting exposures that elicit oxidative stress. This is an important approach, given that as suggested above, oxidative stress is being identified as part of a common pathway in the effects of diverse chemicals (Wells et al. 2009). The incorporation of GFP allows the detection of organ specific activation of the EpRE, and the luciferase construct allows the effect to be quantified. Carvan and coworkers used the construct in both stable and transient transfections and to assess exposure to HgCl, and observed that expression was detected at doses below those eliciting morphological defects. However, there were differences between the two reporters as regards the behavior found in the transient as compared to the stable line. This is a valuable approach that no doubt will see greater application.

Zebrafish have been engineered with reporter genes linked to promoters or response elements for the estrogen receptor (Chen et al. 2010), for heat shock proteins (Hsp 70; Krone et al. 2003), for sonic hedgehog (shh; Neumann and Nuesslein-Volhard 2000; Hill et al. 2003), for retinoic acid receptors (Carvan et al. 2000) and for the AHR (Mattingly et al. 2001). In each case there has been success in identifying effects in particular organs or cells (Blechinger et al. 2007). These successes indicate that the utility of transgenic fish in detecting environmental exposures, and in assessing toxic and mutagenic mechanisms, will be possible.

Studies with transgenic fish have shown that heterologous promoters and proteins are able to function in zebrafish, with meaningful expression patterns and ability to rescue deleterious mutant phenotypes (Udvadia and Linney 2003). The generation of humanized zebrafish, created by expressing human defensome target genes in zebrafish, and the converse generation of human cell lines expressing zebrafish homologs, will help to determine functional differences between the proteins of the two species, and further refine the utility of zebrafish as a model

### 6.3. High-throughput Techniques

Technological approaches that are applicable to high-throughput testing in zebrafish are changing the pace and possibilities in assessing the effects of chemicals. The logistical advantages of zebrafish as a developmental model are particularly evident in the context of high-throughput applications, as the small size and rapid development of the zebrafish embryo allow for the use of 96-well microplates in performing many assays in parallel, and the transparency of the embryos allows direct observation of chemical effects. High throughput techniques are generally used for screening hundreds or even thousands of chemicals (e.g. Kitambi et al. 2009), and will be important in assessing the effects of mixtures as well.

Several different types of high-throughput assays have been used in zebrafish embryos. Most studies have used automated microscopic analysis of phenotype for specific organ endpoints (Burns et al. 2005; Milan and Macrae 2008), an approach that can be especially powerful in transgenic lines (Vogt et al. 2009). Other assays include immunohistochemical staining or in situ hybridization to look at alterations in the expression or localization of particular molecular targets (Sabaliauskas et al. 2006; Kaufman et al. 2009). In addition, behavioral assays using image processing techniques to extract data from videomicroscopy has proved to be an efficient method for assaying the effects of compounds on neural function (Kokel and Peterson 2008; Egan et al. 2009; Guo 2009).

The utility of zebrafish in high-throughput assays is not limited to developmental toxicology, however, as Lam et al. (2008) have convincingly demonstrated the utility of adult zebrafish for predictive chemogenomics. They created prediction models that could robustly classify the gene expression signatures of different classes of compounds, and identified biomarkers for AHR and estrogen receptor agonists (Lam et al. 2008).

One present limitation of the use of zebrafish is the current fragmentary knowledge of basic biological and endocrine processes in zebrafish (Segner 2009; and see also Chapter 5). Another limitation arises from the often complicated homologous relationships between a mammalian gene and its zebrafish counterpart (or counterparts; see Section 4.3). The decreased cost and increased sensitivity of DNA sequencing will soon allow for the identification of altered transcriptional activity in high-throughput screening. However, proteomic approaches remain expensive, although the analysis of single zebrafish embryos is now possible (Lin et al. 2009).

*Small molecule screening:* Small molecule screening in a whole organism with defined biological endpoints is termed chemical genetics, and some progress in zebrafish has recently been reviewed (Kaufman et al. 2009). The "one compound/one target" paradigm that once limited chemical biology is giving way to a more integrated view of a compound's actions (Peterson 2008). The need to screen huge numbers of compounds during the search for and evaluation of drugs will continue to be the major driver of rapid assay development, and there are growing numbers of commercial concerns offering rapid testing. At present, there is still substantial effort required to validate assays for such use, however (e.g. Eimon and Rubinstein 2009). Technical progress, such as flow-through systems (Lammer et al. 2009b) will further help make rapid approaches commonplace in testing of environmental matrices. However, many high throughput approaches are somewhat limited, as they often do not consider the mechanisms underlying the endpoint.

## 7. AREAS OF LOOMING SIGNIFICANCE AND PROMISING APPROACHES

As we look ahead to consider how zebrafish will be utilized in toxicology, it is obvious that this will be driven both by the continued advancement in technical approaches for the zebrafish model generally and by the emergence of new toxicological questions for which zebrafish are well suited as a model. We mention here a few examples of these emerging questions and techniques that are likely to be the focus of much research activity in the coming years.

### 7.1. Zinc Finger Nuclease Technology for Gene Knockout

The use of morpholino technology has resulted in an exceptional understanding of the role of many genes, including genes involved in chemical–biological interactions (Section 5.3; and see also Chapter 1). However, there are limitations, in that the MO injected at the 2–4-cell stage is eliminated or diluted as the embryo grows, with the consequence that translation can ensue, in effect resulting in only a partial knockdown. Knockout technology such as used commonly in mice has proven difficult in zebrafish. A recent technology, involving zinc-finger nuclease targeting of specific genes, is an exciting alternative (Amacher 2008; Ekker 2008). This approach now has been successfully applied to zebrafish, in knocking out the VEGF-2 receptor (Meng et al. 2008) and golden and no-tail genes (Doyon et al. 2008). Using Oligomerized Pool Engineering, from the Zinc Finger Nuclease Consortium, Foley et al. (2009) successfully generated ZFN pairs for five endogenous zebrafish genes: tfr2, dopamine transporter, telomerase, hif1aa, and gridlock. Importantly, Foley et al. also used an in silico approach to estimate that there are > 25 000 zebrafish genes that may be amenable to ZFN knockdown using OPEN libraries currently in hand (Foley et al. 2009). As ZFN technology is applied to genes in the chemical defense array we should finally begin to get a clear definition of the role of particular genes in the toxicity of a given chemical, in embryos and adults. Look for a rapid growth in application of ZFN technology in mechanistic toxicology in zebrafish. The technology is likely to become employed aggressively also in other species, in which knockout using stem cell technology has proven difficult (Geurts et al. 2009).

### 7.2. Defining Ligand Space and Active Sites: Modeling and Docking Studies

One of the questions often left unanswered, is whether a putative ligand for a receptor, or a substrate for an enzyme binds to a particular protein, and with what affinity, and how such binding relates to the effects that may be inferred from gene expression or substrate turnover studies. Approaches to these questions can be empirical or computational. Empirical studies with expressed or in vitro translated ER and AHR proteins are often applied to assess binding of ligands to these receptors. Competitive binding assays show, for example, that 6-formylindolo-3,2b-carbazole (FICZ), a possible endogenous ligand for the AHR, binds to both AHR2 and AHR1b of zebrafish (Jonsson et al. 2009b). Likewise, questions regarding substrate specificity of CYP or other xenobiotic metabolizing enzymes can be addressed with heterologously expressed proteins. As indicated in earlier sections, there still are few studies that report on activities of expressed zebrafish proteins with more than a few substrates.

Computational approaches can predict ligand or substrate specificities in advance of biochemical confirmation. Homology modeling and docking can be an informative approach yielding strong inference about how different ligands interact with particular proteins. While at present there are few studies that have addressed ligand or substrate accessibility of zebrafish defensome enzymes or receptors, the examples from other species, and the questions in zebrafish, are compelling. Examples (below) include recent modeling and docking studies suggesting not only which ligands might access the ligand binding pocket or the active site, but in what orientation, and with what frequency.

Pandini et al. (2007) developed a two-template model (based on Hif2a and ARNT solution structures) of the mouse AHR, and through mutagenesis and binding studies, determined that the model gave reliable results (Pandini et al. 2007). This study was extended by modeling the structures of the ligand binding domains of six additional high-affinity mammalian AhRs (Pandini et al. 2009). With site-directed mutagenesis and AhR functional analysis, they described the "TCDD binding-fingerprint" of conserved residues within the ligand-binding cavity necessary for high-affinity TCDD binding and TCDD-dependent AhR transformation and

DNA binding. Applying similar approaches to the zebrafish (and other model species AHRs) should help to further establish zebrafish as a model for understanding AHR-dependent signaling.

In addition to studies with zebrafish, modeling of ligand binding to AHRs from different taxa may help to explain susceptibility of different groups to dioxins and related compounds. Hahn and colleagues suggested that AHR diversity in vertebrates arose via both gene and whole-genome duplications, combined with lineage-specific gene loss (Hahn 2002; Hahn et al. 2005). They also suggested that sensitivity to the developmental toxicity of PHAH originated in the chordate lineage, through evolution of the ligand-binding capacity of the AHR. Addressing such hypotheses, and determining whether novel compounds may be potential agonists, may be aided by modeling studies. Modeling and docking studies with AHR from various taxa may indicate the evolution of the "ligand space" and may contribute to an understanding of the physiological roles of endogenous ligands.

An example of ensemble modeling and docking to CYP structures illustrates the potential of this approach to evaluate functional properties of CYP enzymes (Prasad et al. 2007). A large number of models, based on six related P450 crystal structures as templates, were constructed for human and rat CYP1A1, and killifish (*F. heteroclitus*) and scup (*S. chrysops*) CYP1A. Multiple dockings to these families of models were used to assess structural features associated with differences in catalytic efficiency between the teleost and mammalian CYP1A orthologs with the substrate 3,3′,4,4′-tetrachlorobiphenyl (TCB) (e.g., Fig. 10.7). The results provided a structural explanation for the observation that fish CYP1As metabolized the TCB hundreds of times slower than the mammalian CYP1A1 proteins. Coupling modeling of all zebrafish CYP1s (CYP1A, 1B1, 1Cs and 1D1) with high throughput docking of libraries of small chemicals would be an important adjunct to empirical studies of substrate specificity and rates. Such computational screening, in concert with empirical screening using small molecule arrays to detect protein binding, could quickly provide substrate information for CYPs for which there is currently none.

Modeling studies of other proteins also are informative. Thorsteinson et al. (2009) employed modeling, molecular dynamics, and QSAR approaches and identified chemical structures that could bind to zebrafish steroid hormone binding globulin (SHBG). Among the compounds that they identified was dihydrobenzo(a)pyren-7(8H)-one. In determining that dihydrobenzo(a)pyren-7(8H)-one binds to the SHBG, this modeling and docking study provides additional information that suggests a mechanism by which BaP could disrupt reproduction, via metabolite(s) that bind to steroid-binding pockets, at least on one protein, but possibly others (Thorsteinson et al. 2009).

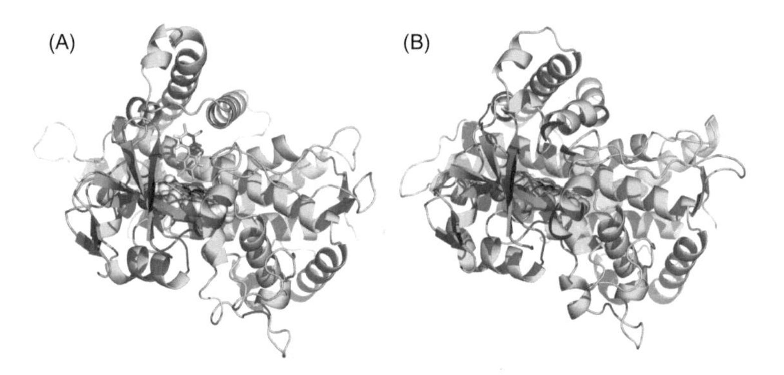


**Fig. 10.7.** Homology models of teleost CYP1A. These homology models depict (A) killifish (*Fundulus heteroclitus*) CYP1A with 3,3′,4,4′-tetrachlorobiphenyl (TCB) docked in the active site, in a productive orientation, above the heme; and (B) a model of zebrafish CYP1A, without TCB docked. Multiple dockings of TCB to homology models of human, rat, and fish CYP1A provide a structural explanation for the observed differences in TCB metabolism between these species. Such computational screening could quickly provide information on ligand binding for protein targets for which there is currently little or no information. (The killifish model is based on data from Prasad et al. 2007; the zebrafish model is unpublished). See color plate section

Modeling and docking studies in theory may be applied to any protein for which there is a suitable crystal structure for use as a template. Additional studies of receptor ligand-binding pockets and enzyme active sites should begin to appear soon, as crystal structures are accruing and as modeling programs are now robust enough to produce reliable structures (Qu et al. 2009).

### 7.3. Emerging Issues

*Nanomaterials:* Application of zebrafish in assessing the effects of nanomaterials has grown in the past two years, with more than about 30 papers currently identified. Studies on most of the types of nanoparticles have been undertaken and published (Cu, Zn, Ce, silica, Au, Ag, $C_{60}$ fullerenes). While this is still a developing area, some important observations are appearing. For example, several studies have used microarray analysis of zebrafish exposed to nanomaterials and have indicated that fullerenes elicit oxidative stress (Henry et al. 2007; Zhu et al. 2007; Usenko et al. 2008). It is clear that exposure conditions and the nature of the preparation used in experiments are critical to the interpretations. These are both growth areas and a detailed consideration would be very useful, but is beyond our scope.

*Pharmaceuticals and personal care products:* There is increasing awareness of the environmental impact of pharmaceuticals and other personal care products, including fragrances (musks), UV sunscreen compounds, and biocides such as triclosan. In vitro and in vivo transgenic zebrafish assays have shown that several polycyclic musks and UV filters have estrogenic or anti-estrogenic activities (Schreurs et al. 2005) and are toxic to developing zebrafish at environmental concentrations (Carlsson and Norrgren 2004). Zebrafish have been used not only to examine specific pollutant effects (e.g. van der Ven et al. 2006b), but also to screen multiple drugs for specific developmental effects (Raldua and Babin 2009), and to evaluate overall transcriptional effects of pharmaceutical mixtures (Pomati et al. 2007). Future work with zebrafish will likely also include investigating sublethal toxic effects of complex mixtures assayed both with embryos and with adults (e.g. via neuromast impairment assays [Froehlicher et al. 2009b]).

### 7.4. Areas Promising New Insight into Gene Regulation and Protein Function

Developments over the past several years have shown that gene regulation and protein function can be strongly affected by factors and processes, some recently discovered and some long known but with renewed appreciation of their significance. These genetic and epigenetic factors and processes are certain to alter the panorama of gene expression linked to adaptive and adverse consequences of chemical exposure. Epigenetic regulation of gene expression patterns involves DNA methylation, histone modifications, and non-coding RNAs. Genetic changes involve mutations, SNPs and allelic variation that affect regulation and function.

*microRNAs (miRNAs):* miRNAs are involved in the regulation of gene expression primarily as post-transcriptional silencers, by base pairing to the 3′-UTR (and sometimes other parts) of mRNAs. miRNAs are critical regulators of gene expression, with demonstrated and substantial importance in a variety of processes including embryonic development, the response to environmental stressors, and the progression of diseases, including cancer (Bushati and Cohen 2007; Hobert 2008; Hudder and Novak 2008). Thus, it seems highly likely that miRNAs may be involved in mediating chemical toxicity, either directly or indirectly through altered xenobiotic metabolism. For example, miRNA genes could be regulated by xenobiotics if their promoters contain response elements for ligand-activated transcription factors (Shah et al. 2007). Conversely, miRNAs may regulate the expression of xenobiotic-responsive transcription factors such as PXR (Takagi et al. 2008) or ER (Adams et al. 2007; Pandey and Picard 2009), and thus potentially affect (enhance or reduce) xenobiotic metabolism or the sensitivity to xenobiotic effects. MicroRNAs also may regulate the translation of proteins induced by xenobiotics (Tsuchiya et al. 2006).

The studies cited as examples above were all in mammals. What is the role for zebrafish in understanding mechanisms of toxicity involving miRNAs? Zebrafish have a similar complement of miRNAs as compared to mammals (Thatcher et al. 2008a; Soares et al. 2009) and the zebrafish model has been instrumental in some of the fundamental advances in understanding the roles of miRNAs in embryonic development in vertebrates (Giraldez et al. 2006; Plasterk 2006). Thus, the zebrafish embryo should be an excellent model with which to investigate the role of miRNAs in developmental toxicity and teratogenicity. Currently, however, there are few published reports concerning the involvement of miRNAs in chemical effects in zebrafish. In one study, differential expression of miR-124a and miR-133a was reported in zebrafish embryos treated with sodium thiosulfate (Hu et al. 2009). However, whether the altered miRNA expression was a cause or effect of the embryotoxicity was not ascertained. Thatcher et al. (2007) showed that inhibiting the hedgehog signaling with cyclopamine in zebrafish embryos caused rapid and widespread changes in miRNA expression.

In light of the dramatic effects of TCDD on development and the many studies using zebrafish to address this, it is of great interest to determine whether miRNAs might be involved. It is intriguing that miRNAs affect processes such as cardiac and vascular development and fin regeneration that also are affected by TCDD (Fish et al. 2008; Morton et al. 2008; Thatcher et al. 2008b). Despite these possible connections between miRNA and TCDD-affected processes, Moffat et al. (2007) found that TCDD had little effect on miR expression in rodent liver and concluded that miRNAs are not likely to play a significant role in dioxin toxicity. Initial results in zebrafish embryos suggest a similar absence of dramatic changes in miRNA expression, at least at the level of the whole embryo, after exposure to TCDD (Jenny and Hahn 2009). However, miRNAs could still have important roles in TCDD toxicity, for example in tissue-specific effects of TCDD on the cardiovascular system or on fin regeneration.

Clearly, research on the roles of miRNAs in toxicology, whether in zebrafish or other models, is in its infancy. A thorough, systematic assessment of these roles and the underlying mechanisms is sorely needed. Investigation of RNA targets of miRNAs, and also the regulation of miRNAs (and their putative promoter regions (Fujita and Iba 2008)) are expected to open the door to fundamentally new insights into the mechanisms of chemical effect.

*Epigenetics:* An area of potentially great importance in toxicology that has so far received little attention in zebrafish concerns the effects of chemical exposure on epigenetic regulation of gene expression. The study of epigenetic effects of toxicants is growing rapidly and increasingly is viewed as highly significant (Jirtle and Skinner 2007; Heindel 2008) and there are a few key examples documenting such effects in mammals (Anway et al. 2005; Dolinoy et al. 2007; Skinner et al. 2008). Much of the interest is in DNA methylation, which is involved in embryonic development and potentially can be modified by chemical exposure, leading to adult disease and transgenerational effects.

Zebrafish embryos would appear to be an ideal model for investigating epigenetic effects of toxicants. However, there is a need first to establish some of the fundamental features of epigenetic regulation in zebrafish, including a better understanding of how it may differ from that in mammals. For example, there has been controversy concerning whether methylation patterns are reprogrammed in fish embryos, as they are in mammals (Macleod et al. 1999; Martin et al. 1999; Mhanni and McGowan 2004; MacKay et al. 2007; Rai et al. 2008).

Studies involving analysis of DNA methylation in relation to toxicant exposure in zebrafish are just beginning to appear. For example, an abnormal genomic DNA methylation pattern was detected in arsenite-treated embryos (Li et al. 2009). New approaches now being developed will greatly enhance the ability to use zebrafish as models to study epigenetic effects of toxicants (Goll et al. 2009).

Major unanswered questions in toxicology involve the nature and significance of developmental origins of adult disease attributed to developmental exposures and transgenerational effects of exposure. A study by King Heiden et al. (2009) suggests that zebrafish may be an important model for addressing these. Adult fish surviving developmental exposure to TCDD had altered craniofacial structures (i.e., operculum and jaw), heart, swim bladder, and ovary, and/or impaired egg production, fertility, and gamete quality. The same study also identified transgenerational effects, with offspring from parents that had been exposed to TCDD as embryos showing increases in mortality, and reduced egg production (King Heiden et al. 2009).

*Post-translational protein modification:* As with DNA methylation, post-translational modification of proteins relevant in toxicology is an area not well explored in zebrafish. Various studies document the importance of phosphorylation of nuclear receptors, and its influence on gene regulation. For example, phosphorylation of the PXR can affect expression of CYP3A (Pondugula et al. 2009). Suzawa and Ingraham proposed that an effect of atrazine on CYP19A1, CYP11A1 and other genes involved in steroidogenesis could involve phosphorylation of nuclear receptor SF-1 (NR5A1 [Suzawa and Ingraham 2008]).

Direct phosphorylation of CYP proteins has long been known (Koch and Waxman 1991), and could be important in modulating enzyme activity, targeting to specific subcellular compartments, and tagging for ubiquitination and degradation in the proteasome. The potential significance of this is being re-examined (Oesch-Bartlomowicz and Oesch 2009), and recent application of proteomic approaches has identified potential phosphorylation sites on human CYPs and determined that at least eight are phosphorylated in human liver (Redlich et al. 2008). Whether homologous CYP in zebrafish will be subject to phosphorylation that affects function is not known.

*Allelic variation and SNPs:* Allelic variation has profound implications for human disease. The implications for therapeutics also are marked. Individual differences in drug metabolism often reside in those same defensome genes and enzymes that metabolize xenobiotics and the receptors that regulate them. Thus, the same individual differences could contribute to differences in response to environmental chemicals. While single nucleotide variation in human drug and xenobiotic metabolizing enzymes is becoming well known in humans, we know little about such variation in the homologous genes in zebrafish. Allelic variation resulting in amino acid changes could affect results with in vitro translation or heterologous expression of proteins for functional analysis, depending on which variant was expressed. However, differences in alleles also can be exploited in identifying those domains important to function, or regulation.

The extent of variation in zebrafish has been estimated in some studies. Guryev et al. (2006) estimated the variation in seven commonly used strains of zebrafish and found that inter-strain variation ranged from 7% (in inbred) to 37% (in wild-derived) of polymorphic sites being heterozygous. They concluded that "…levels and organization of genetic variation between and within commonly used zebrafish strains are markedly different from other laboratory model organisms". Less variation in inbred strains than in wild zebrafish has been reported (Coe et al. 2009).

*Gene networks and cross talk:* Identification of how xenobiotic response and defensome genes fit in gene regulatory networks (e.g., Jupiter et al. 2009) will provide a link to fundamental processes, and which pathways are conserved, which are redundant, and which are involved in compensatory as compared to adverse effects of chemical exposure, and could indicate mechanisms involved in cross-talk.

It has become increasingly apparent in recent years that signaling pathways activated by toxicants do not function in isolation, but rather as part of networks of transcription factors exhibiting substantial cross-talk (Kohle and Bock 2009; Puga et al. 2009). As an example, we highlight a few of the interactions involving AHR. Interactions between AHR and hypoxia signaling are well known, although the mechanisms are not yet well understood (Gradin et al. 1996; Chan et al. 1999; Lee et al. 2006). Similarly, AHR and ER exhibit substantial cross-talk, through several mechanisms (Safe and Wormke 2003; Ohtake et al. 2009). AHR also interacts in several ways with NRF2 signaling; AHR controls expression of NRF2 (Miao et al. 2005), NRF2 regulates expression of AHR (Shin et al. 2007), and AHR and NRF2 co-regulate gene expression (Kohle and Bock 2007; Yeager et al. 2009). AHR–PXR interactions also have been demonstrated (Maglich et al. 2002; Gu et al. 2007). Most of these studies have been in mammalian systems. However, such cross-talk has been demonstrated also in zebrafish and other fish models (Bemanian et al. 2004; Prasch et al. 2004; Mathew et al. 2008, 2009). Zebrafish is likely to emerge soon as an important system for studying such receptor cross-talk, especially as it relates to developmental toxicology.

## 8. SOME CONCLUSIONS

The rapid pace of studies with zebrafish is sure to render a summary such as this obsolete soon. So what lessons have we learned, and how has environmental toxicology been changed by the research to date, and what areas require greater attention?

Unbiased transcription profiling and gene knockdown studies have been informative, in supporting inferences regarding participation of genes and proteins, especially as related to developmental defects. Most prominent are those attributed to dioxin-like compounds and other AHR agonists. These studies are uncovering the involvement of new genes, and pointing to the possibility of common mechanisms, that are shared at least partly in different processes, such as wnt signaling involved in craniofacial defects in embryos and fin regeneration in adults. Other pathways are less well known. Placing these genes in gene regulatory networks is an open area, rich with possibilities.

Morpholino knockdown technology has been of immense value in toxicology studies. The addition of knockout technology, with zinc finger nucleases, TILLING, or technology yet to be developed, perhaps in conjuction with MO, will be important in defining involvement of particular genes/protcins.

The metabolism and disposition of chemicals is strongly related to their toxicity. While we are learning much about the expression of genes that are implicated in these processes, study of the proteins themselves has lagged substantially. Strong inferences regarding effects of particular chemicals from results in this model require knowledge of the function of these proteins, whether enzyme, receptor or other transcription factor. Studies of many of the relevant genes and proteins in zebrafish are still emerging from the descriptive phase, relative to the studies in mammals. Cloning of genes, examining molecular phylogeny, determining developmental expression and response to chemicals, and determining protein function all are essential, and necessary to eventually understand fully their role in toxic effects. Knowledge of such similarities and differences between zebrafish and other species will lead to new fundamental understanding of the defense against and susceptibility to chemicals.

## ACKNOWLEDGMENTS

We gratefully acknowledge the U.S. National Institutes of Health for support in preparation of this chapter and in many of the studies considered here; grants R01ES015912 (JJS), R01ES006272 (MEH), R01ES016366 (MEH), R21ES017304 (MEH), and P42ES007381 (Superfund Basic Research Program at Boston University) (JJS and MEH). We also acknowledge the support of the Andrew W. Mellon Fund for Innovative Research.

## REFERENCES

Adams, B. D., Furneaux, H., and White, B. (2007). The micro-RNA miR-206 targets the human estrogen receptor-{alpha}, and represses ER{alpha} mRNA and protein expression in breast cancer cell lines. *Mol. Endocrinol.* **21**(5), 1132–1147.

Almeida, D. V., da Silva Nornberg, B. F., Geracitano, L. A., Barros, D. M., Monserrat, J. M., and Marins, L. F. (2009). Induction of phase II enzymes and hsp70 genes by copper sulfate through the electrophile-responsive element (EpRE): Insights obtained from a transgenic zebrafish model carrying an orthologous EpRE sequence of mammalian origin. *Fish Physiol. Biochem.* DOI 10.1007/s10695-008-9299-X.

Amacher, S. L. (2008). Emerging gene knockout technology in zebrafish: Zinc-finger nucleases. *Brief Funct. Genomics Proteomics* **7**(6), 460–464.

Amanuma, K., Takeda, H., Amanuma, H., and Aoki, Y. (2000). Transgenic zebrafish for detecting mutations caused by compounds in aquatic environments. *Nat. Biotechnol.* **18**(1), 62–65.

Amanuma, K., Tone, S., Nagaya, M., Matsumoto, M., Watanabe, T., Totsuka, Y., Wakabayashi, K., and Aoki, Y. (2008). Mutagenicity of 2-[2-(acetylamino)-4-[bis(2-hydroxyethyl)amino]-5-methoxyphenyl]-5-amino-7-bromo-4-chloro-2H-benzotriazole (PBTA-6) and benzo[a]pyrene (BaP) in the gill and hepatopancreas of rpsL transgenic zebrafish. *Mutat. Res.* **656**(1-2), 36–43.

Amanuma, K., Tone, S., Saito, H., Shigeoka, T., and Aoki, Y. (2002). Mutational spectra of benzo[a]pyrene and MeIQx in rpsL transgenic zebrafish embryos. *Mutat. Res.* **513**(1-2), 83–92.

Amores, A., Force, A., Yan, Y.-L., Joly, L., Amemiya, C., Fritz, A., Ho, R. K., Langeland, J., Prince, V., Wang, Y.-L., et al. (1998). Zebrafish hox clusters and vertebrate genome evolution. *Science* **282**, 1711–1714.

Andreasen, E. A., Hahn, M. E., Heideman, W., Peterson, R. E., and Tanguay, R. L. (2002a). The zebrafish (Danio rerio) aryl hydrocarbon receptor type 1 is a novel vertebrate receptor. *Mol. Pharmacol.* **62**(2), 234–249.

Andreasen, E. A., Mathew, L. K., and Tanguay, R. L. (2006). Regenerative growth is impacted by TCDD: Gene expression analysis reveals extracellular matrix modulation. *Toxicol. Sci.* **92**(1), 254–269.

Andreasen, E. A., Spitsbergen, J. M., Tanguay, R. L., Stegeman, J. J., Heideman, W., and Peterson, R. E. (2002b). Tissue-specific expression of AHR2, ARNT2, and CYP1A in zebrafish embryos and larvae: Effects of developmental stage and 2,3,7,8- tetrachlorodibenzo-p-dioxin exposure. *Toxicol. Sci.* **68**(2), 403–419.

Ankley, G. T., Bencic, D. C., Breen, M. S., Collette, T. W., Conolly, R. B., Denslow, N. D., Edwards, S. W., Ekman, D. R., Garcia-Reyero, N., Jensen, K. M., et al. (2009). Endocrine disrupting chemicals in fish: Developing exposure indicators and predictive models of effects based on mechanism of action. *Aquat. Toxicol.* **92**(3), 168–178.

Annilo, T., Chen, Z. Q., Shulenin, S., Costantino, J., Thomas, L., Lou, H., Stefanov, S., and Dean, M. (2006). Evolution of the vertebrate ABC gene family: Analysis of gene birth and death. *Genomics* **88**(1), 1–11.

Antkiewicz, D. S., Peterson, R. E., and Heideman, W. (2006). Blocking expression of AHR2 and ARNT1 in zebrafish larvae protects against cardiac toxicity of 2,3,7,8-Tetrachlorodibenzo-p-dioxin. *Toxicol. Sci.* **94**(1), 175–182.

Anway, M. D., Cupp, A. S., Uzumcu, M., and Skinner, M. K. (2005). Epigenetic transgenerational actions of endocrine disruptors and male fertility. *Science* **308**(5727), 1466–1469.

Baker, M. E., Ruggeri, B., Sprague, L. J., Eckhardt-Ludka, C., Lapira, J., Wick, I., Soverchia, L., Ubaldi, M., Polzonetti-Magni, A. M., Vidal-Dorsch, D., et al. (2009). Analysis of endocrine disruption in Southern California coastal fish using an aquatic multispecies microarray. *Environ. Health Perspect.* **117**(2), 223–230.

Bardet, P. L., Horard, B., Robinson-Rechavi, M., Laudet, V., and Vanacker, J. M. (2002). Characterization of oestrogen receptors in zebrafish (*Danio rerio*). *J. Mol. Endocrinol.* **28** (3), 153–163.

Begay, V., Falcon, J., Cahill, G. M., Klein, D. C., and Coon, S. L. (1998). Transcripts encoding two melatonin synthesis enzymes in the teleost pineal organ: Circadian regulation in pike and zebrafish, but not in trout. *Endocrinology* **139**(3), 905–912.

Bello, S. M., Franks, D. G., Stegeman, J. J., and Hahn, M. E. (2001). Acquired resistance to aryl hydrocarbon receptor agonists in a population of *Fundulus heteroclitus* from a marine superfund site: In vivo and in vitro studies on the induction of xenobiotic-metabolizing enzymes. *Toxicol. Sci.* **60**(1), 77–91.

Bemanian, V., Male, R., and Goksoyr, A. (2004). The aryl hydrocarbon receptor-mediated disruption of vitellogenin synthesis in the fish liver: Cross-talk between AHR- and ERalpha-signalling pathways. *Comp. Hepatol.* **3**(1), 2.

Bentley, D. R., Balasubramanian, S., Swerdlow, H. P., Smith, G. P., Milton, J., Brown, C. G., Hall, K. P., Evers, D. J., Barnes, C. L., Bignell, H. R., et al. (2008). Accurate whole human genome sequencing using reversible terminator chemistry. *Nature* **456**(7218), 53–59.

Bertrand, S., Thisse, B., Tavares, R., Sachs, L., Chaumot, A., Bardet, P. L., Escriva, H., Duffraisse, M., Marchand, O., Safi, R., et al. (2007). Unexpected novel relational links uncovered by extensive developmental profiling of nuclear receptor expression. *PLoS Genet.* **3**(11), e188.

Billiard, S. M., Timme-Laragy, A. R., Wassenberg, D. M., Cockman, C., and Di Giulio, R. T. (2006). The role of the aryl hydrocarbon receptor pathway in mediating synergistic developmental toxicity of polycyclic aromatic hydrocarbons to zebrafish. *Toxicol. Sci.* **92** (2), 526–536.

Blechinger, S. R., Kusch, R. C., Haugo, K., Matz, C., Chivers, D. P., and Krone, P. H. (2007). Brief embryonic cadmium exposure induces a stress response and cell death in the developing olfactory system followed by long-term olfactory deficits in juvenile zebrafish. *Toxicol. Appl. Pharmacol.* **224**(1), 72–80.

Bohonowych, J. E., Zhao, B., Timme-Laragy, A., Jung, D., Di Giulio, R. T., and Denison, M. S. (2008). Newspapers and newspaper ink contain agonists for the ah receptor. *Toxicol. Sci.* **102**(2), 278–290.

Bolton, J. L., Trush, M. A., Penning, T. M., Dryhurst, G., and Monks, T. J. (2000). Role of quinones in toxicology. *Chem. Res. Toxicol.* **13**(3), 135–160.

Bosio, A., Binczek, E., Le Beau, M. M., Fernald, A. A., and Stoffel, W. (1996). The human gene CGT encoding the UDP-galactose ceramide galactosyl transferase (cerebroside synthase): Cloning, characterization, and assignment to human chromosome 4, band q26. *Genomics* **34**(1), 69–75.

Braunbeck, T., Boettcher, M., Hollert, H., Kosmehl, T., Lammer, E., Leist, E., Rudolf, M., and Seitz, N. (2005). Towards an alternative for the acute fish LC(50) test in chemical assessment: The fish embryo toxicity test goes multi-species – an update. *Altex* **22**(2), 87–102.

Bresell, A., Weinander, R., Lundqvist, G., Raza, H., Shimoji, M., Sun, T. H., Balk, L., Wiklund, R., Eriksson, J., Jansson, C., et al. (2005). Bioinformatic and enzymatic characterization of the MAPEG superfamily. *FEBS J.* **272**(7), 1688–1703.

Bresolin, T., de Freitas Rebelo, M., and Celso Dias Bainy, A. (2005). Expression of PXR, CYP3A and MDR1 genes in liver of zebrafish. *Comp. Biochem. Physiol. C Toxicol. Pharmacol.* **140**(3-4), 403–407.

Burnett, K. G., Bain, L. J., Baldwin, W. S., Callard, G. V., Cohen, S., Di Giulio, R. T., Evans, D. H., Gomez-Chiarri, M., Hahn, M. E., Hoover, C. A., et al. (2007). Fundulus as the premier teleost model in environmental biology: Opportunities for new insights using genomics. *Comp. Biochem. Physiol. Part D Genomics Proteomics* **2**(4), 257–286.

Burns, C. G., Milan, D. J., Grande, E. J., Rottbauer, W., MacRae, C. A., and Fishman, M. C. (2005). High-throughput assay for small molecules that modulate zebrafish embryonic heart rate. *Nat. Chem. Biol.* **1**(5), 263–264.

Bushati, N., and Cohen, S. M. (2007). microRNA functions. *Annu. Rev. Cell Dev. Biol.* **23**, 175–205.

Cantrell, S. M., Lutz, L. H., Tillitt, D. E., and Hannink, M. (1996). Embryotoxicity of 2,3,7,8-tetrachlorodibenzo-p-dioxin (TCDD): The embryonic vasculature is a physiological target for TCDD-induced DNA damage and apoptotic cell death in Medaka (*Orizias latipes*). *Toxicol. Appl. Pharmacol.* **141**, 23–34.

Carlsson, G., and Norrgren, L. (2004). Synthetic musk toxicity to early life stages of zebrafish (Danio rerio). *Arch. Environ. Contam. Toxicol.* **46**(1), 102–105.

Carney, S. A., Chen, J., Burns, C. G., Xiong, K. M., Peterson, R. E., and Heideman, W. (2006a). Aryl hydrocarbon receptor activation produces heart-specific transcriptional and toxic responses in developing zebrafish. *Mol. Pharmacol.* **70**(2), 549–561.

Carney, S. A., Peterson, R. E., and Heideman, W. (2004). 2,3,7,8-Tetrachlorodibenzo-p-dioxin activation of the aryl hydrocarbon receptor/aryl hydrocarbon receptor nuclear translocator pathway causes developmental toxicity through a CYP1A-independent mechanism in zebrafish. *Mol. Pharmacol.* **66**(3), 512–521.

Carney, S. A., Prasch, A. L., Heideman, W., and Peterson, R. E. (2006b). Understanding dioxin developmental toxicity using the zebrafish model. *Birth Defects Res. A Clin. Mol. Teratol.* **76**(1), 7–18.

Carson, R. (1962). *Silent Spring*. Houghton Mifflin, Boston.

Carvan, M. J., III, Sonntag, D. M., Cmar, C. B., Cook, R. S., Curran, M. A., and Miller, G. L. (2001). Oxidative stress in zebrafish cells: Potential utility of transgenic zebrafish as a deployable sentinel for site hazard ranking. *Sci. Total Environ.* **274**(1-3), 183–196.

Carvan, M. J., 3rd, Dalton, T. P., Stuart, G. W., and Nebert, D. W. (2000). Transgenic zebrafish as sentinels for aquatic pollution. *Ann. N. Y. Acad. Sci.* **919**, 133–147.

Carvan, M. J., 3rd, Gallagher, E. P., Goksoyr, A., Hahn, M. E., and Larsson, D. G. (2007). Fish models in toxicology. *Zebrafish* **4**(1), 9–20.

Celander, M., Buhler, D. R., Forlin, L., Goksoyr, A., Miranda, C. L., Woodin, B. R., and Stegeman, J. J. (1996). Immunochemical relationships of cytochrome P4503A-like proteins in teleost fish. *Fish Physiol. Biochem.* **15**(4), 323–332.

Chan, K. M., Ku, L. L., Chan, P. C., and Cheuk, W. K. (2006). Metallothionein gene expression in zebrafish embryo-larvae and ZFL cell-line exposed to heavy metal ions. *Mar. Environ. Res.* **62**(Suppl.), S83–S87.

Chan, W. K., Yao, G., Gu, Y.-Z., and Bradfield, C. A. (1999). Cross-talk between the aryl hydrocarbon receptor and hypoxia inducible factor signaling pathways. Demonstration of competition and compensation. *J. Biol. Chem.* **274**, 12115–12123.

Chen, H., Hu, J., Yang, J., Wang, Y., Xu, H., Jiang, Q., Gong, Y., Gu, Y., and Song, H. (2010). Generation of a fluorescent transgenic zebrafish for detection of environmental estrogens. *Aquat. Toxicol.* **96**, 53–61.

Chen, J., Carney, S. A., Peterson, R. E., and Heideman, W. (2008). Comparative genomics identifies genes mediating cardiotoxicity in the embryonic zebrafish heart. *Physiol. Genomics* **33**(2), 148–158.

Chen, W. Y., John, J. A., Lin, C. H., and Chang, C. Y. (2007). Expression pattern of metallothionein, MTF-1 nuclear translocation, and its dna-binding activity in zebrafish (Danio rerio) induced by zinc and cadmium. *Environ. Toxicol. Chem.* **26**(1), 110–117.

Chen, W. Y., John, J. A., Lin, C. H., Lin, H. F., Wu, S. C., Lin, C. H., and Chang, C. Y. (2004). Expression of metallothionein gene during embryonic and early larval development in zebrafish. *Aquat. Toxicol.* **69**(3), 215–227.

Chen, X., Li, L., Wong, C. K., and Cheng, S. H. (2009). Rapid adaptation of molecular resources from zebrafish and medaka to develop an estuarine/marine model. *Comp. Biochem. Physiol. C Toxicol. Pharmacol.* **149**(4), 647–655.

Cheshenko, K., Brion, F., Le Page, Y., Hinfray, N., Pakdel, F., Kah, O., Segner, H., and Eggen, R. I. (2007). Expression of zebrafish aromatase CYP19a amd CYP19b genes in response to the ligands of estrogen receptor and aryl hydrocarbon receptor. *Toxicol. Sci.* **96**(2), 255–267.

Christoffels, A., Koh, E. G., Chia, J. M., Brenner, S., Aparicio, S., and Venkatesh, B. (2004). Fugu genome analysis provides evidence for a whole-genome duplication early during the evolution of ray-finned fishes. *Mol. Biol. Evol.* **21**(6), 1146–1151.

Coe, T. S., Hamilton, P. B., Griffiths, A. M., Hodgson, D. J., Wahab, M. A., and Tyler, C. R. (2009). Genetic variation in strains of zebrafish (Danio rerio) and the implications for ecotoxicology studies. *Ecotoxicology* **18**(1), 144–150.

Coughtrie, M. W. (2002). Sulfation through the looking glass – recent advances in sulfotransferase research for the curious. *Pharmacogenomics J.* **2**(5), 297–308.

Craig, P. M., Hogstrand, C., Wood, C. M., and McClelland, G. B. (2009). Gene expression endpoints following chronic waterborne copper exposure in a genomic model organism, the zebrafish, Danio rerio. *Physiol. Genomics* **40**(1), 23–33.

de Souza, F. S., Bumaschny, V. F., Low, M. J., and Rubinstein, M. (2005). Subfunctionalization of expression and peptide domains following the ancient duplication of the proopiomelanocortin gene in teleost fishes. *Mol. Biol. Evol.* **22**(12), 2417–2427.

De Souza, A. G., MacCormack, T. J., Wang, N., Li, L., and Goss, G. G. (2009). Large-scale proteome profile of the zebrafish (Danio rerio) gill for physiological and biomarker discovery studies. *Zebrafish* **6**(3), 229–238.

De Wit, M., Keil, D., Remmerie, N., van der Ven, K., van den Brandhof, E. J., Knapen, D., Witters, E., and De Coen, W. (2008). Molecular targets of TBBPA in zebrafish analysed through integration of genomic and proteomic approaches. *Chemosphere* **74**(1), 96–105.

Dean, M., and Annilo, T. (2005). Evolution of the ATP-binding cassette (ABC) transporter superfamily in vertebrates. *Annu. Rev. Genomics. Hum. Genet.* **6**, 123–142.

Dehal, P., and Boore, J. L. (2005). Two rounds of whole genome duplication in the ancestral vertebrate. *PLoS Biol.* **3**(10), e314.

Denison, M. S., and Nagy, S. R. (2003). Activation of the aryl hydrocarbon receptor by structurally diverse exogenous and endogenous chemicals. *Annu. Rev. Pharmacol. Toxicol.* **43**, 309–334.

Dobbs-McAuliffe, B., Zhao, Q., and Linney, E. (2004). Feedback mechanisms regulate retinoic acid production and degradation in the zebrafish embryo. *Mech. Dev.* **121**(4), 339–350.

Dolinoy, D. C., Huang, D., and Jirtle, R. L. (2007). Maternal nutrient supplementation counteracts bisphenol A-induced DNA hypomethylation in early development. *Proc. Natl Acad. Sci. USA* **104**(32), 13056–13061.

Dong, W., Teraoka, H., Yamazaki, K., Tsukiyama, S., Imani, S., Imagawa, T., Stegeman, J. J., Peterson, R. E., and Hiraga, T. (2002). 2,3,7,8-tetrachlorodibenzo-p-dioxin toxicity in the zebrafish embryo: Local circulation failure in the dorsal midbrain is associated with increased apoptosis. *Toxicol. Sci.* **69**(1), 191–201.

Douard, V., Brunet, F., Boussau, B., Ahrens-Fath, I., Vlaeminck-Guillem, V., Haendler, B., Laudet, V., and Guiguen, Y. (2008). The fate of the duplicated androgen receptor in fishes: A late neofunctionalization event? *BMC Evol. Biol.* **8**, 336.

Doyon, Y., McCammon, J. M., Miller, J. C., Faraji, F., Ngo, C., Katibah, G. E., Amora, R., Hocking, T. D., Zhang, L., Rebar, E. J., et al. (2008). Heritable targeted gene disruption in zebrafish using designed zinc-finger nucleases. *Nat. Biotechnol.* **26**(6), 702–708.

Dragin, N., Shi, Z., Madan, R., Karp, C. L., Sartor, M. A., Chen, C., Gonzalez, F. J., and Nebert, D. W. (2008). Phenotype of the Cyp1a1/1a2/1b1-/- triple-knockout mouse. *Mol. Pharmacol.* **73**(6), 1844–1856.

Egan, R. J., Bergner, C. L., Hart, P. C., Cachat, J. M., Canavello, P. R., Elegante, M. F., Elkhayat, S. I., Bartels, B. K., Tien, A. K., Tien, D. H., et al. (2009). Understanding behavioral and physiological phenotypes of stress and anxiety in zebrafish. *Behav. Brain Res.* **205**(1), 38–44.

Eimon, P. M., and Rubinstein, A. L. (2009). The use of in vivo zebrafish assays in drug toxicity screening. *Expert Opin. Drug Metab. Toxicol.* **5**(4), 393–401.

Ekins, S., Reschly, E. J., Hagey, L. R., and Krasowski, M. D. (2008). Evolution of pharmacologic specificity in the pregnane X receptor. *BMC Evol. Biol.* **8**, 103.

Ekker, S. C. (2008). Zinc finger-based knockout punches for zebrafish genes. *Zebrafish* **5**(2), 121–123.

Elskus, A. A., Monosson, E., McElroy, A. E., Stegeman, J. J., and Woltering, D. S. (1999). Altered CYP1A expression in *Fundulus heteroclitus* adults and larvae: A sign of pollutant resistance?. *Aquat. Toxicol.* **45**, 99–113.

Evans, B. R., Karchner, S. I., Allan, L. L., Pollenz, R. S., Tanguay, R. L., Jenny, M. J., Sherr, D. H., and Hahn, M. E. (2008). Repression of aryl hydrocarbon receptor (AHR) signaling by AHR repressor: Role of DNA binding and competition for AHR nuclear translocator. *Mol. Pharmacol.* **73**(2), 387–398.

Evans, B. R., Karchner, S. I., Franks, D. G., and Hahn, M. E. (2005). Duplicate aryl hydrocarbon receptor repressor genes (ahrr1 and ahrr2) in the zebrafish Danio rerio: Structure, function, evolution, and AHR-dependent regulation in vivo. *Arch. Biochem. Biophys.* **441**(2), 151–167.

Fish, J. E., Santoro, M. M., Morton, S. U., Yu, S., Yeh, R. F., Wythe, J. D., Ivey, K. N., Bruneau, B. G., Stainier, D. Y., and Srivastava, D. (2008). miR-126 regulates angiogenic signaling and vascular integrity. *Dev. Cell* **15**(2), 272–284.

Fishman, M. C. (2001). Genomics. Zebrafish–the canonical vertebrate. *Science* **294**(5545), 1290–1291.

Fitch, W. M. (1970). Distinguishing homologous from analogous proteins. *Syst. Zool.* **19**, 99–113.

Foley, J. E., Yeh, J. R., Maeder, M. L., Reyon, D., Sander, J. D., Peterson, R. T., and Joung, J. K. (2009). Rapid mutation of endogenous zebrafish genes using zinc finger nucleases made by Oligomerized pool ENgineering (OPEN). *PLoS ONE* **4**(2), e4348.

Force, A., Lynch, M., Pickett, F. B., Amores, A., Yan, Y.-L., and Postlethwait, J. H. (1999). Preservation of duplicate genes by complementary, degenerative mutations. *Genetics* **151**, 1531–1545.

Froehlicher, M., Liedtke, A., Groh, K., Lopez-Schier, H., Neuhauss, S. C., Segner, H., and Eggen, R. I. (2009a). Estrogen receptor subtype beta2 is involved in neuromast development in zebrafish (Danio rerio) larvae. *Dev. Biol.* **330**(1), 32–43.

Froehlicher, M., Liedtke, A., Groh, K. J., Neuhauss, S. C., Segner, H., and Eggen, R. I. (2009b). Zebrafish (Danio rerio) neuromast: Promising biological endpoint linking developmental and toxicological studies. *Aquat. Toxicol.* **95**(4), 307–319.

Fujita, S., and Iba, H. (2008). Putative promoter regions of miRNA genes involved in evolutionarily conserved regulatory systems among vertebrates. *Bioinformatics* **24**(3), 303–308.

Garcia-Reyero, N., Villeneuve, D. L., Kroll, K. J., Liu, L., Orlando, E. F., Watanabe, K. H., Sepulveda, M. S., Ankley, G. T., and Denslow, N. D. (2009). Expression signatures for a model androgen and antiandrogen in the fathead minnow (Pimephales promelas) ovary. *Environ. Sci. Technol.* **43**(7), 2614–2619.

George, S. G., and Taylor, B. (2002). Molecular evidence for multiple UDP-glucuronosyltransferase gene familes in fish. *Mar. Environ. Res.* **54**(3-5), 253–257.

Geurts, A. M., Cost, G. J., Freyvert, Y., Zeitler, B., Miller, J. C., Choi, V. M., Jenkins, S. S., Wood, A., Cui, X., Meng, X., et al. (2009). Knockout rats via embryo microinjection of zinc-finger nucleases. *Science* **325**(5939), 433.

Giraldez, A. J., Mishima, Y., Rihel, J., Grocock, R. J., Van Dongen, S., Inoue, K., Enright, A. J., and Schier, A. F. (2006). Zebrafish MiR-430 promotes deadenylation and clearance of maternal mRNAs. *Science* **312**(5770), 75–79.

Godard, C. A., Goldstone, J. V., Said, M. R., Dickerson, R. L., Woodin, B. R., and Stegeman, J. J. (2005). The new vertebrate CYP1C family: Cloning of new subfamily members and phylogenetic analysis. *Biochem. Biophys. Res. Commun.* **331**(4), 1016–1024.

Goldstone, H. M., and Stegeman, J. J. (2006). Molecular mechanisms of 2,3,7,8-tetrachlorodibenzo-p-dioxin cardiovascular embryotoxicity. *Drug Metab. Rev.* **38**(1-2), 261–289.

Goldstone, J. V. (2008). Environmental sensing and response genes in cnidaria: The chemical defensome in the sea anemone *Nematostella vectensis*. *Cell Biol. Toxicol.* **24**(6), 483–502.

Goldstone, J. V., Hamdoun, A., Cole, B. J., Howard-Ashby, M., Nebert, D. W., Scally, M., Dean, M., Epel, D., Hahn, M. E., and Stegeman, J. J. (2006). The chemical defensome: Environmental sensing and response genes in the *Strongylocentrotus purpuratus* genome. *Dev. Biol.* **300**(1), 366–384.

Goldstone, J. V., Jonsson, M. E., Behrendt, L., Woodin, B. R., Jenny, M. J., Nelson, D. R., and Stegeman, J. J. (2009). Cytochrome P450 1D1: A novel CYP1A-related gene that is not transcriptionally activated by PCB126 or TCDD. *Arch. Biochem. Biophys.* **482**(1-2), 7–16.

Goll, M. G., Anderson, R., Stainier, D. Y., Spradling, A. C., and Halpern, M. E. (2009). Transcriptional silencing and reactivation in transgenic zebrafish. *Genetics* **182**(3), 747–755.

Gradin, K., McGuire, J., Wenger, R. H., Kvietikova, I., Whitelaw, M. L., Toftgård, R., Tora, L., Gassmann, M., and Poellinger, L. (1996). Functional interference between hypoxia and dioxin signal transduction pathways: Competition for recruitment of the Arnt transcription factor. *Mol. Cell. Biol.* **16**(10), 5221–5231.

Gu, X., Ke, S., Liu, D., Sheng, T., and Tian, Y. (2007). Cross-talk between aryl hydrocarbon receptor and pregnane-X-receptor pathways. *Toxicol. Sci. (The Toxicologist Supplement)* **96**(Suppl.), 294. (Abstract #1420)

Guo, S. (2009). Using zebrafish to assess the impact of drugs on neural development and function. *Expert Opin. Drug Discov.* **4**(7), 715–726.

Gurvich, N., Berman, M. G., Wittner, B. S., Gentleman, R. C., Klein, P. S., and Green, J. B. (2005). Association of valproate-induced teratogenesis with histone deacetylase inhibition *in vivo*. *FASEB J.* **19**(9), 1166–1168.

Guryev, V., Koudijs, M. J., Berezikov, E., Johnson, S. L., Plasterk, R. H., van Eeden, F. J., and Cuppen, E. (2006). Genetic variation in the zebrafish. *Genome Res.* **16**(4), 491–497.

Hahn, M. E. (2002). Aryl hydrocarbon receptors: Diversity and evolution. *Chem. Biol. Interact.* **141**(1/2), 131–160.

Hahn, M. E., and Hestermann, E. V. (2007). Chapter 5. Receptor-mediated Mechanisms of Toxicity. In: *The Toxicology of Fishes* (R.T. Di Giulio and D.E. Hinton, eds), pp. 235–272. Taylor & Francis.

Hahn, M. E., Karchner, S. I., Evans, B. R., Franks, D. G., Merson, R. R., and Lapseritis, J. M. (2006). Unexpected diversity of aryl hydrocarbon receptors in non-mammalian vertebrates: Insights from comparative genomics. *J. Exp. Zool. A Comp. Exp. Biol.* **305**(9), 693–706.

Hahn, M. E., Karchner, S. I., Franks, D. G., Woodin, B. R., Barott, K. L., Cipriano, M. J., and McArthur, A. G. (2007). The transcriptional response to oxidative stress in zebrafish embryos. *Toxicol. Sci. (The Toxicologist Supplement)* **96**, 326–327. (Abstract #1578)

Hahn, M. E., Merson R. R., and Karchner S. I. (2005). Xenobiotic receptors in fishes: Structural and functional diversity and evolutionary insights. In *Biochemistry and Molecular Biology of Fishes*, Vol. 6 – *Environmental Toxicology* (T. W. Moon and T. P. Mommsen, eds.), pp. 191–228.

Handley-Goldstone, H. M., Grow, M. W., and Stegeman, J. J. (2005). Cardiovascular gene expression profiles of dioxin exposure in zebrafish embryos. *Toxicol. Sci.* **85**(1), 683–693.

Hassoun, E. A., Walter, A. C., Alsharif, N. Z., and Stohs, S. J. (1997). Modulation of TCDD-induced fetotoxicity and oxidative stress in embryonic and placental tissues of C57BL/6J mice by vitamin E succinate and ellagic acid. *Toxicology* **124**(1), 27–37.

Haudry, Y., Berube, H., Letunic, I., Weeber, P. D., Gagneur, J., Girardot, C., Kapushesky, M., Arendt, D., Bork, P., Brazma, A., et al. (2008). 4DXpress: A database for cross-species expression pattern comparisons. *Nucleic Acids Res.* **36**(Database issue), D847–853.

Hawkins, M. B., Godwin, J., Crews, D., and Thomas, P. (2005). The distributions of the duplicate oestrogen receptors ER-beta a and ER-beta b in the forebrain of the Atlantic croaker (Micropogonias undulatus): Evidence for subfunctionalization after gene duplication. *Proc. R. Soc. B* **272**(1563), 633–641.

He, X., and Ma, Q. (2009). Induction of metallothionein I by arsenic via metal-activated transcription factor 1: Critical role of C-terminal cysteine residues in arsenic sensing. *J. Biol. Chem.* **284**(19), 12609–12621.

Hegedus, Z., Zakrzewska, A., Agoston, V. C., Ordas, A., Racz, P., Mink, M., Spaink, H. P., and Meijer, A. H. (2009). Deep sequencing of the zebrafish transcriptome response to mycobacterium infection. *Mol. Immunol.* **46**(15), 2918–2930.

Heideman, W., Antkiewicz, D. S., Carney, S. A., and Peterson, R. E. (2005). Zebrafish and cardiac toxicology. *Cardiovasc. Toxicol.* **5**(2), 203–214.

Heiden, T. C., Struble, C. A., Rise, M. L., Hessner, M. J., Hutz, R. J., and Carvan, M. J., 3rd (2008). Molecular targets of 2,3,7,8-tetrachlorodibenzo-p-dioxin (TCDD) within the zebrafish ovary: Insights into TCDD-induced endocrine disruption and reproductive toxicity. *Reprod. Toxicol.* **25**(1), 47–57.

Heindel, J. J. (2008). Animal models for probing the developmental basis of disease and dysfunction paradigm. *Basic Clin. Pharmacol. Toxicol.* **102**(2), 76–81.

Henry, T. B., McPherson, J. T., Rogers, E. D., Heah, T. P., Hawkins, S. A., Layton, A. C., and Sayler, G. S. (2009). Changes in the relative expression pattern of multiple vitellogenin genes in adult male and larval zebrafish exposed to exogenous estrogens. *Comp. Biochem. Physiol. A Mol. Integr. Physiol.* **154**(1), 119–126.

Henry, T. B., Menn, F. M., Fleming, J. T., Wilgus, J., Compton, R. N., and Sayler, G. S. (2007). Attributing effects of aqueous C60 nano-aggregates to tetrahydrofuran decomposition products in larval zebrafish by assessment of gene expression. *Environ. Health Perspect.* **115** (7), 1059–1065.

Hernandez, R. E., Putzke, A. P., Myers, J. P., Margaretha, L., and Moens, C. B. (2007). Cyp26 enzymes generate the retinoic acid response pattern necessary for hindbrain development. *Development* **134**(1), 177–187.

Hill, A., Howard, C. V., Strahle, U., and Cossins, A. (2003). Neurodevelopmental defects in zebrafish (Danio rerio) at environmentally relevant dioxin (TCDD) concentrations. *Toxicol. Sci.* **76**(2), 392–399.

Hinton, D. E., Kullman, S. W., Hardman, R. C., Volz, D. C., Chen, P. J., Carney, M., and Bencic, D. C. (2005). Resolving mechanisms of toxicity while pursuing ecotoxicological relevance? *Mar. Pollut. Bull.* **51**(8-12), 635–648.

Hobert, O. (2008). Gene regulation by transcription factors and microRNAs. *Science* **319**(5871), 1785–1786.

Hoffmann, J. L., and Oris, J. T. (2006). Altered gene expression: A mechanism for reproductive toxicity in zebrafish exposed to benzo[a]pyrene. *Aquat. Toxicol.* **78**(4), 332–340.

Hoffmann, J. L., Thomason, R. G., Lee, D. M., Brill, J. L., Price, B. B., Carr, G. J., and Versteeg, D. J. (2008). Hepatic gene expression profiling using GeneChips in zebrafish exposed to 17alpha-methyldihydrotestosterone. *Aquat. Toxicol.* **87**(2), 69–80.

Hoffmann, J. L., Torontali, S. P., Thomason, R. G., Lee, D. M., Brill, J. L., Price, B. B., Carr, G. J., and Versteeg, D. J. (2006). Hepatic gene expression profiling using Genechips in zebrafish exposed to 17alpha-ethynylestradiol. *Aquat. Toxicol.* **79**(3), 233–246.

Hogstrand, C., Zheng, D., Feeney, G., Cunningham, P., and Kille, P. (2008). Zinc-controlled gene expression by metal-regulatory transcription factor 1 (MTF1) in a model vertebrate, the zebrafish. *Biochem. Soc. Trans.* **36**(Pt 6), 1252–1257.

Holth, T. F., Nourizadeh-Lillabadi, R., Blaesbjerg, M., Grung, M., Holbech, H., Petersen, G. I., Alestrom, P., and Hylland, K. (2008). Differential gene expression and biomarkers in zebrafish (Danio rerio) following exposure to produced water components. *Aquat. Toxicol.* **90**(4), 277–291.

Hu, W., Cheng, L., Xia, H., Sun, D., Li, D., Li, P., Song, Y., and Ma, X. (2009). Teratogenic effects of sodium thiosulfate on developing zebrafish embryos. *Front. Biosci.* **14**, 3680–3687.

Huang, H., and Wu, Q. (2010). Cloning and comparative analyses of the zebrafish Ugt repertoire reveal its evolutionary diversity. *PLoS ONE* **5**(2), e9144.

Hudder, A., and Novak, R. F. (2008). miRNAs: Effectors of environmental influences on gene expression and disease. *Toxicol. Sci.* **103**(2), 228–240.

Incardona, J. P., Carls, M. G., Teraoka, H., Sloan, C. A., Collier, T. K., and Scholz, N. L. (2005). Aryl hydrocarbon receptor-independent toxicity of weathered crude oil during fish development. *Environ. Health Perspect.* **113**(12), 1755–1762.

Incardona, J. P., Day, H. L., Collier, T. K., and Scholz, N. L. (2006). Developmental toxicity of 4-ring polycyclic aromatic hydrocarbons in zebrafish is differentially dependent on AH receptor isoforms and hepatic cytochrome P4501A metabolism. *Toxicol. Appl. Pharmacol.* **217**(3), 308–321.

Jenny, M. J., and Hahn, M. E. (2009). MicroRNAs in developmental toxicology: Effects of TCDD on MicroRNA expression in embryos. *Toxicological Sciences (The Toxicologist Supplement)* **108**(1), 20. (Abstract #104)

Jenny, M. J., Karchner, S. I., Franks, D. G., Woodin, B. R., Stegeman, J. J., and Hahn, M. E. (2009). Distinct roles of two zebrafish AHR repressors (AHRRa and AHRRb) in embryonic development and regulating the response to 2,3,7,8-tetrachlorodibenzo-p-dioxin. *Toxicol. Sci.* **110**(2), 426–441.

Jin, Y., Chen, R., Sun, L., Qian, H., Liu, W., and Fu, Z. (2009). Induction of estrogen-responsive gene transcription in the embryo, larval, juvenile and adult life stages of zebrafish as biomarkers of short-term exposure to endocrine disrupting chemicals. *Comp. Biochem. Physiol. C Toxicol. Pharmacol.* **150**(3), 414–420.

Jin, Y., and Penning, T. M. (2007). Aldo-keto reductases and bioactivation/detoxication. *Annu. Rev. Pharmacol. Toxicol.* **47**, 263–292.

Jirtle, R. L., and Skinner, M. K. (2007). Environmental epigenomics and disease susceptibility. *Nat. Rev. Genet.* **8**(4), 253–262.

Jones, R. W., and Huffman, M. N. (1957). Fish embryos as bio-assay material in testing chemicals for effects on cell division and differentiation. *Trans. Am. Microsc. Soc.* **76**, 177–183.

Jonsson, M. E., Brunstrom, B., and Brandt, I. (2009a). The zebrafish gill model: Induction of CYP1A, EROD and PAH adduct formation. *Aquat. Toxicol.* **91**(1), 62–70.

Jonsson, M. E., Franks, D. G., Woodin, B. R., Jenny, M. J., Garrick, R. A., Behrendt, L., Hahn, M. E., and Stegeman, J. J. (2009b). The tryptophan photoproduct 6-formylindolo [3,2-b]carbazole (FICZ) binds multiple AHRs and induces multiple CYP1 genes via AHR2 in zebrafish. *Chem. Biol. Interact.* **181**(3), 447–454.

Jonsson, M. E., Jenny, M. J., Woodin, B. R., Hahn, M. E., and Stegeman, J. J. (2007). Role of AHR2 in the expression of novel cytochrome P450 1 family genes, cell cycle genes, and morphological defects in developing zebra fish exposed to 3,3′,4,4′,5-pentachlorobiphenyl or 2,3,7,8-tetrachlorodibenzo-p-dioxin. *Toxicol. Sci.* **100**(1), 180–193.

Jupiter, D., Chen, H., and VanBuren, V. (2009). STARNET 2: A web-based tool for accelerating discovery of gene regulatory networks using microarray co-expression data. *BMC Bioinformatics* **10**, 332.

Karchner, S. I., Franks, D. G., and Hahn, M. E. (2005). AHR1B, a new functional aryl hydrocarbon receptor in zebrafish: Tandem arrangement of ahr1b and ahr2 genes. *Biochem. J.* **392**(Pt 1), 153–161.

Kaufman, C. K., White, R. M., and Zon, L. (2009). Chemical genetic screening in the zebrafish embryo. *Nat. Protoc.* **4**(10), 1422–1432.

Kausch, U., Alberti, M., Haindl, S., Budczies, J., and Hock, B. (2008). Biomarkers for exposure to estrogenic compounds: Gene expression analysis in zebrafish (Danio rerio). *Environ. Toxicol.* **23**(1), 15–24.

Kensler, T. W., and Wakabayashi, N. (2010). Nrf2: Friend or Foe for Chemoprevention? *Carcinogenesis* **31**(1), 90–99.

Kily, L. J., Cowe, Y. C., Hussain, O., Patel, S., McElwaine, S., Cotter, F. E., and Brennan, C. H. (2008). Gene expression changes in a zebrafish model of drug dependency suggest conservation of neuro-adaptation pathways. *J. Exp. Biol.* **211**(Pt 10), 1623–1634.

King Heiden, T. C., Spitsbergen, J., Heideman, W., and Peterson, R. E. (2009). Persistent adverse effects on health and reproduction caused by exposure of zebrafish to 2,3,7,8-tetrachlorodibenzo-p-dioxin during early development and gonad differentiation. *Toxicol. Sci.* **109**(1), 75–87.

Kishida, M., and Callard, G. V. (2001). Distinct cytochrome P450 aromatase isoforms in zebrafish (Danio rerio) brain and ovary are differentially programmed and estrogen regulated during early development. *Endocrinology* **142**(2), 740–750.

Kitambi, S. S., McCulloch, K. J., Peterson, R. T., and Malicki, J. J. (2009). Small molecule screen for compounds that affect vascular development in the zebrafish retina. *Mech. Dev.* **126**(5-6), 464–477.

Kleiner, H. E., Vulimiri, S. V., Hatten, W. B., Reed, M. J., Nebert, D. W., Jefcoate, C. R., and DiGiovanni, J. (2004). Role of cytochrome p4501 family members in the metabolic activation of polycyclic aromatic hydrocarbons in mouse epidermis. *Chem. Res. Toxicol.* **17** (12), 1667–1674.

Kling, P., and Forlin, L. (2009). Proteomic studies in zebrafish liver cells exposed to the brominated flame retardants HBCD and TBBPA. *Ecotoxicol. Environ. Saf.* **72**(7), 1985–1993.

Kling, P., Norman, A., Andersson, P. L., Norrgren, L., and Forlin, L. (2008). Gender-specific proteomic responses in zebrafish liver following exposure to a selected mixture of brominated flame retardants. *Ecotoxicol. Environ. Saf.* **71**(2), 319–327.

Kobayashi, M., Itoh, K., Suzuki, T., Osanai, H., Nishikawa, K., Katoh, Y., Takagi, Y., and Yamamoto, M. (2002). Identification of the interactive interface and phylogenic conservation of the Nrf2-Keap1 system. *Genes Cells* **7**(8), 807–820.

Kobayashi, M., Li, L., Iwamoto, N., Nakajima-Takagi, Y., Kaneko, H., Nakayama, Y., Eguchi, M., Wada, Y., Kumagai, Y., and Yamamoto, M. (2009). The antioxidant defense system Keap1-Nrf2 comprises a multiple sensing mechanism for responding to a wide range of chemical compounds. *Mol. Cell. Biol.* **29**(2), 493–502.

Koch, J. A., and Waxman, D. J. (1991). P450 phosphorylation in isolated hepatocytes and in vivo. *Methods Enzymol.* **206**, 305–315.

Kohle, C., and Bock, K. W. (2007). Coordinate regulation of Phase I and II xenobiotic metabolisms by the Ah receptor and Nrf2. *Biochem. Pharmacol.* **73**(12), 1853–1862.

Kohle, C., and Bock, K. W. (2009). Coordinate regulation of human drug-metabolizing enzymes, and conjugate transporters by the Ah receptor, pregnane X receptor and constitutive androstane receptor. *Biochem. Pharmacol.* **77**(4), 689–699.

Kokel, D., and Peterson, R. T. (2008). Chemobehavioural phenomics and behaviour-based psychiatric drug discovery in the zebrafish. *Brief Funct. Genomics Proteomics* **7**(6), 483–490.

Krasowski, M. D., Yasuda, K., Hagey, L. R., and Schuetz, E. G. (2005). Evolutionary selection across the nuclear hormone receptor superfamily with a focus on the NR1I subfamily (vitamin D, pregnane X, and constitutive androstane receptors). *Nucl. Recept.* **3**, 2.

Kreiling, J. A., Creton, R., and Reinisch, C. (2007). Early embryonic exposure to polychlorinated biphenyls disrupts heat-shock protein 70 cognate expression in zebrafish. *J. Toxicol. Environ. Health A.* **70**(12), 1005–1013.

Krone, P. H., Evans, T. G., and Blechinger, S. R. (2003). Heat shock gene expression and function during zebrafish embryogenesis. *Semin. Cell Dev. Biol.* **14**(5), 267–274.

Krueger, S. K., and Williams, D. E. (2005). Mammalian flavin-containing monooxygenases: Structure/function, genetic polymorphisms and role in drug metabolism. *Pharmacol. Ther.* **106**(3), 357–387.

Kusik, B. W., Carvan, M. J., 3rd, and Udvadia, A. J. (2008). Detection of mercury in aquatic environments using EPRE reporter zebrafish. *Mar. Biotechnol. (NY)* **10**(6), 750–757.

Lam, S. H., Mathavan, S., Tong, Y., Li, H., Karuturi, R. K., Wu, Y., Vega, V. B., Liu, E. T., and Gong, Z. (2008). Zebrafish whole-adult-organism chemogenomics for large-scale predictive and discovery chemical biology. *PLoS Genet.* **4**(7), e1000121.

Lam, S. H., Winata, C. L., Tong, Y., Korzh, S., Lim, W. S., Korzh, V., Spitsbergen, J., Mathavan, S., Miller, L. D., Liu, E. T., et al. (2006). Transcriptome kinetics of arsenic-induced adaptive response in zebrafish liver. *Physiol. Genomics* **27**(3), 351–361.

Lammer, E., Carr, G. J., Wendler, K., Rawlings, J. M., Belanger, S. E., and Braunbeck, T. (2009a). Is the fish embryo toxicity test (FET) with the zebrafish (Danio rerio) a potential alternative for the fish acute toxicity test? *Comp. Biochem. Physiol. C Toxicol. Pharmacol.* **149**(2), 196–209.

Lammer, E., Kamp, H. G., Hisgen, V., Koch, M., Reinhard, D., Salinas, E. R., Wendler, K., Zok, S., and Braunbeck, T. (2009b). Development of a flow-through system for the fish embryo toxicity test (FET) with the zebrafish (Danio rerio). *Toxicol In Vitro* **23**(7), 1436–1442.

Lassen, N., Estey, T., Tanguay, R. L., Pappa, A., Reimers, M. J., and Vasiliou, V. (2005). Molecular cloning, baculovirus expression, and tissue distribution of the zebrafish aldehyde dehydrogenase 2. *Drug Metab. Dispos.* **33**(5), 649–656.

Latchoumycandane, C., and Mathur, P. P. (2002). Effects of vitamin E on reactive oxygen species-mediated 2,3,7,8-tetrachlorodi-benzo-p-dioxin toxicity in rat testis. *J. Appl. Toxicol.* **22**(5), 345–351.

Le Page, Y., Scholze, M., Kah, O., and Pakdel, F. (2006). Assessment of xenoestrogens using three distinct estrogen receptors and the zebrafish brain aromatase gene in a highly responsive glial cell system. *Environ. Health Perspect.* **114**(5), 752–758.

Leaver, M. J., Wright, J., Hodgson, P., Boukouvala, E., and George, S. G. (2007). Piscine UDP-glucuronosyltransferase 1B. *Aquat. Toxicol.* **84**(3), 356–365.

Lee, K., Burgoon, L. D., Lamb, L., Dere, E., Zacharewski, T. R., Hogenesch, J. B., and LaPres, J. J. (2006). Identification and characterization of genes susceptible to transcriptional cross-talk between the hypoxia and dioxin signaling cascades. *Chem. Res. Toxicol.* **19**(10), 1284–1293.

Lee, M. O., Manthey, C. L., and Sladek, N. E. (1991). Identification of mouse liver aldehyde dehydrogenases that catalyze the oxidation of retinaldehyde to retinoic acid. *Biochem. Pharmacol.* **42**(6), 1279–1285.

Lefebvre, K. A., Tilton, S. C., Bammler, T. K., Beyer, R. P., Srinouanprachan, S., Stapleton, P. L., Farin, F. M., and Gallagher, E. P. (2009). Gene expression profiles in zebrafish brain after acute exposure to domoic acid at symptomatic and asymptomatic doses. *Toxicol. Sci.* **107**(1), 65–77.

Li, D., Lu, C., Wang, J., Hu, W., Cao, Z., Sun, D., Xia, H., and Ma, X. (2009). Developmental mechanisms of arsenite toxicity in zebrafish (Danio rerio) embryos. *Aquat. Toxicol.* **91**(3), 229–237.

Li, L., Kobayashi, M., Kaneko, H., Nakajima-Takagi, Y., Nakayama, Y., and Yamamoto, M. (2008). Molecular Evolution of Keap1: Two Keap1 molecules with distinctive intervening region structures are conserved among fish. *J. Biol. Chem.* **283**(6), 3248–3255.

Lin, Y., Chen, Y., Yang, X., Xu, D., and Liang, S. (2009). Proteome analysis of a single zebrafish embryo using three different digestion strategies coupled with liquid chromatography-tandem mass spectrometry. *Anal. Biochem.* **394**(2), 177–185.

Liu, M. Y., Yang, Y. S., Sugahara, T., Yasuda, S., and Liu, M. C. (2005). Identification of a novel zebrafish SULT1 cytosolic sulfotransferase: Cloning, expression, characterization, and developmental expression study. *Arch. Biochem. Biophys.* **437**(1), 10–19.

Lynch, M., and Conery, J. S. (2000). The evolutionary fate and consequences of duplicate genes. *Science* **290**(5494), 1151–1155.

Ma, Q. (2008). Xenobiotic-activated receptors: From transcription to drug metabolism to disease. *Chem. Res. Toxicol.* **21**(9), 1651–1671.

MacKay, A. B., Mhanni, A. A., McGowan, R. A., and Krone, P. H. (2007). Immunological detection of changes in genomic DNA methylation during early zebrafish development. *Genome* **50**(8), 778–785.

Mackenzie, P. I., Bock, K. W., Burchell, B., Guillemette, C., Ikushiro, S., Iyanagi, T., Miners, J. O., Owens, I. S., and Nebert, D. W. (2005). Nomenclature update for the mammalian UDP glycosyltransferase (UGT) gene superfamily. *Pharmacogenet. Genomics* **15**(10), 677–685.

Mackenzie, P. I., Rogers, A., Treloar, J., Jorgensen, B. R., Miners, J. O., and Meech, R. (2008). Identification of UDP glycosyltransferase 3A1 as a UDP N-acetylglucosaminyltransferase. *J. Biol. Chem.* **283**(52), 36205–36210.

Macleod, D., Clark, V. H., and Bird, A. (1999). Absence of genome-wide changes in DNA methylation during development of the zebrafish. *Nat. Genet.* **23**(2), 139–140.

Maglich, J. M., Stoltz, C. M., Goodwin, B., Hawkins-Brown, D., Moore, J. T., and Kliewer, S. A. (2002). Nuclear pregnane x receptor and constitutive androstane receptor regulate overlapping but distinct sets of genes involved in xenobiotic detoxification. *Mol. Pharmacol.* **62**(3), 638–646.

Margulies, M., Egholm, M., Altman, W. E., Attiya, S., Bader, J. S., Bemben, L. A., Berka, J., Braverman, M. S., Chen, Y. J., Chen, Z., et al. (2005). Genome sequencing in microfabricated high-density picolitre reactors. *Nature* **437**(7057), 376–380.

Martin, C. C., Laforest, L., Akimenko, M. A., and Ekker, M. (1999). A role for DNA methylation in gastrulation and somite patterning. *Dev. Biol.* **206**(2), 189–205.

Martyniuk, C. J., Gerrie, E. R., Popesku, J. T., Ekker, M., and Trudeau, V. L. (2007). Microarray analysis in the zebrafish (Danio rerio) liver and telencephalon after exposure to low concentration of 17alpha-ethinylestradiol. *Aquat. Toxicol.* **84**(1), 38–49.

Mathew, L. K., Andreasen, E. A., and Tanguay, R. L. (2006). Aryl hydrocarbon receptor activation inhibits regenerative growth. *Mol. Pharmacol.* **69**(1), 257–265.

Mathew, L. K., Sengupta, S. S., Ladu, J., Andreasen, E. A., and Tanguay, R. L. (2008). Crosstalk between AHR and Wnt signaling through R-Spondin1 impairs tissue regeneration in zebrafish. *FASEB J.* **22**(8), 3087–3096.

Mathew, L. K., Simonich, M. T., and Tanguay, R. L. (2009). AHR-dependent misregulation of Wnt signaling disrupts tissue regeneration. *Biochem. Pharmacol.* **77**(4), 498–507.

Mattingly, C. J., Hampton, T. H., Brothers, K. M., Griffin, N. E., and Planchart, A. (2009). Perturbation of defense pathways by low-dose arsenic exposure in zebrafish embryos. *Environ. Health Perspect.* **117**(6), 981–987.

Mattingly, C. J., McLachlan, J. A., and Toscano, W. A., Jr. (2001). Green fluorescent protein (GFP) as a marker of aryl hydrocarbon receptor (AhR) function in developing zebrafish (Danio rerio). *Environ. Health Perspect.* **109**(8), 845–849.

McArthur, A. G., Hegelund, T., Cox, R. L., Stegeman, J. J., Liljenberg, M., Olsson, U., Sundberg, P., and Celander, M. C. (2003). Phylogenetic analysis of the cytochrome P450 3 (CYP3) gene family. *J. Mol. Evol.* **57**(2), 200–211.

Meier-Abt, F., Mokrab, Y., and Mizuguchi, K. (2005). Organic anion transporting polypeptides of the OATP/SLCO superfamily: Identification of new members in nonmammalian species, comparative modeling and a potential transport mode. *J. Membr. Biol.* **208**(3), 213–227.

Meng, X., Noyes, M. B., Zhu, L. J., Lawson, N. D., and Wolfe, S. A. (2008). Targeted gene inactivation in zebrafish using engineered zinc-finger nucleases. *Nat. Biotechnol.* **26**(6), 695–701.

Menuet, A., Pellegrini, E., Anglade, I., Blaise, O., Laudet, V., Kah, O., and Pakdel, F. (2002). Molecular characterization of three estrogen receptor forms in zebrafish: Binding characteristics, transactivation properties, and tissue distributions. *Biol. Reprod.* **66**(6), 1881–1892.

Mhanni, A. A., and McGowan, R. A. (2004). Global changes in genomic methylation levels during early development of the zebrafish embryo. *Dev. Genes. Evol.* **214**(8), 412–417.

Miao, W., Hu, L., Scrivens, P. J., and Batist, G. (2005). Transcriptional regulation of NF-E2 p45-related factor (NRF2) expression by the aryl hydrocarbon receptor-xenobiotic response element signaling pathway: Direct cross-talk between phase I and II drug-metabolizing enzymes. *J. Biol. Chem.* **280**(21), 20340–20348.

Milan, D. J., and Macrae, C. A. (2008). Zebrafish genetic models for arrhythmia. *Prog. Biophys. Mol. Biol.* **98**(2-3), 301–308.

Mindnich, R., and Adamski, J. (2009). Zebrafish 17beta-hydroxysteroid dehydrogenases: An evolutionary perspective. *Mol. Cell Endocrinol.* **301**(1-2), 20–26.

Miranda, C. L., Chung, W. G., Wang-Buhler, J. L., Musafia-Jeknic, T., Baird, W. M., and Buhler, D. R. (2006). Comparative in vitro metabolism of benzo[a]pyrene by recombinant zebrafish CYP1A and liver microsomes from beta-naphthoflavone-treated rainbow trout. *Aquat. Toxicol.* **80**(2), 101–108.

Moffat, I. D., Boutros, P. C., Celius, T., Linden, J., Pohjanvirta, R., and Okey, A. B. (2007). microRNAs in adult rodent liver are refractory to dioxin treatment. *Toxicol. Sci.* **99**(2), 470–487.

Moore, L. B., Maglich, J. M., McKee, D. D., Wisely, B., Willson, T. M., Kliewer, S. A., Lambert, M. H., and Moore, J. T. (2002). Pregnane X receptor (PXR), constitutive androstane receptor (CAR), and benzoate X receptor (BXR) define three pharmacologically distinct classes of nuclear receptors. *Mol. Endocrinol.* **16**(5), 977–986.

Morrison, H. G., Oleksiak, M. F., Cornell, N. W., Sogin, M. L., and Stegeman, J. J. (1995). Identification of cytochrome P-450 1A (CYP1A) genes from two teleost fish, toadfish (Opsanus tau) and scup (Stenotomus chrysops), and phylogenetic analysis of CYP1A genes. *Biochem. J.* **308**, 97–104.

Morrison, H. G., Weil, E. J., Karchner, S. I., Sogin, M. L., and Stegeman, J. J. (1998). Molecular cloning of CYP1A from the estuarine fish Fundulus heteroclitus and phylogenetic analysis of CYP1 genes: Update with new sequences. *Comp. Biochem. Physiol. C. Pharmacol. Toxicol. Endocrinol.* **121**(1-3), s231–s240.

Morton, S. U., Scherz, P. J., Cordes, K. R., Ivey, K. N., Stainier, D. Y., and Srivastava, D. (2008). microRNA-138 modulates cardiac patterning during embryonic development. *Proc. Natl Acad. Sci. USA* **105**(46), 17830–17835.

Nagel, R. (2002). DarT: The embryo test with the Zebrafish Danio rerio – a general model in ecotoxicology and toxicology. *Altex* **19**(Suppl. 1), 38–48.

Nasevicius, A., and Ekker, S. C. (2000). Effective targeted gene 'knockdown' in zebrafish. *Nat. Genet.* **26**(2), 216–220.

NCI (1984). Use of small fish in carcinogenicity testing. Proceedings of a symposium. Bethesda, Maryland, December 8–10, 1981. *Natl Cancer Inst. Monogr.* **65**: 1–409.

Nebert, D. W., and Dalton, T. P. (2006). The role of cytochrome P450 enzymes in endogenous signalling pathways and environmental carcinogenesis. *Nat. Rev. Cancer* **6**(12), 947–960.

Neumann, C. J., and Nuesslein-Volhard, C. (2000). Patterning of the zebrafish retina by a wave of sonic hedgehog activity. *Science* **289**(5487), 2137–2139.

Nguyen, T., Nioi, P., and Pickett, C. B. (2009). The Nrf2-antioxidant response element signaling pathway and its activation by oxidative stress. *J. Biol. Chem.* **284**(20), 13291–13295.

Nourizadeh-Lillabadi, R., Lyche, J. L., Almaas, C., Stavik, B., Moe, S. J., Aleksandersen, M., Berg, V., Jakobsen, K. S., Stenseth, N. C., Skare, J. U., et al. (2009). Transcriptional regulation in liver and testis associated with developmental and reproductive effects in male zebrafish exposed to natural mixtures of persistent organic pollutants (POP). *J. Toxicol. Environ. Health A.* **72**(3-4), 112–130.

NRC (2000). *Scientific Frontiers in Developmental Toxicology and Risk Assessment*. National Academy Press, Washington, DC.

Oesch-Bartlomowicz, B., and Oesch, F. (2009). Role of cAMP in mediating AHR signaling. *Biochem. Pharmacol.* **77**(4), 627–641.

Ohtake, F., Fujii-Kuriyama, Y., and Kato, S. (2009). AhR acts as an E3 ubiquitin ligase to modulate steroid receptor functions. *Biochem. Pharmacol.* **77**(4), 474–484.

Okey, A. B. (2007). An aryl hydrocarbon receptor odyssey to the shores of toxicology: The Deichmann lecture, international congress of toxicology-XI. *Toxicol. Sci.* **98**(1), 5–38.

Oleksiak, M. F., Wu, S., Parker, C., Qu, W., Cox, R., Zeldin, D. C., and Stegeman, J. J. (2003). Identification and regulation of a new vertebrate cytochrome P450 subfamily, the CYP2Ps, and functional characterization of CYP2P3, a conserved arachidonic acid epoxygenase/19-hydroxylase. *Arch. Biochem. Biophys.* **411**(2), 223–234.

Pandey, D. P., and Picard, D. (2009). miR-22 inhibits estrogen signaling by directly targeting the estrogen receptor alpha mRNA. *Mol. Cell. Biol.* **29**(13), 3783–3790.

Pandini, A., Denison, M. S., Song, Y., Soshilov, A. A., and Bonati, L. (2007). Structural and functional characterization of the aryl hydrocarbon receptor ligand binding domain by homology modeling and mutational analysis. *Biochemistry* **46**(3), 696–708.

Pandini, A., Soshilov, A. A., Song, Y., Zhao, J., Bonati, L., and Denison, M. S. (2009). Detection of the TCDD binding-fingerprint within the Ah receptor ligand binding domain by structurally driven mutagenesis and functional analysis. *Biochemistry* **48**(25), 5972–5983.

Pang, Y., Dong, J., and Thomas, P. (2008). Estrogen signaling characteristics of Atlantic croaker G protein-coupled receptor 30 (GPR30) and evidence it is involved in maintenance of oocyte meiotic arrest. *Endocrinology* **149**(7), 3410–3426.

Pang, Y., and Thomas, P. (2009). Involvement of estradiol-17beta and its membrane receptor, G protein coupled receptor 30 (GPR30) in regulation of oocyte maturation in zebrafish, Danio rario. *Gen. Comp. Endocrinol.* **161**(1), 58–61.

Parng, C. (2005). In vivo zebrafish assays for toxicity testing. *Curr. Opin. Drug Discov. Devel.* **8** (1), 100–106.

Pearson, W. R. (2005). Phylogenies of glutathione transferase families. *Methods Enzymol.* **401**, 186–204.

Penning, T. M., and Drury, J. E. (2007). Human aldo-keto reductases: Function, gene regulation, and single nucleotide polymorphisms. *Arch. Biochem. Biophys.* **464**(2), 241–250.

Peterson, R. T. (2008). Chemical biology and the limits of reductionism. *Nat. Chem. Biol.* **4**(11), 635–638.

Pittlik, S., Domingues, S., Meyer, A., and Begemann, G. (2008). Expression of zebrafish aldh1a3 (raldh3) and absence of aldh1a1 in teleosts. *Gene Expr. Patterns* **8**(3), 141–147.

Plasterk, R. H. (2006). Micro RNAs in animal development. *Cell* **124**(5), 877–881.

Pomati, F., Cotsapas, C. J., Castiglioni, S., Zuccato, E., and Calamari, D. (2007). Gene expression profiles in zebrafish (Danio rerio) liver cells exposed to a mixture of pharmaceuticals at environmentally relevant concentrations. *Chemosphere* **70**(1), 65–73.

Pondugula, S. R., Dong, H., and Chen, T. (2009). Phosphorylation and protein-protein interactions in PXR-mediated CYP3A repression. *Expert Opin. Drug Metab. Toxicol.* **5**(8), 861–873.

Popovic, M., Zaja, R., Loncar, J., and Smital, T. (2009). A novel ABC transporter: The first insight into zebrafish (Danio rerio) Abch1. *Mar. Environ. Res.* PMID 19926124.

Postlethwait, J., Amores, A., Cresko, W., Singer, A., and Yan, Y. L. (2004). Subfunction partitioning, the teleost radiation and the annotation of the human genome. *Trends Genet.* **20**(10), 481–490.

Postlethwait, J. H. (2006). The zebrafish genome: A Review and msx Gene Case Study. In *Vertebrate Genomes*, Vol. 2. Karger, Basel, pp. 183–197.

Postlethwait, J. H. (2007). The zebrafish genome in context: Ohnologs gone missing. *J. Exp. Zool. B Mol. Dev. Evol.* **308**(5), 563–577.

Prasad, J. C., Goldstone, J. V., Camacho, C. J., Vajda, S., and Stegeman, J. J. (2007). Ensemble modeling of substrate binding to cytochromes P450: Analysis of catalytic differences between CYP1A orthologs. *Biochemistry* **46**(10), 2640–2654.

Prasch, A. L., Andreasen, E. A., Peterson, R. E., and Heideman, W. (2004). Interactions between 2,3,7,8-tetrachlorodibenzo-p-dioxin (TCDD) and hypoxia signaling pathways in zebrafish: Hypoxia decreases responses to TCDD in zebrafish embryos. *Toxicol. Sci.* **78**(1), 68–77.

Prasch, A. L., Tanguay, R. L., Mehta, V., Heideman, W., and Peterson, R. E. (2006). Identification of zebrafish ARNT1 homologs: 2,3,7,8-tetrachlorodibenzo-p-dioxin toxicity in the developing zebrafish requires ARNT1. *Mol. Pharmacol.* **69**(3), 776–787.

Prasch, A. L., Teraoka, H., Carney, S. A., Dong, W., Hiraga, T., Stegeman, J. J., Heideman, W., and Peterson, R. E. (2003). Aryl hydrocarbon receptor 2 mediates 2,3,7,8-tetrachlorodibenzo-p-dioxin developmental toxicity in zebrafish. *Toxicol. Sci.* **76**, 138–150.

Puga, A., Ma, C., and Marlowe, J. L. (2009). The aryl hydrocarbon receptor cross-talks with multiple signal transduction pathways. *Biochem. Pharmacol.* **77**(4), 713–722.

Qiu, H., Taudien, S., Herlyn, H., Schmitz, J., Zhou, Y., Chen, G., Roberto, R., Rocchi, M., Platzer, M., and Wojnowski, L. (2008). CYP3 phylogenomics: Evidence for positive selection of CYP3A4 and CYP3A7. *Pharmacogenet. Genomics* **18**(1), 53–66.

Qu, X., Swanson, R., Day, R., and Tsai, J. (2009). A guide to template based structure prediction. *Curr. Protein Pept. Sci.* **10**(3), 270–285.

Rai, K., Huggins, I. J., James, S. R., Karpf, A. R., Jones, D. A., and Cairns, B. R. (2008). DNA demethylation in zebrafish involves the coupling of a deaminase, a glycosylase, and gadd45. *Cell* **135**(7), 1201–1212.

Raldua, D., and Babin, P. J. (2009). Simple, rapid zebrafish larva bioassay for assessing the potential of chemical pollutants and drugs to disrupt thyroid gland function. *Environ. Sci. Technol.* **43**(17), 6844–6850.

Reddy, C. M. (2008). A cautionary tale about evaluating analytical methods to assess contamination after oil spills. *Mar. Pollut. Bull.* **56**(7), 1380.

Redlich, G., Zanger, U. M., Riedmaier, S., Bache, N., Giessing, A. B., Eisenacher, M., Stephan, C., Meyer, H. E., Jensen, O. N., and Marcus, K. (2008). Distinction between human cytochrome P450 (CYP) isoforms and identification of new phosphorylation sites by mass spectrometry. *J. Proteome. Res.* **7**(11), 4678–4688.

Reimers, M. J., Hahn, M. E., and Tanguay, R. L. (2004a). Two zebrafish alcohol dehydrogenases share common ancestry with mammalian class I, II, IV, and V ADH genes but have distinct functional characteristics. *J. Biol. Chem.* **279**, 38303–38312.

Reimers, M. J., Hahn, M. E., and Tanguay, R. L. (2004b). Two zebrafish alcohol dehydrogenases share common ancestry with mammalian class I, II, IV, and V alcohol dehydrogenase genes but have distinct functional characteristics. *J. Biol. Chem.* **279**(37), 38303–38312.

Reschly, E. J., Bainy, A. C., Mattos, J. J., Hagey, L. R., Bahary, N., Mada, S. R., Ou, J., Venkataramanan, R., and Krasowski, M. D. (2007). Functional evolution of the vitamin D and pregnane X receptors. *BMC Evol. Biol.* **7**, 222.

Ruggeri, B., Ubaldi, M., Lourdusamy, A., Soverchia, L., Ciccocioppo, R., Hardiman, G., Baker, M. E., Palermo, F., and Polzonetti-Magni, A. M. (2008). Variation of the genetic expression pattern after exposure to estradiol-17beta and 4-nonylphenol in male zebrafish (Danio rerio). *Gen. Comp. Endocrinol.* **158**(1), 138–144.

Sabaliauskas, N. A., Foutz, C. A., Mest, J. R., Budgeon, L. R., Sidor, A. T., Gershenson, J. A., Joshi, S. B., and Cheng, K. C. (2006). High-throughput zebrafish histology. *Methods* **39**(3), 246–254.

Safe, S., and Wormke, M. (2003). Inhibitory aryl hydrocarbon receptor-estrogen receptor-α cross-talk and mechanisms of action. *Chem. Res. Toxicol.* **16**(7), 807–816.

Santos, E. M., Workman, V. L., Paull, G. C., Filby, A. L., Van Look, K. J., Kille, P., and Tyler, C. R. (2007). Molecular basis of sex and reproductive status in breeding zebrafish. *Physiol. Genomics.* **30**(2), 111–122.

Schlezinger, J. J., Keller, J., Verbrugge, L. A., and Stegeman, J. J. (2000). 3,3′,4,4′-Tetrachlorobiphenyl oxidation in fish, bird and reptile species: Relationship to cytochrome P450 1A inactivation and reactive oxygen production. *Comp. Biochem. Physiol. C Toxicol. Pharmacol.* **125**(3), 273–286.

Schlezinger, J. J., Struntz, W. D., Goldstone, J. V., and Stegeman, J. J. (2006). Uncoupling of cytochrome P450 1A and stimulation of reactive oxygen species production by co-planar polychlorinated biphenyl congeners. *Aquat. Toxicol.* **77**(4), 422–432.

Schlezinger, J. J., White, R. D., and Stegeman, J. J. (1999). Oxidative inactivation of cytochrome P450 1A (CYP1A) stimulated by 3,3′,4,4′-tetrachlorobiphenyl: Production of reactive oxygen by vertebrate CYP1As. *Mol. Pharmacol.* **56**, 588–597.

Schreurs, R. H., Sonneveld, E., Jansen, J. H., Seinen, W., and van der Burg, B. (2005). Interaction of polycyclic musks and UV filters with the estrogen receptor (ER), androgen receptor (AR), and progesterone receptor (PR) in reporter gene bioassays. *Toxicol. Sci.* **83**(2), 264–272.

Segner, H. (2009). Zebrafish (Danio rerio) as a model organism for investigating endocrine disruption. *Comp. Biochem. Physiol. C Toxicol. Pharmacol.* **149**(2), 187–195.

Shah, Y. M., Morimura, K., Yang, Q., Tanabe, T., Takagi, M., and Gonzalez, F. J. (2007). Peroxisome proliferator-activated receptor alpha regulates a microRNA-mediated signaling cascade responsible for hepatocellular proliferation. *Mol. Cell. Biol.* **27**(12), 4238–4247.

Shi, X., Yeung, L. W., Lam, P. K., Wu, R. S., and Zhou, B. (2009). Protein profiles in zebrafish (Danio rerio) embryos exposed to perfluorooctane sulfonate. *Toxicol. Sci.* **110**(2), 334–340.

Shin, S., Wakabayashi, N., Misra, V., Biswal, S., Lee, G. H., Agoston, E. S., Yamamoto, M., and Kensler, T. W. (2007). NRF2 modulates aryl hydrocarbon receptor signaling: Influence on adipogenesis. *Mol. Cell. Biol.* **27**(20), 7188–7197.

Sim, E., Lack, N., Wang, C. J., Long, H., Westwood, I., Fullam, E., and Kawamura, A. (2008). Arylamine N-acetyltransferases: Structural and functional implications of polymorphisms. *Toxicology* **254**(3), 170–183.

Skinner, M. K., Anway, M. D., Savenkova, M. I., Gore, A. C., and Crews, D. (2008). Transgenerational epigenetic programming of the brain transcriptome and anxiety behavior. *PLoS ONE* **3**(11), e3745.

Soares, A. R., Pereira, P. M., Santos, B., Egas, C., Gomes, A. C., Arrais, J., Oliveira, J. L., Moura, G. R., and Santos, M. A. (2009). Parallel DNA pyrosequencing unveils new zebrafish microRNAs. *BMC Genomics* **10**, 195.

Song, W., Zou, Z., Xu, F., Gu, X., Xu, X., and Zhao, Q. (2006). Molecular cloning and expression of a second zebrafish aldehyde dehydrogenase 2 gene (aldh2b). *DNA Seq.* **17**(4), 262–269.

Spitsbergen, J. M., and Kent, M. L. (2003). The state of the art of the zebrafish model for toxicology and toxicologic pathology research – advantages and current limitations. *Toxicol. Pathol.* **31**(Suppl.), 62–87.

Stegeman, J. J. (1981). Polynuclear aromatic hydrocarbons and their metabolism in the marine environment. In: *Polycyclic Hydrocarbons and Cancer* (H.V. Gelboin and P.O.P. Ts'o, eds), Vol. 3, pp. 1–60. Academic Press, New York.

Stegeman, J. J., Brouwer, M., DiGiulio, R. T., Forlin, L., Fowler, B. M., Sanders, B. M., and Van Veld, P. (1992). Molecular responses to environmental contamination: Enzyme and protein systems as indicators of contaminant exposure and effect. In: *Biomarkers for Chemical Contaminants* (R.J. Huggett, ed.), pp. 237–339. CRC Press.

Stegeman, J. J., Smolowitz, R. M., and Hahn, M. E. (1991). Immunohistochemical localization of environmentally induced cytochrome P4501A1 in multiple organs of the marine teleost *Stenotomus chrysops* (scup). *Toxicol. Appl. Pharmacol.* **110**, 486–504.

Stegeman, J. J., Woodin, B. R., and Binder, R. L. (1984). Patterns of benzo[a]pyrene metabolism by varied species, organs, and developmental stages of fish. *Natl Cancer Inst. Monogr.* **65**, 371–377.

Suzawa, M., and Ingraham, H. A. (2008). The herbicide atrazine activates endocrine gene networks via non-steroidal NR5A nuclear receptors in fish and mammalian cells. *PLoS ONE* **3**(5), e2117.

Suzuki, T., Takagi, Y., Osanai, H., Li, L., Takeuchi, M., Katoh, Y., Kobayashi, M., and Yamamoto, M. (2005). Pi class glutathione S-transferase genes are regulated by Nrf2 through an evolutionarily conserved regulatory element in zebrafish. *Biochem. J.* **388**(Pt 1), 65–73.

Takagi, S., Nakajima, M., Mohri, T., and Yokoi, T. (2008). Post-transcriptional regulation of human pregnane X receptor by micro-RNA affects the expression of cytochrome P450 3A4. *J. Biol. Chem.* **283**(15), 9674–9680.

Takagi, Y., Kobayashi, M., Li, L., Suzuki, T., Nishikawa, K., and Yamamoto, M. (2004). MafT, a new member of the small Maf protein family in zebrafish. *Biochem. Biophys. Res. Commun.* **320**(1), 62–69.

Tanguay, R. L., Abnet, C. C., Heideman, W., and Peterson, R. E. (1999). Cloning and characterization of the zebrafish (Danio rerio) aryl hydrocarbon receptor. *Biochim. Biophys. Acta* **1444**(1), 35–48.

Tanguay, R. L., Andreasen, E., Heideman, W., and Peterson, R. E. (2000). Identification and expression of alternatively spliced aryl hydrocarbon nuclear translocator 2 (ARNT2) cDNAs from zebrafish with distinct functions. *Biochim. Biophys. Acta* **1494**(1-2), 117–128.

Teraoka, H., Dong, W., and Hiraga, T. (2003). Zebrafish as a novel experimental model for developmental toxicology. *Congenit. Anom. (Kyoto)* **43**(2), 123–132.

Thatcher, E. J., Bond, J., Paydar, I., and Patton, J. G. (2008a). Genomic Organization of Zebrafish microRNAs. *BMC Genomics* **9**, 253.

Thatcher, E. J., Flynt, A. S., Li, N., Patton, J. R., and Patton, J. G. (2007). MiRNA expression analysis during normal zebrafish development and following inhibition of the hedgehog and notch signaling pathways. *Dev. Dyn.* **236**(8), 2172–2180.

Thatcher, E. J., Paydar, I., Anderson, K. K., and Patton, J. G. (2008b). Regulation of zebrafish fin regeneration by microRNAs. *Proc. Natl Acad. Sci. USA* **105**(47), 18384–18389.

Thorsteinson, N., Ban, F., Santos-Filho, O., Tabaei, S. M., Miguel-Queralt, S., Underhill, C., Cherkasov, A., and Hammond, G. L. (2009). In silico identification of anthropogenic chemicals as ligands of zebrafish sex hormone binding globulin. *Toxicol. Appl. Pharmacol.* **234**(1), 47–57.

Tiboni, G. M., Marotta, F., and Carletti, E. (2009). Fluconazole alters CYP26 gene expression in mouse embryos. *Reprod. Toxicol.* **27**(2), 199–202.

Timme-Laragy, A. R., Van Tiem, L. A., Linney, E. A., and Di Giulio, R. T. (2009). Antioxidant responses and NRF2 in synergistic developmental toxicity of PAHs in zebrafish. *Toxicol. Sci.* **109**(2), 217–227.

Ton, C., Stamatiou, D., and Liew, C. C. (2003). Gene expression profile of zebrafish exposed to hypoxia during development. *Physiol. Genomics* **13**(2), 97–106.

Tsuchiya, Y., Nakajima, M., Takagi, S., Taniya, T., and Yokoi, T. (2006). MicroRNA regulates the expression of human cytochrome P450 1B1. *Cancer Res.* **66**(18), 9090–9098.

Tukey, R. H., and Strassburg, C. P. (2000). Human UDP-glocuronosyltransferases: metabolism, expression, and disease. *Annu. Rev. Pharmacol. Toxicol.* **40**, 581–616.

Udvadia, A. J., and Linney, E. (2003). Windows into development: Historic, current, and future perspectives on transgenic zebrafish. *Dev. Biol.* **256**(1), 1–17.

Usenko, C. Y., Harper, S. L., and Tanguay, R. L. (2008). Fullerene $C_{60}$ exposure elicits an oxidative stress response in embryonic zebrafish. *Toxicol. Appl. Pharmacol.* **229**(1), 44–55.

van Boxtel, A. L., Kamstra, J. H., Cenijn, P. H., Pieterse, B., Wagner, J. M., Antink, M., Krab, K., van der Burg, B., Marsh, G., Brouwer, A., et al. (2008). Microarray analysis reveals a mechanism of phenolic polybrominated diphenylether toxicity in zebrafish. *Environ. Sci. Technol.* **42**(5), 1773–1779.

van der Ven, K., De Wit, M., Keil, D., Moens, L., Van Leemput, K., Naudts, B., and De Coen, W. (2005). Development and application of a brain-specific cDNA microarray for effect evaluation of neuro-active pharmaceuticals in zebrafish (Danio rerio). *Comp. Biochem. Physiol. B Biochem. Mol. Biol.* **141**(4), 408–417.

van der Ven, K., Keil, D., Moens, L. N., Hummelen, P. V., van Remortel, P., Maras, M., and De Coen, W. (2006a). Effects of the antidepressant mianserin in zebrafish: Molecular markers of endocrine disruption. *Chemosphere* **65**(10), 1836–1845.

van der Ven, K., Keil, D., Moens, L. N., Van Leemput, K., van Remortel, P., and De Coen, W. M. (2006b). Neuropharmaceuticals in the environment: Mianserin-induced neuroendocrine disruption in zebrafish (Danio rerio) using cDNA microarrays. *Environ. Toxicol. Chem.* **25**(10), 2645–2652.

Van Veld, P. A., Vogelbein, W. K., Cochran, M. K., Goksoyr, A., and Stegeman, J. J. (1997). Route-specific cellular expression of cytochrome P4501A (CYP1A) in fish (Fundulus heteroclitus) following exposure to aqueous and dietary benzo[a]pyrene. *Toxicol. Appl. Pharmacol.* **142**(2), 348–359.

Vascotto, S. G., Beckham, Y., and Kelly, G. M. (1997). The zebrafish's swim to fame as an experimental model in biology. *Biochem. Cell Biol.* **75**(5), 479–485.

Verkade, H., and Heath, J. K. (2008). Wnt signaling mediates diverse developmental processes in zebrafish. *Methods Mol. Biol.* **469**, 225–251.

Vezina, C. M., Lin, T. M., and Peterson, R. E. (2009). AHR signaling in prostate growth, morphogenesis, and disease. *Biochem. Pharmacol.* **77**(4), 566–576.

Villeneuve, L., Wang, R. L., Bencic, D. C., Biales, A. D., Martinovic, D., Lazorchak, J. M., Toth, G., and Ankley, G. T. (2009). Altered gene expression in the brain and ovaries of zebrafish (Danio rerio) exposed to the aromatase inhibitor fadrozole: Microarray analysis and hypothesis generation. *Environ. Toxicol. Chem.* **28**(8), 1767–1782.

Voelker, D., Vess, C., Tillmann, M., Nagel, R., Otto, G. W., Geisler, R., Schirmer, K., and Scholz, S. (2007). Differential gene expression as a toxicant-sensitive endpoint in zebrafish embryos and larvae. *Aquat. Toxicol.* **81**(4), 355–364.

Vogt, A., Cholewinski, A., Shen, X., Nelson, S. G., Lazo, J. S., Tsang, M., and Hukriede, N. A. (2009). Automated image-based phenotypic analysis in zebrafish embryos. *Dev. Dyn.* **238** (3), 656–663.

Wang, L., Yao, J., Chen, L., Chen, J., Xue, J., and Jia, W. (2007). Expression and possible functional roles of cytochromes P450 2J1 (zfCyp 2J1) in zebrafish. *Biochem. Biophys. Res. Commun.* **352**(4), 850–855.

Wang, M., Chan, L. L., Si, M., Hong, H., and Wang, D. (2010). Proteomic analysis of hepatic tissue of zebrafish (Danio rerio) experimentally exposed to chronic microcystin-LR. *Toxicol. Sci.* **113**(1), 60–69.

Wang, R. L., Bencic, D., Biales, A., Lattier, D., Kostich, M., Villeneuve, D., Ankley, G. T., Lazorchak, J., and Toth, G. (2008a). DNA microarray-based ecotoxicological biomarker discovery in a small fish model species. *Environ. Toxicol. Chem.* **27**(3), 664–675.

Wang, R. L., Biales, A., Bencic, D., Lattier, D., Kostich, M., Villeneuve, D., Ankley, G. T., Lazorchak, J., and Toth, G. (2008b). DNA microarray application in ecotoxicology: Experimental design, microarray scanning, and factors affecting transcriptional profiles in a small fish species. *Environ. Toxicol. Chem.* **27**(3), 652–663.

Wang, Y., Wimmer, U., Lichtlen, P., Inderbitzin, D., Stieger, B., Meier, P. J., Hunziker, L., Stallmach, T., Forrer, R., Rulicke, T., et al. (2004). Metal-responsive transcription factor-1 (MTF-1) is essential for embryonic liver development and heavy metal detoxification in the adult liver. *FASEB J.* **18**(10), 1071–1079.

Watanabe, K. H., Jensen, K. M., Orlando, E. F., and Ankley, G. T. (2007). What is normal? A characterization of the values and variability in reproductive endpoints of the fathead minnow, Pimephales promelas. *Comp. Biochem. Physiol. C Toxicol. Pharmacol.* **146**(3), 348–356.

Waxman, D. J. (1999). P450 gene induction by structurally diverse xenochemicals: Central role of nuclear receptors CAR, PXR, and PPAR. *Arch. Biochem. Biophys.* **369**(1), 11–23.

Webb, K. J., Norton, W. H., Trumbach, D., Meijer, A. H., Ninkovic, J., Topp, S., Heck, D., Marr, C., Wurst, W., Theis, F. J., et al. (2009). Zebrafish reward mutants reveal novel transcripts mediating the behavioral effects of amphetamine. *Genome Biol.* **10**(7), R81.

Wei, P., Zhang, J., Dowhan, D. H., Han, Y., and Moore, D. D. (2002). Specific and overlapping functions of the nuclear hormone receptors CAR and PXR in xenobiotic response. *Pharmacogenom J.* **2**(2), 117–126.

Wells, P. G., McCallum, G. P., Chen, C. S., Henderson, J. T., Lee, C. J., Perstin, J., Preston, T. J., Wiley, M. J., and Wong, A. W. (2009). Oxidative stress in developmental origins of disease: Teratogenesis, neurodevelopmental deficits, and cancer. *Toxicol. Sci.* **108**(1), 4–18.

White, J. A., Beckett-Jones, B., Guo, Y. D., Dilworth, F. J., Bonasoro, J., Jones, G., and Petkovich, M. (1997). cDNA cloning of human retinoic acid-metabolizing enzyme (hP450RAI) identifies a novel family of cytochromes P450. *J. Biol. Chem.* **272**(30), 18538–18541.

Wolfe, K. (2000). Robustness – it's not where you think it is. *Nat. Genet.* **25**(1), 3–4.

Wu, S. M., Zheng, Y. D., and Kuo, C. H. (2008). Expression of mt2 and smt-B upon cadmium exposure and cold shock in zebrafish (Danio rerio). *Comp. Biochem. Physiol. C Toxicol. Pharmacol.* **148**(2), 184–193.

Xie, W., Barwick, J. L., Simon, C. M., Pierce, A. M., Safe, S., Blumberg, B., Guzelian, P. S., and Evans, R. M. (2000). Reciprocal activation of xenobiotic response genes by nuclear receptors SXR/PXR and CAR. *Genes. Dev.* **14**(23), 3014–3023.

Xiong, K. M., Peterson, R. E., and Heideman, W. (2008). Aryl hydrocarbon receptor-mediated down-regulation of sox9b causes jaw malformation in zebrafish embryos. *Mol. Pharmacol.* **74**(6), 1544–1553.

Yamamoto, Y., Kawamoto, T., and Negishi, M. (2003). The role of the nuclear receptor CAR as a coordinate regulator of hepatic gene expression in defense against chemical toxicity. *Arch. Biochem. Biophys.* **409**(1), 207–211.

Yan, C. H., and Chan, K. M. (2002). Characterization of zebrafish metallothionein gene promoter in a zebrafish caudal fin cell-line, SJD. 1. *Mar. Environ. Res.* **54**(3-5), 335–339.

Yan, C. H., and Chan, K. M. (2004). Cloning of zebrafish metallothionein gene and characterization of its gene promoter region in HepG2 cell line. *Biochim. Biophys. Acta* **1679**(1), 47–58.

Yang, L., Kemadjou, J. R., Zinsmeister, C., Bauer, M., Legradi, J., Muller, F., Pankratz, M., Jakel, J., and Strahle, U. (2007). Transcriptional profiling reveals barcode-like toxicogenomic responses in the zebrafish embryo. *Genome Biol.* **8**(10), R227.

Yasuda, S., Kumar, A. P., Liu, M. Y., Sakakibara, Y., Suiko, M., Chen, L., and Liu, M. C. (2005a). Identification of a novel thyroid hormone-sulfating cytosolic sulfotransferase, SULT1 ST5, from zebrafish. *FEBS J.* **272**(15), 3828–3837.

Yasuda, S., Liu, C. C., Takahashi, S., Suiko, M., Chen, L., Snow, R., and Liu, M. C. (2005b). Identification of a novel estrogen-sulfating cytosolic SULT from zebrafish: Molecular cloning, expression, characterization, and ontogeny study. *Biochem. Biophys. Res. Commun.* **330**(1), 219–225.

Yasuda, T., Yasuda, S., Williams, F. E., Liu, M. Y., Sakakibara, Y., Bhuiyan, S., Snow, R., Carter, G., and Liu, M. C. (2008). Characterization and ontogenic study of novel steroid-sulfating SULT3 sulfotransferases from zebrafish. *Mol. Cell. Endocrinol.* **294**(1-2), 29–36.

Yeager, R. L., Reisman, S. A., Aleksunes, L. M., and Klaassen, C. D. (2009). Introducing the 'TCDD Inducible AhR-Nrf2 Gene Battery'. *Toxicol. Sci.* **11**(2), 238–246.

Yin, H. C., Tseng, H. P., Chung, H. Y., Ko, C. Y., Tzou, W. S., Buhler, D. R., and Hu, C. H. (2008). Influence of TCDD on zebrafish CYP1B1 transcription during development. *Toxicol. Sci.* **103**(1), 158–168.

Zheng, D., Feeney, G. P., Kille, P., and Hogstrand, C. (2008). Regulation of ZIP and ZnT zinc transporters in zebrafish gill: Zinc repression of ZIP10 transcription by an intronic MRE cluster. *Physiol. Genomics* **34**(2), 205–214.

Zhu, X., Zhu, L., Li, Y., Duan, Z., Chen, W., and Alvarez, P. J. (2007). Developmental toxicity in zebrafish (Danio rerio) embryos after exposure to manufactured nanomaterials: Buckminsterfullerene aggregates ($nC_{60}$) and fullerol. *Environ. Toxicol. Chem.* **26**(5), 976–979.

Ziegler, D. M. (2002). An overview of the mechanism, substrate specificities, and structure of FMOs. *Drug Metab. Rev.* **34**(3), 503–511.

Zodrow, J. M., Stegeman, J. J., and Tanguay, R. L. (2004). Histological analysis of acute toxicity of 2,3,7,8-tetrachlorodibenzo-p-dioxin (TCDD) in zebrafish. *Aquat. Toxicol.* **66**(1), 25–38.

# INDEX

**D**

F



G

**H**



**I**

## O



## P

# OTHER VOLUMES IN THE FISH PHYSIOLOGY SERIES

VOLUME 1 Excretion, Ionic Regulation, and Metabolism
*Edited by W. S. Hoar and D. J. Randall*

VOLUME 2 The Endocrine System
*Edited by W. S. Hoar and D. J. Randall*

VOLUME 3 Reproduction and Growth: Bioluminescence, Pigments, and Poisons
*Edited by W. S. Hoar and D. J. Randall*

VOLUME 4 The Nervous System, Circulation, and Respiration
*Edited by W. S. Hoar and D. J. Randall*

VOLUME 5 Sensory Systems and Electric Organs
*Edited by W. S. Hoar and D. J. Randall*

VOLUME 6 Environmental Relations and Behavior
*Edited by W. S. Hoar and D. J. Randall*

VOLUME 7 Locomotion
*Edited by W. S. Hoar and D. J. Randall*

VOLUME 8 Bioenergetics and Growth
*Edited by W. S. Hoar, D. J. Randall, and J. R. Brett*

VOLUME 9A Reproduction: Endocrine Tissues and Hormones
*Edited by W. S. Hoar, D. J. Randall, and E. M. Donaldson*

VOLUME 9B Reproduction: Behavior and Fertility Control
*Edited by W. S. Hoar, D. J. Randall, and E. M. Donaldson*

VOLUME 10A Gills: Anatomy, Gas Transfer, and Acid-Base Regulation
*Edited by W. S. Hoar and D. J. Randall*

VOLUME 10B Gills: Ion and Water Transfer
*Edited by W. S. Hoar and D. J. Randall*

VOLUME 11A The Physiology of Developing Fish: Eggs and Larvae
*Edited by W. S. Hoar and D. J. Randall*

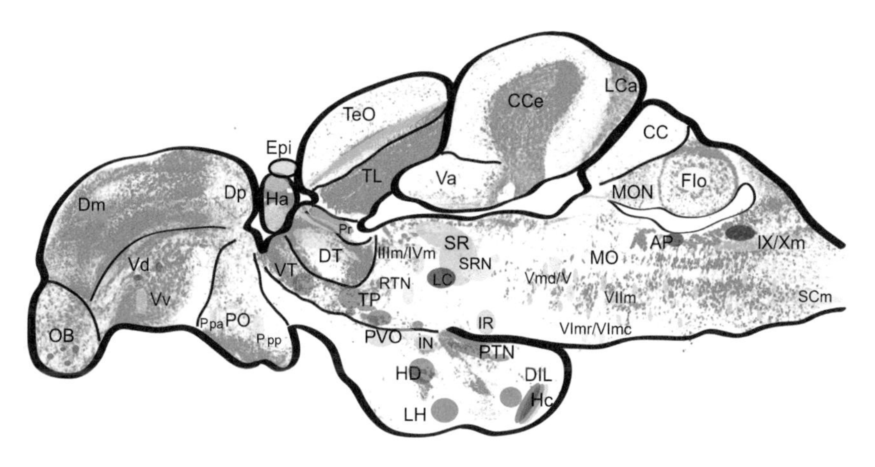
TeO
CCe
LCa
CC
Epi
TL
Va
Flo
MON
Dm
Dp
Ha
Pr
IIIm/IVm
SR
SRN
AP
IX/Xm
Vd
DT
VT
RTN
LC
Vmd/V
MO
Vv
TP
VIIm
SCm
OB
Ppa
PO
Ppp
PVO
IR
IN
PTN
VImr/VImc
HD
DIL
Hc
LH

Fig. 2.4.

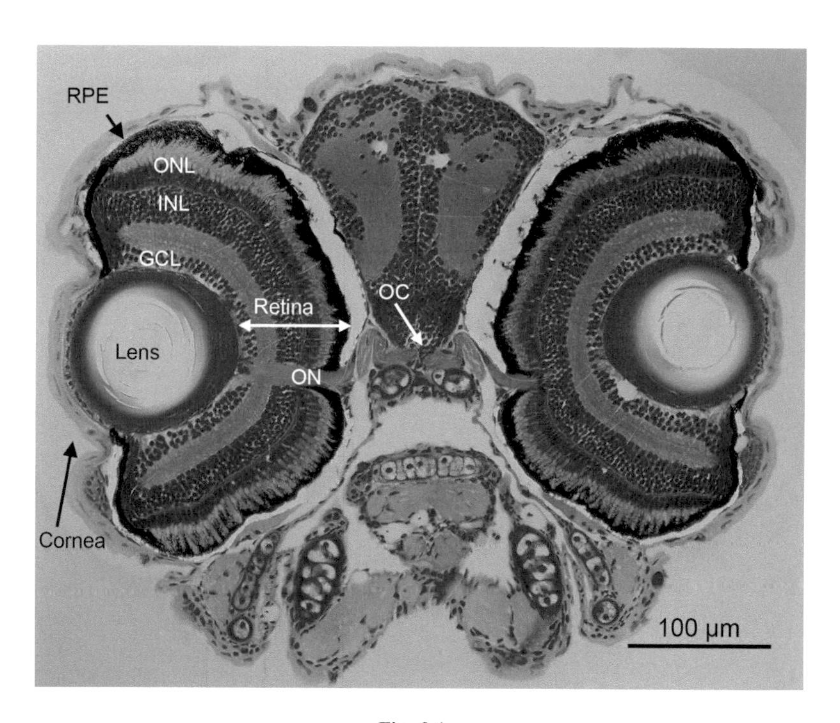
RPE
ONL
INL
GCL
Retina
OC
Lens
ON
Cornea
100 µm

Fig. 3.1.

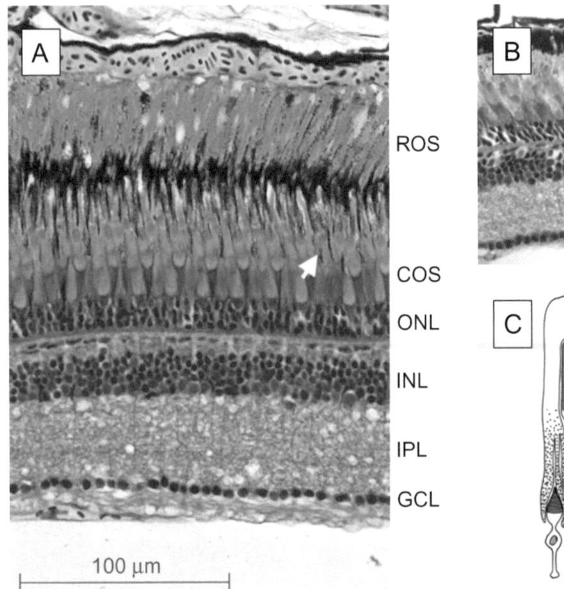


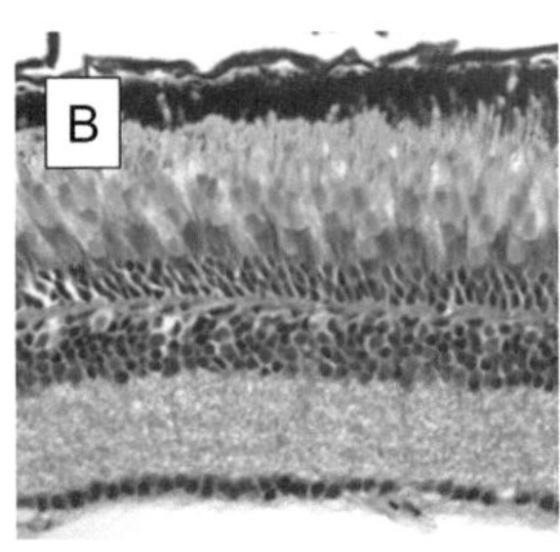


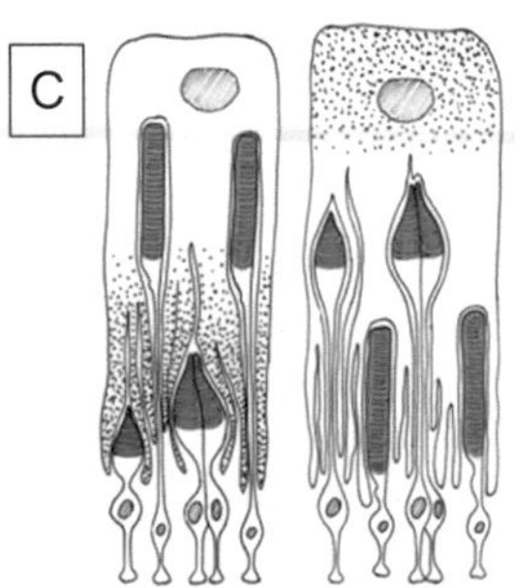


Fig. 3.2.

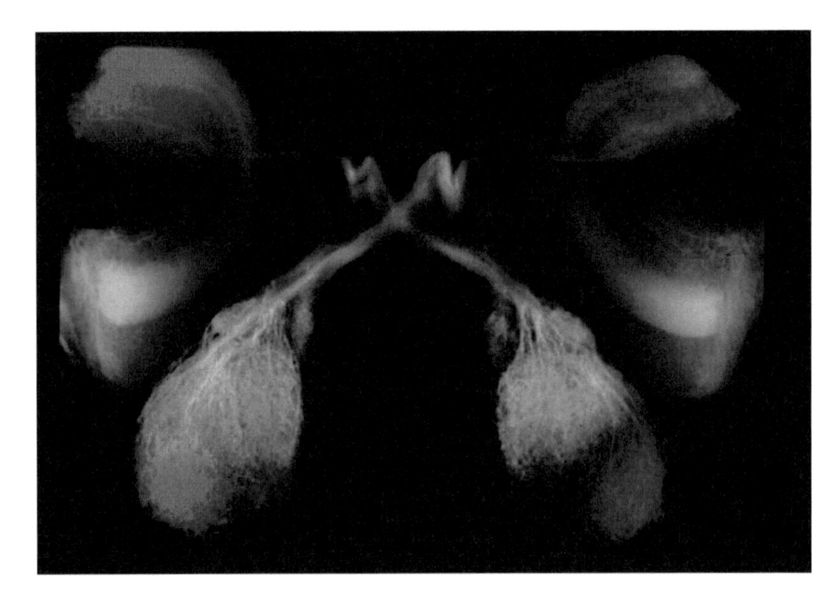

Fig. 3.3.

Fig. 3.4.

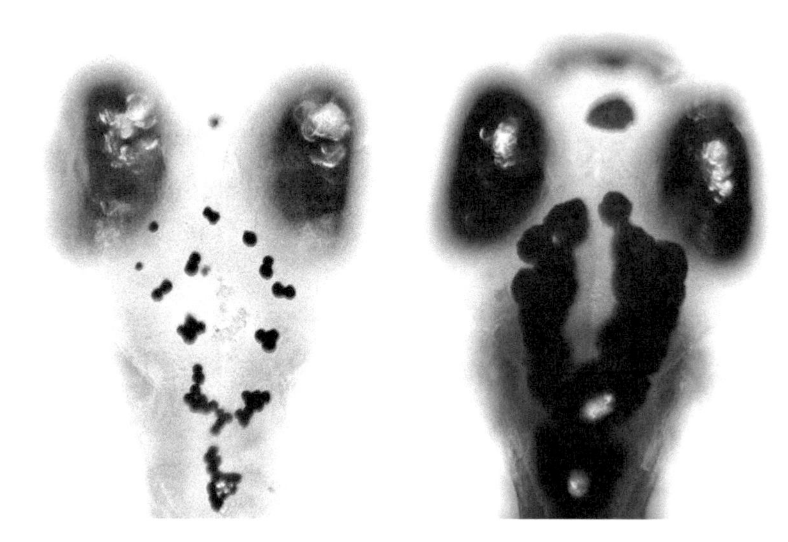

Fig. 3.5.

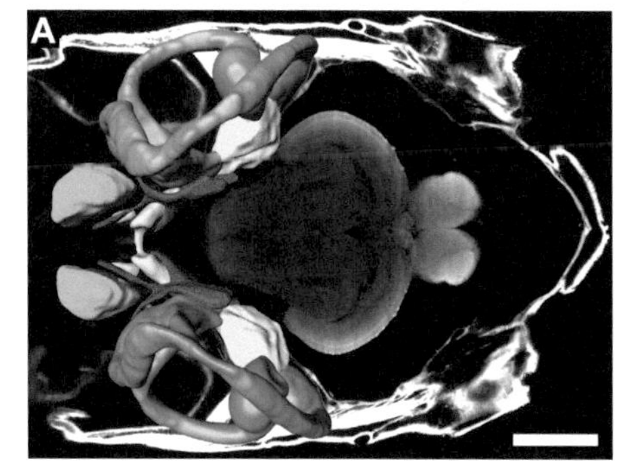


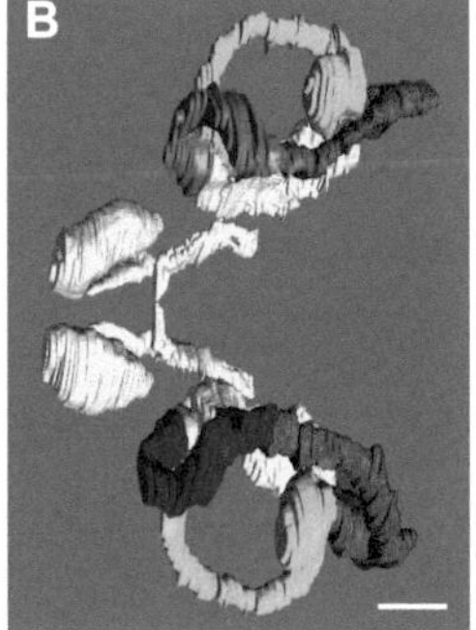


Fig. 4.1.

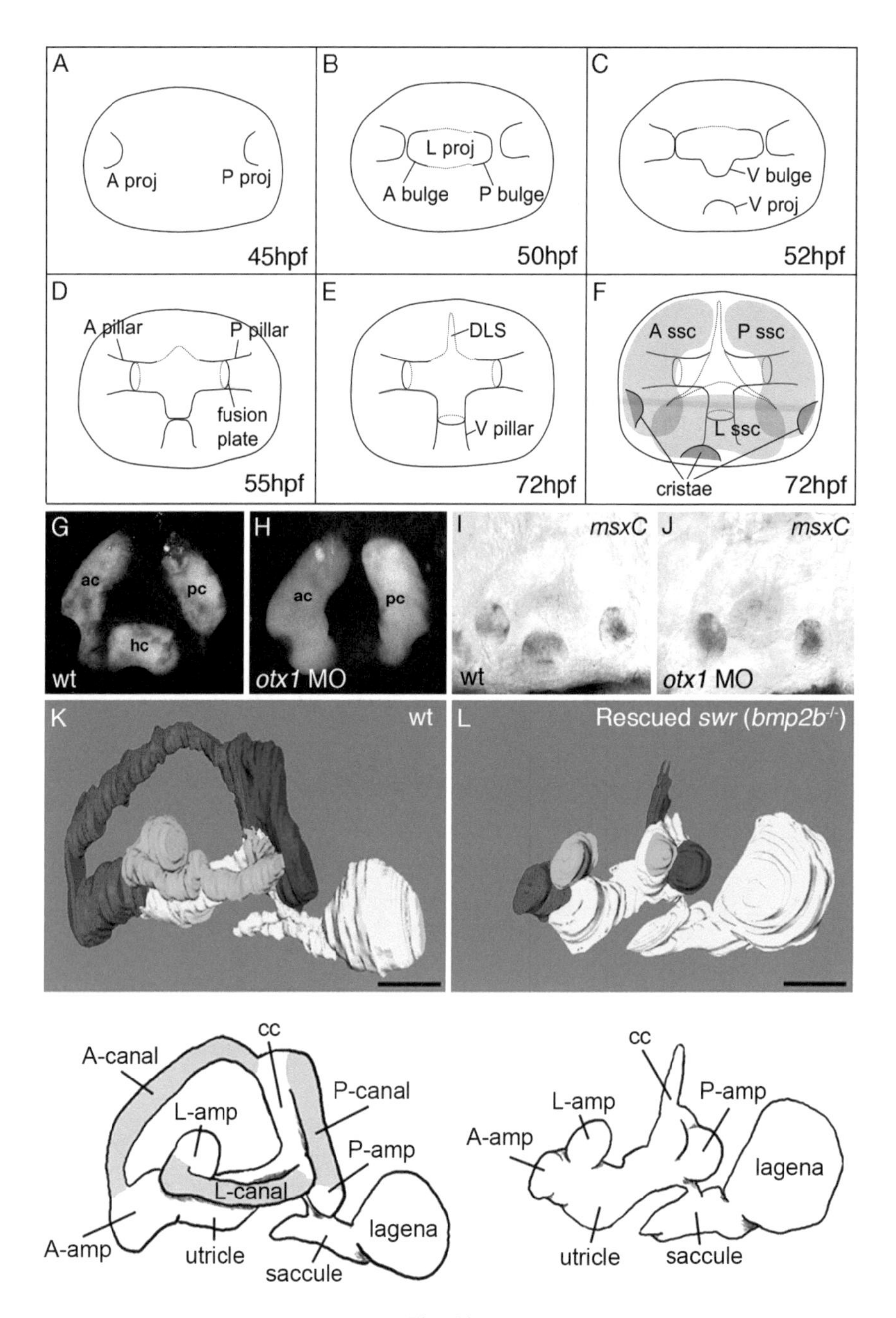

A
A proj
P proj
45hpf
B
L proj
A bulge
P bulge
50hpf
C
V bulge
V proj
52hpf
D
A pillar
P pillar
fusion plate
55hpf
E
DLS
V pillar
72hpf
F
A ssc
P ssc
L ssc
cristae
72hpf
G
ac
pc
hc
wt
H
ac
pc
otx1 MO
I
msxC
wt
J
msxC
otx1 MO
K
wt
L
Rescued swr (bmp2b-/-)
cc
A-canal
L-amp
P-canal
P-amp
L-canal
A-amp
utricle
saccule
lagena
cc
L-amp
P-amp
A-amp
lagena
utricle
saccule


**Fig. 4.2.**

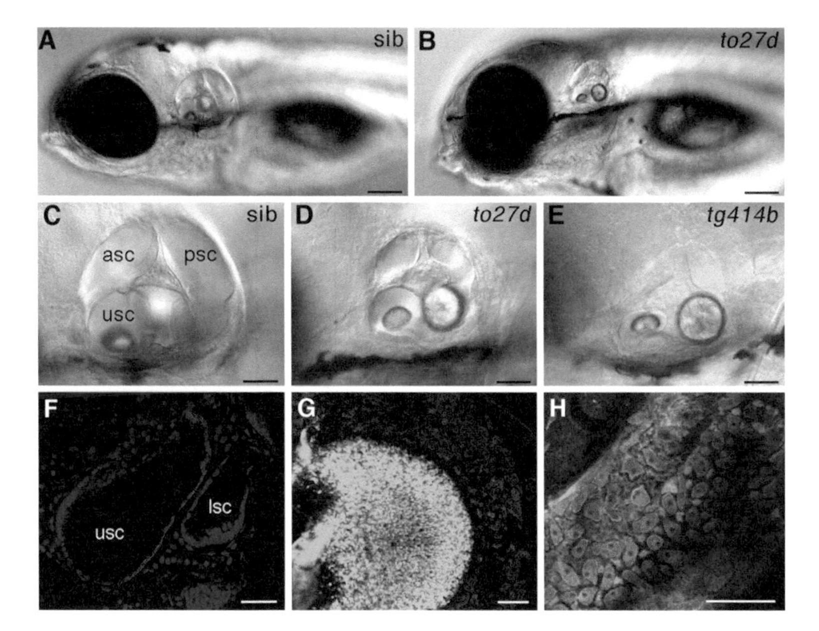


**Fig. 4.3.**

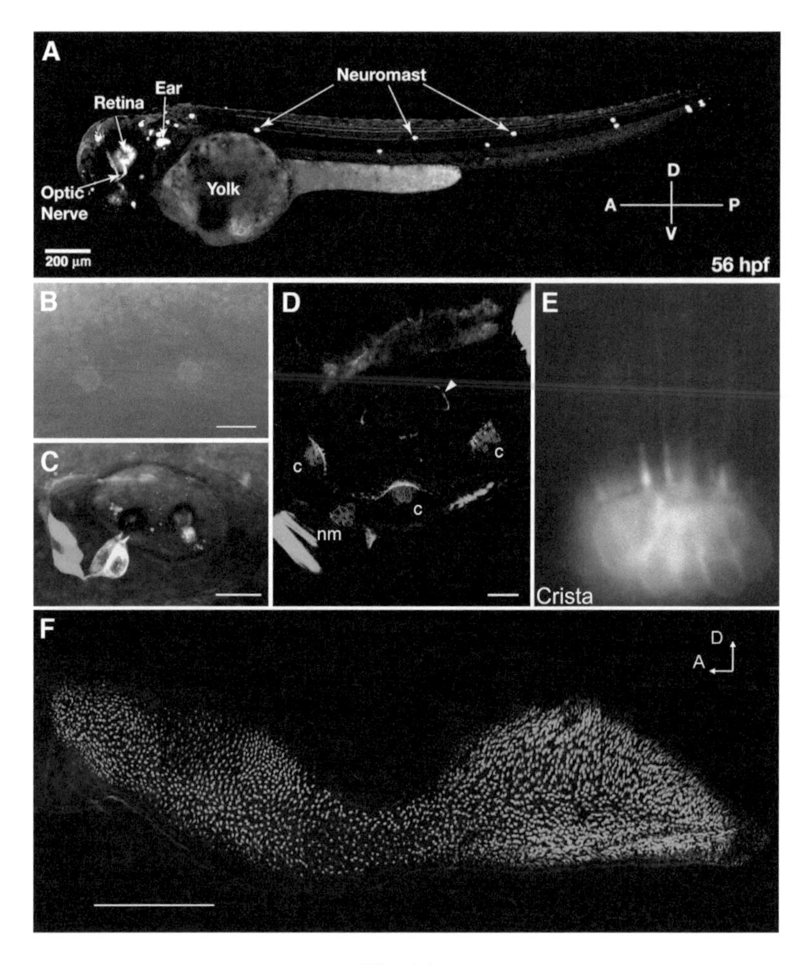
A
Neuromast
Ear
Retina
Optic
Nerve
Yolk
D
A
P
V
200 μm
56 hpf
B
C
D
c
c
c
nm
E
Crista
F
D
A

**Fig. 4.4.**

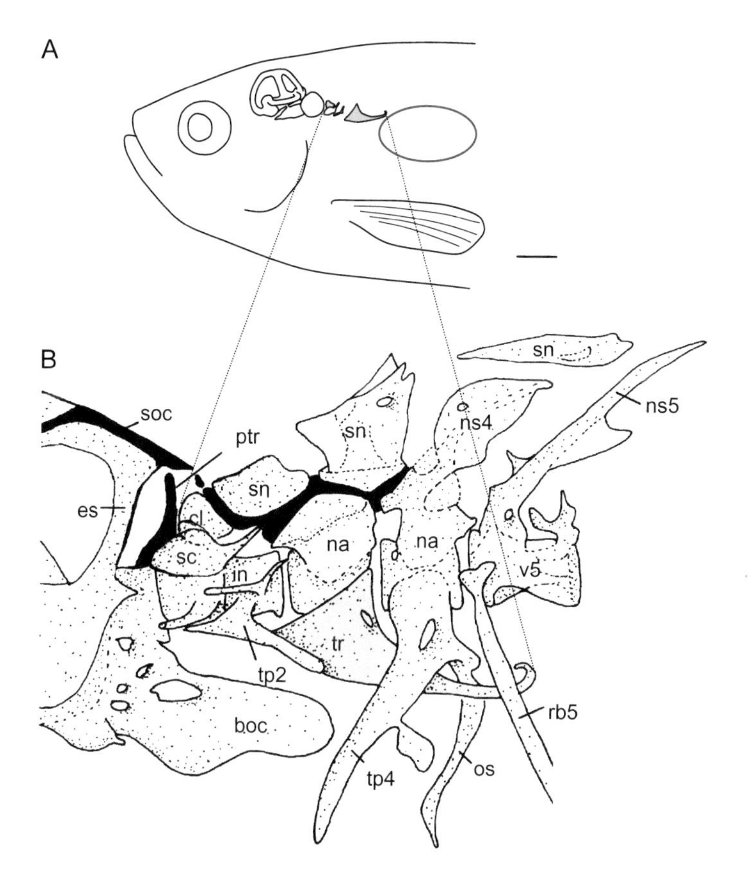


**Fig. 4.6.**

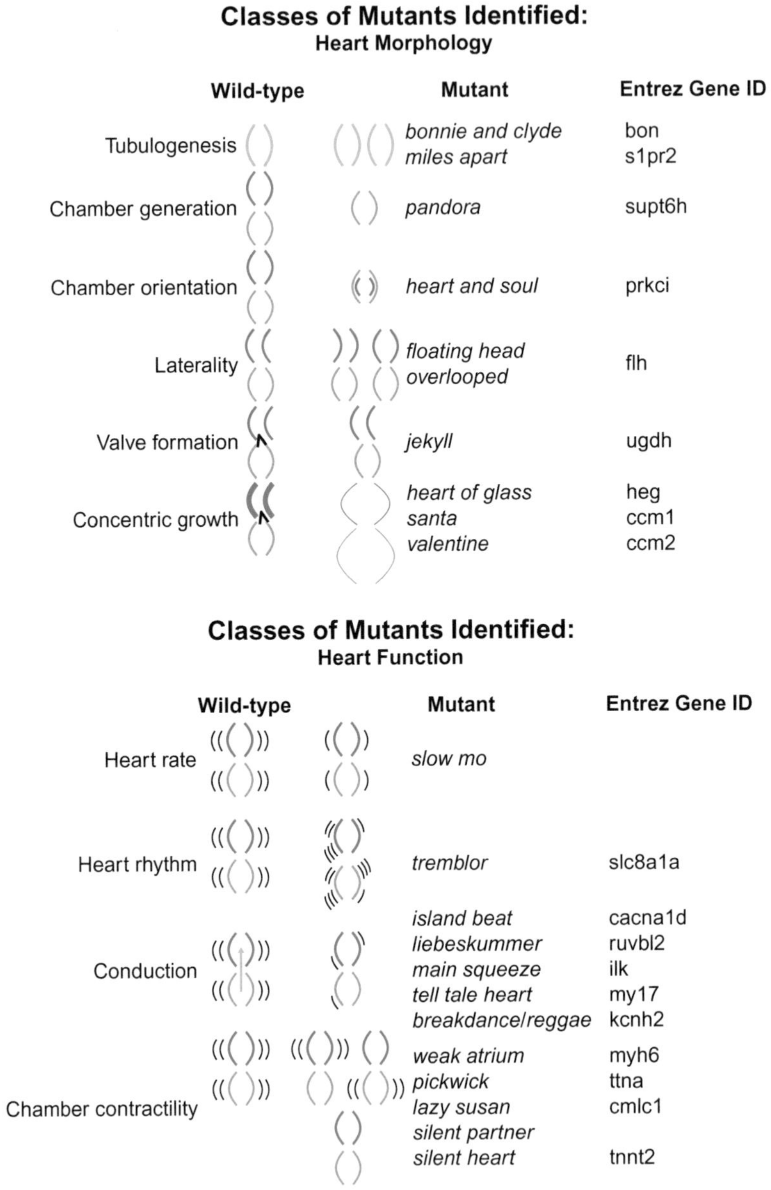

## Classes of Mutants Identified:
### Heart Morphology

| | Wild-type | Mutant | Entrez Gene ID |
|---|---|---|---|
| Tubulogenesis | | *bonnie and clyde* | bon |
| | | *miles apart* | s1pr2 |
| Chamber generation | | *pandora* | supt6h |
| Chamber orientation | | *heart and soul* | prkci |
| Laterality | | *floating head* | flh |
| | | *overlooped* | |
| Valve formation | | *jekyll* | ugdh |
| Concentric growth | | *heart of glass* | heg |
| | | *santa* | ccm1 |
| | | *valentine* | ccm2 |

## Classes of Mutants Identified:
### Heart Function

| | Wild-type | Mutant | Entrez Gene ID |
|---|---|---|---|
| Heart rate | | *slow mo* | |
| Heart rhythm | | *tremblor* | slc8a1a |
| Conduction | | *island beat* | cacna1d |
| | | *liebeskummer* | ruvbl2 |
| | | *main squeeze* | ilk |
| | | *tell tale heart* | my17 |
| | | *breakdance/reggae* | kcnh2 |
| Chamber contractility | | *weak atrium* | myh6 |
| | | *pickwick* | ttna |
| | | *lazy susan* | cmlc1 |
| | | *silent partner* | |
| | | *silent heart* | tnnt2 |

**Fig. 6.1.**

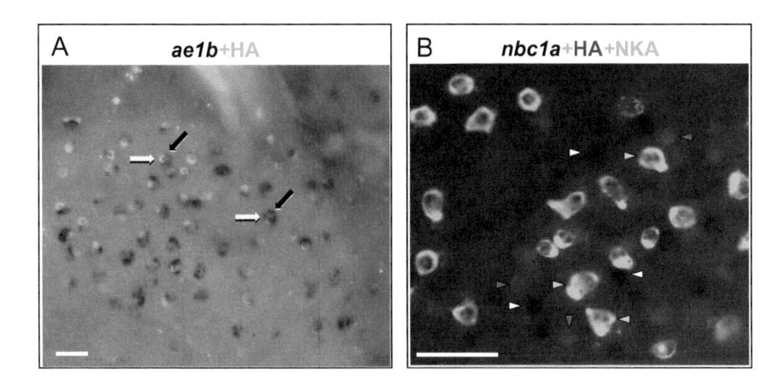
A
ae1b+HA
B
nbc1a+HA+NKA

**Fig. 8.1.**

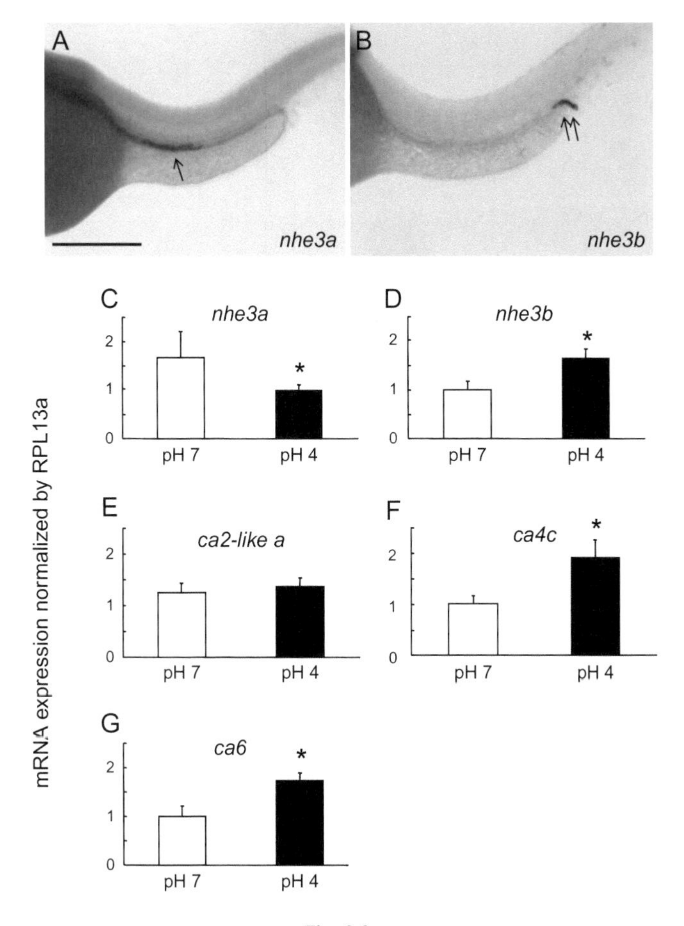
A
nhe3a
B
nhe3b
mRNA expression normalized by RPL13a
C
nhe3a
D
nhe3b
E
ca2-like a
F
ca4c
G
ca6
pH 7
pH 4
0
1
2
*

**Fig. 8.3.**

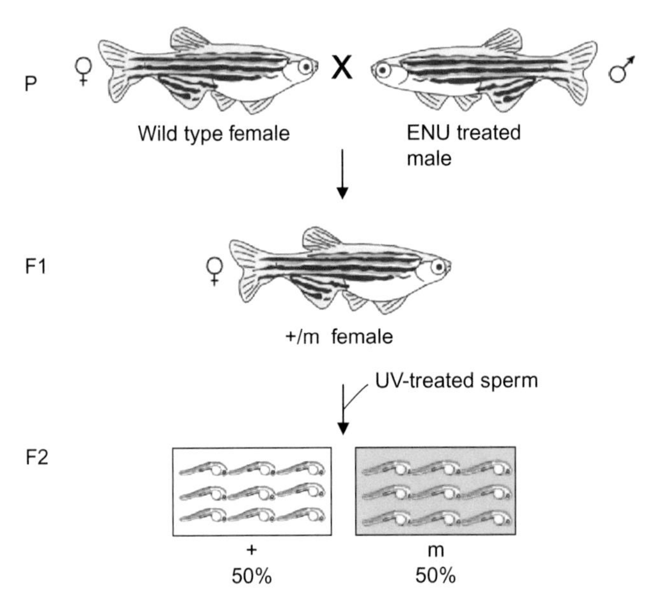
P
♀
Wild type female
X
ENU treated
male
♂
F1
♀
+/m female
UV-treated sperm
F2
+
50%
m
50%

**Fig. 9.1.**

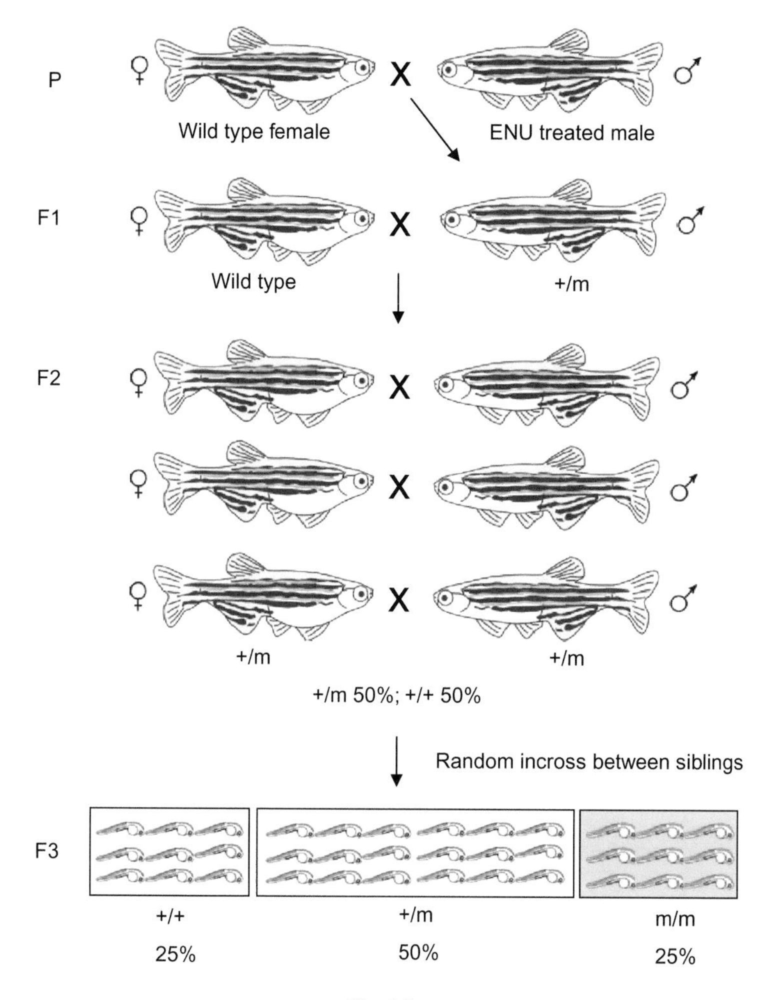
P
Wild type female
X
ENU treated male
F1
X
Wild type
+/m
F2
X
X
X
+/m
+/m
+/m 50%; +/+ 50%
Random incross between siblings
F3
+/+
25%
+/m
50%
m/m
25%

**Fig. 9.2.**

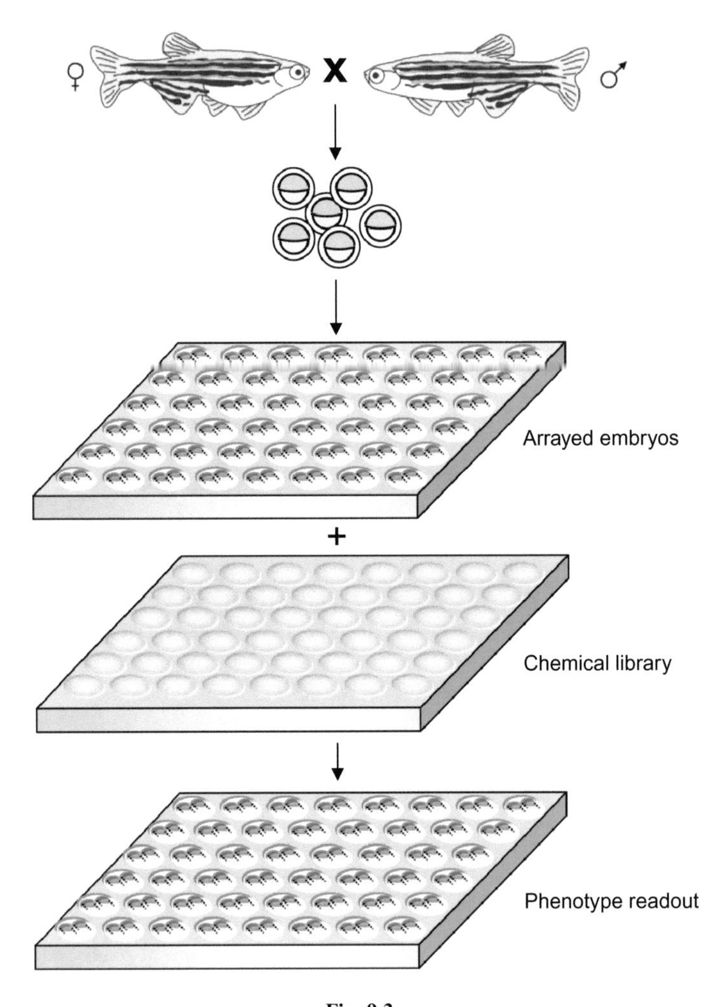


**Fig. 9.3.**

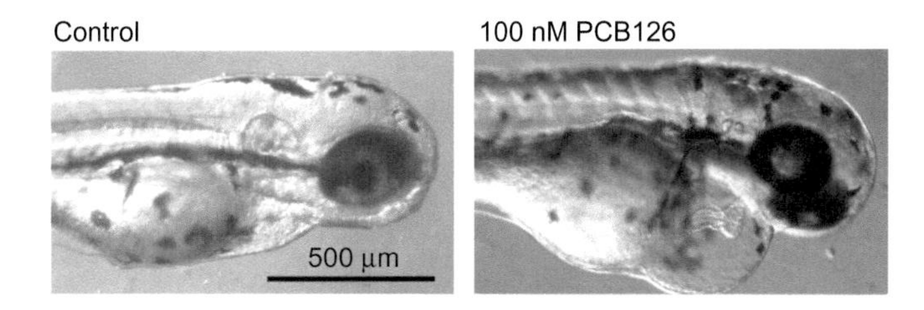


**Fig. 10.6.**

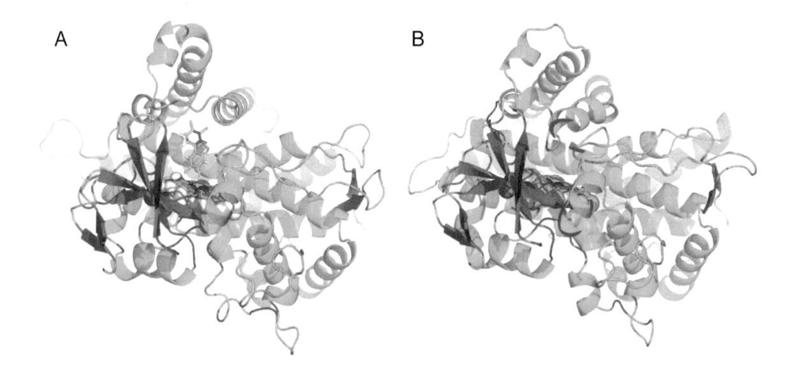


**Fig. 10.7.**

**Table 6.1**

| Phenotype | Wild type | Mutant | Mutant name | Gene name |
|---|---|---|---|---|
| Endothelial specification | 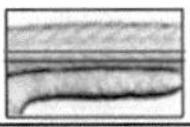 | 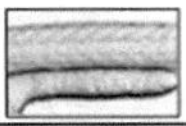 | *cloche* | ? |
| Endothelial specification & ISV sprouting |  |  | *Y11* | Ets related protein (etsrp) |
| Artery specification |  | 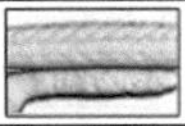 | *gridlock* | Hairy/Enhancer of split with YRPW motif 2 (Hey2) |
| Arteriovenous shunting |  | 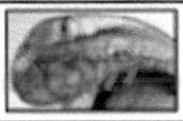 | *kurzschluss* | Unc45a |
| ISV sprouting |  |  | *t20257* | Flk (VEGFR1) |
| ISV sprouting |  |  | *Y10* | Phospholipase Cγ1 |
| Venous and lymphatic sprouting | 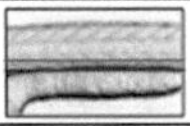 | 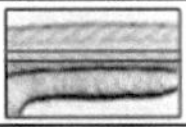 | *full of fluid* | Collagen and calcium-binding EGF domain-1 (ccbe1) |
| Endothelial cell number | 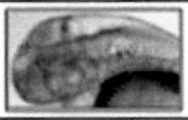 |  | *violet beauregarde* | Activin like kinase 1 (acvrl1) |
| Vessel patterning | 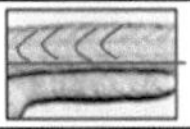 | 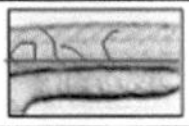 | *out of bounds* | Piexin D1 |
| Vessel dilation & patterning |  |  | *adrasteia* | Seryl-tRNA synthetase (sars) |
| Vascular stabilization |  |  | *bubblehead* | Pak-interacting exchange factor (bPix) |
| Vascular stabilization | |  | *redhead* | P21 activated kinase 2a (Pak2a) |
| Vascular stabilization | 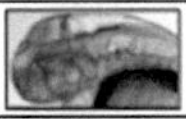 | 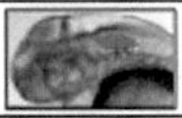 | *tomato* | Birc2 (ciap 1) |
| Vascular regeneration |  |  | *reg6* | ? |